VISUAL ENCYCLOPEDIA — OF — SCIENCE

VISUAL ENCYCLOPEDIA OF SCIENCE

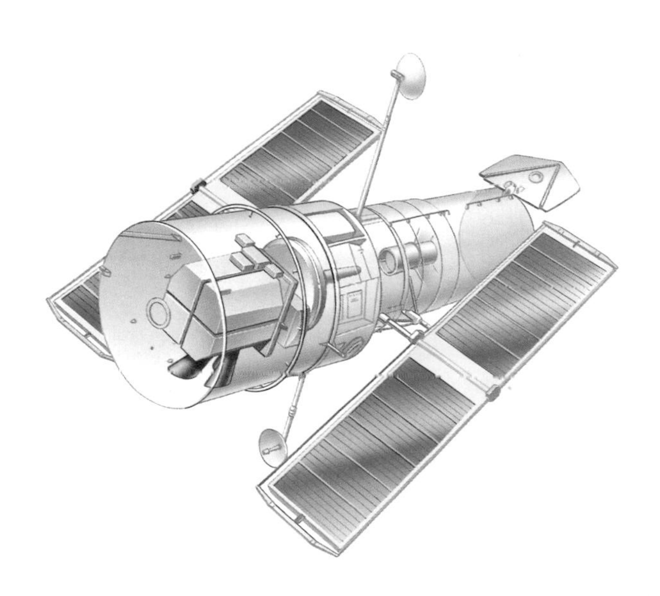

Kingfisher
NEW YORK

Authors

PLANET EARTH
Michael Allaby, Neil Curtis

THE LIVING WORLD
Brian Williams

STARS AND PLANETS
James Muirden

SCIENCE AND TECHNOLOGY
Brian Williams

Series Editor
Michèle Byam

Series Designer and Art Editor
Ralph Pitchford

Editors
Andrea Moran, Cynthia O'Neill

Designers and Art Editors
Shaun Barlow, Sandra Begnor, Nigel Bradley, John Kelly,
Cathy Tincknell, Steve Woosnam-Savage

Picture Research
Elaine Willis, Su Alexander

Production Managers
Julia Mather, Oonagh Phelan

Additional editorial help from
Joan Angelbeck, Stuart Atkinson, Nicky Barber, Hilary Bird,
Catherine Bradley, Catherine Headlam

Additional art preparation by
Matthew Gore, Andy Archer, Julian Ewart, Mustafa Sidki,
Martin Wilson, Janet Woronkowicz

KINGFISHER
Larousse Kingfisher Chambers Inc.
95 Madison Avenue
New York, New York 10016

First American edition 1994
10 9 8 7 6 5 4 3 2 1

The material in this edition was previously published
in four individual volumes in 1993.

LIBRARY OF CONGRESS CATALOGING-IN-PUBLICATION DATA
Visual encyclopedia of science: planet earth. the living world.
stars and planets. science and technology.—1st U.S. ed.
p. cm.—(Visual encyclopedia)
Includes index.
1. Curiosities and wonders—Juvenile literature. [1. Curiosities
and wonders.] I. Kingfisher Books. II. Series.
AG243.K53 1994
031.02—dc20 93-43118 CIP AC

ISBN 1-85697-998-9

Printed in Portugal.

CONTENTS

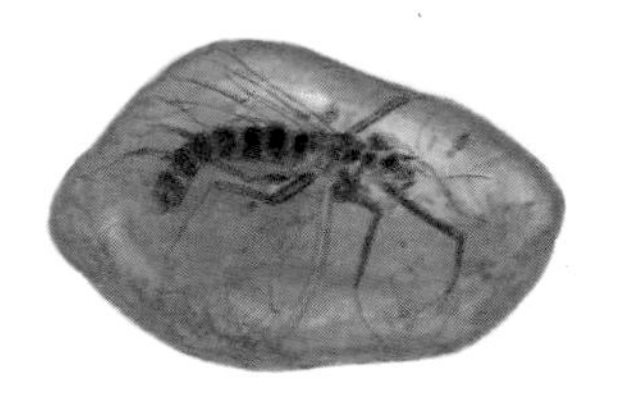

PLANET EARTH

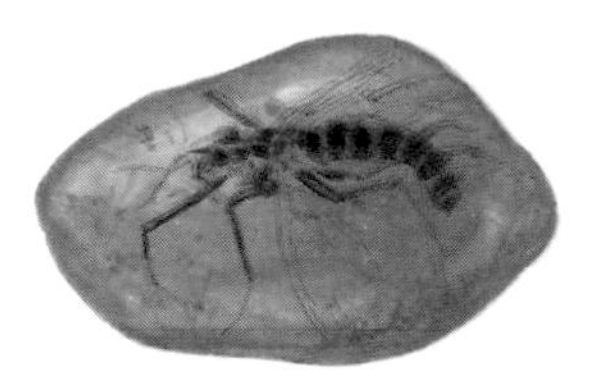

Our Planet 8
The Land 16
Water 38
Weather and Climate 48
Landscapes 60
Past, Present, Future 72

THE LIVING WORLD

The Living Earth 84
The Plant Kingdom 90
The Animal Kingdom 108
The Human Body 142

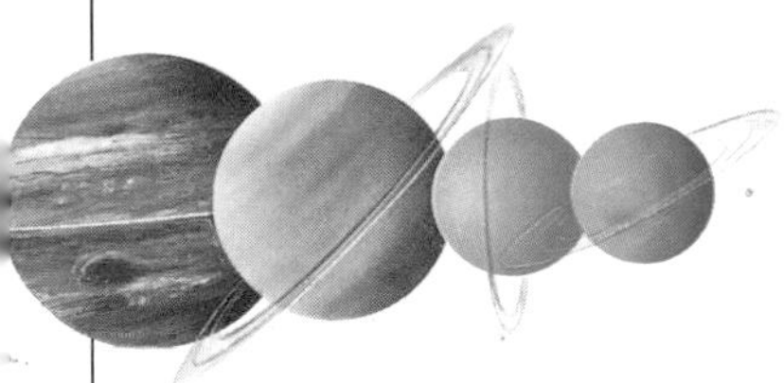

STARS AND PLANETS

Cosmic Time 160
The Solar System 162
Beyond the Solar System 196
The Stars 202
Observing the Skies 214
Space Exploration 222

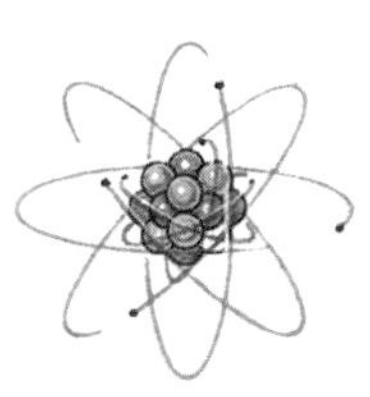

SCIENCE AND TECHNOLOGY

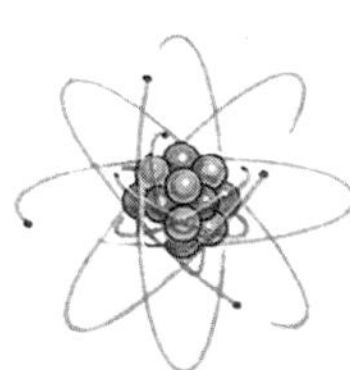

Discovering Science 236
Matter and Energy 238
Force and Motion 252
Space and Time 260
Light and Sound 266
Electricity 274
Technology 282

This landscape on the island of Sardinia, Italy, shows one of the many pressures that have shaped the Earth's surface.

PLANET EARTH

The Earth in Space **8** Gravity and the Earth **10** The Structure of the Earth **12**
Land, Water, and Air **14** Earthquakes **16** Volcanoes **18** Drifting Continents **20** Plate Tectonics **22**
Mountains **24** Bending and Breaking **26** Rocks **28** Minerals and Gems **30** Frozen in Time **32**
Shaping the Earth **34** The Work of Ice **36** Oceans and Seas **38** The Life of the Ocean **40**
The Seashore **42** Rivers **44** Lakes and Swamps **46** The Atmosphere **48** Climate **50**
Winds and Storms **52** The Types of Cloud **54** Rain and Snow **56**
Weather Forecasting **58** The Changing Scene **60**
Polar Regions and the Tundra **62** Temperate Woodland **64**
Grasslands **66** Deserts **68** Tropical Rain Forests **70**
Natural Resources **72** Air Pollution **76**
Environmental Problems **78** Conservation **80**

Why is there day and night? Why do things fall down? What is lightning? For thousands of years questions such as these have puzzled people. Early humans were superstitious about these confusing and sometimes frightening conditions. Later, philosophers and other classical thinkers tried to find a logical explanation for natural events. Although a number of their theories may seem foolish to us with hindsight, many form the foundations of modern scientific thought.

In the past few centuries we have developed the technology to gather vast amounts of information about the Earth. It is in part due to this abundance of information that we are able to address problems, such as pollution, that our ancestors never encountered. Perhaps the same technologies we have used to explain how the world evolved will eventually help us to preserve it.

OUR PLANET

The Earth in Space

Viewed from space, the Earth appears as a round ball that shines bright and blue. People have not always seen the Earth in this way. Aristotle (384–322 B.C.), a philosopher in ancient Greece, and other scholars, believed that any problem could be solved by thinking carefully about it. He believed that the Earth was at the center of the universe, and that the Moon, Sun, planets, and stars orbited around it. You might come to the same conclusion if you lived on a desert island, with no radio, television, computers, or books, and no telescope to study the night sky. Indeed, you might not even guess you were on a spherical planet floating in space. At best, if you watched a ship disappearing beneath the horizon, you might figure out that the world's surface was curved. Today, we know that the Earth is one of a system of planets orbiting the Sun.

► *The Earth is not completely round. It is slightly flattened at the poles and bulges slightly at the equator. Clouds swirl continuously above the surface. Over two-thirds of the Earth is covered by water. Most of this water is contained in the oceans.*

EARTH DATAFILE

Diameter at the poles: 7,900 mi.
Diameter at the equator: 7,926 mi.
Circumference (distance around the Earth at the equator): 24,912 mi.
Volume: about 240 billion cu. mi.
Mass: 6.5 sextillion tons
Average density: 5.5 (water = 1)
Surface area: 197 million sq. mi.
Percentage of surface area covered by water: 71 percent
Age: 4.6 billion years
Age of the oldest known rocks: 3.7 billion years
Distance to the Moon: *maximum* 252,717 mi., *minimum* 221,462 mi.
Average distance to the Sun: 94 million mi.
Average thickness of the crust: 12 mi.
Average thickness of the mantle: 1,740 mi.
Average diameter of the core: 4,327 mi.
Temperature at the center: 8,132°F

► *Before modern mapping techniques, people had many different ideas about the Earth. These ideas included the belief that the Earth (and the Sun) were gods and that the Earth was flat.*

Egyptian Sun disk

15th century world map

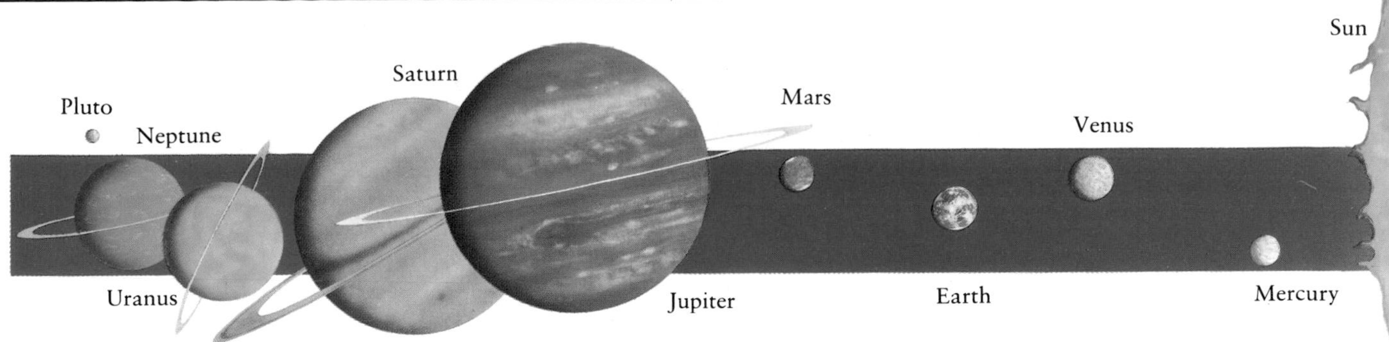

THE EARTH'S POSITION

Our Sun has a solar system of nine planets around it. The planets' characteristics are related to their distance from the Sun. Earth is one of the small inner planets, along with Mercury, Venus, and Mars. Closest to the Sun, Mercury and Venus are extremely hot. Each day on Mercury lasts over two Earth-months, during which its surface heats up to 842°F (450°C), and then cools to −274°F (−170°C) during the long night. Venus is even hotter, because of the greenhouse effect of its carbon dioxide atmosphere *(see page 77)*. Next is Earth—the only planet with liquid water. Beyond Earth, Mars resembles a cold, red, stony desert. Colder still are the giant outer planets—Jupiter, Saturn, Uranus, and Neptune. They are made up partly of gases such as hydrogen and helium. Pluto is the smallest planet and is thought to be ice.

THE BIRTH OF THE EARTH

The Earth was formed at the same time as the Sun and the other planets in the Solar System. About 4.6 billion years ago, a rotating cloud of dust and gas, called a nebula, contracted under the pull of gravity. The pressure and temperature at the center of the nebula became so great it triggered a nuclear reaction. Some of the hydrogen in the cloud fused into helium, releasing great amounts of energy. This was the birth of the Sun. Farther from the center, material surrounding the Sun cooled and collided, building up into larger bodies. These eventually became the planets.

LINES ON EARTH

The Earth spins on an axis. The imaginary points on the Earth's surface where this axis projects are called the geographical poles.

The equator is an imaginary circle drawn around the Earth at an equal distance from the North and South poles. Imaginary circles drawn parallel to the equator are called lines of latitude. They are measured in degrees. The latitude of the equator is 0°, and the poles are 90° north and 90° south. Great circles drawn through the poles give lines of longitude. Zero longitude runs through Greenwich, London.

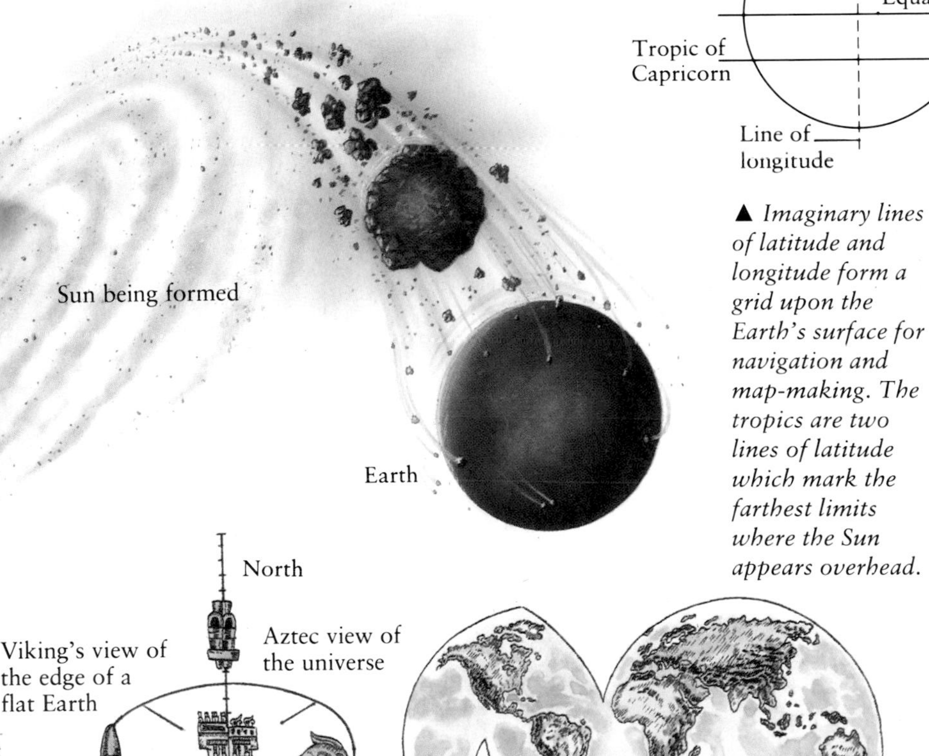

▲ *Imaginary lines of latitude and longitude form a grid upon the Earth's surface for navigation and map-making. The tropics are two lines of latitude which mark the farthest limits where the Sun appears overhead.*

Satellite image of the Earth

Gravity and the Earth

Gravity is the force of attraction that every object in the universe exerts on every other object. It is the weakest force known in physics, but its effects extend across the huge distances of space. The greater the mass of an object, the larger its gravitational attraction, but the farther away it is, the smaller the force. Gravitational forces are largely responsible for the orbits of bodies around the Sun and the orbit of our sat[illegible] Moon, around the Earth. These forces also g[illegible] the seasons and the tides.

THE SEASONS

Since the Earth's axis is tilted, the hemispheres are at different angles to the Sun. This phenomenon gives rise to the seasons. In winter the days are short and cold and in summer they are long and warm. At the equator, daylength varies little throughout the year, and the seasons are linked to rainfall.

THE EARTH'S DAY

The Earth's day is the time it takes to spin once on its axis (a little under 24 hours). If the Earth's axis were at right angles to its orbit, the length of a day would be the same all over the Earth. In fact, the Earth's axis is tilted by 23° 27′. This means that people in the hemisphere tilted toward the Sun see the Sun passing higher across the sky, and daylength is longer. The hemisphere tilted away from the Sun has a shorter day.

Sun

21 December
Winter begins in Northern Hemisphere, summer begins in Southern Hemisphere

21 March
Spring begins in Northern Hemisphere, autumn in Southern Hemisphere

21 June
Summer begins in Northern Hemisphere, winter begins in Southern Hemisphere

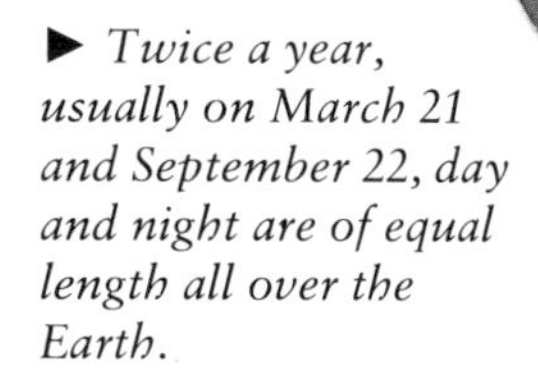

► *Twice a year, usually on March 21 and September 22, day and night are of equal length all over the Earth.*

▲ *In the Northern Hemisphere, summer begins officially on June 21, the longest day of the year. In the Southern Hemisphere June 21 is the shortest day.*

THE EARTH'S YEAR

The Earth's year is the time it takes for the Earth to complete one orbit of the Sun (365 days, 5 hours, 48 minutes, and 46 seconds). The year is divided into 365 days, but every fourth year, a leap year, has 366 days to make up the extra time. In a leap year February has 29 days.

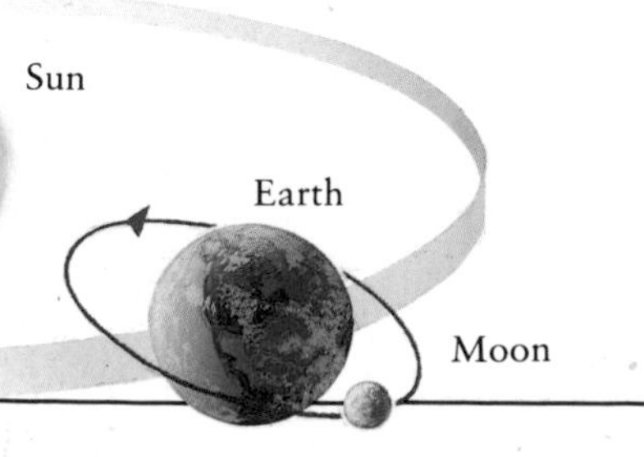

THE PULL OF GRAVITY

The sea rises and falls every 12 hours and 26 minutes. These tides are caused by the gravity of the Moon, and also the Sun. The Moon's gravity pulls the oceans on the side of the Earth nearer to it more than it pulls the Earth itself, causing a bulge of water, or tide. On the opposite side, the Earth is pulled more than the oceans, and this leaves behind a slightly smaller bulge of water.

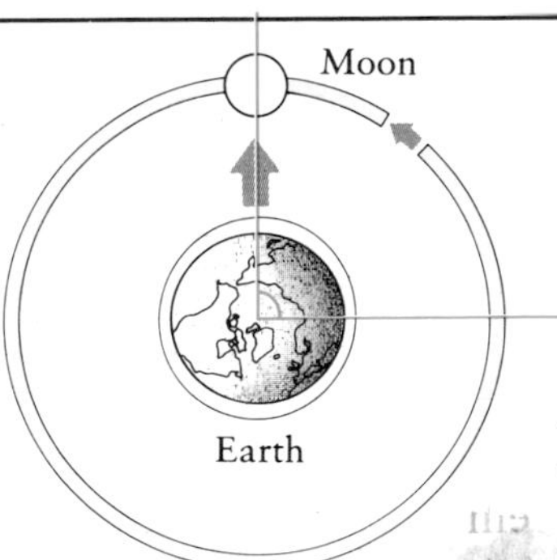

▲ *When the Sun and Moon are at right angles and ...lling in different ...ons, the lowest ...ake place. These are called neap tides.*

23 September
Autumn begins in Northern Hemisphere, spring begins in Southern Hemisphere

Moon
Sun
Earth

▲ *When the Sun and Moon are in line, pulling together—at new or full moon—high tides, known as spring tides, occur.*

◄ *The Bay of Fundy, Canada, boasts the world's greatest spring tidal range—47.5 ft. (14.5 m) between high and low tides.*

THE EARTH'S ENERGY BALANCE

The Sun radiates (gives out) energy and can be thought of as the Earth's power plant. It supplies the energy that drives the Earth's climate and weather. It is the energy source for all plant life and, therefore, for all animal life as well. At a temperature of some 27 million °F (15 million °C), the Sun's fuel source, hydrogen, fuses into helium (the hydrogen atoms combine to form helium atoms) and releases energy. Our planet receives just one two-billionth of the energy that the Sun generates.

► *Thirty percent of the available radiation is reflected directly back into space by clouds, or by the surface of the planet.*

Radiation reflected back to space

▼ *Radiation reflected by the Earth is absorbed by gases in the atmosphere such as carbon dioxide. The energy is then radiated again, some of it returning to Earth.*

Radiation from Sun

Radiation reflected to Earth's surface

Absorption by atmosphere

► *About 70 percent of the Sun's radiation that reaches the Earth is absorbed and then radiated back again. Of this 70 percent, 25 percent is absorbed by the atmosphere and the rest by the planet itself.*

The Structure of the Earth

Although earthquakes can have catastrophic effects, they can also reveal a great deal about the Earth's structure. The shock waves pass through the Earth in different ways and can be recorded by scientists using sensitive instruments called seismometers. Scientists have been able to identify three main zones according to their densities: the thin outer crust, the mantle, and the core in the center. The upper layers of the Earth are also categorized as the lithosphere, hydrosphere, and asthenosphere.

▼ *The Earth's crust is divided into oceanic crust and continental crust, both of which originate in the mantle. The thicker continental crust can vary from about 22 mi. (35 km) thick to as much as 31 mi. (50 km) beneath mountain ranges. It is made mainly from pale, granitelike rocks.*

▼ *The lithosphere is the upper, rocky layer of the Earth. It includes the crust and the top, brittle part of the mantle; it can be up to 186 mi. (300 km) thick.*

▼ *The hydrosphere is the water (mainly the oceans and seas) on the Earth's surface. After the Earth was formed, water vapor in the atmosphere condensed to form the hydrosphere.*

► *The thinner rocky crust beneath the oceans is only about 3 mi. (5 km) thick. It is made up mainly of dark rocks called basalt and gabbro and has been formed in the last 200 million years.*

► *The asthenosphere is the layer of the mantle beneath the lithosphere which is in an almost fluid or "plastic" state, so that it behaves like an extremely thick liquid.*

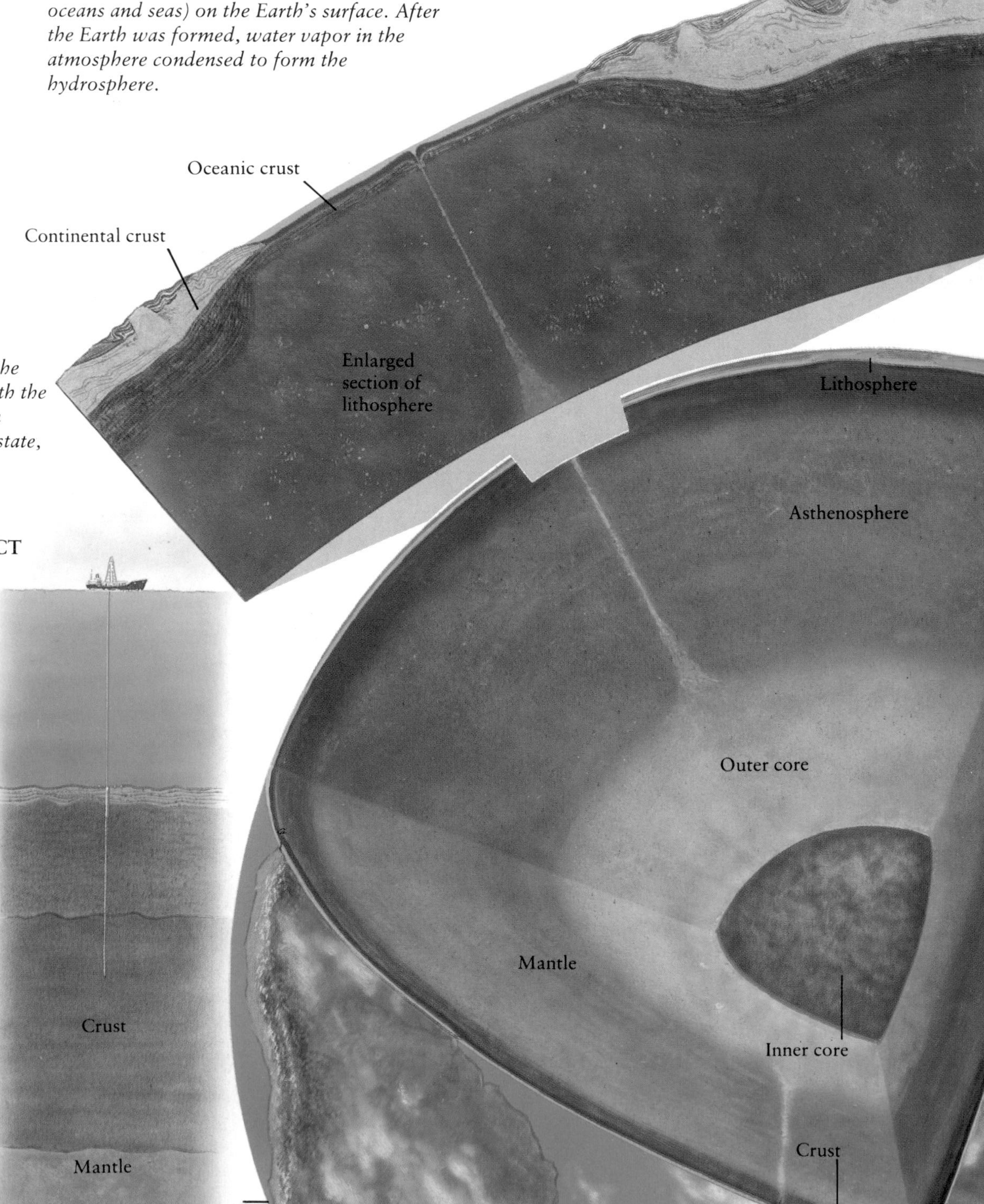

THE MOHOLE PROJECT
Rocks on the surface of the Earth can easily be examined. But the inside of the planet is hidden from scientists, except when volcanic eruptions spew out material. The boundary between the crust and the mantle is called the Mohorovičić discontinuity, after a Croatian scientist. In the 1960s, an attempt was made to drill through the ocean crust to take samples of the mantle. The "Mohole" project was abandoned because of increasing costs.

Crust

Mantle

MAGNETIC ATTRACTION

The Earth's magnetic field extends far into space. The Sun showers the Earth with charged atomic particles —the solar wind—which are affected by the Earth's magnetic field, or the magnetosphere. In places where the magnetosphere is dense, the atomic particles are trapped in two layers around the Earth. These layers are called the Van Allen belts.

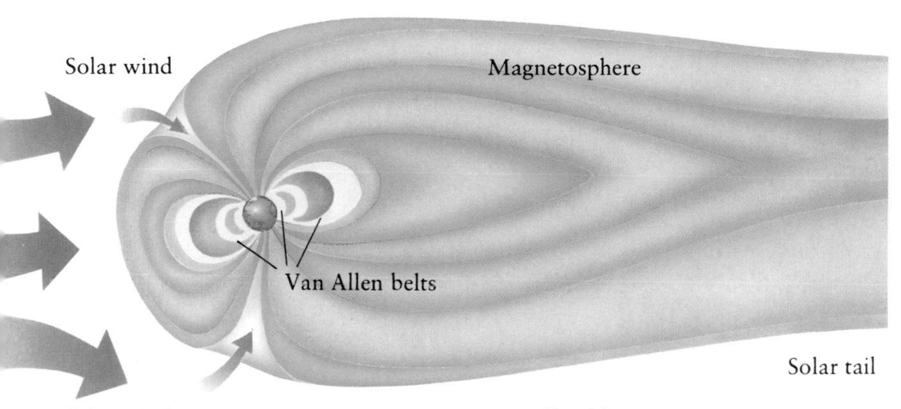

Earth's magnetic field

Lines of force

◀ *The Earth is magnetic and behaves somewhat like a bar magnet with a magnetic field. The Earth's magnetic poles and the geographic poles do not exactly coincide.*

THE EARTH'S MAGNETISM

Homing pigeon

Magnetism is a property of materials such as iron, and of electric currents, in which moving charged particles develop a force of attraction. The Earth is magnetic and has magnetic poles and a magnetic field. If a magnetized needle is floated in a bowl of water it will align itself with the Earth's magnetic poles. This is a simple compass and can be used for navigation, although the difference between the geographic poles and the magnetic poles (magnetic variation) must be taken into account. Magnetic variation changes over time. It appears that the magnetic poles are "wandering," but it is actually the continents that are moving over them.

Some animals, such as pigeons, have been shown to use the Earth's magnetic field to navigate. If they are released in areas where there is some kind of magnetic disturbance, homing pigeons seem to lose their way.

Magnetic compass

◀ *The aurora australis, or the southern lights, is a "curtain" of spectacular lights in the sky at the South Pole. They are caused by the collision of the solar wind with the Earth's atmosphere. The aurora borealis occurs at the North Pole.*

THE CORE AND MANTLE

Lying beneath the Earth's crust, the mantle is about 1,740 mi. (2,800 km) thick. It is made of rock. The central zone of the Earth is called the core. It is divided into inner and outer zones. The inner core is solid, but the outer core is liquid. Both are dense and hot and consist mainly of nickel and iron.

THE CRUST

These are the main elements that make up the crust:

Oxygen	Calcium
Silicon	Sodium
Aluminum	Potassium
Iron	Magnesium

Land, Water, and Air

The elements that make up the land, water, and air of the Earth are constantly being recycled as they combine, break up, and recombine. For example, the air around the Earth is unstable. It is too rich in the reactive gases oxygen, nitrogen, and methane, and too poor in carbon dioxide to be in balance chemically. But Earth's air is ideal for life. The British scientist James Lovelock has developed the Gaia hypothesis, in which he suggests that it is life itself which maintains the conditions that it needs.

THE WATER CYCLE
Almost 80 percent of the rain that falls goes into the oceans, the rest falls onto land. When the Sun's rays heat the Earth's surface, water evaporates back into the atmosphere. Over 80 percent of this water evaporates from the oceans. Some of the rest is returned to the atmosphere by transpiration from plants. As the water in the atmosphere cools, it condenses to form clouds. Some of this water falls to Earth again as rain.

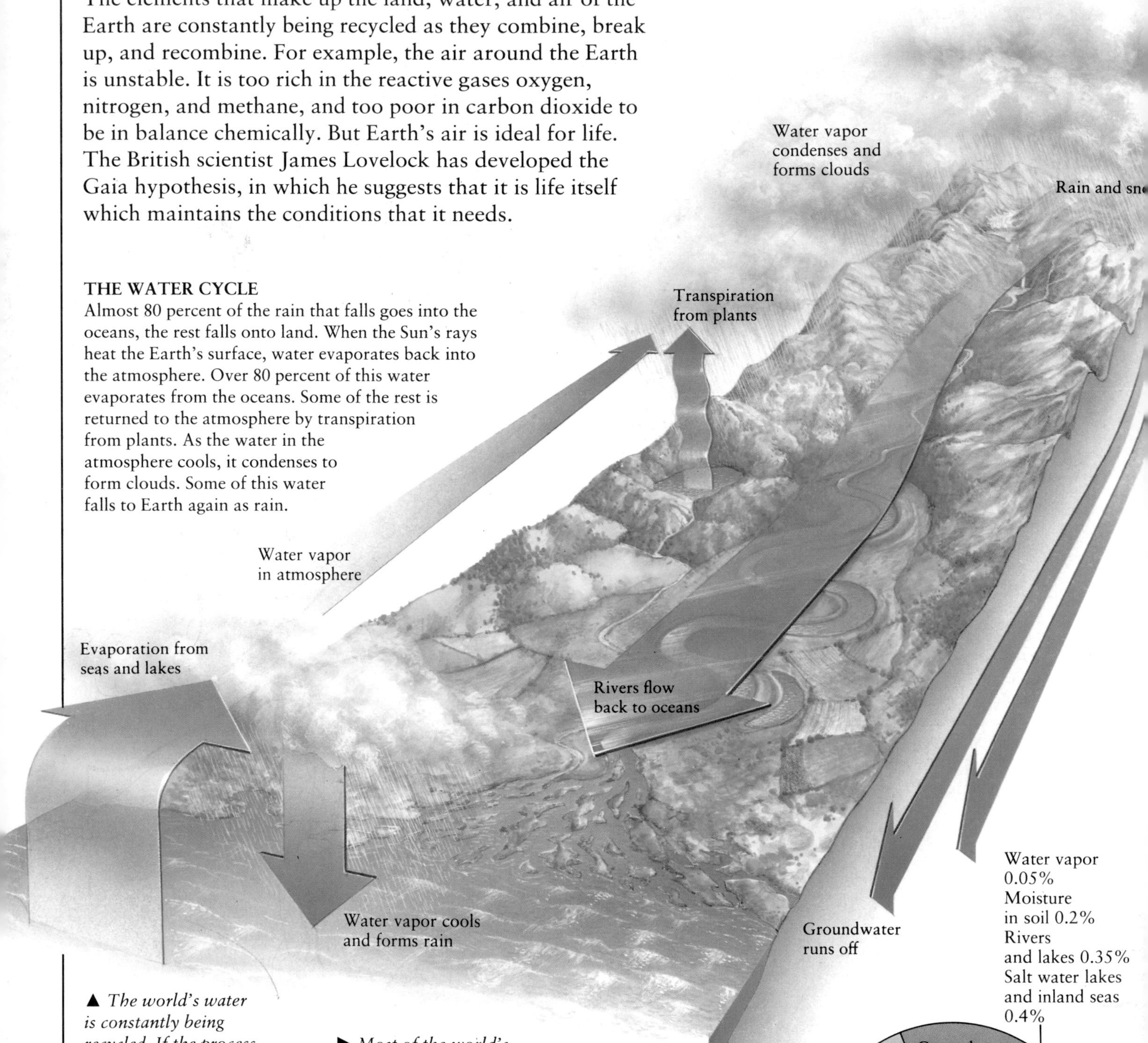

▲ *The world's water is constantly being recycled. If the process of evaporation from the oceans and the ground, and transpiration from the Earth's plants, suddenly stopped, there would be enough rain held in the atmosphere to last for 10 days.*

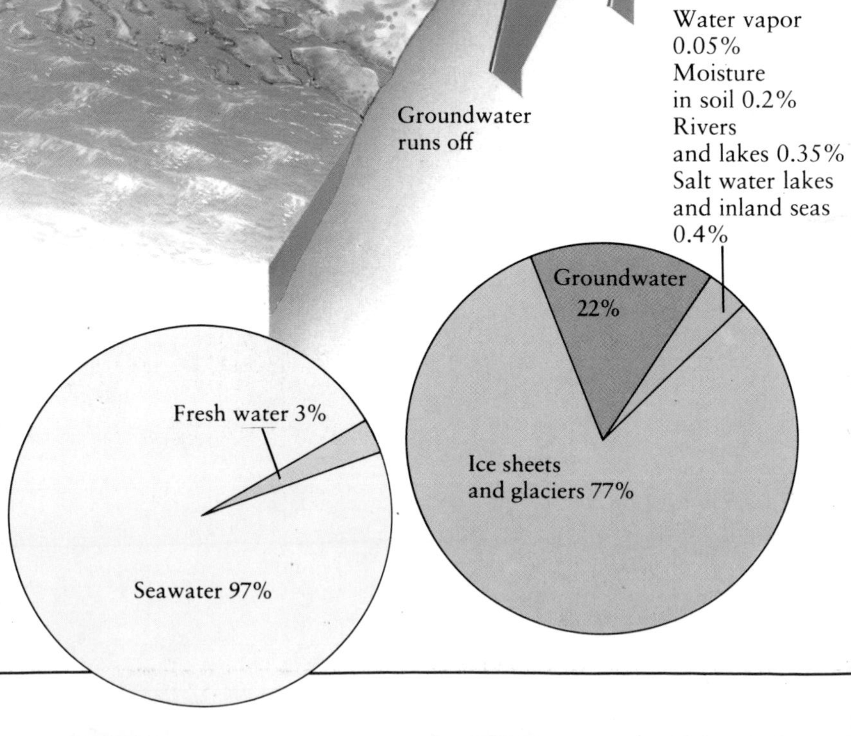

► *Most of the world's water is found in the oceans. Of the 3 percent of water that is not in the sea, over 75 percent is locked up in the ice caps and glaciers of the Arctic and Antarctic.*

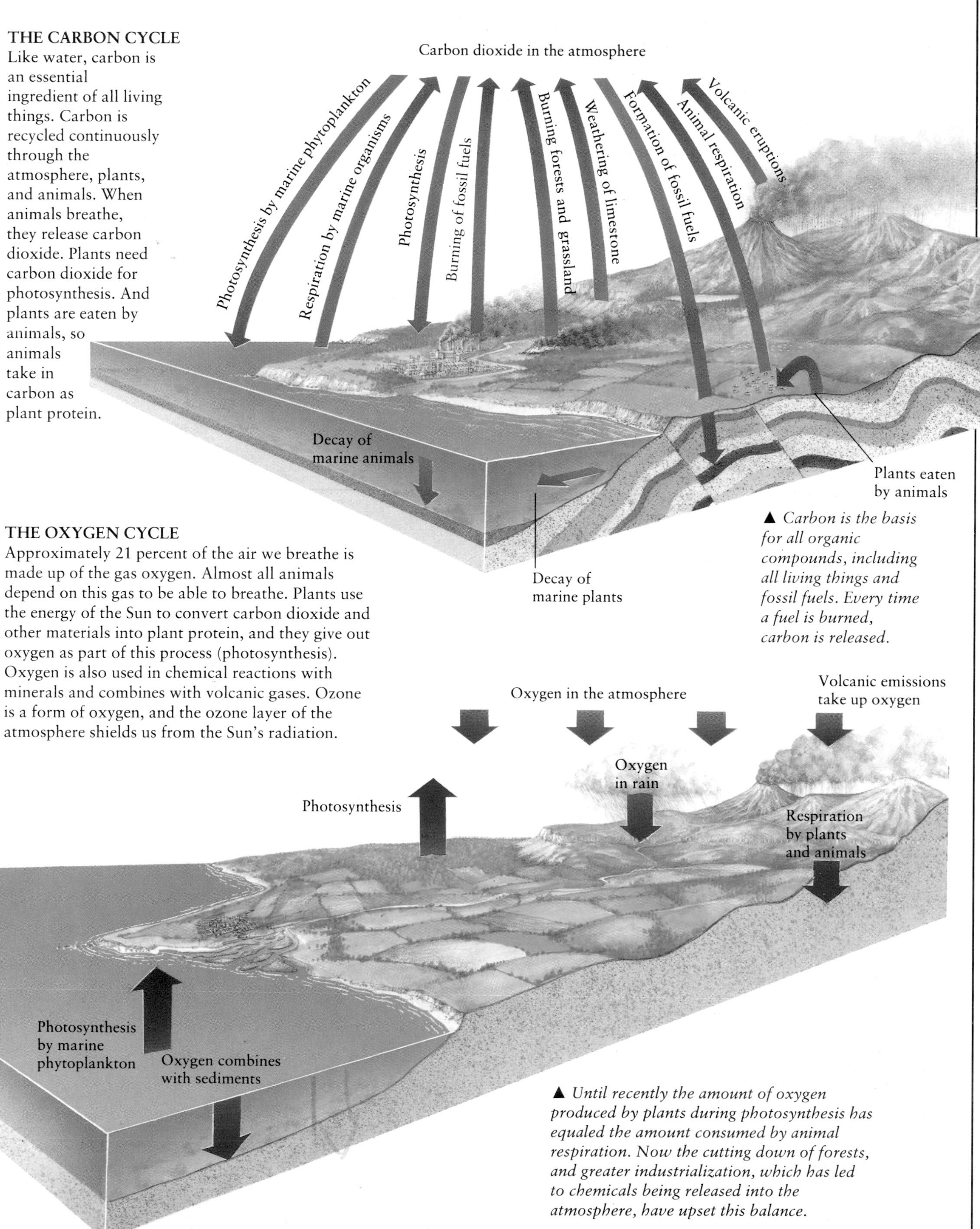

THE CARBON CYCLE

Like water, carbon is an essential ingredient of all living things. Carbon is recycled continuously through the atmosphere, plants, and animals. When animals breathe, they release carbon dioxide. Plants need carbon dioxide for photosynthesis. And plants are eaten by animals, so animals take in carbon as plant protein.

▲ *Carbon is the basis for all organic compounds, including all living things and fossil fuels. Every time a fuel is burned, carbon is released.*

THE OXYGEN CYCLE

Approximately 21 percent of the air we breathe is made up of the gas oxygen. Almost all animals depend on this gas to be able to breathe. Plants use the energy of the Sun to convert carbon dioxide and other materials into plant protein, and they give out oxygen as part of this process (photosynthesis). Oxygen is also used in chemical reactions with minerals and combines with volcanic gases. Ozone is a form of oxygen, and the ozone layer of the atmosphere shields us from the Sun's radiation.

▲ *Until recently the amount of oxygen produced by plants during photosynthesis has equaled the amount consumed by animal respiration. Now the cutting down of forests, and greater industrialization, which has led to chemicals being released into the atmosphere, have upset this balance.*

THE LAND

Earthquakes

Earthquakes are among the most serious natural disasters: the ground shakes and may crack; poorly constructed buildings collapse; dams are destroyed; huge ocean waves flood the land; gas mains rupture; and fires break out. Many hundreds or even thousands of people may be killed or left homeless as a result.

Earthquakes occur because within the Earth's asthenosphere, stress causes the semiplastic rocks to move very slowly. This builds up strain within the more brittle rocks of the lithosphere above. Eventually, the brittle rocks break and the stress is released as shock waves. Earthquakes can take place at depths of up to 450 mi. (720 km). But those that have effects at the surface usually occur no deeper than 45 mi. (70 km). Every year there are about 1,000 earthquakes around the world that are strong enough to cause some damage.

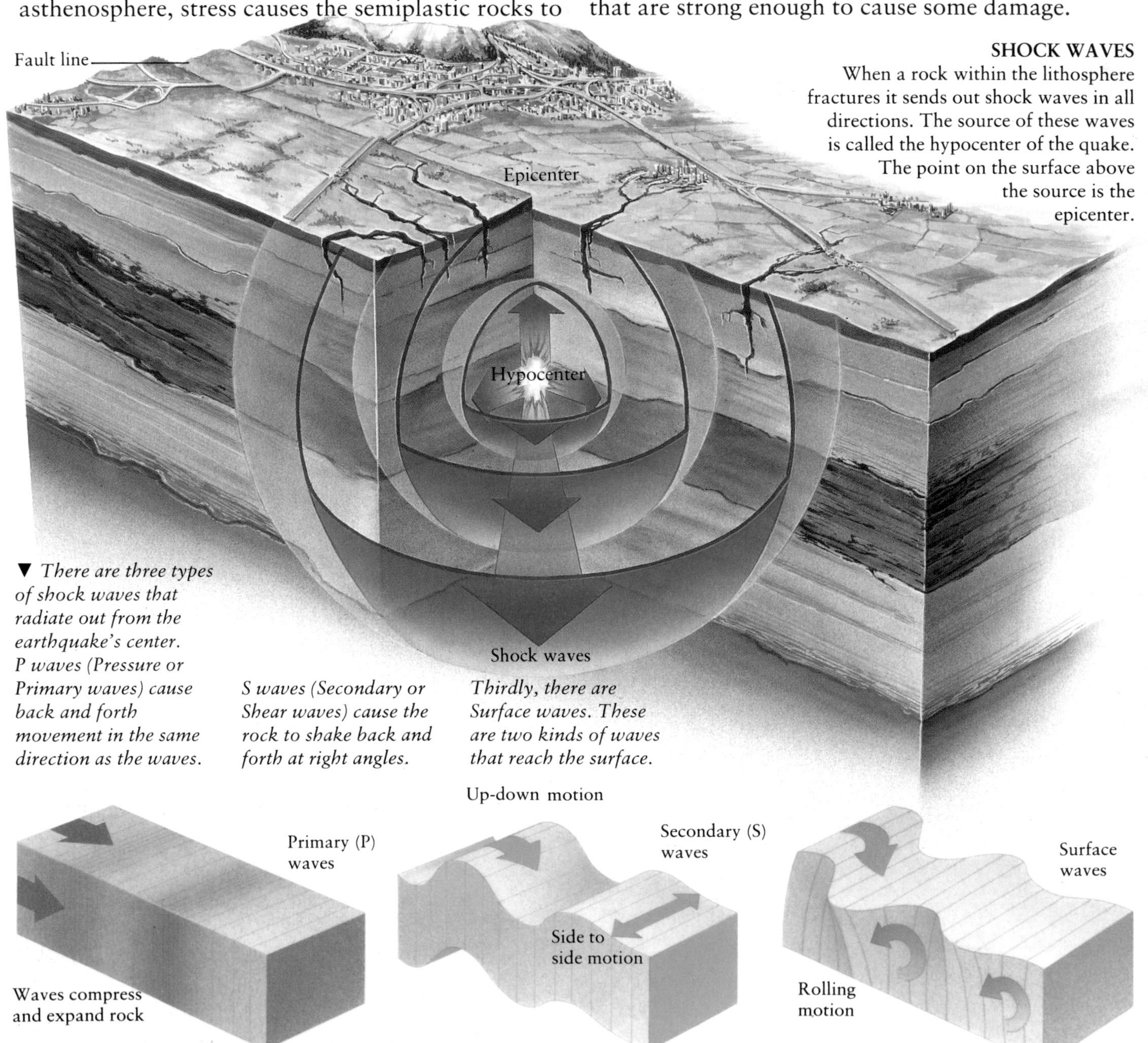

SHOCK WAVES
When a rock within the lithosphere fractures it sends out shock waves in all directions. The source of these waves is called the hypocenter of the quake. The point on the surface above the source is the epicenter.

▼ *There are three types of shock waves that radiate out from the earthquake's center. P waves (Pressure or Primary waves) cause back and forth movement in the same direction as the waves.*

S waves (Secondary or Shear waves) cause the rock to shake back and forth at right angles.

Thirdly, there are Surface waves. These are two kinds of waves that reach the surface.

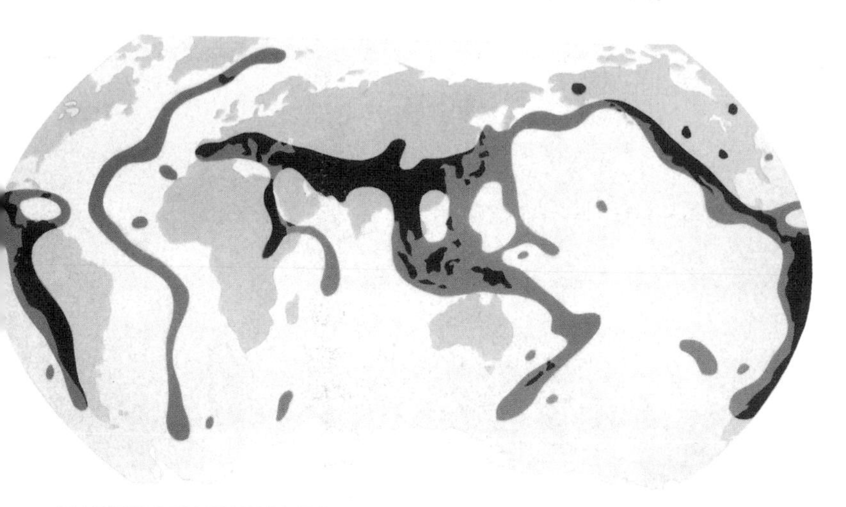

EARTHQUAKE ZONES

Scientists continuously monitor earthquakes and plot the sites of major quakes on a map. A distinct pattern has emerged showing that, with a few exceptions, earthquakes take place in a number of definite zones. Earthquakes happen where there are deep trenches in the ocean bed with groups of islands nearby, such as around the Pacific.

In earthquake areas, seismologists try to figure out whether stress is building up in the rocks. If the area is quiet for a long time, it may be that stress is building up and may eventually cause a major earthquake. Seismologists can also use one or more seismometers to detect the tiny shock waves that occur before an earthquake.

▲ *When an earthquake takes place under the seabed, it may produce a wave in the sea. In the open ocean the wave may hardly be noticed. But if it reaches shallow water near a coast, the wave may rise to heights of 100 ft. (30 m) or more. These giant waves are called tsunamis.*

EARTHQUAKE DAMAGE

If a large earthquake occurs where people live then there may be casualties as buildings collapse. Fire has always been a hazard, originally as buildings fell on domestic fires and burned, and now as gas mains break and catch fire.

▲ *In 1906, the San Andreas fault moved 21.3 ft. (6.5 m), causing an earthquake that destroyed San Francisco. Much of the city burned down as gas mains broke and caught fire.*

FACTS ABOUT EARTHQUAKES

- The strongest earthquake ever recorded on the Richter scale took place in Japan on March 2, 1933. The earthquake killed 2,990 people and measured 8.9.
- Historical records indicate that in July of the year 1201, more than a million people were killed during an earthquake which affected the eastern Mediterranean area.
- The highest tsunami ever recorded hit the island of Ishigaki to the west of Taiwan in April 1771. It is thought to have reached a height of 280 ft. (85 m).
- In 1750 an earthquake caused widespread panic and confusion in London, a city where earthquakes are virtually unknown.
- The costliest earthquake in terms of property damage destroyed 570,000 buildings on the Kwanto plain, Japan, 1923.

EARTHQUAKE SCALES

The intensity of earthquakes is measured on two scales. The Mercalli scale, based on observable effects, ranges from "not felt" at 1 to "total devastation" at scale 12. The Richter scale is based on the size of the shock waves produced.

Mercalli and Richter Scales

Mercalli	Richter	Effects
1	<3	Very slight: detected by instruments only
2	3–3.4	Feeble: felt by people resting
3	3.5–4	Slight: like heavy trucks passing
4	4.1–4.4	Moderate: windows rattle
5	4.5–4.8	Rather strong: wakes sleeping people
6	4.9–5.4	Strong: trees sway, walls crack
7	5.5–6	Very strong: people fall over, buildings crack
8	6.1–6.5	Destructive: chimneys fall, buildings move
9	6.6–7	Ruinous: heavy damage to buildings, ground cracks
10	7.1–7.3	Disastrous: most buildings destroyed, landslides
11	7.4–8.1	Very disastrous: railroads and pipelines break
12	>8.1	Catastrophic: total devastation

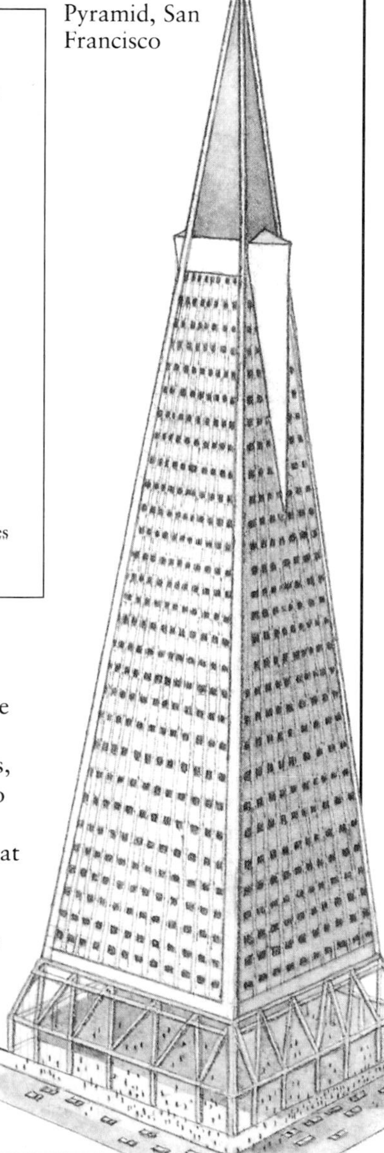

PREVENTING COLLAPSE

When an earthquake strikes, badly built houses collapse. It is, however, possible to construct buildings, even skyscrapers, that will resist collapse. Earthquake-proof buildings must have their foundations built into the solid rock, and they must be able to bend without shattering.

Volcanoes

A volcano is any kind of fissure, or natural opening, in the Earth's crust through which hot molten rock (called lava), ash, steam, gas, and other material is spewed. The heat comes from within the Earth's mantle. The word "volcano" (after Vulcan, the Roman god of fire and metalworking) is also used to describe the cone of lava and ash that builds up around the opening. The shape depends on the type of eruption. Volcanic activity may take place under the sea as well as on land, and it sometimes creates new land.

► *Sometimes, a volcano may explode very violently, emptying the lava chamber that feeds it. The roof and walls may then collapse, leaving a hole called a caldera.*

Ash and smoke

TYPES OF VOLCANOES

▲ *Mount Pelé is a volcano on Martinique, West Indies. In an explosive Peléan eruption the lava is thick and the volcano gives out glowing, gas-charged clouds.*

▲ *The cones created by a Hawaiian eruption are big at the bottom and slope gently because the lava, which pours out in fountains, is quite runny.*

▲ *Stromboli is an active volcano on an island north of Sicily. A Strombolian eruption occurs regularly, but with small explosions.*

▲ *Pliny the Elder died in* A.D. *79 when Vesuvius erupted and the Roman city of Pompeii was destroyed. A Plinian eruption is explosive with great clouds of ash and volcanic rock, or pumice.*
◄ *A vulcanian eruption* (left), *after Vulcano, near Stromboli, is characterized by rare explosions of almost solid magma thrown long distances.*

INSIDE A VOLCANO

A typical volcano, with a crater and a cone of solidified lava and ash, is called a central type. The activity takes place in a chimneylike "pipe," or vent, through which the material erupts. Far below the vent is a chamber of molten rock (magma) containing dissolved gases. Gas bubbling out of the magma keeps the vent open. Sometimes, lava and other material may break through farther down the sides of the volcano, and a secondary or parasite core is formed.

Caldera

Lava flow

FAMOUS ERUPTIONS

Eruptions occur when the pressure in the chamber and vent has built up and the material breaks through.

Vesuvius, Italy A.D. 79

Mt. Etna, Sicily, 1669

Cotopaxi, Ecuador, 1877

Mt. St. Helens, Wash., 1980

Mauna, Loa, Hawaii, 1872

Taupa, New Zeala A.D. 130

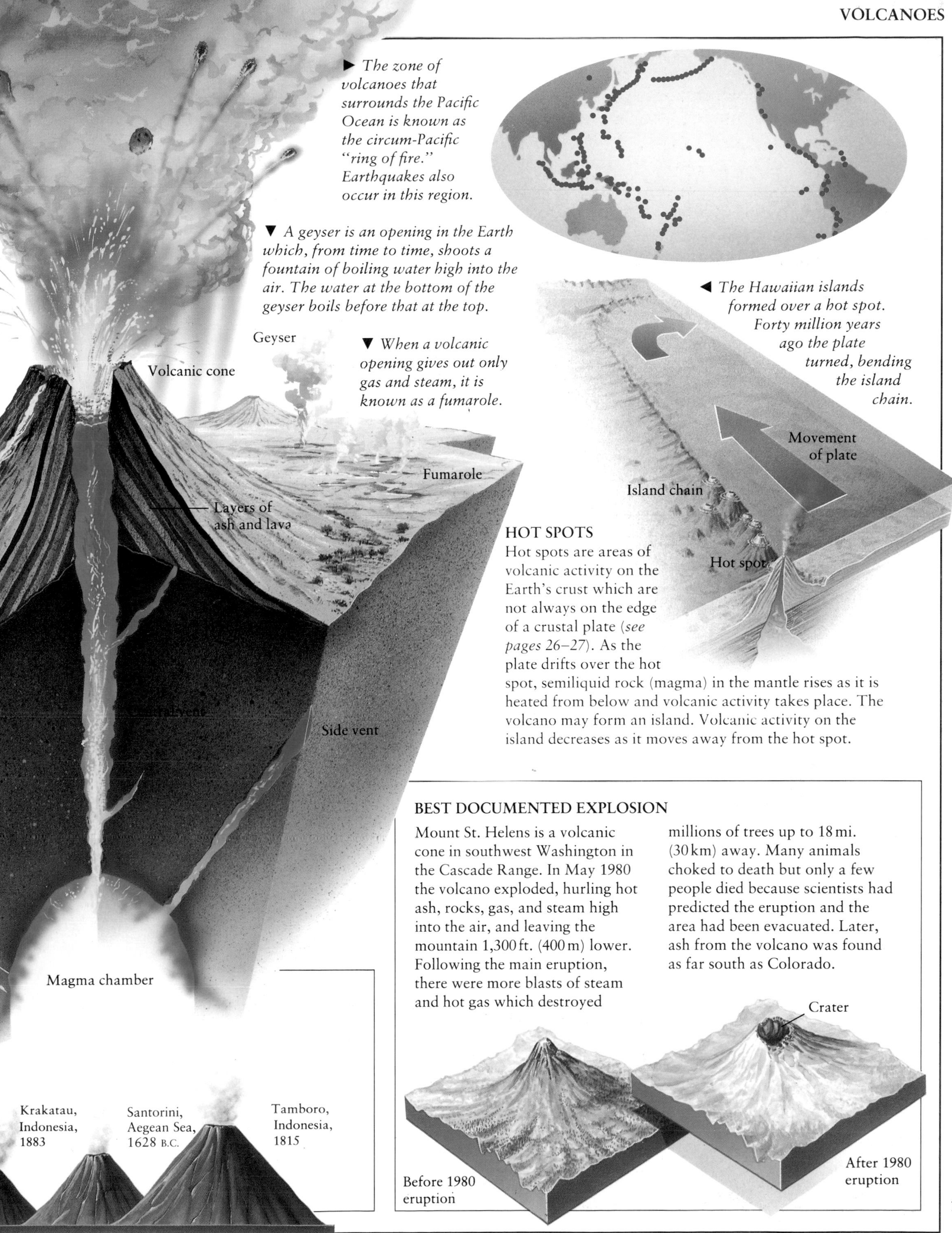

▶ *The zone of volcanoes that surrounds the Pacific Ocean is known as the circum-Pacific "ring of fire." Earthquakes also occur in this region.*

▼ *A geyser is an opening in the Earth which, from time to time, shoots a fountain of boiling water high into the air. The water at the bottom of the geyser boils before that at the top.*

▼ *When a volcanic opening gives out only gas and steam, it is known as a fumarole.*

◀ *The Hawaiian islands formed over a hot spot. Forty million years ago the plate turned, bending the island chain.*

HOT SPOTS

Hot spots are areas of volcanic activity on the Earth's crust which are not always on the edge of a crustal plate (*see pages 26–27*). As the plate drifts over the hot spot, semiliquid rock (magma) in the mantle rises as it is heated from below and volcanic activity takes place. The volcano may form an island. Volcanic activity on the island decreases as it moves away from the hot spot.

BEST DOCUMENTED EXPLOSION

Mount St. Helens is a volcanic cone in southwest Washington in the Cascade Range. In May 1980 the volcano exploded, hurling hot ash, rocks, gas, and steam high into the air, and leaving the mountain 1,300 ft. (400 m) lower. Following the main eruption, there were more blasts of steam and hot gas which destroyed millions of trees up to 18 mi. (30 km) away. Many animals choked to death but only a few people died because scientists had predicted the eruption and the area had been evacuated. Later, ash from the volcano was found as far south as Colorado.

Drifting Continents

The coastlines of eastern South America and West Africa could fit together like pieces in a jigsaw. This match was noticed in the 17th century. However, it was not until 1912 that Alfred Wegener proposed that all the land masses of the world had originally formed one super-continent, which he called Pangaea. This could not be explained until the early 1960s, when scientists discovered that the rocky plates of the Earth's lithosphere were moving, floating on the more mobile rock below.

► *As well as the matching coastlines, there is other evidence that there was once a single continent. There are remains of an ancient mountain belt, between 470 and 350 million years old, which are now separated by the Atlantic Ocean. These mountains were created by what was a continuous belt of geological activity.*

EVOLVING EARTH
The supercontinent, Pangaea, is thought to have evolved some 280 million years ago, at the end of the Carboniferous Period. By mid-Jurassic times, 150 million years ago, Pangaea had split into a northern continent, Laurasia, and a southern continent, Gondwanaland. By the end of the Cretaceous, about 65 million years ago, Gondwanaland was breaking up, although North America had not yet split from Eurasia.

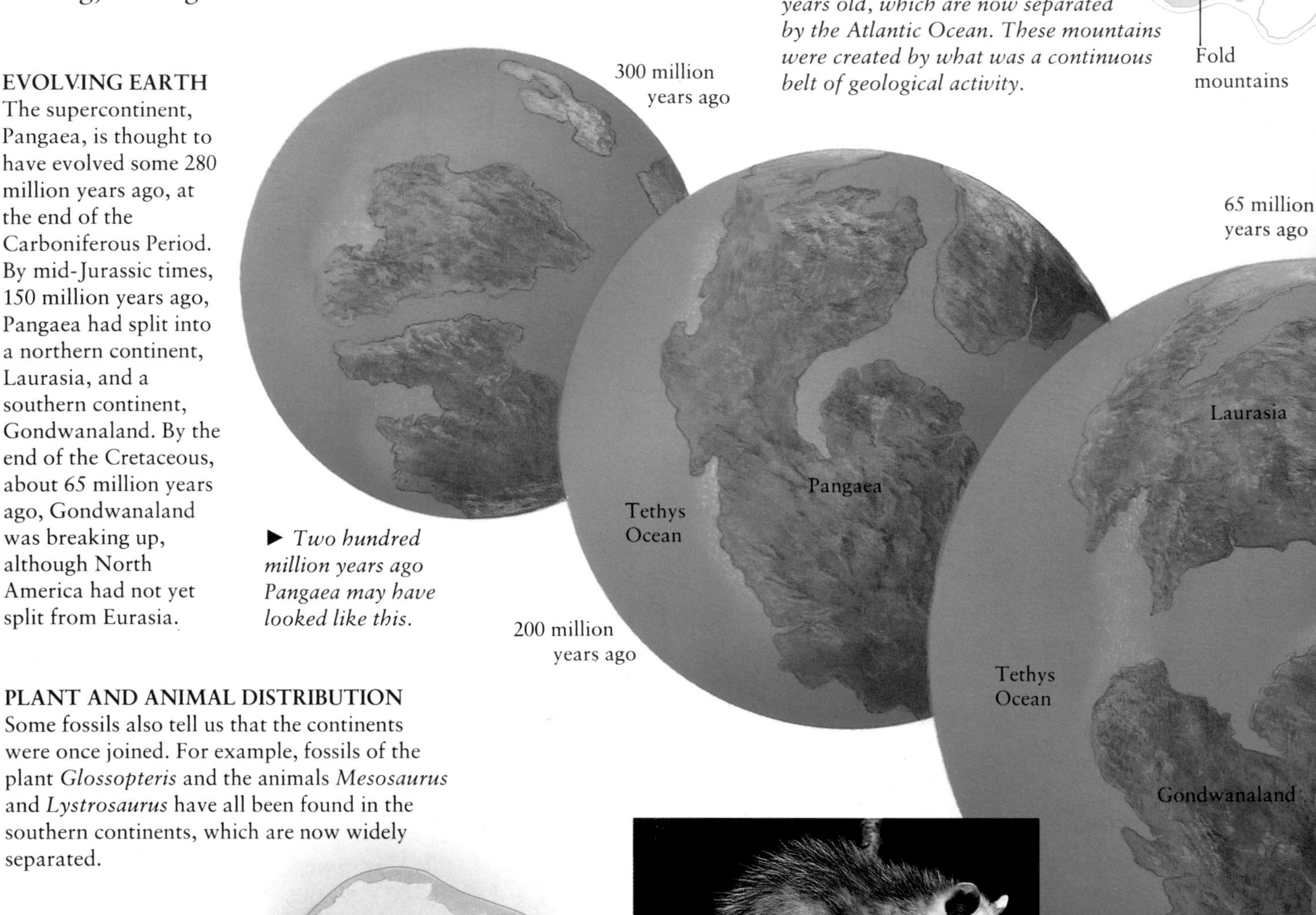

► *Two hundred million years ago Pangaea may have looked like this.*

PLANT AND ANIMAL DISTRIBUTION
Some fossils also tell us that the continents were once joined. For example, fossils of the plant *Glossopteris* and the animals *Mesosaurus* and *Lystrosaurus* have all been found in the southern continents, which are now widely separated.

▲ *Some very similar species exist far apart. This marsupial opposum lives in North America, whereas most marsupials live in Australasia.*

Mesosaurus

TODAY

The present distribution of the continents has taken place in the last 65 million years. Today, the drift still continues. The Atlantic Ocean is getting wider by a few inches a year, the Pacific Ocean is getting smaller, and the Red Sea is part of a crack in the crust that will widen to produce a new ocean millions of years in the future.

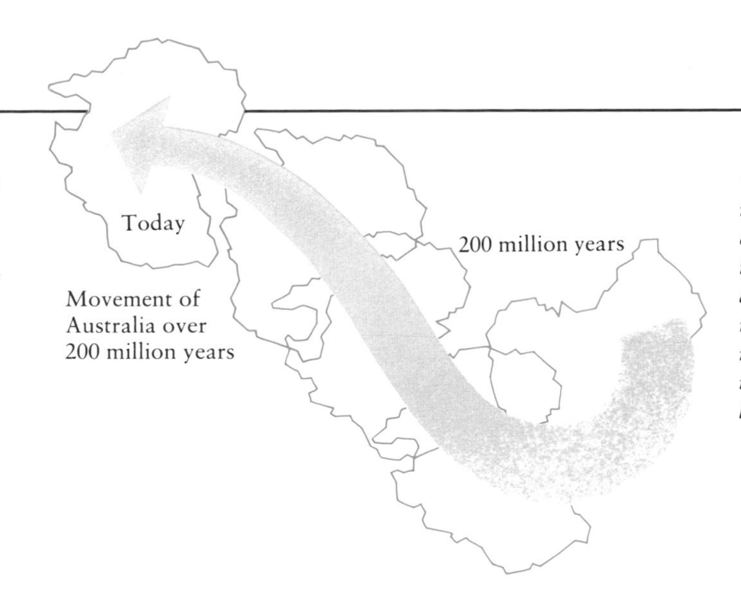

◀ *Australia has nearly turned completely around from its original position and is now moving northward. In 50 million years time it will be touching the landmass of Eurasia.*

Today

▼ *The world begins to look more familiar 65 million years ago. The widening South Atlantic Ocean has separated Africa and South America, and Madagascar has split from Africa, but Australia and Antarctica are still joined.*

25 million years ago

North America

Eurasia

Africa

South America

ALFRED WEGENER (1880–1930)

Alfred Wegener was born in Berlin, Germany. He studied meteorology at the Universities of Heidelberg, Innsbruck in Austria, and in Berlin. In 1924 he returned to Graz, Austria, to become professor of meteorology and geophysics.

In 1910 the American scientist F. B. Taylor put forward the idea that whole continents could have drifted across the surface of the planet. This theory was taken up in 1911 by H. B. Baker and then by Wegener in 1912, who developed his theory of continental drift known as the Wegener hypothesis. He published *Origins of Continents and Oceans* in which he explained his ideas. Wegener also suggested that, with the shifting land masses, the magnetic poles moved around. He died in Greenland on his fourth expedition there.

Plate Tectonics

The word "tectonics" comes from the Greek *tekton*, meaning "builder." The theory suggests that the surface of the Earth is made up of rigid plates of lithosphere which "float" on the more mobile asthenosphere. Owing to movements in the asthenosphere, the plates are in constant motion. It explains many of the major processes of the Earth, such as the drifting of continents, mountain building, and earthquake and volcanic activity. Much of this activity occurs at the edges, or margins, of the plates.

► *Movements of the plates of lithosphere may be driven by convection currents in the asthenosphere. Hot currents rise, then cool as they reach the surface. At the same time, cooler currents sink. This movement carries the crustal plates.*

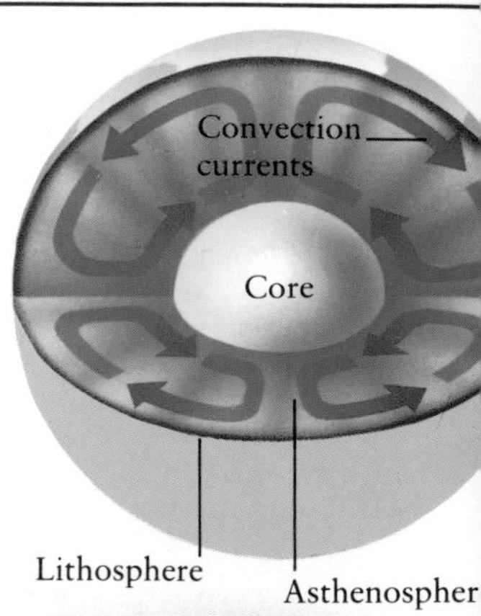

SEAFLOOR SPREADING
Down the middle of the Atlantic Ocean floor is a ridge where volcanic activity takes place. This ridge marks a plate margin. Along the ridge are cracks where molten rocks push up to form new crust. The crust spreads away from the ridge, and the ocean basin widens.

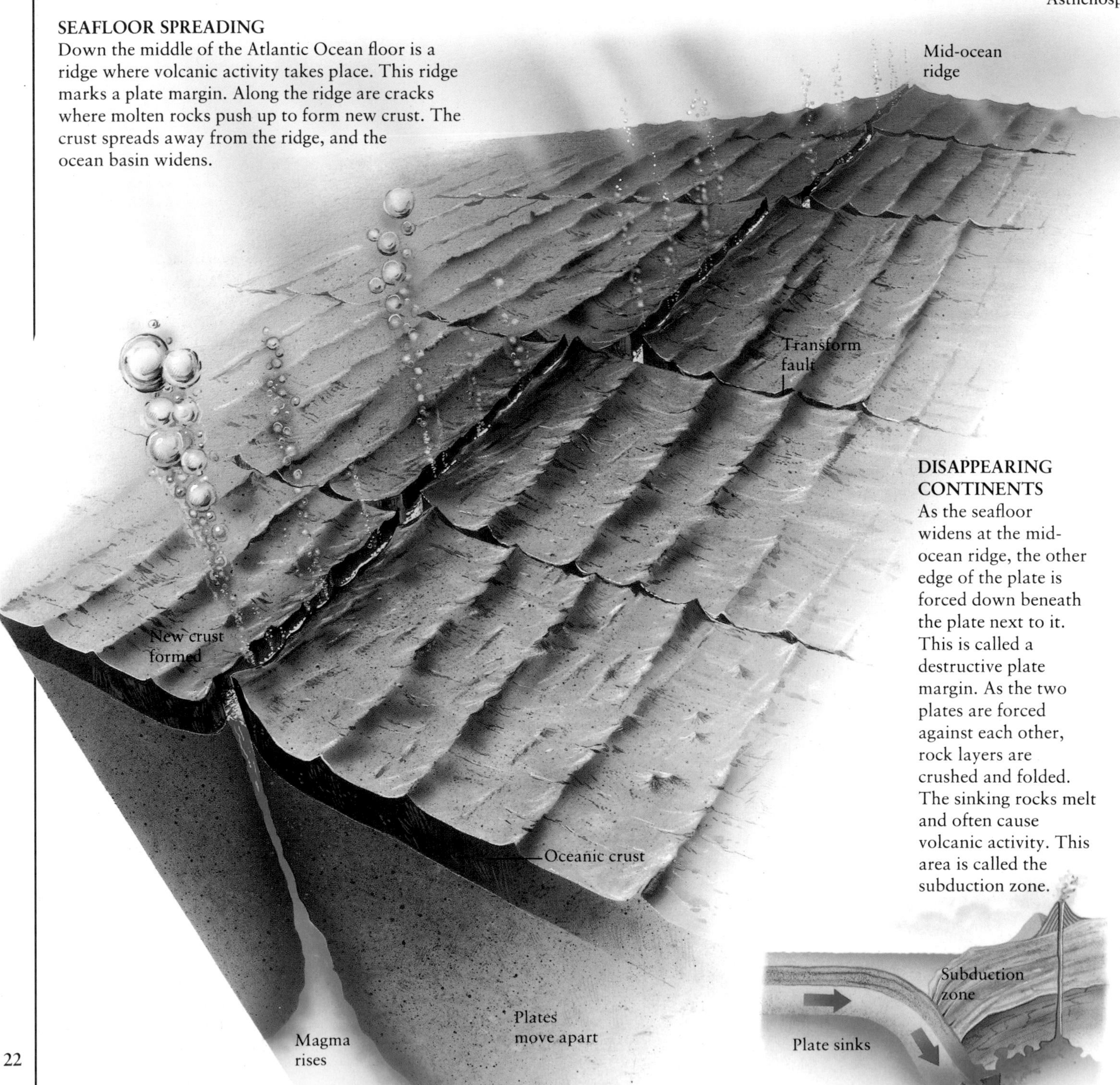

DISAPPEARING CONTINENTS
As the seafloor widens at the mid-ocean ridge, the other edge of the plate is forced down beneath the plate next to it. This is called a destructive plate margin. As the two plates are forced against each other, rock layers are crushed and folded. The sinking rocks melt and often cause volcanic activity. This area is called the subduction zone.

ICELAND

Iceland is situated on the mid-Atlantic ridge between Greenland and Scandinavia. It was formed from eruptions from the ridge and is still getting wider by 1 in. (2.5 cm) a year. This activity means that it is a volcanic island, with geysers and hot springs.

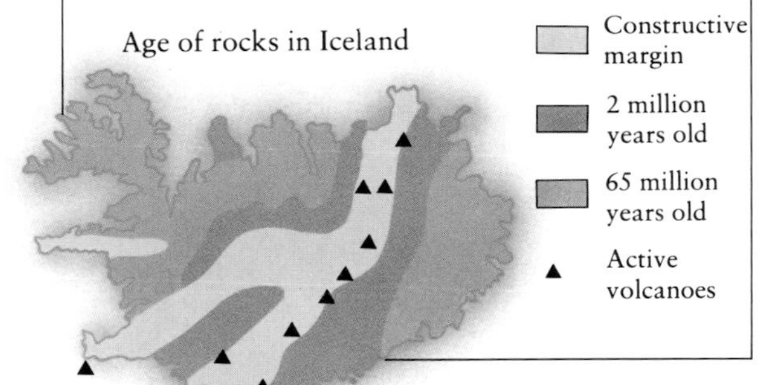

OCEANS AND CONTINENTS

All oceanic crust is less than 200 million years old. It has four main layers. The top layer is made up of sand and mud laid down in the world's seas. Beneath it is a layer of basalt. Then comes a layer of another dark rock called gabbro, and finally a thin layer above the mantle. Oceanic crust is about 7 mi. (11 km) thick.

The thicker continental crust has only two layers: the upper one is mainly granite, with gabbro beneath. On average, continental crust is about 22 mi. (35 km) thick, but may be 31 mi. (50 km) thick under recently built mountains.

► *The crust that forms the continents is far thicker than that under the oceans.*

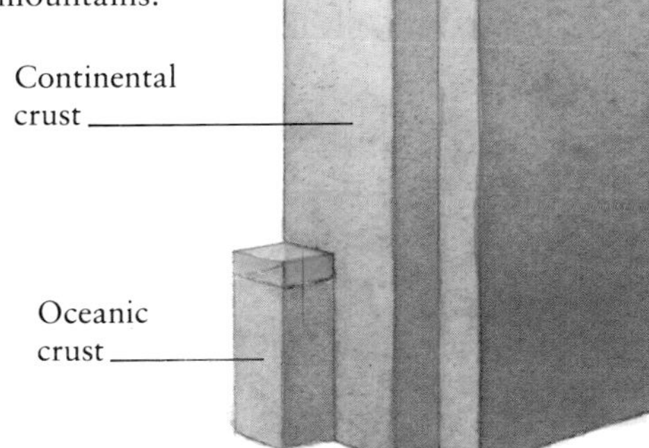

Plate margin

Transform faults

◄ *The Earth's crust is made up of eight main plates and several smaller ones. The edges of these plates are marked by ridges and trenches. New crust is formed at the ridges and destroyed at the trenches. Large faults, known as transform faults, occur at right angles to the mid-oceanic ridges. As new crust is created in the mid-Atlantic, the North American and South American plates are moving westward. India continues to move northward; as it collides with the Eurasian plate, which is traveling eastward, the Himalayas get higher. The African plate continues its drift to the northwest as the Red Sea opens up.*

FACTS ABOUT PLATE TECTONICS

- When lavas cool and harden into rock, some minerals are magnetized in the direction of the Earth's magnetic poles at that time. About every 400,000–500,000 years, the Earth's magnetic poles reverse.
- By measuring the ages and magnetism of rocks on either side of the mid-Atlantic ridge, geologists have proved that the Atlantic is widening.
- The Atlantic widens by 0.4–2 in. (1–5 cm) a year.
- It has taken about 200 million years to separate South America from Africa and create the Atlantic Ocean, and about 40 million years for Australia and the Antarctic to move apart to their present positions.

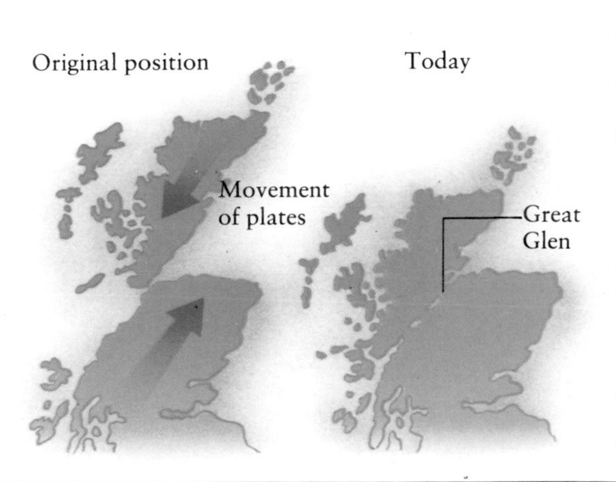

► *Running from the northeast of Scotland to the southwest, a series of lochs follows the line of the Great Glen. This valley is a good example of a tear fault* (see page 26). *Some 400 million years ago, the land north of the fault slid southwest by 60 mi. (100 km).*

Mountains

A mountain is an area of high ground that is higher (over 1,000 ft. [300 m]) than a hill. A group of mountains is called a range. The greatest mountain ranges are the European Alps, the Andes of South America, the Rockies of North America, and—the highest of all—the Himalayas of Asia. It takes millions of years for mountains to be formed. The process is going on continuously as sections of the Earth's crust are thrust, folded, and broken, pushing up rocks to make new mountains.

HOW MOUNTAINS ARE FORMED

The surface of the Earth is made up of giant slow-moving plates of crust and mantle. Where two continental plates collide, a mountain belt is thrust slowly upward. The sediments of the ocean floor are squeezed into folds, which may be tens of thousands of feet high. The folds may then be overturned, one on top of the other. Beneath the chain of mountains is a thick layer of continental crust.

THE HIMALAYAS

The high mountains of the Himalayas mark a region of the Earth's crust where two continental plates are colliding. The Indo-Australian plate is pushing north into the Eurasian plate. Eventually the two continents will become locked together.

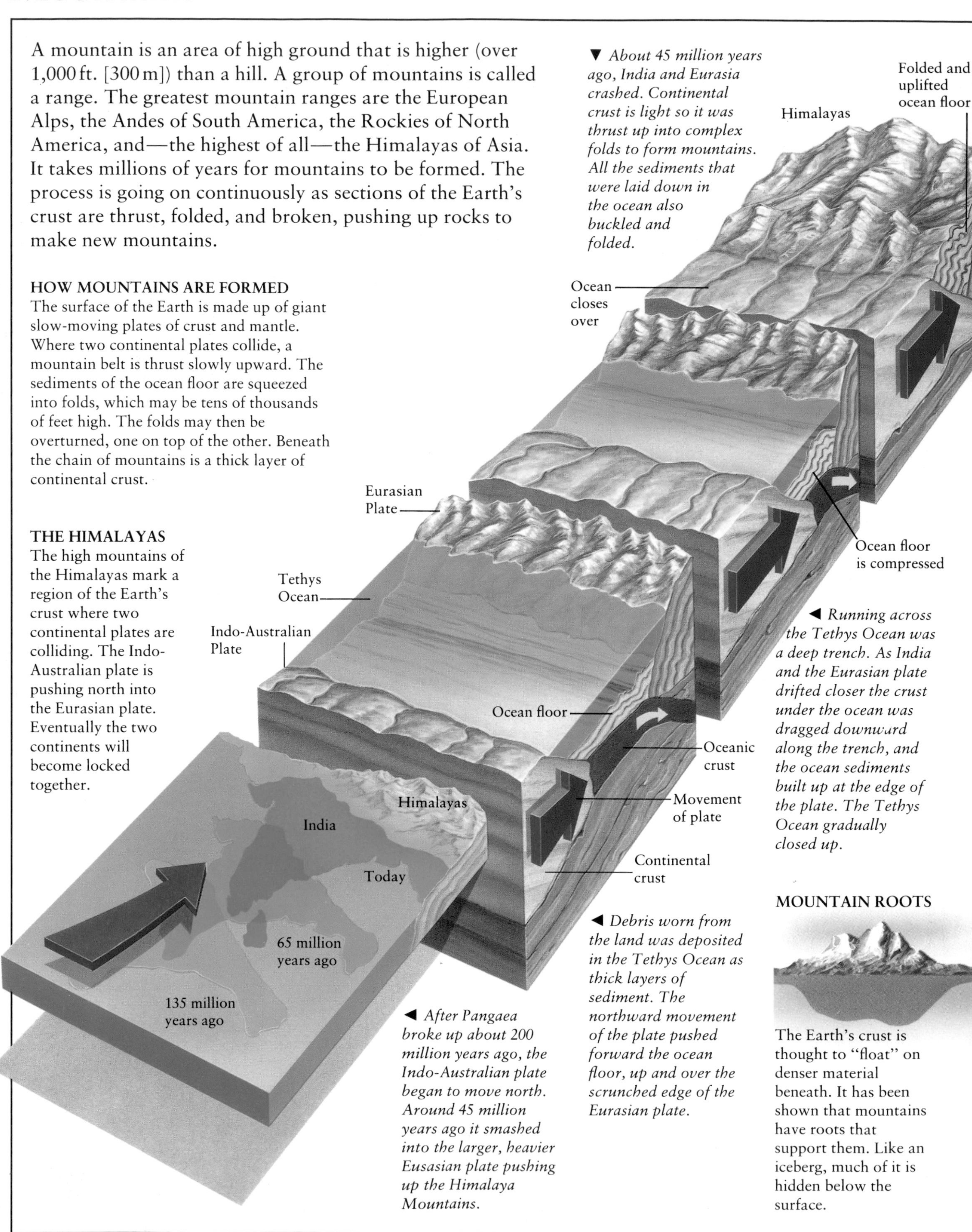

▼ *About 45 million years ago, India and Eurasia crashed. Continental crust is light so it was thrust up into complex folds to form mountains. All the sediments that were laid down in the ocean also buckled and folded.*

◀ *Running across the Tethys Ocean was a deep trench. As India and the Eurasian plate drifted closer the crust under the ocean was dragged downward along the trench, and the ocean sediments built up at the edge of the plate. The Tethys Ocean gradually closed up.*

◀ *After Pangaea broke up about 200 million years ago, the Indo-Australian plate began to move north. Around 45 million years ago it smashed into the larger, heavier Eusasian plate pushing up the Himalaya Mountains.*

◀ *Debris worn from the land was deposited in the Tethys Ocean as thick layers of sediment. The northward movement of the plate pushed forward the ocean floor, up and over the scrunched edge of the Eurasian plate.*

MOUNTAIN ROOTS

The Earth's crust is thought to "float" on denser material beneath. It has been shown that mountains have roots that support them. Like an iceberg, much of it is hidden below the surface.

▲ *The shape of a mountain depends on how it was formed, its age, and how much it has been eroded and worn. Young mountains, such as the Himalayas, are high and rugged. Old mountains are smoother and lower.*

MOUNTAIN LIFE

Different animals and plants live in different zones, or areas, of a mountain. Many cannot survive the extreme cold and thin air at high altitudes, but flourish on the lower slopes. The types of animals and plants that live in each of the zones vary in different mountain ranges. In warm equatorial regions, trees may be able to live at 13,000 ft. (4,000 m). But in the colder Alps, trees can survive only up to 6,000 ft. (1,800 m).

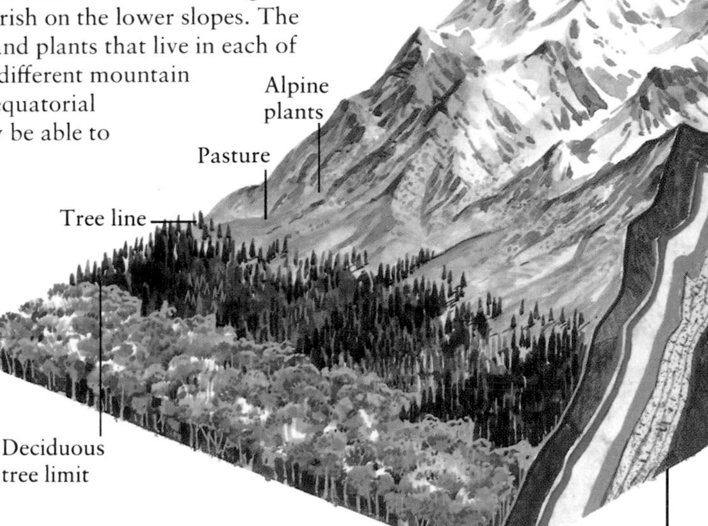

MOUNTAIN BELTS

Eventually, weathering wears away the crumpled and faulted rocks of a mountain range. Some rocks resist weathering better than others. Those that do may survive for longer as high, jagged peaks.

MAJOR MOUNTAIN RANGES

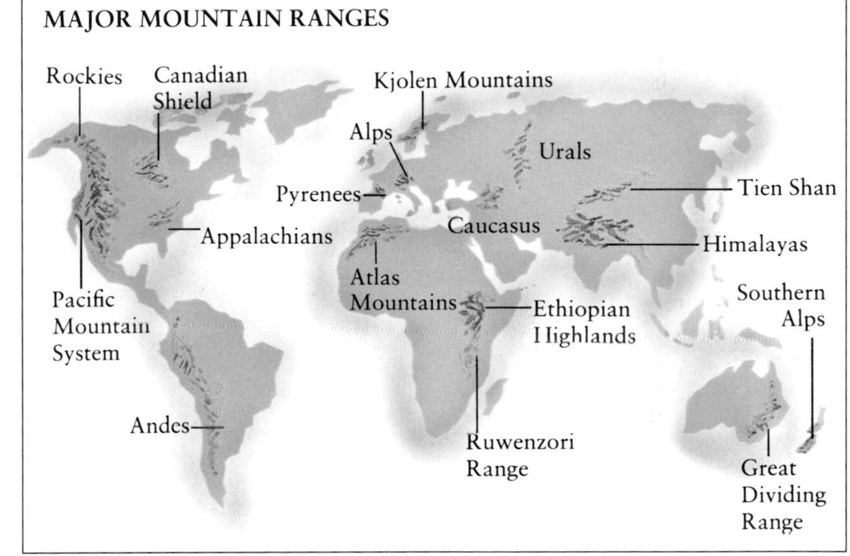

MOUNTAIN DATAFILE

The world's highest peaks (14 over 24,000 ft. [8,000 m]) are in the Himalaya-Karakoram ranges.

Top 5 in Asia

1 Mount Everest 29,028 ft. (8,848 m)
2 Godwin-Austin (K2) 28,250 ft. (8,611 m)
3 Kanchenjunga 28,208 ft. (8,597 m)
4 Makalu 27,824 ft. (8,480 m)
5 Dhaulagiri 26,810 ft. (8,172 m)

Highest in other continents

6 Aconcagua (South America) 22,831 ft. (6,959 m)
7 McKinley (North America) 20,320 ft. (6,194 m)
8 Kilimanjaro (Africa) 19,340 ft. (5,895 m)
9 Elbrus (Europe) 18,481 ft. (5,633 m)
10 Vinson Massif (Antarctica) 16,860 ft. (5,139 m)
11 Mt. Wilhelm (Oceania) 14,793 ft. (4,509 m)

► *Mountaineering began as a sport in the mid-1800s, when British and other European climbers tried to scale the peaks of the Alps. On May 29, 1953, mountaineering history was made when the New Zealander Edmund Hillary (1919–) and his Sherpa guide, Tenzing Norgay (1914–86), reached the summit of Mount Everest, the world's highest peak. In 1975, the Japanese climber Junko Tabei (1939–) became the first woman to climb Everest.*

Bending and Breaking

If you hit a rock hard enough with a hammer, it will break. It is harder to imagine rocks bending. But when movements in the Earth cause stress to build up over a long period of time, the layers of rock may break or bend. A break in the rocks is known as a fault. When rock layers bend into waves, or sometimes overturn until they are upside-down, they are said to have folded. There are different kinds of folds and faults, depending on the type of rock and the force that has pushed or pulled it.

THE PARTS OF A FAULT

The angle made by the fault and the horizontal is called the dip. If the angle is measured from the vertical it is called the hade. The relative movements of the blocks of rock on either side of the fault are known as its slip; the vertical movement when the blocks have slipped up or down the dip is the throw.

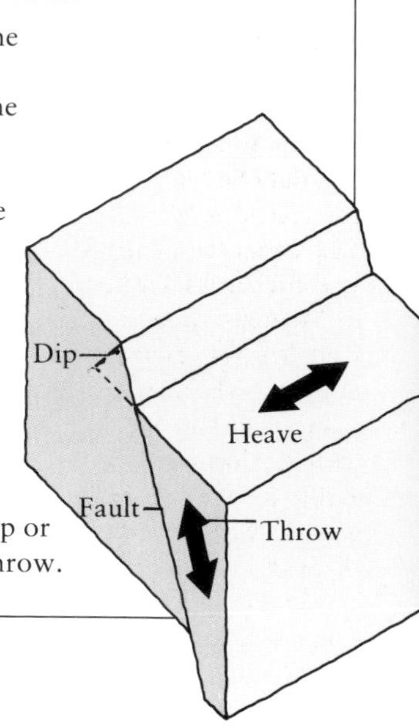

FOLDS BEFORE FAULTS

If you take several layers of paper, hold them by the edges and then push, they will bend into a dome. If you try the same experiment with some cardboard, it will probably not bend as easily, and may break. Rocks can behave more like the paper than the cardboard. When they are pushed or pulled, they may recover if the stress is removed. But if the rock is brittle and is deformed too much it will break. It will bend if it is more elastic.

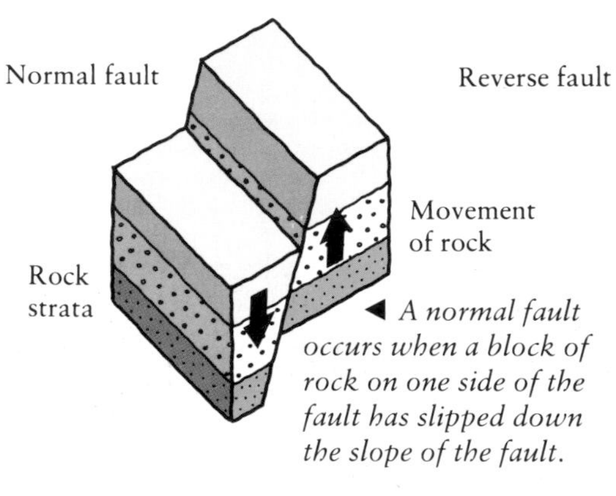

◀ *A normal fault occurs when a block of rock on one side of the fault has slipped down the slope of the fault.*

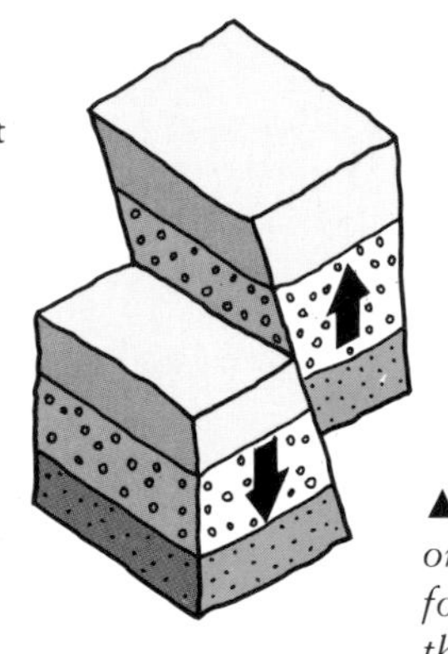

▲ *When a block on one side of the fault is forced up the slope of the fault, it is called a reverse fault.*

Strike-slip fault

▲ *In a transcurrent fault (or strike-slip fault) the main movement is horizontal when the beds of rock are horizontal.*

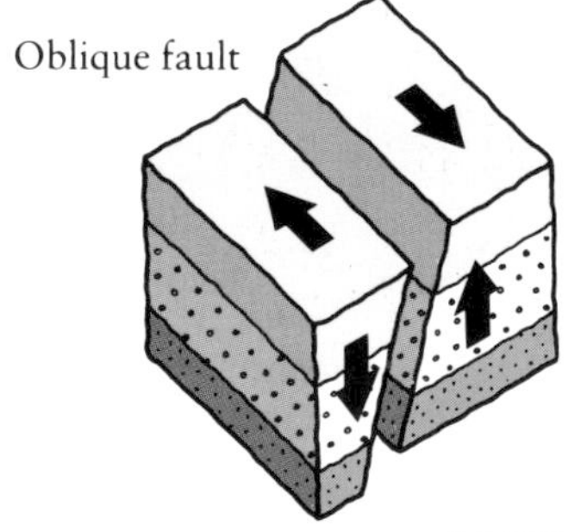

◀ *An oblique fault resembles a combination of a transcurrent fault with either a normal or a reverse fault. The amounts of movement in each direction are similar.*

▼ *When a block of rock has been thrown upward between two steeply angled faults, it is called a horst.*

Horst

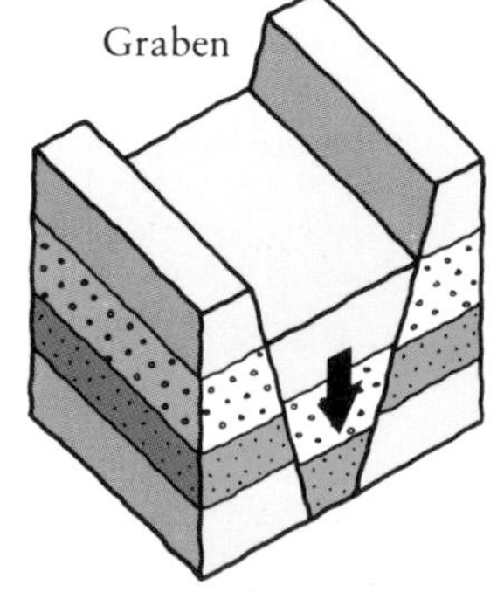

▲ *A graben is a down-thrown block of rock between two steeply angled normal faults.*

▼ *Sandstone and shale beds in this cliff have been crushed into folds, and then shifted onto one end by the movement of the Earth's crust.*

THE GREAT RIFT VALLEY

When a graben occurs on a very large scale, it is called a rift valley. Such valleys are usually quite straight and may be hundreds of miles long. The Great Rift Valley stretches from Turkey to Mozambique. In East Africa it divides in two, with the north-eastern rift running from Ethiopia to Zambia. The south-western rift curves through Uganda and Tanzania and contains some of the great East African lakes.

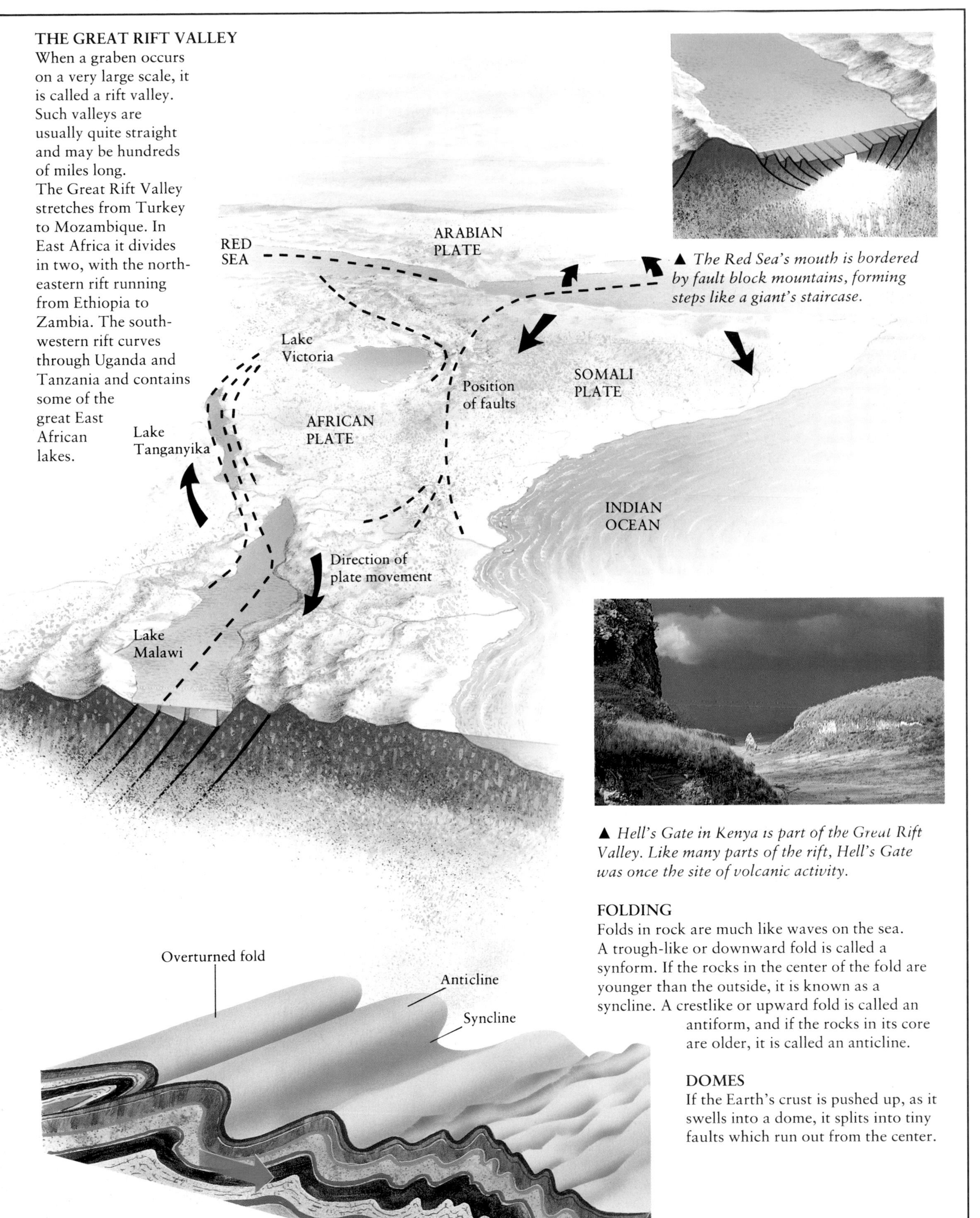

▲ *The Red Sea's mouth is bordered by fault block mountains, forming steps like a giant's staircase.*

▲ *Hell's Gate in Kenya is part of the Great Rift Valley. Like many parts of the rift, Hell's Gate was once the site of volcanic activity.*

FOLDING

Folds in rock are much like waves on the sea. A trough-like or downward fold is called a synform. If the rocks in the center of the fold are younger than the outside, it is known as a syncline. A crestlike or upward fold is called an antiform, and if the rocks in its core are older, it is called an anticline.

DOMES

If the Earth's crust is pushed up, as it swells into a dome, it splits into tiny faults which run out from the center.

Rocks

Granite, sandstone, chalk, marble, and slate are all different types of rock. The pebbles you find on the beach are rocks that have been worn down and smoothed by the action of the sea. The stones that are used to build structures, from small cottages to magnificent cathedrals, are rocks. Not all rocks are hard; clay is a type of rock, but it is quite soft. Petrologists (scientists who study rocks) might define a rock as any natural mass of mineral matter that makes up the Earth's crust.

IGNEOUS ROCKS

Granite is coarse-grained because it has cooled slowly. Syenite is similar to granite but less common. Basalt is almost black, fine-grained, and has cooled quickly.

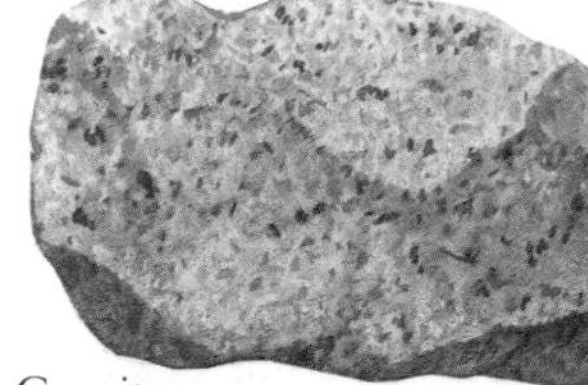

Granite

Basalt

Syenite

HOW ROCKS ARE FORMED

There are three main groups of rocks. Igneous rocks are formed from lava hurled out of a volcano, or from hot magma forced up through the ground. Sedimentary rocks are made from sediments formed by the erosion and weathering of other rocks. The sediments are carried by wind or water to the sea, where they are deposited and harden to rock. Metamorphic rocks are rocks that have been changed by heat and/or pressure.

Sedimentary rocks

Limestone

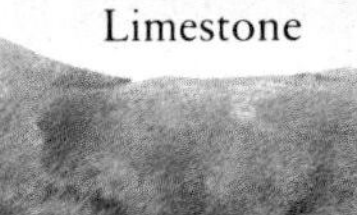

SEDIMENTARY ROCKS

Sandstone is made from grains of sand that have been naturally cemented together. The red rock of Devon, England, is a typical sandstone. Chalk is made up of millions of tiny calcium carbonate (lime) skeletons.

Chalk

Sandstone

GEOLOGICAL TIMESCALE

For the last 150 years, scientists have been figuring out the ages of rocks. Unless they have been overturned, the oldest rocks are deeper than younger ones. Rocks can be related to a scale of different ages.

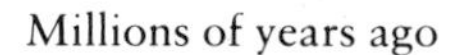

Millions of years ago

4600	570	505	440	410	360
Period	Precambrian	Cambrian	Ordovician	Silurian	Devonian
Era		PALEOZOIC			

METAMORPHIC ROCKS

Metamorphic rocks are changed rocks. When a rock is subjected to heat and/or pressure, new minerals are formed, altering the characteristics of the rock. Slate is a hard rock formed from muds and clays. It splits easily into thin sheets.

MARBLE

Marble is a metamorphic rock. It is a type of limestone (sedimentary rock) which may have been changed by the heat of a lava flow or by contact with molten rocks far below the ground. Marble can be polished, and it may be pure white in color or mottled or banded. It has long been favored by architects and sculptors. The Carrara marble quarry in Italy *(left)* produces some of the world's finest stone.

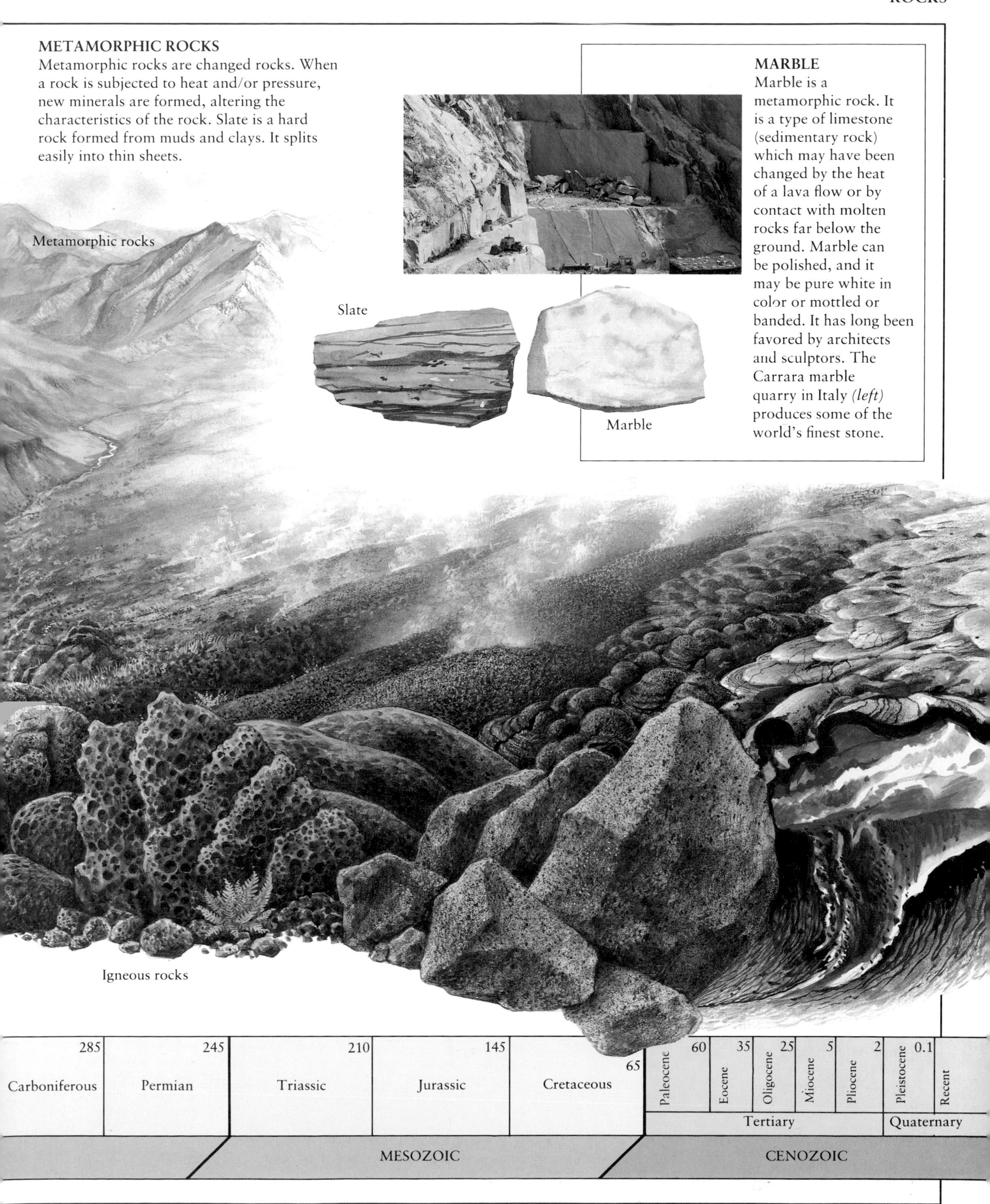

Minerals and Gems

Minerals are the building blocks of rocks. All rocks, igneous, sedimentary, or metamorphic, are composed of minerals. A mineral is a chemical compound that occurs naturally. Each different mineral is made up of crystals of a particular chemical. Minerals can be identified by their hardness, color, density, the way they reflect light, and the way they break. Sometimes, using a hand lens, you can even identify the shape of the individual crystals of a mineral within a rock.

▲ *One of the most common minerals is salt. Common salt is sodium chloride and occurs naturally in the sea. When an inland lake in a hot region evaporates, the salts are left behind as crystals, forming a salt pan, or growing to form pillars such as these, in the Dead Sea.*

ROCK-FORMING MINERALS

Rocks are made up of mineral grains. The most common minerals that make up igneous rocks include quartz, plagioclase, and olivine. Augite is found in metamorphic rocks. Dolomite makes up limestone sedimentary rocks.

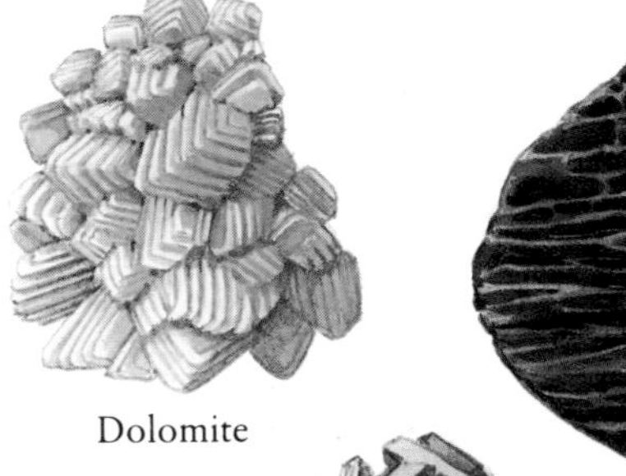

Dolomite

Plagioclase

Olivine

Augite

HOW MINERALS ARE FORMED

All minerals are originally formed from hot magma. When the magma cools, crystals of minerals appear. These first crystals may sink in the magma so that the composition of the magma changes with depth. Thus, a sequence of minerals is formed in the rocks as the magma cools.

WHAT MINERALS ARE MADE OF

About 90 different chemical elements are found naturally in the Earth's crust. But almost 99 percent of all minerals are made of just 8 elements: oxygen, silicon, aluminum, iron, calcium, sodium, potassium, and magnesium. Some minerals, such as gold and diamond, consist of a single element. These minerals are called native elements.

► *In areas of volcanic activity, hot water under pressure may force its way into cracks in the rocks. The water contains dissolved minerals. The minerals may crystallize on the sides of the crack forming a vein. Important ore minerals (containing metals) are formed in this way.*

FACTS ABOUT MINERALS AND GEMS

- There are at least 2,000 minerals that have been named and identified. However, most rocks are made up of no more than 12 different classes of minerals—the rock-forming minerals.
- The most common element in the Earth's crust is oxygen, forming over 46 percent by weight. The second most common element is silicon at just over 27 percent by weight. Quartz is silicon dioxide. Quartz is a very common mineral.
- The largest diamond ever found was the Cullinan diamond, discovered in South Africa in 1905. It was 3,106 carats, which means it weighed more than 21 oz (600 g).
- The highest price ever paid for a diamond was more than $9 million.
- The largest cut emerald was 86,136 carats (well over 37 lb [17 kg]). It was found in Brazil and was valued in 1982 at over $1 million.

CRYSTAL SYSTEMS

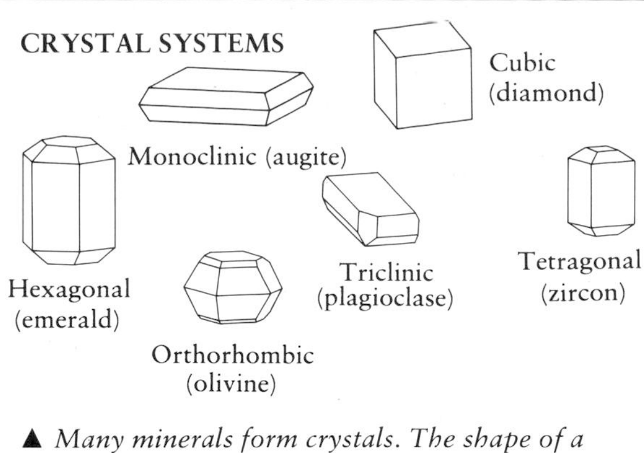

▲ *Many minerals form crystals. The shape of a crystal is determined by the arrangement of its atoms. A crystal has many flat faces. The angles between each face are characteristic of a given type of crystal. The whole crystal is symmetrical. On the basis of its symmetry, any crystal can be included within one of just six (sometimes seven) crystal systems.*

▼ *Most minerals are formed underground when heat and pressure transforms one form of rock into another. The minerals in molten rock or dissolved in very hot water crystallize out of solution as the temperature falls. Lighter minerals occur above denser minerals. If the crystals form slowly then gemstones may form.*

GEMSTONES

A gemstone is a mineral that is especially beautiful and rare. Gems are formed under particular conditions of temperature and pressure, when the right minerals are present and crystals can form. These conditions are found deep in the Earth's crust. A gem may have a beautiful color: deep green emeralds or brilliant red rubies. Or its surface may show a rainbow of colors when it is moved, as in the case of opal.

Silver

Gold

▲ *Gold and silver are elements. Diamond and graphite are both forms of carbon but their atoms are arranged differently. Diamond is the hardest known naturally occurring substance. It may occur as gem-quality stones. Graphite is soft, black, and feels greasy.*

Diamond

Graphite

▼ *Under the sea, minerals dissolved in water crystallize around the vents of faults or fissures in the Earth's crust. They also precipitate (become solid) in the sea water above the sea floor.*

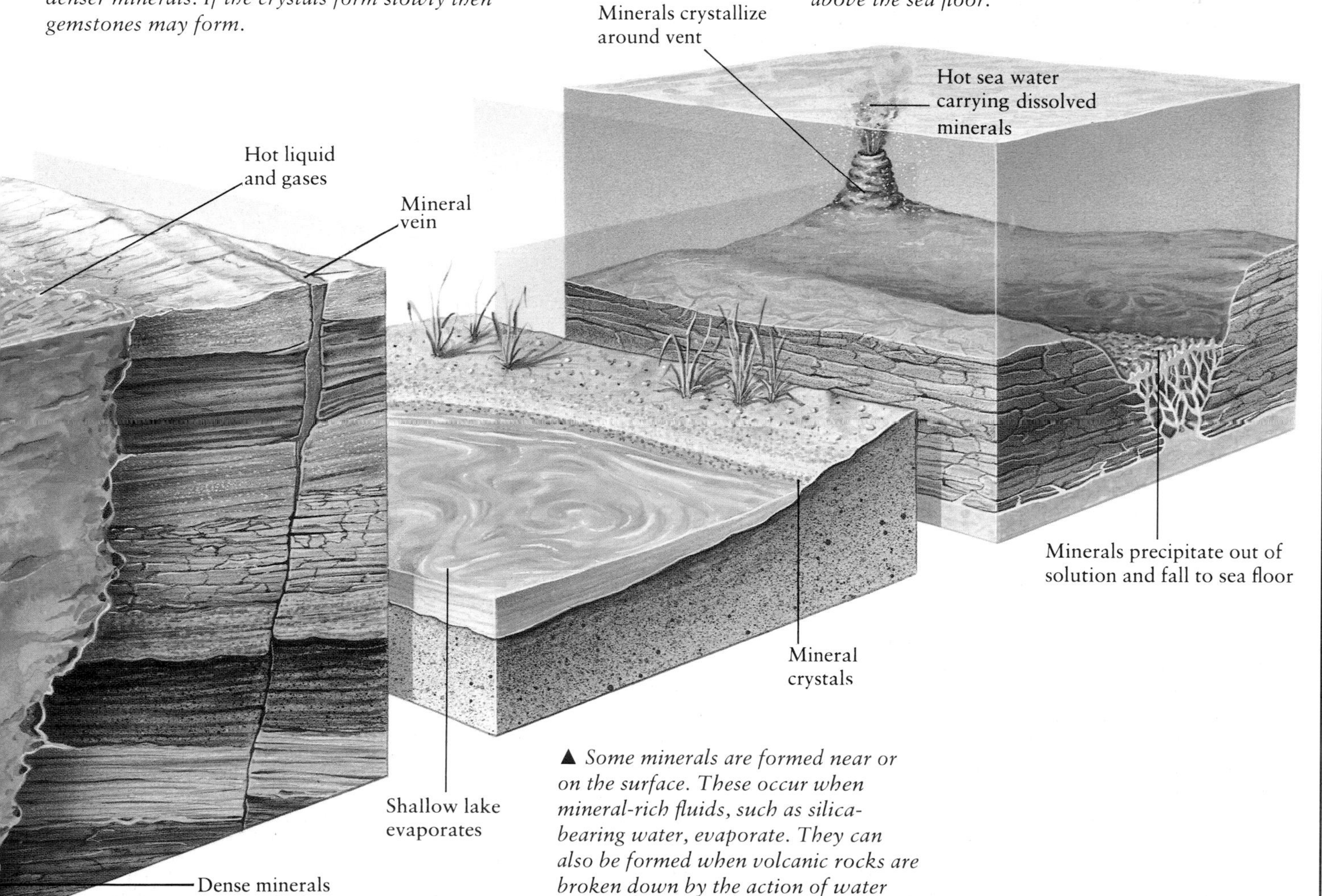

▲ *Some minerals are formed near or on the surface. These occur when mineral-rich fluids, such as silica-bearing water, evaporate. They can also be formed when volcanic rocks are broken down by the action of water and wind.*

Frozen in Time

The word fossil comes from the Latin, *fossilis*, meaning "dug up." Until the beginning of the 18th century, fossil meant anything that was dug out of the ground. Now we use the word to describe any remains of animals or plants that lived before about 10,000 years ago. The term is also used to describe the "fossil fuels"—coal, oil, and gas—because these fuels have been formed from the remains of ancient plants and animals. Fossils include skeletons, teeth, tracks, leaves, and even plant pollen.

▼ When sap runs down a tree trunk, it may trap an insect climbing up. The sap hardens into a golden-brown resin called amber, with the insect preserved inside it.

Oil may seep up through the ground to form tar pits. An animal may fall into a pit and its remains be preserved. Parts of ancient plants are often preserved in coal.

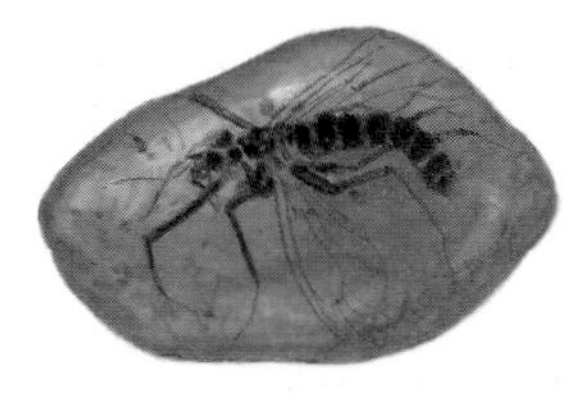

Fossil insect in amber

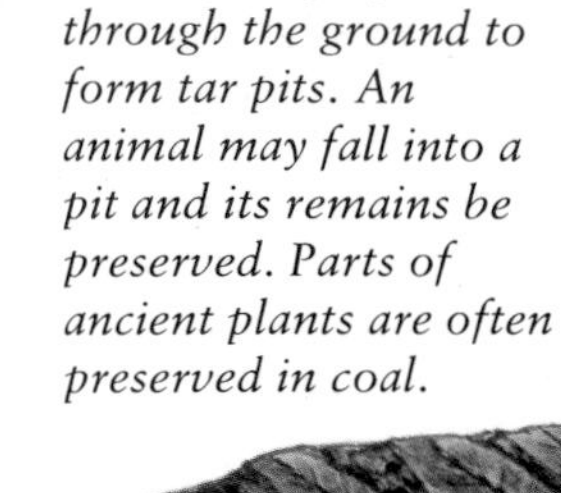

Fossil leaf in coal

Tar pit

HOW FOSSILS ARE FORMED

Most fossils are found in sedimentary rocks, such as limestone or shale, which have been formed in the sea. Fossils of sea creatures are therefore much more common than those of land creatures. Those land animals and plants that have been preserved are usually found in sediments in a lake, river, or estuary.

The soft parts of an animal or plant—the flesh, or a delicate shoot—will decay more quickly than its hard parts after the creature dies. So shells, teeth, and bones are much more likely to be preserved as fossils than skin or organs. But sometimes, even an animal's last meal may be preserved as a fossil.

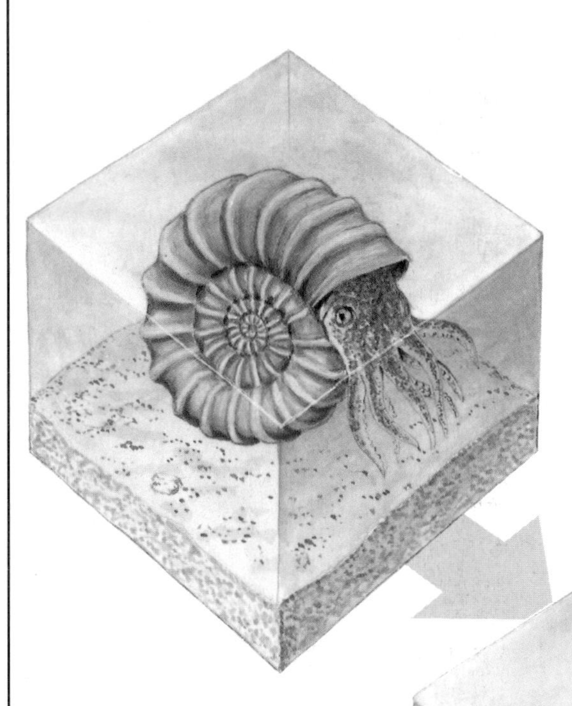

◄ When a sea animal, such as an ammonite, dies, its body will sink to the seabed. If there are swift currents the shell may be swept around. The shell will quickly break up and itself become part of the sediment on the seabed.

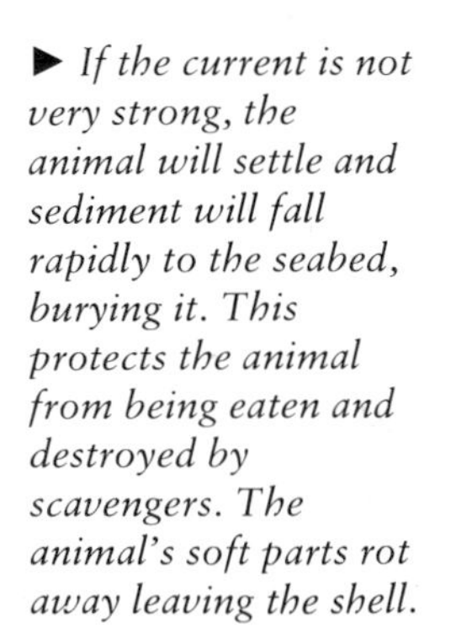

► If the current is not very strong, the animal will settle and sediment will fall rapidly to the seabed, burying it. This protects the animal from being eaten and destroyed by scavengers. The animal's soft parts rot away leaving the shell.

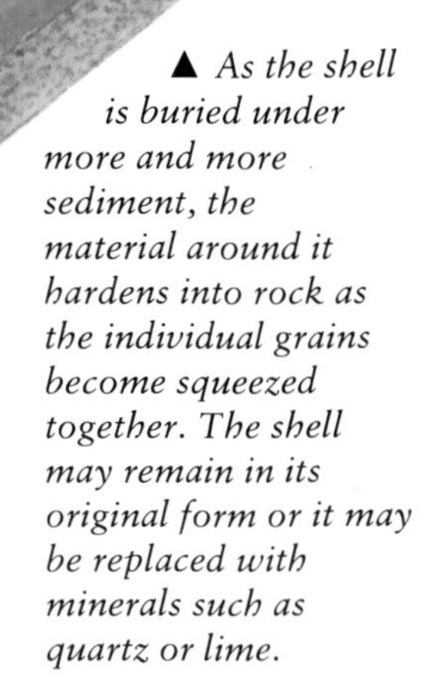

▲ As the shell is buried under more and more sediment, the material around it hardens into rock as the individual grains become squeezed together. The shell may remain in its original form or it may be replaced with minerals such as quartz or lime.

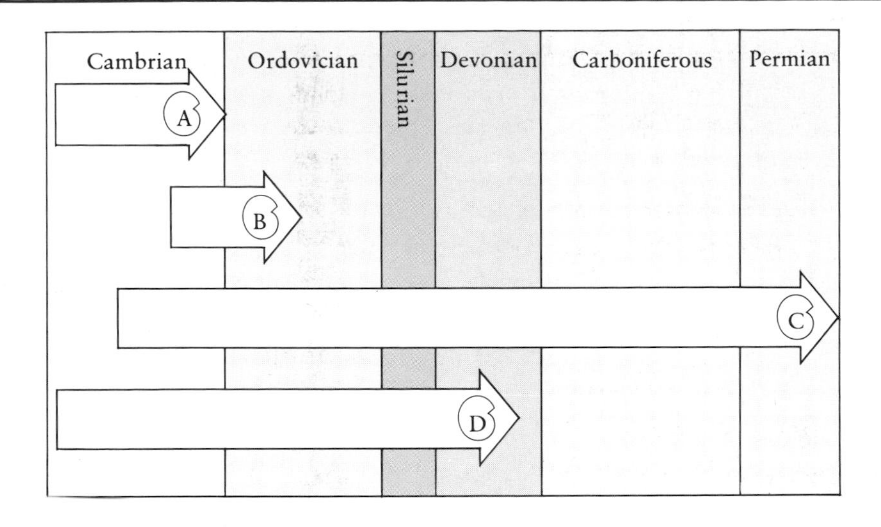

DATING ROCKS FROM FOSSILS

Many types of animals and plants survived on Earth for only a limited period of time. Ammonites, pterosaurs, and dinosaurs all became extinct, or died out. The fossils of extinct animals can be used to date the surrounding rock. This is because these fossils cannot be found in rocks that are younger than the time at which the animal disappeared. Rocks of a known age can also be used to date fossils.

FACTS ABOUT FOSSILS

- The oldest known fossils are called stromatolites. Stromatolites are the remains of matlike structures formed by bacteria. Stromatolites have been found in western North America and Australia. Some are more than 2 billion years old.
- One of the largest fossil bones that has ever been found was the shoulder blade of a giant dinosaur called *Supersaurus*. It measured over 6.5 ft. (2 m) and was found in Colorado in 1972.
- Most of the world's fossils have been found in rocks less than 600 million years old. This is the time when animals with hard parts first evolved.
- Paleontologists (who study fossils) can learn about prehistoric climates from minute fossilized pollen grains. The pollen grains can be used as "fingerprints" to identify the plants living at particular times.

MOLDS AND CASTS

Sometimes a shell may be dissolved away by acid waters seeping through the rock. However, the shape of the shell might be preserved in the surrounding rock—a mold. If the mold is then filled with new mineral material, the resulting fossil is called a cast.

▼ *Eventually, the effects of weathering and erosion may wear away the sedimentary rocks and expose the fossil. Or the fossil may be found in a quarry face, or because of road cutting. It is as though the sea animal has been frozen in time.*

▲ *Over many millions of years, the sea may retreat. The rocks in which the fossil lies may become faulted and folded because of movements within the Earth's crust. What was once the seabed may be thrown up into a newly formed mountain range.*

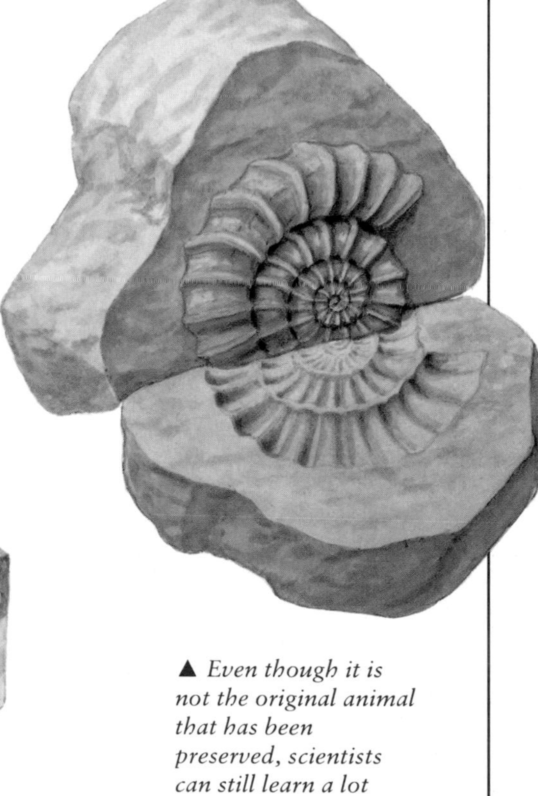

▲ *Even though it is not the original animal that has been preserved, scientists can still learn a lot from the fossil.*

Shaping the Earth

The world's highest mountain is Mount Everest in the Himalaya range, soaring to 29,028 ft. (8,848 m). Britain's highest mountain, Ben Nevis, is 4,406 ft. (1,343 m). Both mountains were created by similar processes of mountain building, so why should one be nearly seven times the height of the other? Part of the answer lies in the effects of weathering and erosion. Ben Nevis is a very old mountain and has been worn down over hundreds of millions of years by the action of wind, rain, ice, and snow.

▲ Plant roots can also break down rock. They work their way into tiny cracks in a rock. As the roots grow, they widen the cracks. Eventually, the force of the roots can shatter the rock. Worms eat large amounts of soil, passing out the waste as casts which can then be washed away by rain.

WEATHERING AND EROSION

Rocks are often formed inside the Earth at high pressures and temperatures. At the Earth's surface, when rocks are exposed, the conditions are quite different. It is this change that causes rocks and minerals to break up. Physical weathering breaks down the rocks into smaller particles, such as sands or grits. In chemical weathering, the minerals that form the rocks are dissolved by the action of water, together with oxygen and carbon dioxide in the air. Once the rocks have been broken down by weathering, the bits are transported elsewhere by ice, wind, running water, or gravity. This is called erosion. Sometimes the actions of people, such as removing plants that hold the soil together, can increase erosion.

▼ These desert rocks have been scoured by sand-carrying wind into strange pillars appearing from the dunes.

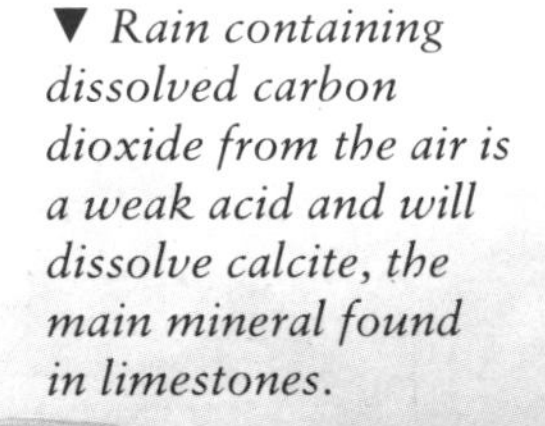

▼ Rain containing dissolved carbon dioxide from the air is a weak acid and will dissolve calcite, the main mineral found in limestones.

Scree formed by frost erosion

Erosion of river valley

▼ Rivers carry weathered rock debris to the coast. The finest particles may be carried out to sea.

Erosion and weathering of folded rocks at surface

Rock tower weathered by wind and rain

► When water freezes, it expands. If water finds its way into cracks in rocks and then freezes, it pushes out against the rock. This may cause the rock to shatter.

◄ Rocks may be subjected to high temperatures during the day, and very low ones at night. They expand and contract, which causes them to split.

SLUMPS AND SLIDES

Gravity often causes erosion. Soil may slide slowly down a slope as a result of disturbance caused by wetting and drying. And gravity can cause loose material to slump. If the material is dry, some kind of shock, such as an earthquake, may cause soil and rock to slump.

▲ *Soil may creep slowly downhill, pulled by gravity.*

▲ *A shock may cause clay to liquefy to form a mudflow.*

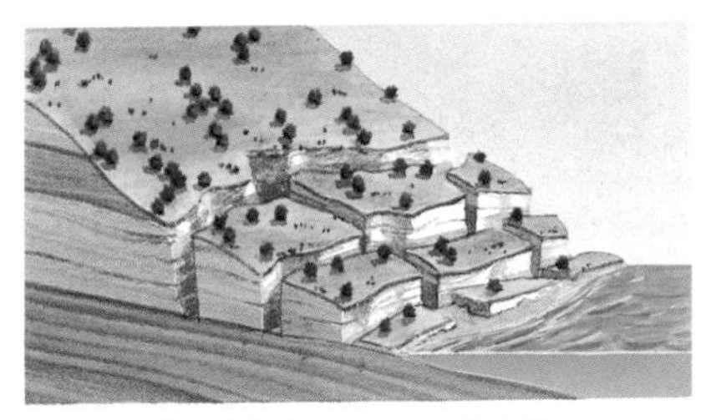

▲ *In a landslide, material falls quite quickly, but the material may fall from a break along a plane or curved surface.*

FACTS ABOUT CAVES AND CAVERNS

- The world's deepest cave is at Rousseau Jean Bernard, in France. It is 5,036 ft. (1,535 m) deep.
- The longest cave system in the world is at least 330 mi. (530 km) long. It is under the Mammoth Cave National Park, Kentucky.
- The world's biggest cavern is the Sarawak Chamber in Sarawak, Indonesia. It is 2,300 ft. (700 m) long, 1,000 ft. (300 m) wide, and 230 ft. (70 m) high, and is supported only by its sides.
- Gaping Gyll, a sinkhole in North Yorkshire, England, descends vertically for more than 360 ft. (110 m).
- The longest stalactite in the world is in a cavern in County Clare in Ireland. It is over 23 ft. (7 m) long.
- Near Lozère in southern France, there is a stalagmite that has now reached 95 ft. (29 m) in height.

CAVES AND CAVERNS

Limestone may be weathered into a pavement of blocks. As acid water works its way down through cracks in limestone, it widens the cracks into passageways. When the water meets a layer it cannot drain through, it runs down until it finds an exit. In this way underground streams and rivers are formed. They dissolve away the limestone to form potholes, caves, and caverns.

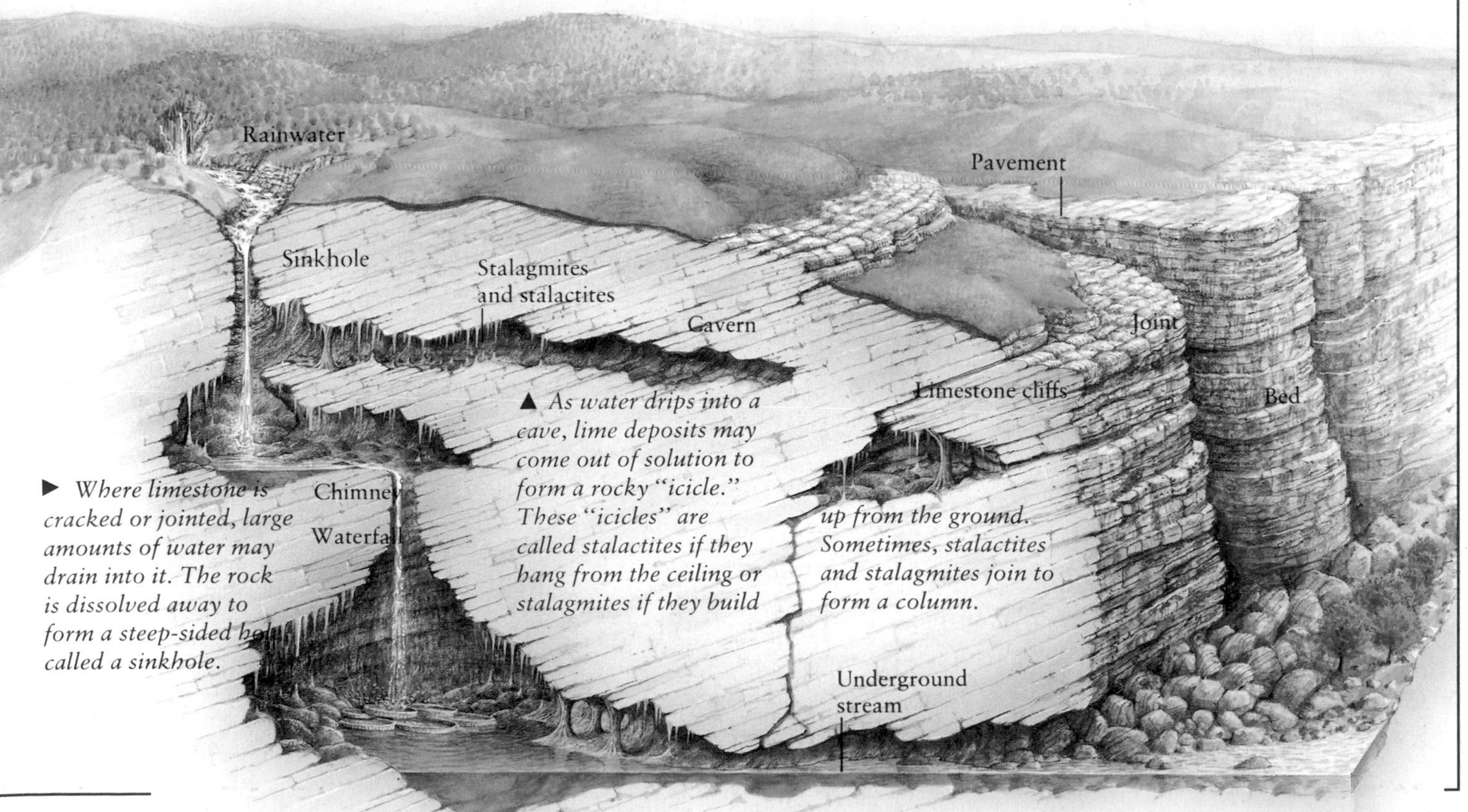

▲ *As water drips into a cave, lime deposits may come out of solution to form a rocky "icicle." These "icicles" are called stalactites if they hang from the ceiling or stalagmites if they build up from the ground. Sometimes, stalactites and stalagmites join to form a column.*

► *Where limestone is cracked or jointed, large amounts of water may drain into it. The rock is dissolved away to form a steep-sided hole called a sinkhole.*

The Work of Ice

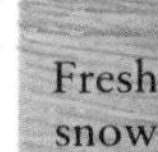

Water freezes into solid ice at 32°F (0°C). As the density of water is greatest at 39.2°F (4°C), ice floats on top of cold water. This is important because it allows animals and plants to survive below the ice on a river, lake, or even the ocean, provided the water does not freeze solid. A large block of ice that floats in the sea is called an iceberg. A large mass of ice on the surface of the land is called a glacier. Glaciers, originally made of snow, are very powerful and their action carves distinctive landscapes.

GLACIERS

Glaciers will form in places where so much snow falls during the winter that not all of it melts or evaporates in the summer. Glaciers are often found high up in mountains at the heads of valleys. Above the snow line, snow accumulates as a permanent snowfield.

Depth (ft.)
Fresh snow
0
3 Névé
Compacted snow
30
Ice
80
150
300 Compressed glacier ice
600 Impermeable ice

► *As the layers of snow build up, they pack close together under the weight of the snow above. Dense snow is called névé.*

► *Air is squeezed out of the névé while water runs into it and freezes. The névé becomes denser.*

▲ *Eventually the white snow is turned into clear, bluish ice, forming a glacier.*

ICE AGES

A period when part of the Earth is permanently covered by ice is called an Ice Age. We are living in an Ice Age at the moment. There are permanent ice sheets at the North and South poles. The extent of the ice varies with the seasons. However, there have been times in the past when larger areas of continents have been covered with ice. This was because the Earth's climate was cooler then.

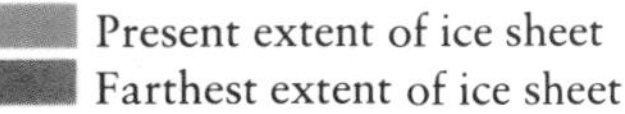

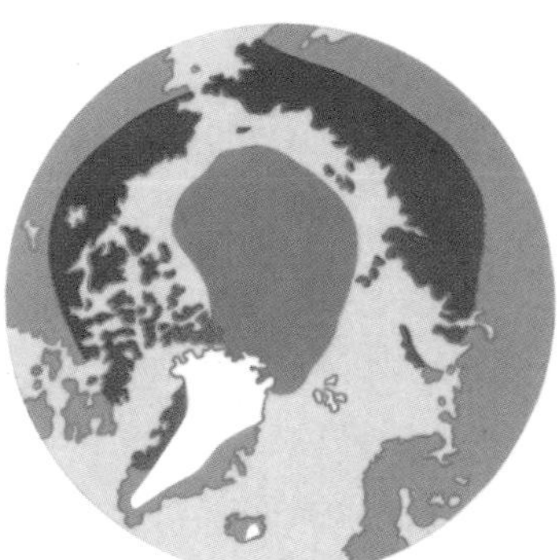

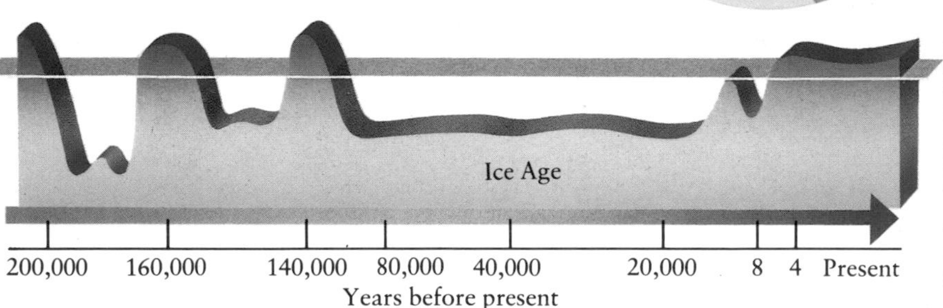

Average July temperatures (°F)

GLACIATION

About 2 million years ago, the Earth's climate cooled and polar ice spread southward, covering Europe as far south as the Severn Estuary in Britain. In places, the European ice sheet was as thick as the Greenland ice sheet is today. The results of glaciation can be seen today. Valleys contained huge glaciers, creeping slowly downhill. As a glacier moved its scouring action wore away the sides and floor of its valley, deepening it. Glaciated valleys can be recognized by their smooth U-shape. And because ice does not meander as the river did, the valley is straightened out and old spurs (projections) cut off.

▼ *A glacier carries with it rock debris from its valley. As the snout of the glacier melts, some debris is dropped, forming a ridge called a terminal moraine.*

Snout
Meltwater
Terminal moraine

FACTS ABOUT ICE

- Today, permanent ice covers more than 10 percent of the Earth's surface.
- During the last Ice Age, over 28 percent of the planet was engulfed in ice. The Scandinavian ice sheet was 10,000 ft. (3,000 m) thick.
- During the Ice Ages, the average temperature of the Earth was only about 5°F (3°C) lower than it is today.
- The world's longest glacier is the Antarctic Lambert Glacier which is over 250 mi. (400 km) in length.
- The fastest moving glacier, in Greenland, flows at up to 79 ft. (24 m) a day.
- The greatest thickness of ice is in the Antarctic and is about 16,400 ft. (5,000 m).
- The biggest iceberg ever sighted was more than 12,000 sq. mi. (31,000 sq. km) in size.

▼ *The head of a glaciated valley is weathered and eroded into an armchair-shape known as a cirque. Where more than one cirque is linked, a knifelike ridge or a pyramid-shaped peak may result.*

► *A typical river valley is a V shape. A moving glacier carries a great deal of rock with it. It works muck like sandpaper, wearing away at the valley until its shape is like a U.*

◄ *Lakes may form in the armchair-shaped cirques made by glaciers.*

► *Sometimes a glacier picks up large blocks of rock and dumps them a long distance away. These are called glacial erratics.*

GLACIAL LAKES

Lakes are a common feature of a landscape that has been glaciated. Indeed, there are more glacial lakes than all other kinds put together. A lake may form in a hollow that the ice has worn in softer rocks, or in holes in the uneven surface where a glacier has deposited a lot of debris.

WATER

Oceans and Seas

Over 70 percent of the Earth's surface is covered by water—the rivers, lakes, oceans, and seas. This watery layer is sometimes referred to as the hydrosphere. There is also water locked up as ice—mainly at the poles—and still more exists as vapor in the atmosphere and as moisture in the soil. But about 97 percent of the planet's water is in the seas, and it is salty. The oceans are not evenly distributed around the globe, and most of the world's land areas are to be found in the Northern Hemisphere. The world's biggest ocean is the Pacific. It occupies almost twice the area of the next biggest ocean, the Atlantic, and is also far deeper. Much of the Pacific coast is bounded by mountain ranges, such as the Andes. This means that there are few major rivers flowing into the Pacific Ocean. Many large rivers flow into the Atlantic, carrying sediment from the land.

BEYOND THE LAND
The ocean floor can be separated into two main zones: the deep ocean basin, and the shallower area at the edge of the land, called the continental margin. The gentle slope from the edge of the land down to about 1,600 ft. (500 m) is the continental shelf. Farther out to sea, the ocean bed falls away more steeply down the continental slope, between 5,000 and 11,500 ft. (1,500–3,500 m) to the abyssal plain. Deep submarine canyons form in the continental slope, where currents of muddy water from the mouths of rivers rush down the slope and deposit their sediments in a fan shape.

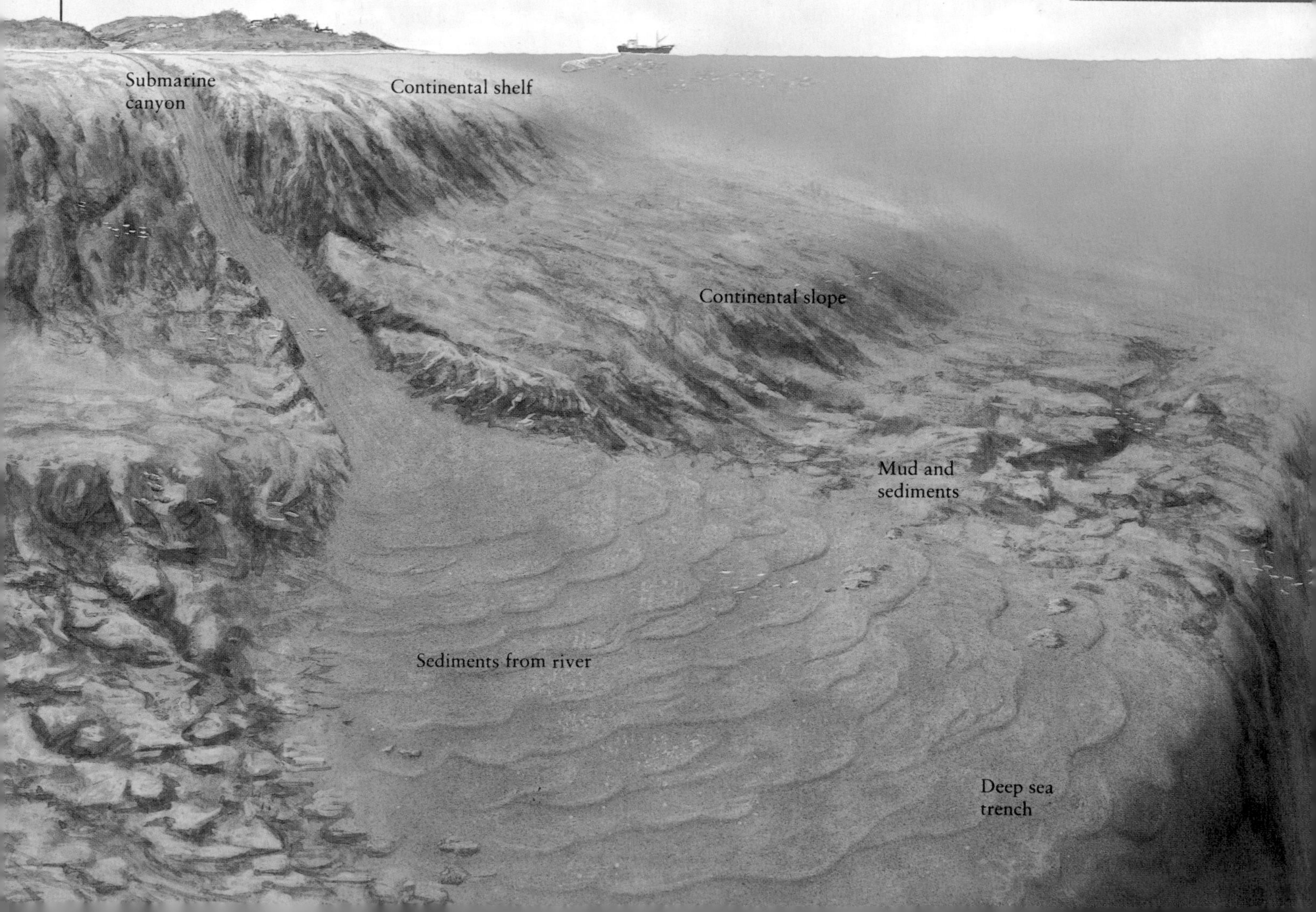

THE OCEANS

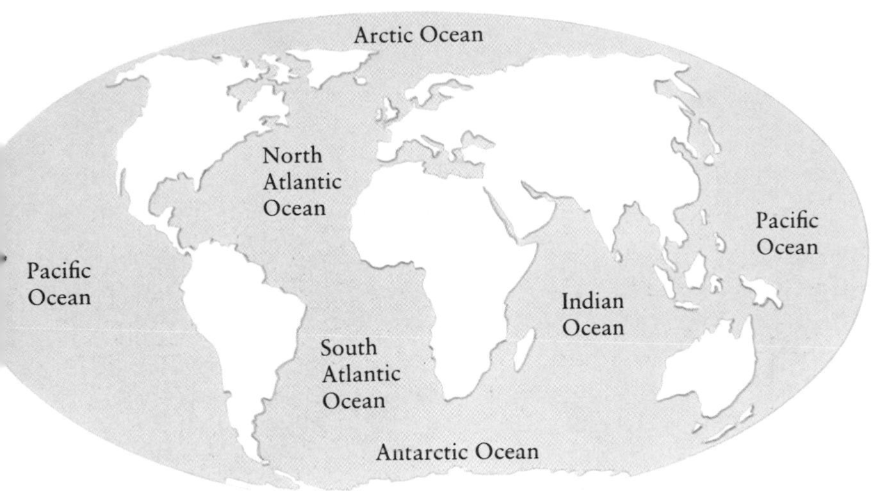

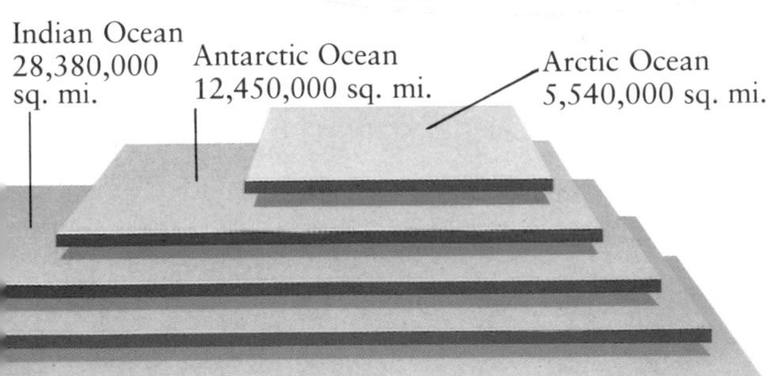

◀ *Two-thirds of the world's surface is covered by the oceans and seas. Of the five oceans, the Pacific is by far the largest.*

FACTS ABOUT OCEANS AND SEAS

- The total area of the oceans is about 140 million sq. mi. (362 million sq. km).
- The Pacific Ocean is about 64 million sq. mi. (166 million sq. km) in area.
- The deepest part of the ocean is the Marianas Trench, in the Pacific Ocean. It is 36,198 ft. (11,033 m) deep.
- The tallest seamount under the ocean is situated between Samoa and New Zealand and is 28,510 ft. (8,690 m) high.
- The largest sea is the South China Sea with an area of 1,148,500 sq. mi. (2,974,600 sq. km).
- Hudson Bay is the world's biggest bay and has an area of 475,000 sq. mi. (1,230,250 sq. km).
- Most of the ocean floors are covered by a layer of loose sediments up to 0.6 mi. (1 km) thick.

MOUNTAINS AND CANYONS

The mountains of the oceans are very tall indeed. Mauna Kea, Hawaii, is really the peak of a mountain that starts on the sea bottom. Measured from there it is taller than Everest. Similarly, the Grand Canyon is dwarfed by the Marianas Trench in the Pacific Ocean.

A volcanic mountain under the sea is called a seamount. If it breaks the surface, it becomes an island. Its top may be eroded and flat so it no longer breaks the sea surface. Then it is known as a guyot.

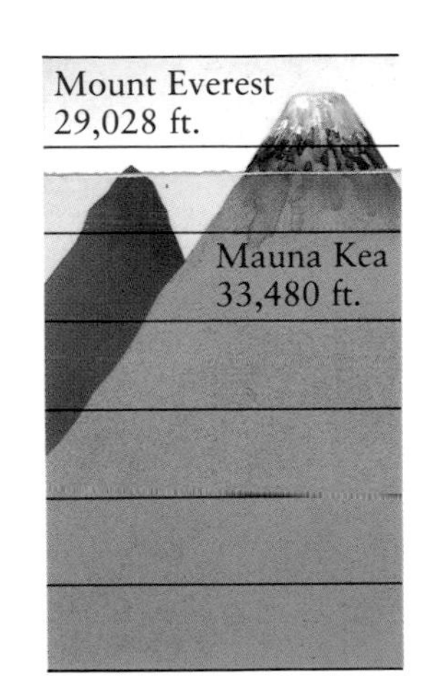

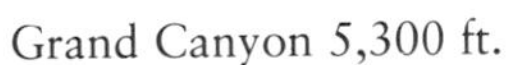

◀ *Close to mid-ocean ridges, there are holes in the seafloor where hot liquids and gases leak into the water from the hot rocks beneath. These liquids include metal oxides and sulfides which give a smokelike effect as they mix with the cold seawater. Some animals live near these "smokers," feeding on the sulfides.*

CORAL REEFS AND ATOLLS

Corals are sea animals related to jellyfish. An individual, known as a polyp, has a cylinder-like trunk which is fixed at one end and has a mouth at the other. Some corals live in vast colonies and build massive limey skeletons which accumulate into reefs.

◀ *Coral reefs form when a new volcanic island rises from the sea over a hot spot. A fringing coral reef grows up around the island.*

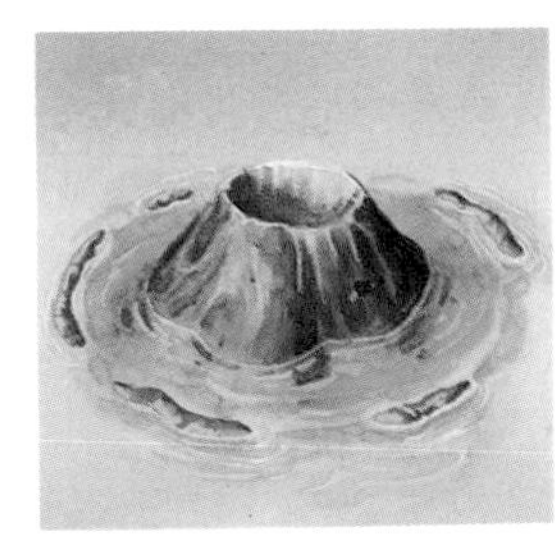

◀ *The coral grows as fast as the seafloor sinks, or the sea level rises. This makes a barrier reef with a lagoon between it and the island.*

◀ *Eventually the island disappears, leaving an atoll enclosing an empty lagoon, or a ring-shaped reef. Finally the coral is also covered by the rising waters, or sinks with the subsiding seabed.*

The Life of the Ocean

The world of the oceans and seas has sometimes been called "inner space." Humans have made use of the sea for thousands of years – for food, transportation, and as a waste dump, for example. However, it has been only in comparatively recent times that people have been able to use machines to explore the fascinating world beneath the surface. The sea also has a considerable effect upon the land and the life that lives there. Water heats and cools quite slowly, and the oceans moderate the world's climate.

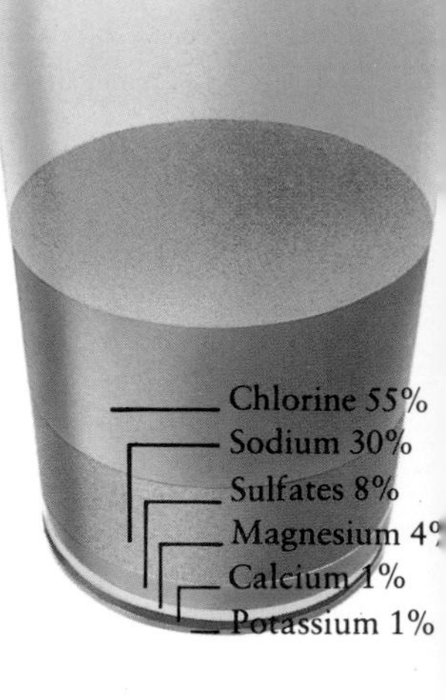

► *The sea tastes salty because it contains common salt and other minerals. On average, the salinity of the sea is 35 parts per thousand. Some inland lakes and seas are saltier than the oceans —the Dead Sea is 250 parts per thousand.*

OCEAN CURRENTS

Warm water is less dense than, and rises through, colder water. In the deep waters of the oceans, these differences in temperature and density create currents. In the top 1,600 ft. (500 m) of water, it is the winds that drive the currents. The ocean's currents follow the prevailing wind directions. Therefore in each ocean basin, there is a roughly circular movement of water called a gyre.

▼ *Near the equator, the ocean's main currents are blown toward the west. Near the poles, they are blown eastward.*

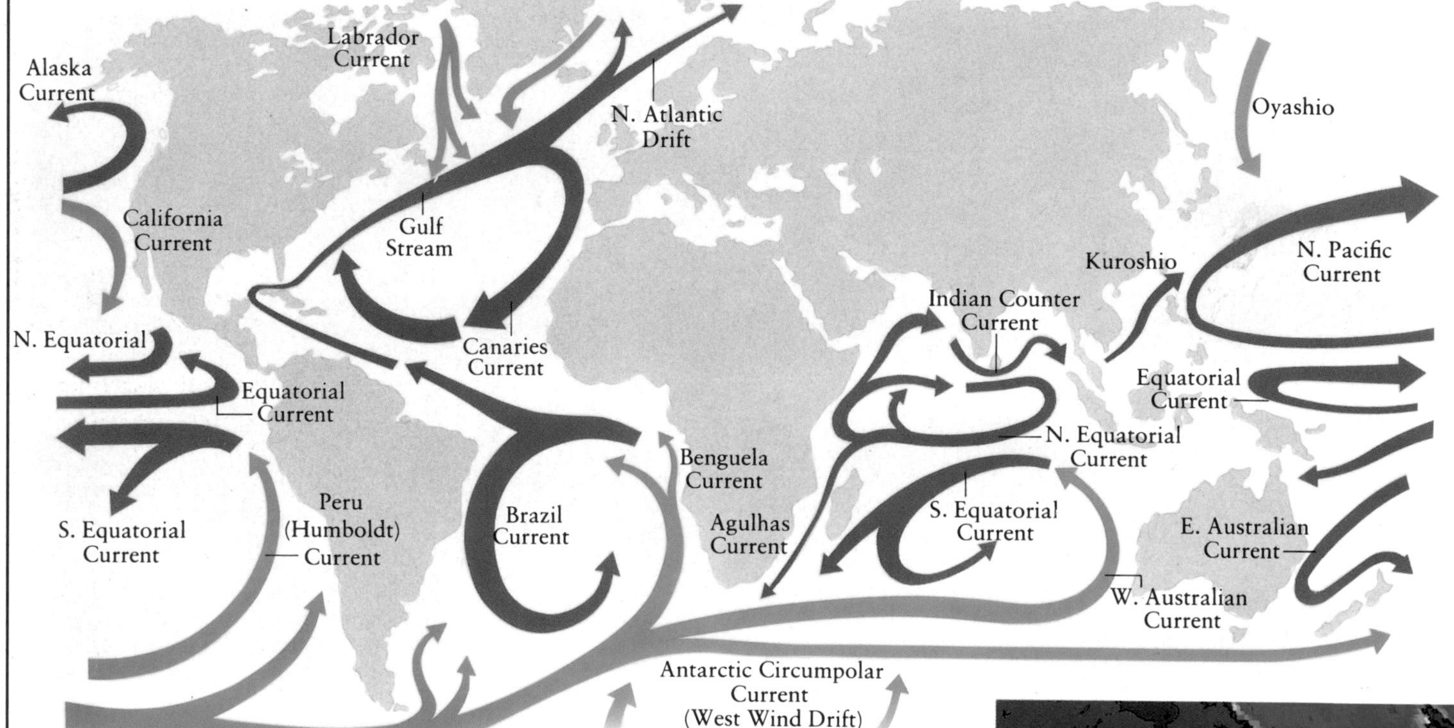

► *In 1947, the Norwegian scientist Thor Heyerdahl built a balsawood raft—the* Kon-Tiki. *The raft was named after an ancient Peruvian. According to legend, this Peruvian had floated from Peru across the Pacific Ocean to settle in the Polynesian Islands. Heyerdahl's raft did reach the islands. Both the Peruvians and Heyerdahl were probably helped on their way by the Humboldt Current.*

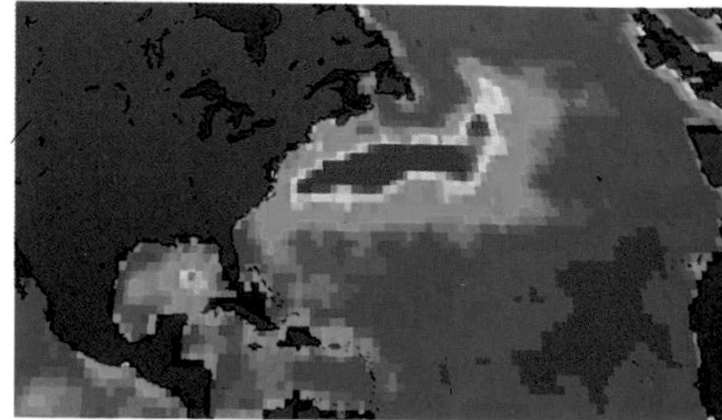

▲ *The Gulf Stream is a current of warm water that begins in the Gulf of Mexico, runs up the coast of the eastern United States, and then turns eastward across the Atlantic toward Europe.*

THE THERMOCLINE

The waters of the top 328 ft. (100 m) of the seas are mixed by the wind. Beneath this, the water temperature falls rapidly. The boundary between the two levels is the thermocline. It prevents nutrients in the water from moving upward.

Depth (miles)
Thermocline
0.5
1.25
2.0
2.5

► *Sunlight penetrates water to about 3,000 ft. (900 m) but only the top 328 ft. (100 m) gets enough light for plants to photosynthesize. Seaweeds that live in upper waters are green and those below are reddish.*

LIFE IN THE SEA

The sea provides support and food for countless animals and plants. Although some kinds of sea creatures may exist in vast numbers, less than 15 percent of the world's species of animals live in the oceans. And 98 percent of these dwell on the seafloor.

Plankton
Small fish
Sharks
Squid
Angler fish
Deep sea shrimp

◄ *Below 3,000 ft. (900 m) the oceans are pitch black. At depths greater than 6,500 ft. (2,000 m), strange creatures survive. Some of them give out their own light signals. Beardworms and fish called rat-tails live here, but there may be other even weirder animals that scientists have yet to discover.*

BROTHER AND SISTER

El Niño (the Spanish word for a boy) is the name given to an occasional change in the world's weather pattern. This change, in turn, affects the circulation of the world's oceans.

Normal Conditions

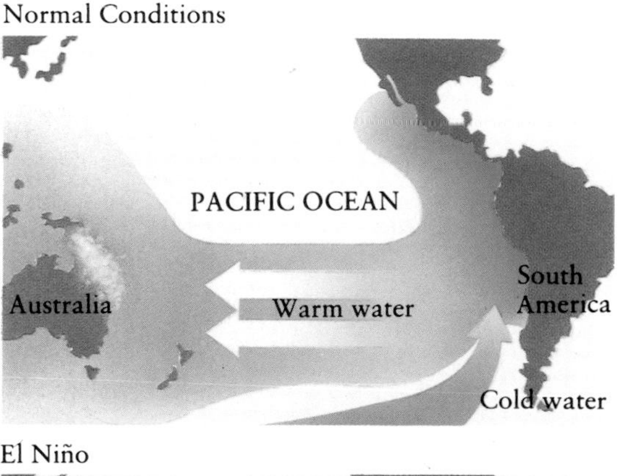

El Niño

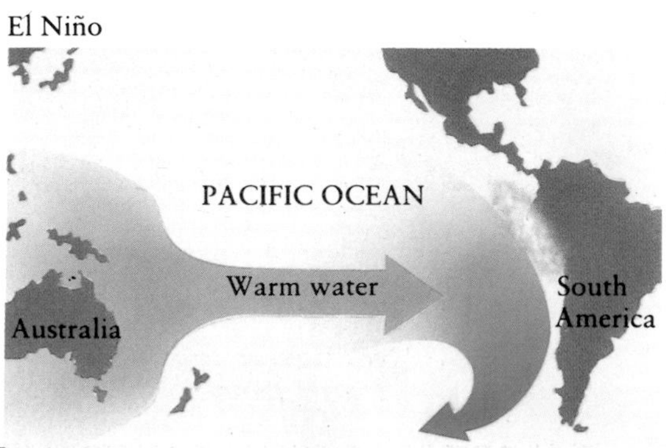

◄ *Usually, food-rich waters well up along the coast of Peru, but during* El Niño *the winds change and the upwelling of cold, rich water is stopped. Instead, warm water arrives at the coast killing much of the wildlife and bringing famine to the people.* La Niña *(the Spanish for a girl) is the name given to a second occasional weather pattern which has the opposite effect to* El Niño. *When it occurs, it brings drought to some parts of the world while others are subjected to freak rains and floods.*

THE SARGASSO SEA

As ocean currents circulate, they sometimes trap areas of relatively calm water. The Sargasso in the western North Atlantic is surrounded by the Florida Current. It is often windless and is choked with seaweed. It is the birthplace of common eels.

The Seashore

Many scientists are worried that, if the greenhouse effect (*see page 77*) continues, there will be a warming of the Earth's climate. If they are correct, then some of the polar ice would melt and sea levels across the world would rise. This could be disastrous for low-lying coastal regions. But it has happened before, about 10,000 to 15,000 years ago, toward the end of the last Ice Age. The effect continued until only a few thousand years ago. So most of the world's coasts are quite young features.

THE CHANGING COASTS
There are different types of coastline: narrow, pebbly, or sandy beaches with a steep cliff at the back; broad beaches which slope down to the sea; or rocky shores. The coast may be cut by bays or deep inlets called fjords. Where parts of the coast are being eroded by the sea, you will see bays and high cliffs. This high coast is said to be retreating before the advance of the sea. At a retreating coastline the beach helps to resist the power of the sea. The sea also builds up parts of the coast. Waves carry sand and pebbles, which are deposited in some places, building up a marsh, a beach, or a spit. This creates a low, sloping coastline.

THE POWER OF THE SEA
Even the everyday action of a relatively calm sea may erode a coastline. Waves, armed with rocks, will hammer their load against the shore. And the sea will pick up and carry off any loose debris. The back-and-forth motion of rocks can smooth and round them into pebbles. Where the sea comes into contact with limestones, it may actually dissolve away the rock.

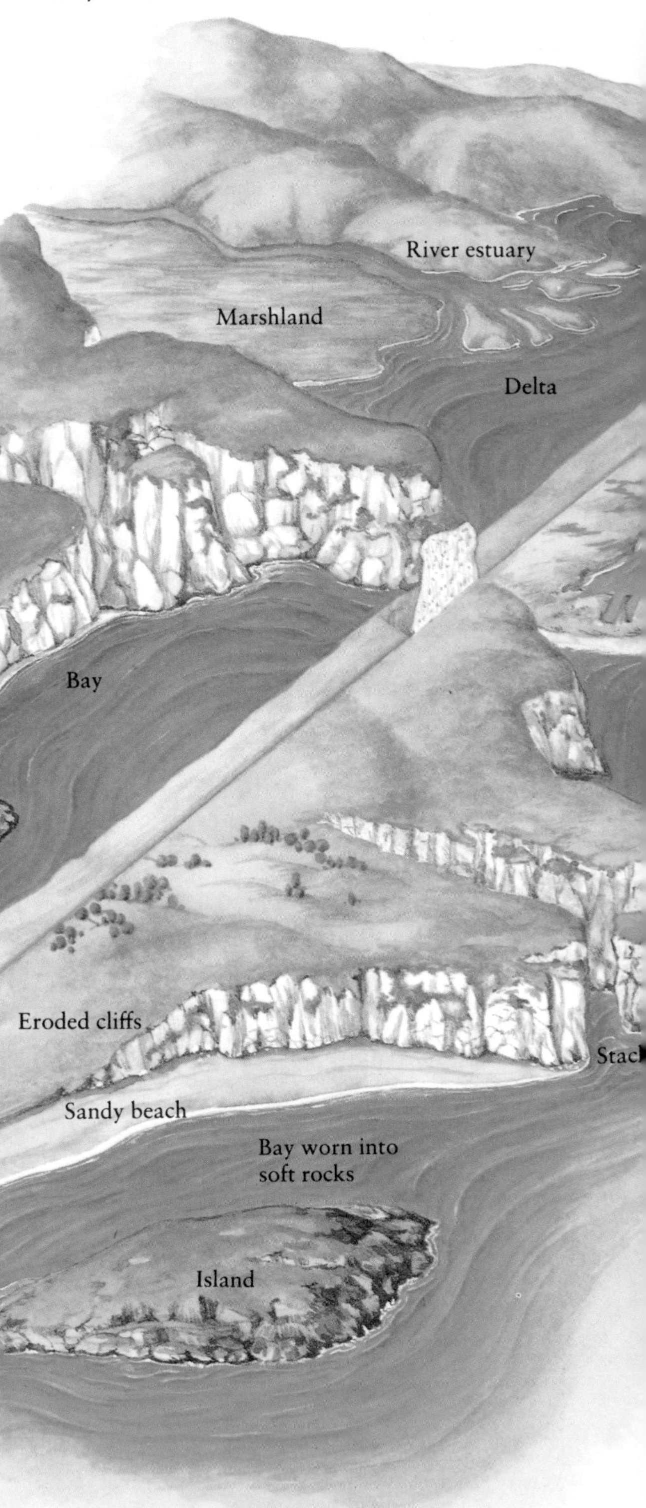

▼ *The seashore is constantly changing. Where coastal rocks are soft the features change quickly, but where the rocks are hard, they erode slowly. A shore may feature cliffs, wide bays, rocky headlands, river deltas, and estuaries.*

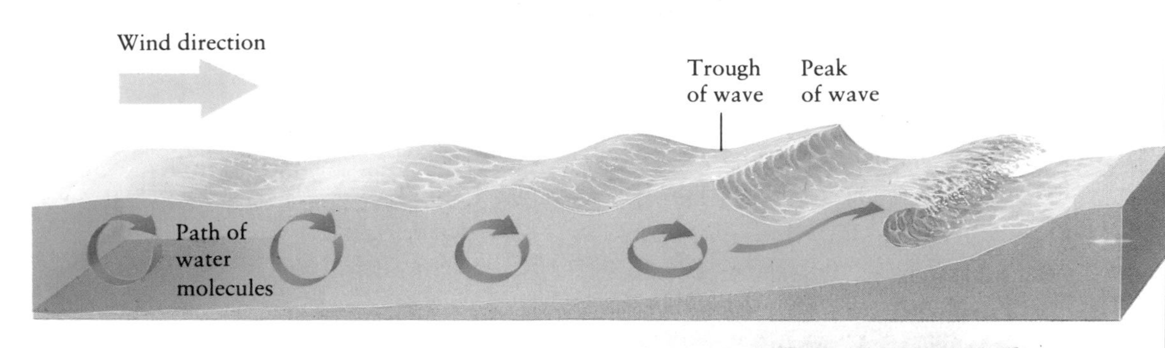

▲ *Waves are caused mainly by the wind. A wave's height depends on the strength of the wind and the area of sea it is blowing over. The motion of a water particle is almost circular until the wave meets the slope of the beach.*

◀ *The steep-sided inlets along the coast of countries such as Norway, Scotland, and New Zealand are called fjords. They are the remains of glaciated river valleys that have been flooded at a retreating coastline.*

SHIFTING SANDS

When a wave breaks on a beach the uprush of water (swash) carries sand and pebbles toward the land. The returning backwash takes some material back down the beach.

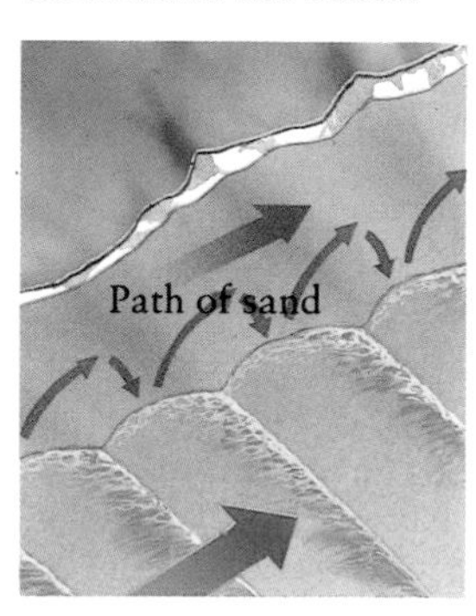

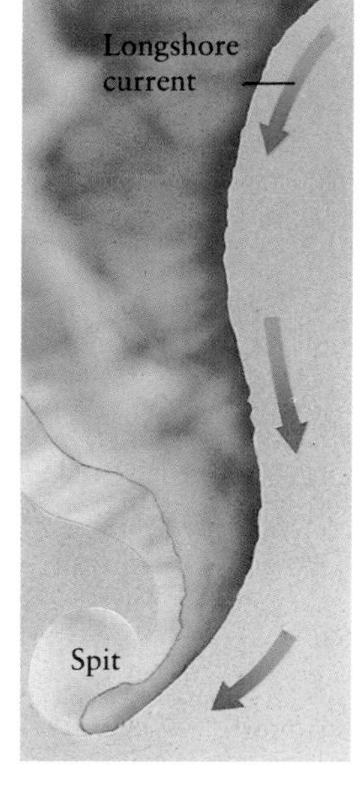

▲ *Waves may hit the beach at an angle, and a longshore current carries sand along the shore.*

SPITS AND BARS

Sometimes, fence-like barriers are used to prevent longshore currents from carrying beach material along the shore. But where longshore drift occurs, the sand and pebbles can be dropped into deeper water. This builds up to form a new area of land called a spit, which will continue to grow (*left*). Sand dunes and marshes often build up behind the spit. Sometimes a spit will grow right across a bay to form a bar.

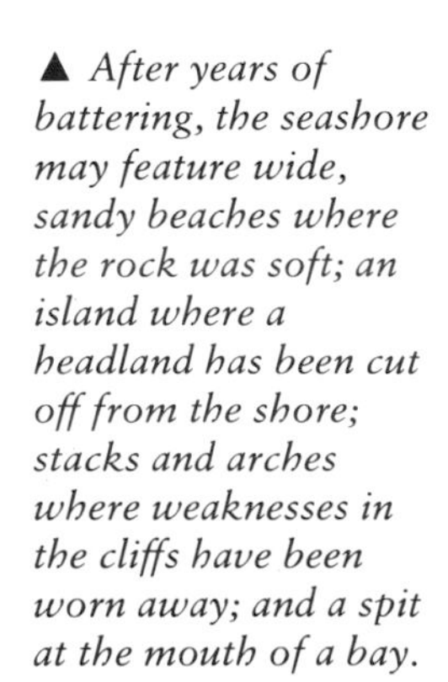

▲ *After years of battering, the seashore may feature wide, sandy beaches where the rock was soft; an island where a headland has been cut off from the shore; stacks and arches where weaknesses in the cliffs have been worn away; and a spit at the mouth of a bay.*

FACTS ABOUT THE SEASHORE

- During a big storm, a wave can crash against the shore with a pressure of almost 66,000 pounds per square yard!
- After an Ice Age, when the ice sheets melt and their weight is removed from the land, the land may rise slowly, like a bobbing cork on water. At the coastline, the original beach will be raised high and dry and a new beach is formed at sea level.

▲ *The sea may sweep away more than the beach. Villages on a clay cliff in Yorkshire, England, have been undercut, and the erosion is still continuing.*

Rivers

Although rivers make up only a tiny percentage of all the surface water on Earth, they are very important. They wear away and form the landscape around us. River valleys have been barriers to the movement of people. Rivers themselves have provided vital transportation links from the sea to inland areas. Where bridges have been built to cross rivers, villages, towns, and cities have grown up. And of course, rivers have supplied food, and water for drinking, washing, and irrigation.

▼ *When a landscape has been newly uplifted, the land surface is steep and irregular and the river naturally follows the pattern of the land. This is known as a consequent river.*

BIRTH AND DEATH OF A RIVER

Some rivers start life as springs. Others are fed by melting glaciers. Most rivers come from the rain and snow that fall on highlands. Water runs along the surface in small rivulets. These rivulets may join to form a small stream. This stream will then begin to erode a valley. The valley sides themselves provide slopes for other tributary streams to flow down. As the amount of water increases, the river grows bigger. At the end of its journey, the river flows to the sea.

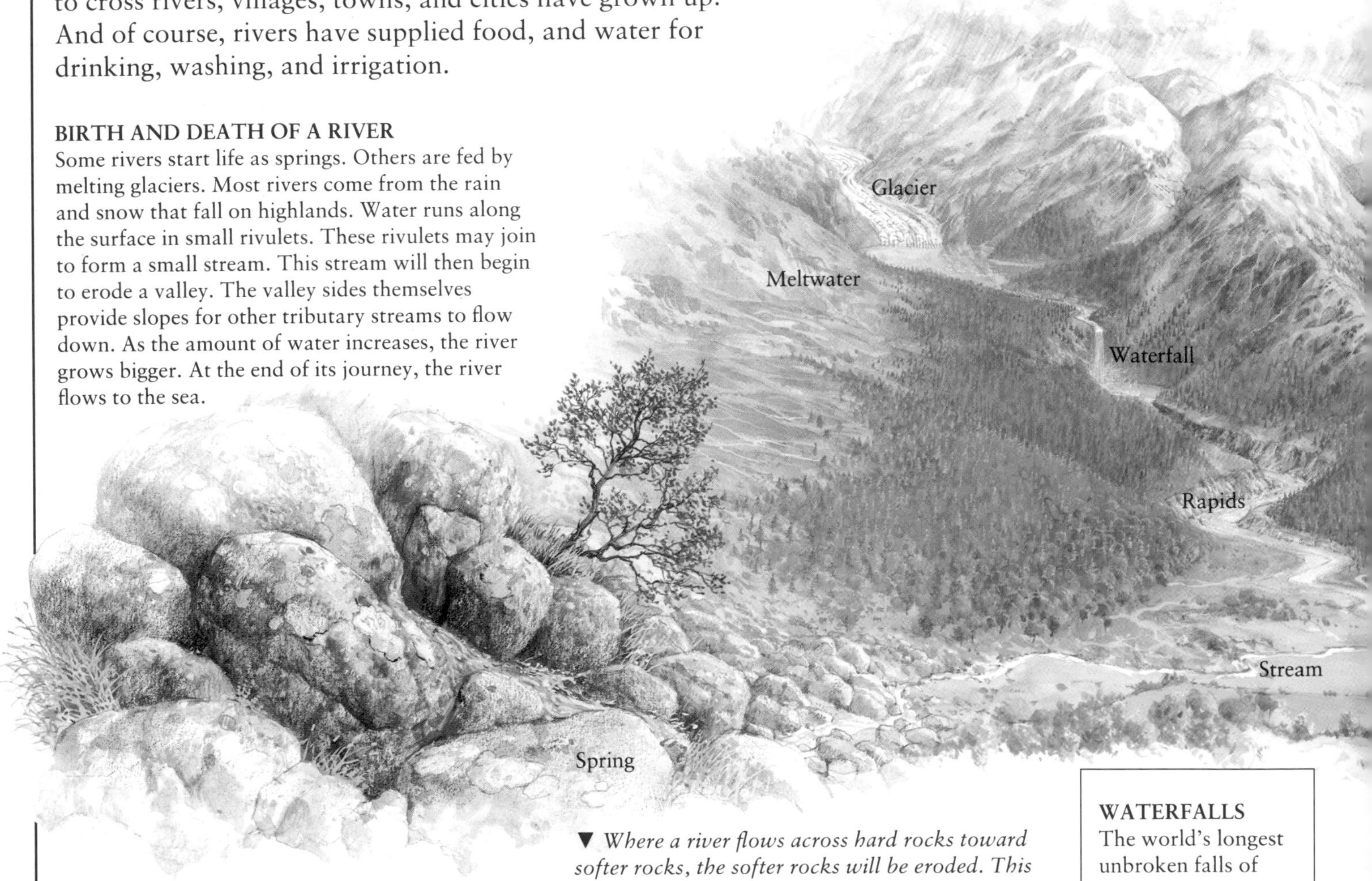

FACTS ABOUT RIVERS

- The longest river in the world is the Nile River in Africa. Running from its source in Burundi to the Mediterranean, the Nile is 4,145 mi. (6,670 km) long.
- The second longest river, the Amazon in South America, is 4,007 mi. (6,448 km). It runs from its source in the Andes Mountains of Peru to the southern Atlantic.
- The Amazon has more than 1,100 tributaries, and it carries much more water than the Nile.
- Rivers can carry amazing amounts of sediment. The Huang He (Yellow River) in China deposits rich silt over 54,690 sq. mi. (141,645 sq. km) in its flood plain and delta.

▼ *Where a river flows across hard rocks toward softer rocks, the softer rocks will be eroded. This erosion creates an abrupt increase in the slope of the river valley. As a result, the river flows more quickly and fiercely, as rapids. If the rock is eroded enough for its face to be vertical, the river will cascade over a waterfall.*

WATERFALLS

The world's longest unbroken falls of water:

RIVER SHAPES

If a river flows quickly down a steep slope, over hard rocks, it will tend to cut a gorge, or deep cleft, in the land.

Where the river flows more slowly, over softer rocks, the valley will be worn back and widened out into an open V shape.

MEANDERS AND OXBOW LAKES

When the channel of a river flows in a snakelike pattern across the broad floor of its valley, it is said to meander. On the inner side of each curve, the river deposits sand and silt. On the outer bend, the bank is eroded away and the channel deepens. Gradually new land is built up on the inner side of the bend, and more land is eroded away on the outer side. This makes the course of the river migrate, or move, into an increasingly wide meander. As the bends migrate, the valley gradually widens and flattens. If a flood occurs, the river may cross the neck of the loop, cutting off the old channel and forming an oxbow lake.

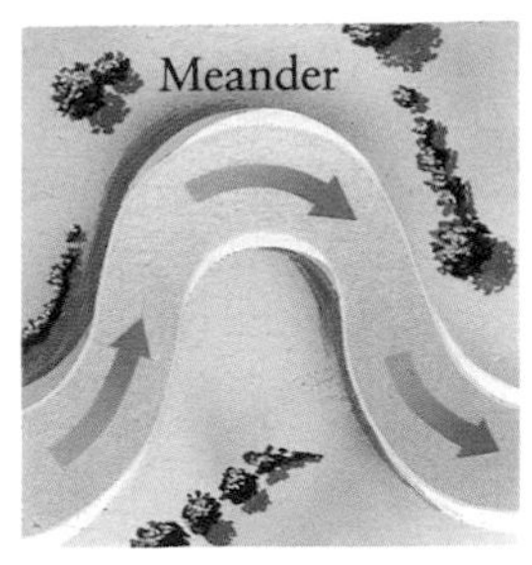

▲ *Some meanders will swell out into much broader loops than others.*

▲ *The neck of the loop may become very narrow as the loop develops.*

▲ *The old channel may be cut off to form an area of water called an oxbow lake.*

▲ *As the river moves farther away from its source, its valley becomes shallower and smoother. The slope of the valley floor decreases, until it is almost flat. The river valley is broader.*

► *As the river reaches its mouth it slows down, dropping the sediments it has been carrying. Its valley is wide, sloping gently to the sea. The river may meander, breaking up into a number of channels.*

Lakes and Swamps

A lake is an area of water completely surrounded by land. Lakes may contain fresh or salt water. Some, such as the Caspian, the salt lake between southeastern Europe and Asia, are big enough to be thought of as inland seas. The water in a lake may seep from its basin, and it may also evaporate. So, for a lake to continue to exist, it must be fed with water at the same rate or faster than the water is being lost. And the lake's bed must be lower than the lowest part of the rim or the water will run off elsewhere.

STILL WATERS

Even though lakes seem still, few are completely stagnant. Water is at its densest at 39.2°F (4°C). So in a cold winter, the lake may be covered with a sheet of ice, while the water at the bottom of the lake is warmer. In spring, the ice melts. As the temperature of the surface water rises, it becomes denser and sinks. The water in the lake circulates until, once again, water at 39.2°F (4°C) settles to the bottom. In summer, the surface waters warm and circulate but stay above the layer of cold water in the lower part of the lake. In autumn the lake's surface cools again and the water circulates once more.

THE LARGEST OR DEEPEST LAKE

There are two main ways that a lake can be measured. The first is the area of its surface and the second is its depth. Lake Baikal in central Siberia, Russia, is the deepest lake in the world. At its deepest point, the Olkhon Crevice, Lake Baikal is 6,365 ft. (1,940 m) deep, and 4,872 ft. (1,485 m) below sea level. The Great Lakes in North America are all linked and so could qualify as the largest by surface area.

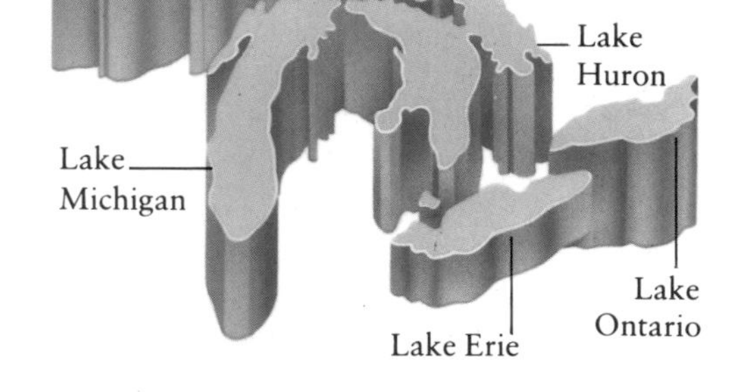

▼ *The five Great Lakes lie on the border of the United States and Canada. They are: Superior, Michigan, Huron, Erie, Ontario.*

▲ *Lake Baikal is the largest freshwater lake in Eurasia, and it is the sixth biggest in the world.*

Volcanic lake

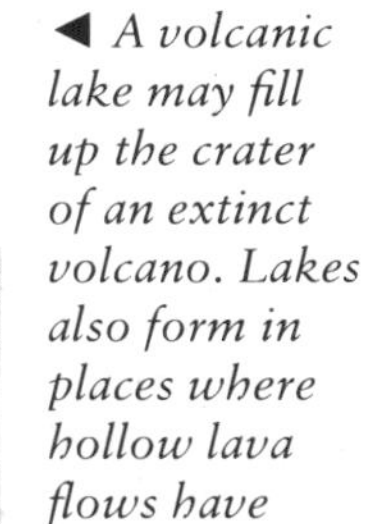

◄ *A volcanic lake may fill up the crater of an extinct volcano. Lakes also form in places where hollow lava flows have collapsed.*

Barrier lake

◄ *If a river valley becomes naturally blocked a lake will form. The barrier might be the result of a landfall, for example, or glacial debris.*

Reservoir

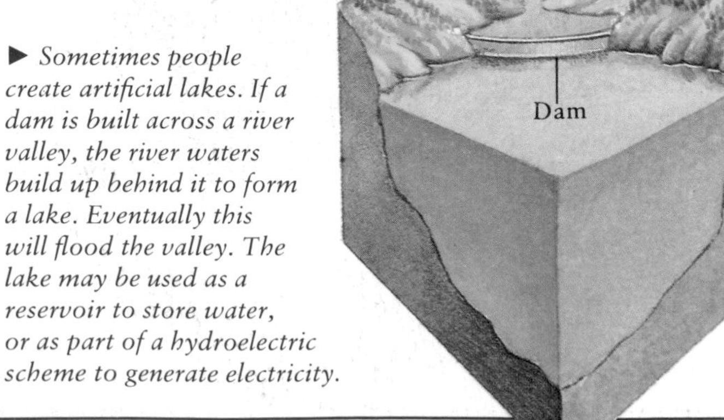

► *Sometimes people create artificial lakes. If a dam is built across a river valley, the river waters build up behind it to form a lake. Eventually this will flood the valley. The lake may be used as a reservoir to store water, or as part of a hydroelectric scheme to generate electricity.*

Rift valley lake

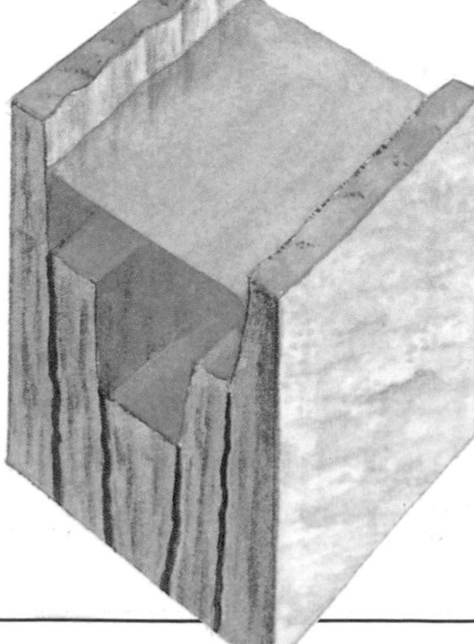

► *Lakes may lie along major fault or fissure lines, such as the long, narrow lochs in the Great Glen fault in Scotland. There are also lakes in the Great Rift Valley of East Africa—the world's largest group of fault-created lakes.*

DISAPPEARING WATERS

Most lakes are quite shallow and are fed by rivers. A river brings with it a great deal of debris and sediment. The larger particles of sediment are dropped soon after the river enters the lake, and a delta-like fan of material gradually spreads out into the water. Very fine material may be carried quite a long way before it settles onto the lake's bed, slowly reducing the depth of water. At the same time, where the river leaves the lake, at the rim, it becomes worn away, and the lake is partly drained.

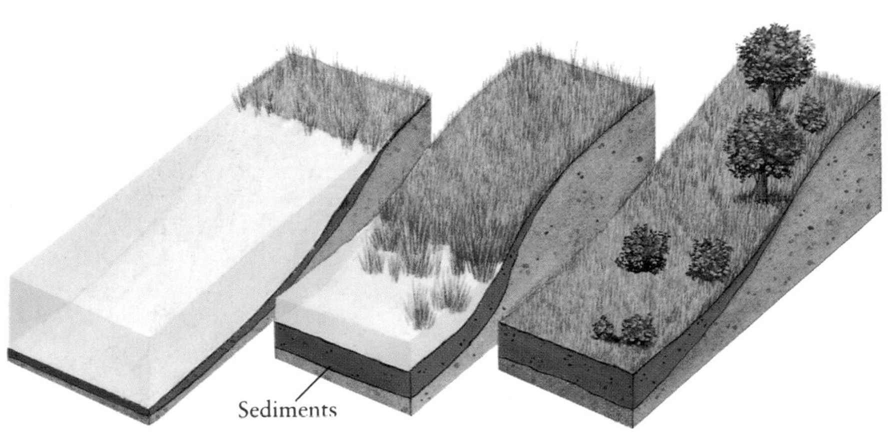

▲ *A fan of sediment builds up where the river feeds the lake and deposits material.*

▲ *Marsh plants colonize the waterlogged soil and trap more sediment.*

▲ *Eventually, the land dries out and other, less marshy kinds of plants move in.*

▼ *To the west of the Uinta Mountains in Utah is the Great Salt Lake, an inland salt lake about 75 mi. (120 km) long and 13 ft. (4 m) deep. It is the remains of a much larger lake, which geologists call Lake Bonneville. The Great Salt Lake is getting larger without getting any less salty.*

▲ *The biggest inland water in the world is the Caspian Sea. It is 761 mi. (1,225 km) long. Estimates suggest that its total surface area is between 140,000 and 170,000 sq. mi. (360,000–440,000 sq. km). The Caspian is shrinking, but it is getting less salty. At its deepest point, it is about 3,280 ft. (1,000 m) deep.*

Great Salt Lake

Extent of Lake Bonneville

FACTS ABOUT LAKES AND SWAMPS

- The world's biggest freshwater lake is Lake Superior, one of the Great Lakes of North America. Its total surface area is 31,700 sq. mi. (82,103 sq. km). Only one of the Great Lakes is wholly in the United States; the others are shared with Canada. The Great Lakes form the largest body of fresh water in the world and with their connecting waterways are the longest inland water transportation system.
- The largest underground lake in the world was discovered in 1986 in the Drachenhauchloch Cave near Grootfontein in Namibia. The lake has a surface area of 10,077 sq. mi. (26,100 sq. km).
- In the Himalayas in Tibet, Lake Nam Tso is at an altitude of 19,290 ft. (5,880 m). The highest navigable lake is Titicaca in South America, at 12,503 ft. (3,811 m).
- Lake Eyre, in south Australia, is normally a dry salty expanse. It has only filled with water three times since it was first visited by Europeans.

SWAMPS AND BOGS

In shallow lakes surrounded by plants, leaves and flowers fall into the water. A layer of peat-like ooze may build up. The ooze accumulates, becoming a bog or swamp.

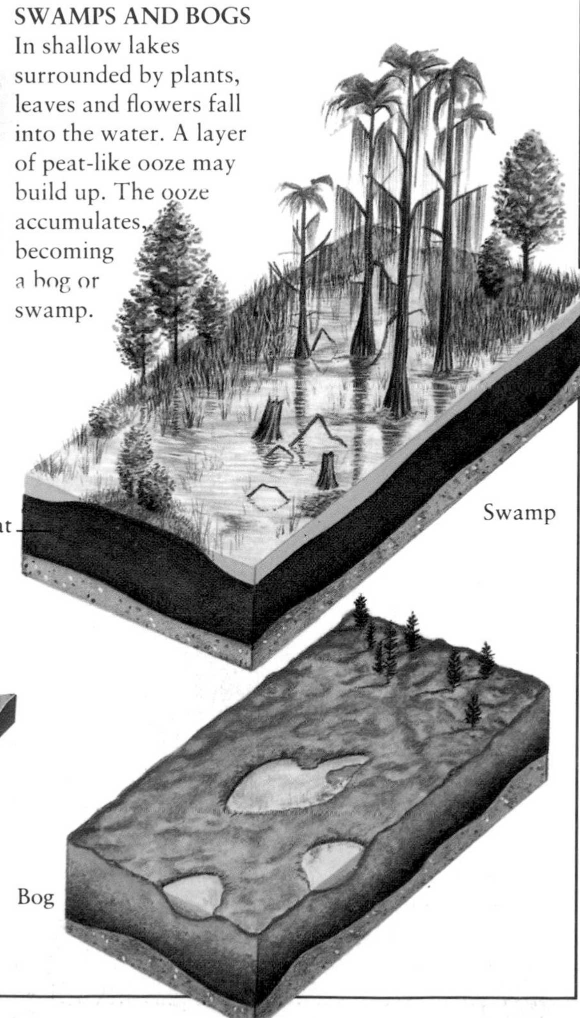

WEATHER AND CLIMATE

The Atmosphere

The atmosphere is the envelope of air that surrounds the Earth. It has four layers: the lowest layer is the troposphere; above this layer are the stratosphere, the mesosphere, and lastly the thermosphere. The atmosphere extends upward to a height of about 300 mi. (500 km). Above this, the air merges with the particles streaming constantly from the Sun. There is no clear boundary where the Earth's atmosphere ends and that of the Sun begins. Air has weight, and the atmosphere nearest the Earth is compressed, or pressed down by the weight of the air above it. At sea level, the atmosphere presses down on every square inch of the Earth's surface with a weight of about 14.7 pounds. This pressure is called "one bar," or 1,000 millibars (mb). Atmospheric pressure decreases with height, until it falls to about one millibar at a height of 19 mi. (30 km) above sea level.

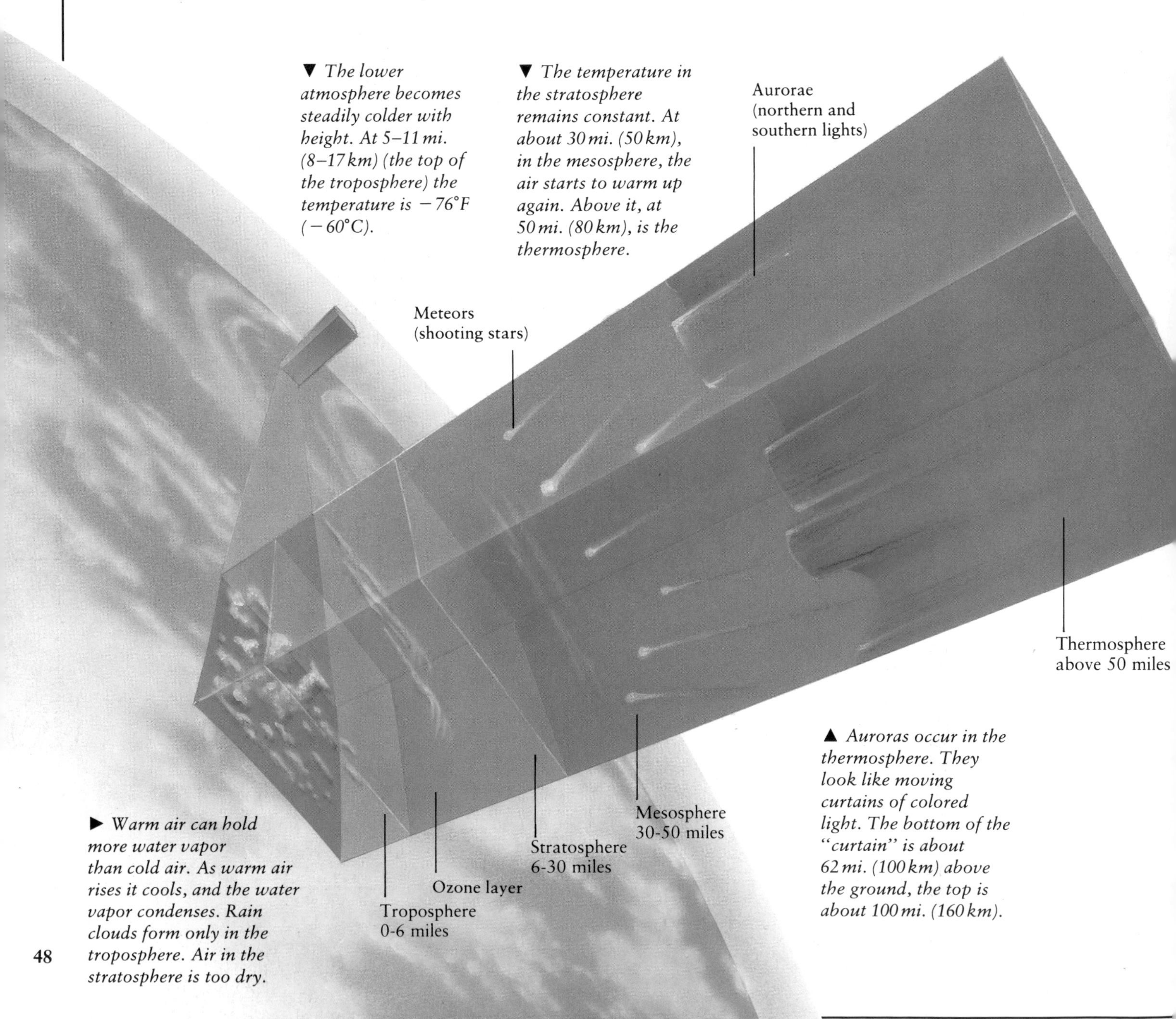

▼ *The lower atmosphere becomes steadily colder with height. At 5–11 mi. (8–17 km) (the top of the troposphere) the temperature is −76°F (−60°C).*

▼ *The temperature in the stratosphere remains constant. At about 30 mi. (50 km), in the mesosphere, the air starts to warm up again. Above it, at 50 mi. (80 km), is the thermosphere.*

▲ *Auroras occur in the thermosphere. They look like moving curtains of colored light. The bottom of the "curtain" is about 62 mi. (100 km) above the ground, the top is about 100 mi. (160 km).*

► *Warm air can hold more water vapor than cold air. As warm air rises it cools, and the water vapor condenses. Rain clouds form only in the troposphere. Air in the stratosphere is too dry.*

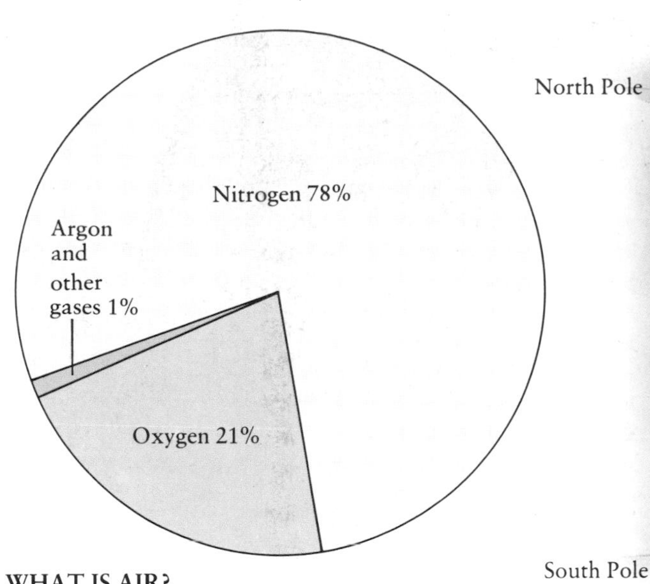

WHAT IS AIR?

The air is a mixture of gases: about 78 percent is nitrogen, 21 percent oxygen, and the rest is mainly argon and carbon dioxide. The air also contains water vapor, but the amount varies.

North Pole

Rays hit Earth at an angle

Rays hit Earth directly

South Pole

THE SUN'S RAYS

As the surface of the Earth is heated by the Sun it warms the air above it. This makes the air less dense and it rises and cools. The air over the equator is heated more strongly than over the poles, since the poles are slightly turned away from the Sun. This difference in temperature produces our climates and weather.

Nitrogen

4 billion years ago

3.5 billion years ago

2 billion years ago

500 million years ago

Today

Oxygen

Carbon dioxide

▲ *The atmosphere was once mainly carbon dioxide. By 600 million years ago, plants had replaced the carbon dioxide with oxygen, through photosynthesis. The amount of nitrogen increased until it was almost 80 percent of the air.*

CLOUDS OF THE UPPER ATMOSPHERE

Noctilucent clouds form at about 50 mi. (80 km). They are visible only on summer nights and are believed to be made up of ice crystals. Nacreous, or "mother-of-pearl," clouds form at about 14 mi. (22 km), and are visible when the Sun is just below the horizon.

Noctilucent clouds

Nacreous clouds

▲ *A view of the Moon setting over the Earth's horizon taken from space. The image of the Moon is distorted by the Earth's atmosphere.*

THE IONOSPHERE

Some radio signals are transmitted long distances by bouncing them off the ionosphere. This extends from about 62–186 mi. (100–300 km). Electrically charged air in the ionosphere reflects some radio waves.

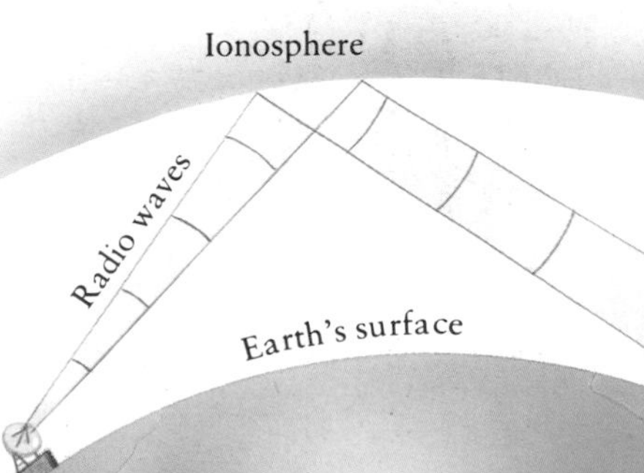

Climate

The Sun warms the surface of the Earth more strongly near the equator than at the poles. The warm, tropical air rises and cooler, denser air from higher latitudes moves in to replace it. Water evaporates into the warm air and condenses in cool air. This constant movement of the air, driven by the warmth of the Sun, produces the world's climates and the weather we experience from day to day. It is called the "general circulation" of the atmosphere. It is influenced by the rotation of the Earth and the oceans.

JET STREAMS

There is a sharp change in temperature between warm tropical air and the cooler air north and south of the tropics, and between this mid-latitude air and polar air. In these two regions, at about 7 mi. (12 km), westerly winds blow at speeds of about 125 mph (200 km/h). These fast-moving winds are called jet streams.

GLOBAL WINDS

Warm air rises and moves away from the equator, cooling and losing moisture as it does so. The trade winds bring cold air in from higher latitudes to replace it. The dry tropical air eventually sinks in the subtropics (about 30° latitude). When it reaches the surface, some of the sinking air is drawn back to the equator, forming the trade winds, and some blows toward the poles. This circulation of the air near the equator is called the Hadley cell. It is named after George Hadley, the meteorologist who described it in 1735.

▼ *At each pole, cold air sinks and spills out, to be replaced by warmer air flowing in from above. This is a polar cell. Ferrel, or mid-latitude, cells form between 30° and 60°. The cold air moving away from the poles meets warm winds from the subtropics and pushes the warm air back to the equator.*

Polar jet stream

Polar cell

Polar front

Ferrel cell

Subtropical jet stream

Wind direction

Hadley cell

WORLD CLIMATES

The world can be divided into several climate zones. The main factors that affect the climate of a particular place are its distance from the equator (latitude) and its distance from the ocean. Climates are cooler farther away from the equator, and drier in places far from the ocean.

Mountain
Tropical wet and dry
Tropical wet
Desert
Subtropical dry summer
Continental moist
Oceanic moist
Subarctic
Polar

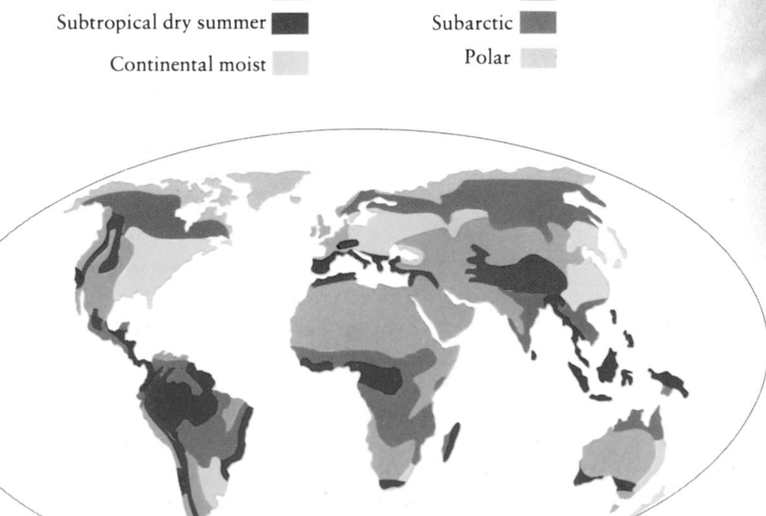

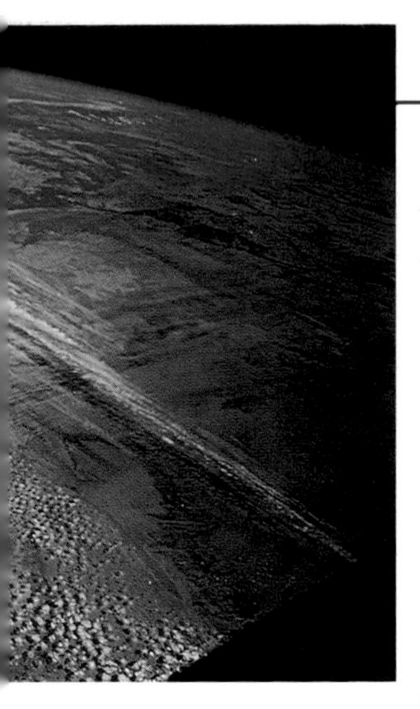

▲ *Jet streams are often revealed by long narrow bands of cirrus cloud.*

THE CORIOLIS EFFECT

The Earth rotates to the east; the surface and the air move faster at the equator. Warm air rises and is replaced by cooler air which is moving eastward more slowly, so the Earth's surface overtakes it. This creates the northeasterly and southeasterly trade winds and is called the Coriolis effect.

Trade winds
Rotation of the Earth
Actual path
Expected path
Equator

▼ *In winter, the sea is warm relative to the land. The air above the sea is heated and rises, and cold, dry winds from the northeast blow toward the Indian Ocean.*

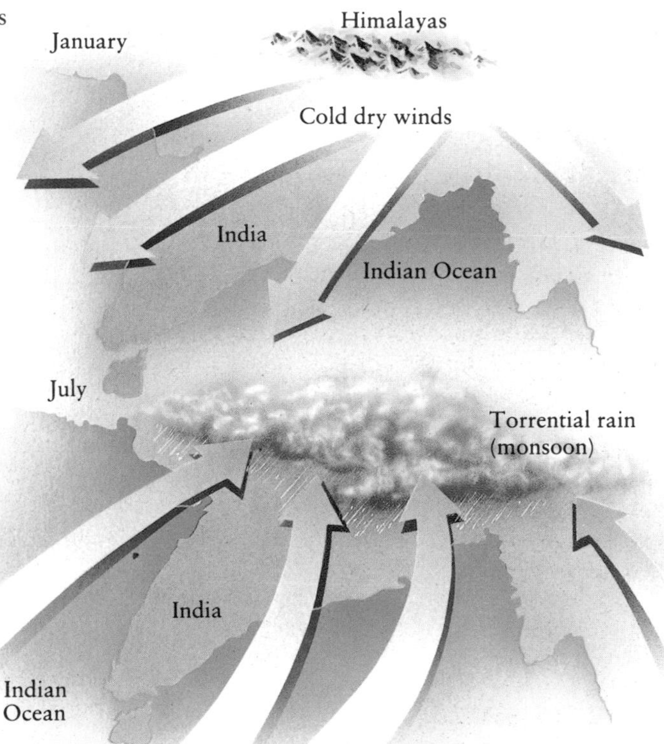

▲ *In summer, the air over northern India becomes very hot and dry. Its pressure falls, causing warm moist air to move to the north, bringing heavy rain to southern Asia.*

MONSOONS

In June, differences in air pressure over land and sea bring thunderstorms and heavy rain to India and southern Asia. The parched ground revives and farm crops flourish. The rainy season is called the monsoon. It lasts until September, and in some places 85 percent of the annual rain falls during this period. Monsoon seasons also occur in West Africa and northern Australia.

AIR MASSES

An air mass is a large body of air that has the same temperature and moisture. Depending on where it forms, it is called "polar" or "tropical," and "maritime" or "continental." Continental air (forming over land) is dry and cool if it is polar, and warm if it forms at the tropics. Maritime air (forming over sea) is moist. As the air mass crosses the land it loses moisture, bringing rain. The weather usually changes when one air mass is replaced by another. The boundary between two air masses is a front.

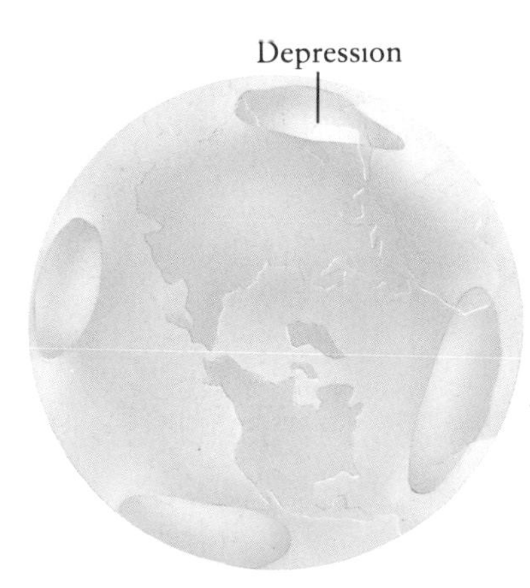

▲ *An area of low air pressure is called a depression. As the air rises its moisture condenses, layers of cloud form, and it rains.*

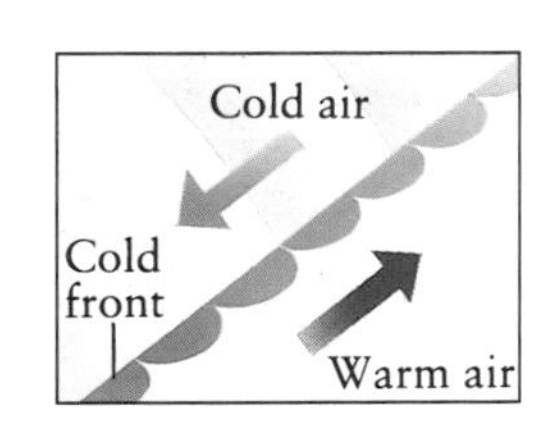

◀ *A cold front has cooler and often drier air behind it.*

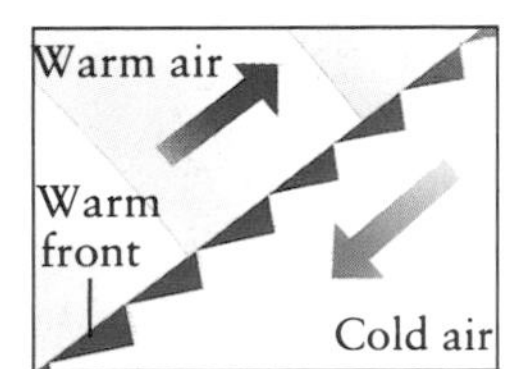

◀ *A warm front brings cloud and rain. Cold air pushes beneath the warm air and lifts it.*

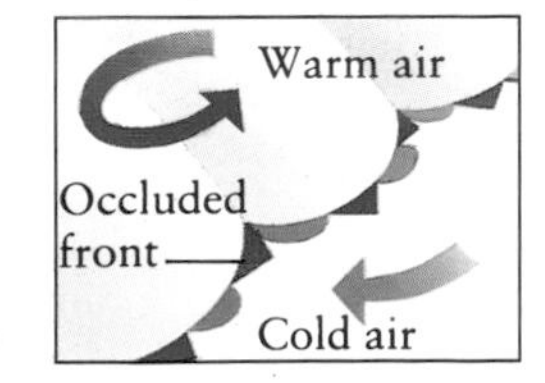

◀ *Where cold and warm air mix, the fronts are "occluded." They weaken and disappear.*

FACTS ABOUT CLIMATES

- The winds drive the ocean currents, which also carry warmth from the equator to cooler regions.
- The ocean currents affect climates by bringing warm or cool water to the shores of continents.
- The climate of western Europe is warmed by the waters of the Gulf Stream and North Atlantic Drift.
- The climate of the northwest coast of North America is cooled by the California Current, which flows south from the Arctic.
- Water warms and cools more slowly than land. A continental air mass that crosses the ocean will be warmed by the water in winter, and cooled in summer.
- The greatest amount of rain ever recorded in a single year was at Cherrapunji in India. It received 67.740 ft. (20.647 m) of rain in the monsoon season of 1861.

Winds and Storms

Air moves from areas of high atmospheric pressure to areas of low pressure, causing winds. It does not flow directly but moves around the centers of high or low pressure because of the Coriolis effect *(see page 51)*. In the Northern Hemisphere, air moves counterclockwise around low pressure and clockwise around high pressure. In the Southern Hemisphere the opposite occurs. The wind always moves "downhill" from high to low pressure. Its speed depends on the difference in pressure.

WINDS OF THE WORLD

Because the Earth spins to the east, the winds either side of the equator are from the northeast and southeast. They are the "trade winds." Between the trade wind belts lie the "doldrums," where winds are light. In middle latitudes (30°–50°), winds are more often from the west than from the east. Easterly winds prevail in the Arctic and Antarctic.

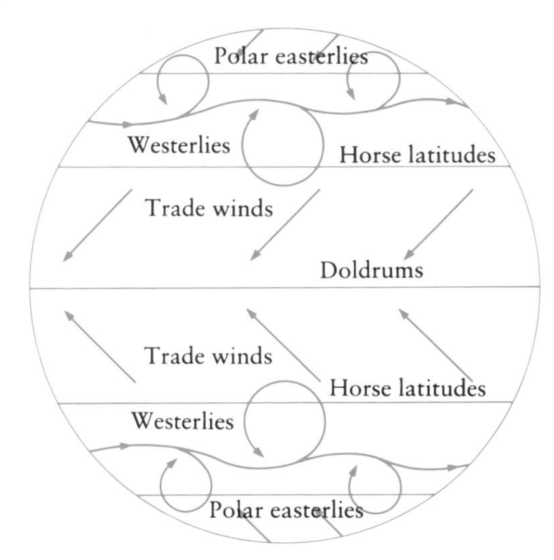

THE BEAUFORT WIND SCALE

In 1806, Sir Francis Beaufort, an English admiral, devised a scale for measuring wind force. It is still used today.

Force	Speed	Definition
0	< 1 mph	calm
1	1–3 mph	light air
2	4–7 mph	light breeze
3	8–12 mph	gentle breeze
4	13–18 mph	moderate breeze
5	19–24 mph	fresh breeze
6	25–31 mph	strong breeze
7	32–38 mph	moderate gale
8	39–46 mph	fresh gale
9	47–54 mph	strong gale
10	55–63 mph	whole gale
11	64–75 mph	storm
12	> 75 mph	hurricane

Wind speed is measured by an anemometer (*see page 60*). It has small cups, mounted on horizontal arms, which spin around on an axis. A wind vane is a flat blade that indicates wind direction.

▶ *The strong upcurrents of air in heavy rainstorms have been known to lift up objects as large as fish and frogs, which then appear to rain down from the sky.*

THUNDER AND LIGHTNING

In a large storm cloud, as water droplets collide, water becomes electrically charged. Positive charges collect at the top of the cloud and negative charges at the bottom. The negative charge creates a positive charge on the ground surface, which builds until lightning sparks from the cloud to the ground and back again.

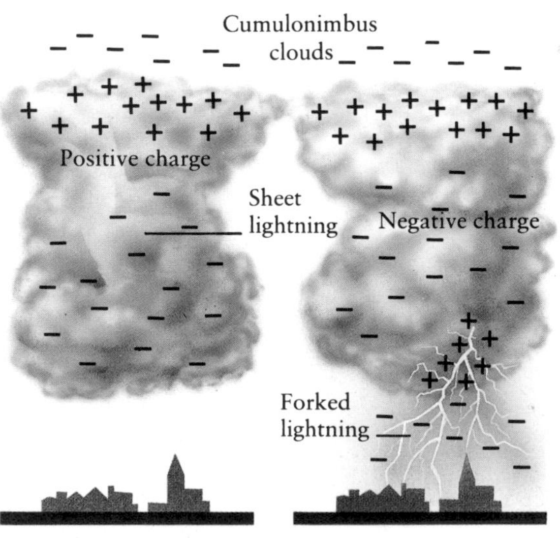

▲ *Sheet lightning flashes inside or between clouds; forked lightning flashes to the ground. Thunder is the sound of hot air exploding.*

FACTS ABOUT WIND

- A warm, dry "föhn" or "chinook" wind is caused by air flowing down the side of a mountain range.
- In southern Europe, the valleys of the Rhône and other rivers funnel the "mistral," a cold, northerly wind.
- In West Africa, a dry easterly wind is called the "harmattan," or "doctor," because it brings relief from very humid conditions.
- Winds on Mt. Washington, New Hampshire, have reached 231 mph (370 km/h).

TORNADOES

A tornado is a twisting funnel, a thousand feet across, in which wind speeds can reach 220 mph (350 km/h). Tornadoes move in a straight line and can cause terrible destruction. The pressure inside is so low that it can cause nearby buildings to explode.

► *A hurricane is a tropical storm, usually about 400 mi. (650 km) across. It brings heavy rain and winds of up to 125 mph (200 km/h). In the northwest Pacific they are called typhoons, and those in the Indian Ocean and north of Australia are called cyclones.*

► *Hurricanes form in late summer and fall over warm water in the Atlantic, Pacific, and Indian oceans. They then move westward, along coasts.*

INSIDE A HURRICANE

In the eye of a hurricane, winds are light and the sky is clear. The descending air is warm and the pressure is very low. Fierce winds bring air swirling around the eye. The air is swept upward by huge cumulonimbus clouds. As air in the clouds rises and cools, the water vapor it carries condenses. This releases heat, which warms the air again, sending it upward into a region of high pressure above the clouds. The big clouds cause heavy rain. Away from the center, cirrus and cirrostratus clouds trace the storm's outline.

The Types of Cloud

Clouds form when water vapor condenses into tiny droplets. The conditions under which this happens vary, producing clouds of different types and shapes. White puffy clouds are often seen on fine days. If they grow bigger, rain showers may fall from them. High above them, there may be small wispy clouds made from ice crystals. Very large dark clouds bring thunderstorms. Layers of cloud, like flat sheets covering the sky, bring dull weather or steady, sometimes heavy, rain.

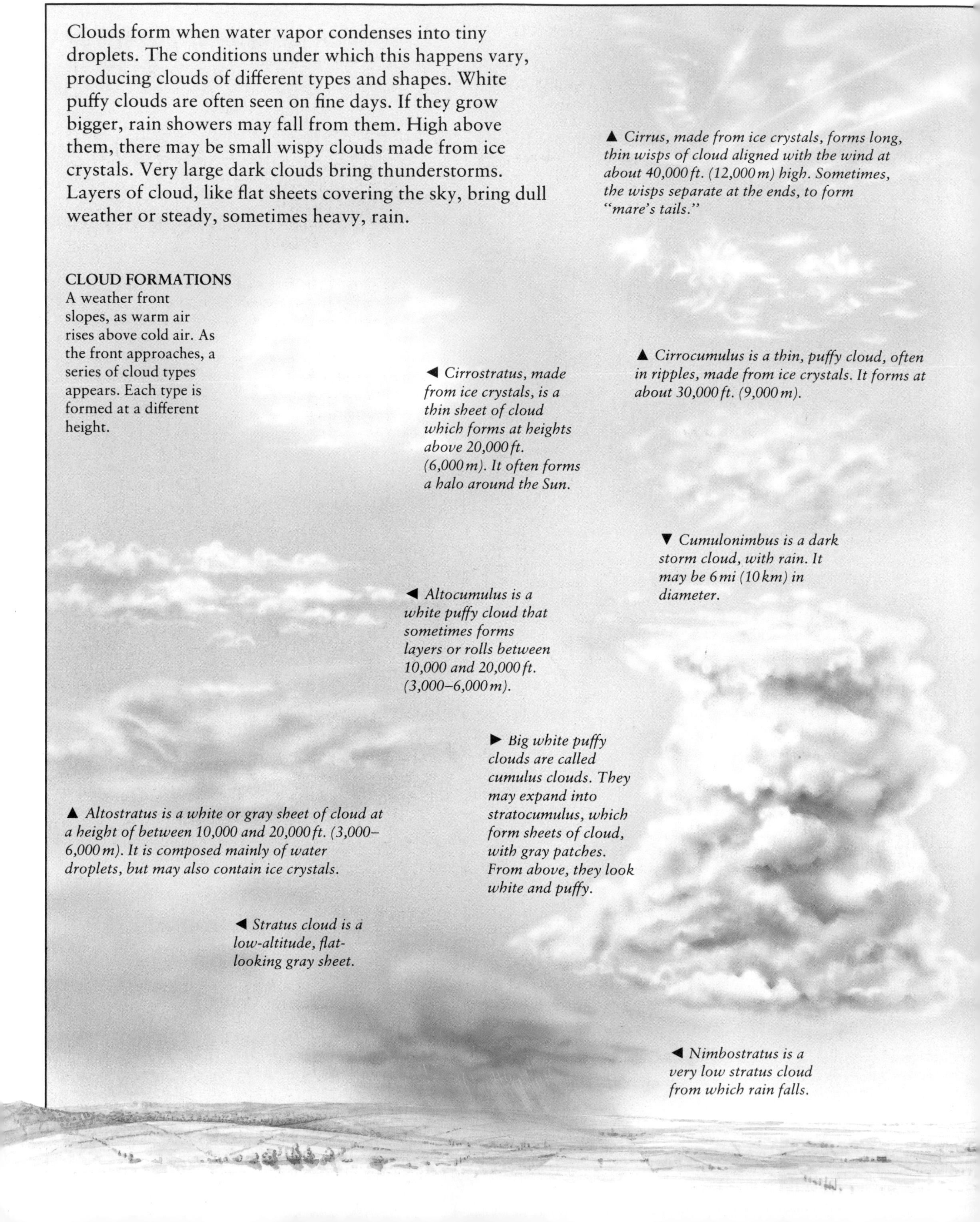

▲ *Cirrus, made from ice crystals, forms long, thin wisps of cloud aligned with the wind at about 40,000 ft. (12,000 m) high. Sometimes, the wisps separate at the ends, to form "mare's tails."*

CLOUD FORMATIONS
A weather front slopes, as warm air rises above cold air. As the front approaches, a series of cloud types appears. Each type is formed at a different height.

◀ *Cirrostratus, made from ice crystals, is a thin sheet of cloud which forms at heights above 20,000 ft. (6,000 m). It often forms a halo around the Sun.*

▲ *Cirrocumulus is a thin, puffy cloud, often in ripples, made from ice crystals. It forms at about 30,000 ft. (9,000 m).*

▼ *Cumulonimbus is a dark storm cloud, with rain. It may be 6 mi (10 km) in diameter.*

◀ *Altocumulus is a white puffy cloud that sometimes forms layers or rolls between 10,000 and 20,000 ft. (3,000–6,000 m).*

▶ *Big white puffy clouds are called cumulus clouds. They may expand into stratocumulus, which form sheets of cloud, with gray patches. From above, they look white and puffy.*

▲ *Altostratus is a white or gray sheet of cloud at a height of between 10,000 and 20,000 ft. (3,000–6,000 m). It is composed mainly of water droplets, but may also contain ice crystals.*

◀ *Stratus cloud is a low-altitude, flat-looking gray sheet.*

◀ *Nimbostratus is a very low stratus cloud from which rain falls.*

▶ *The Sun heats some parts of the ground, such as rock or bare soil, more than others. On warm days, bubbles of hot air form over these areas, and they rise up through the cooler air around them.*

▶ *The warm air rises into low-pressure air and expands and cools. The air cools so much that the water vapor condenses into droplets, and a small cumulus cloud is formed.*

▶ *The cloud grows if it is fed by a series of air bubbles, and the wind detaches it. Fair-weather cumulus looks like cotton balls. It does not carry enough water to cause rain.*

▼ *Moist air that rises over a mountain and then sinks again may set up waves in the air downwind of the mountain. As the air rises to the crest of a wave, it cools and clouds are formed in the shape of a lens. They are called "lenticular" clouds.*

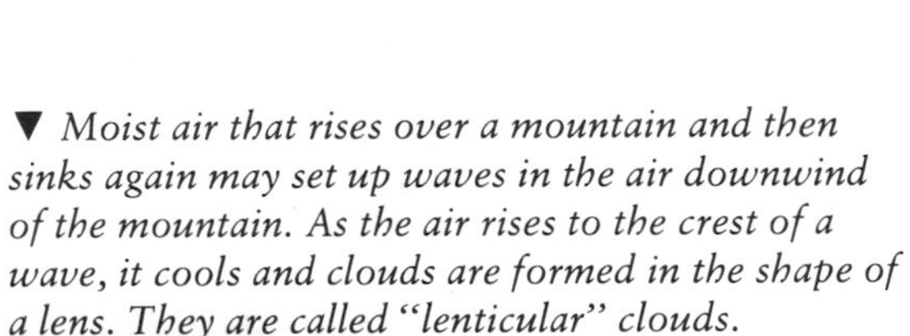

HOW CLOUDS FORM

The amount of water vapor air can hold depends on the temperature of the air. If warm, moist air is cooled, its water vapor will condense. This is why water condenses on a cold window. Clouds form when the water vapor condenses around tiny solid particles.

▲ *At night, fog forms as warm air is chilled by contact with a cold valley floor. As the ground warms again the fog lifts to form low cloud.*

FOG

Fog is cloud that forms close to the ground. The different types of fog are named after the ways in which they are formed. Advection fog forms when warm, moist air passes over cold ground or water. Radiation fog forms at night, as the land cools and the air above it is chilled.

▼ *Advection fogs are common in San Francisco, rolling in from the Pacific to envelop the Golden Gate Bridge. They form when warm, moist air from the south meets cold ocean currents from the Arctic. In the daytime, the air over the warm land is at low pressure and a sea breeze carries the fog ashore.*

FACTS ABOUT CLOUDS

Cloud names are easy to remember:
- Those that begin with **alto-** form at medium height, between 6,500 and 20,000 ft. (2,000–6,000 m).
- Those beginning **cirr-** form above 20,000 ft. (6,000 m). Clouds without these prefixes form below 6,500 ft. (2,000 m).
- **Strat-** clouds form flat-looking layers; those with **cumu-** form heaps.
- **Nimb-** means a rain cloud.
- The highest standard cloud formation is cirrus, up to 40,000 ft. (12,000 m), but nacreous cloud may form at almost 80,000 ft. (24,000 m).
- Sea-level fogs on the Grand Banks in Newfoundland persist for 120 days of the year.
- Aircraft often produce trails of cloud at high altitudes, when water vapor from the hot engine exhaust condenses.

Rain and Snow

Water that falls from a cloud is called "precipitation" and may take the form of rain, drizzle, hail, sleet, or snow. Not all precipitation reaches the ground. In warm weather, rain may fall from a cloud only to evaporate again in midair.

Whether precipitation falls as water, ice, or a mixture of the two, depends on the conditions inside the cloud and the temperature of the air outside it. In summer, most of the ice that forms inside a cloud melts as it falls, except during the occasional hailstorm.

SNOW CRYSTALS
When water freezes, its molecules bind together into flat, six-sided crystals with four long sides and two short ones. The crystal grows as other water molecules attach themselves to its six sides. Each snowflake is unique.

RAIN AND SNOW
There are two main types of rain. In the tropics, rain is formed when air currents cause the tiny water droplets in a cloud to bump into each other. These droplets join together to form larger drops which fall as rain. Most rain outside the tropics is caused by snowflakes melting as they fall. The height at which water freezes as it condenses out of a cloud is called the "freezing level." The ice crystals grow rapidly into snowflakes as water droplets freeze onto them. If the freezing level is below 1,000 ft. (300 m), the ice crystals will not have time to melt before reaching the ground and will fall as snow.

Temperature
Wet snow
Dry snow
Sleet
Rain
Drizzle

▲ *If the base of a stratus cloud is low enough, small droplets of rain may fall as a fine drizzle. Dry snow falls when the ground temperature is cold, but if snow falls from a cloud into air that is just above freezing, some of it will melt. The resulting mixture of snow and rain is called "sleet."*

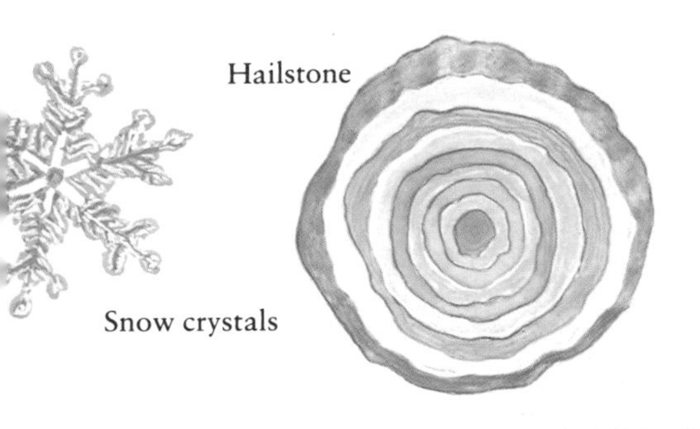

HAILSTONES

Hailstones form around a small ice crystal. Alternate layers of clear and milky ice build up as the hailstones are swept up and float down inside the cloud.

FACTS ABOUT RAIN AND SNOW

- Cherrapunji, India, is one of the world's wettest places, receiving nearly 430 in. (11,000 mm) of rain a year. The South Pole is the driest, with the frozen equivalent of only 1.6 in. (40 mm).
- Libreville, in Gabon, Africa, receives more than 100 in. (2,500 mm) of rain a year, with a short dry season. Chicago has about 33 in. (840 mm).
- As maritime air crosses mountains it loses moisture. The area on the far side of the mountains may lie in a "rain shadow," with relatively little rain. The annual rainfall in Vancouver, Canada, is 57 in. (1,440 mm). In Calgary, on the other side of the Canadian Rockies, it is 17 in. (440 mm). Calgary lies in a rain shadow.

► *Some very cold clouds can be made to release rain by dropping crystals of silver iodide or dry ice into them. Water then freezes onto the crystals. This is called "seeding" the cloud, and has been used to end droughts.*

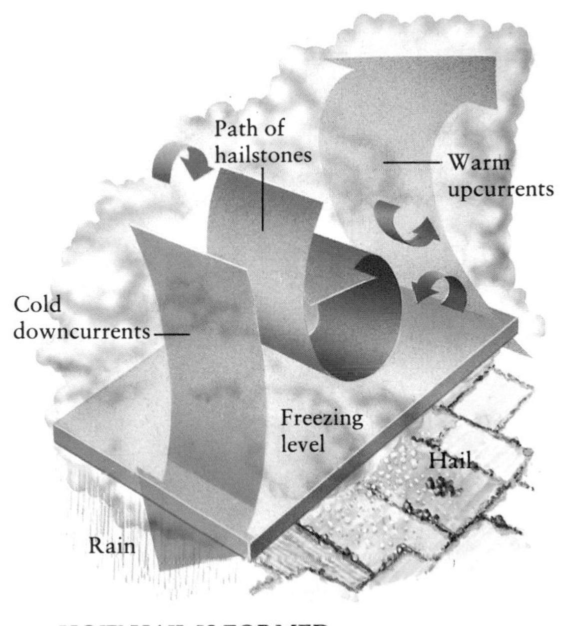

HOW HAIL IS FORMED

Inside a storm cloud, raindrops may be carried up by air currents and frozen high in the cloud. An opaque layer of ice builds up as water vapor freezes onto them. They fall to warmer levels, where the outside melts and is then refrozen into a clear layer of ice as the hailstone is carried up again. The hailstone rises and falls until it is heavy enough to fall out of the cloud.

► *In 1970, hailstones weighing 27 oz. (760 g) fell in Kansas. In 1928 a hailstone measuring 5.5 in. (14 cm) across and weighing 25 oz. (700 g) fell in Nebraska.*

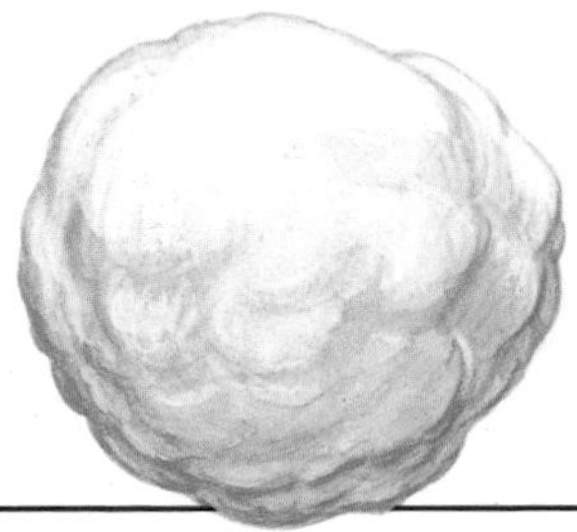

SUN'S HALO

When a thin veil of cloud partly covers the Sun, light rays may be refracted (bent) by the ice crystals. This creates a "halo," a ring of white light around the Sun, sometimes with a faint tinge of red on the inside and violet on the outside. Small water droplets in such clouds as altocumulus can also refract light, making a colored "corona," usually blue on the inside and red on the outside. A white halo can also occur when cirrus clouds obscure the Moon.

RAINBOWS AND FOGBOWS

A rainbow is caused by sunlight, or even moonlight, shining on a screen of water droplets. Rays are refracted as they enter a droplet, reflected from the back of the droplet, and then refracted a second time. Light of different wavelengths is refracted by different amounts, which splits the white light into its rainbow colors. A secondary rainbow, with the order of its colors reversed, may appear outside the primary rainbow. A fogbow is similar, but its colors overlap and mix to produce white.

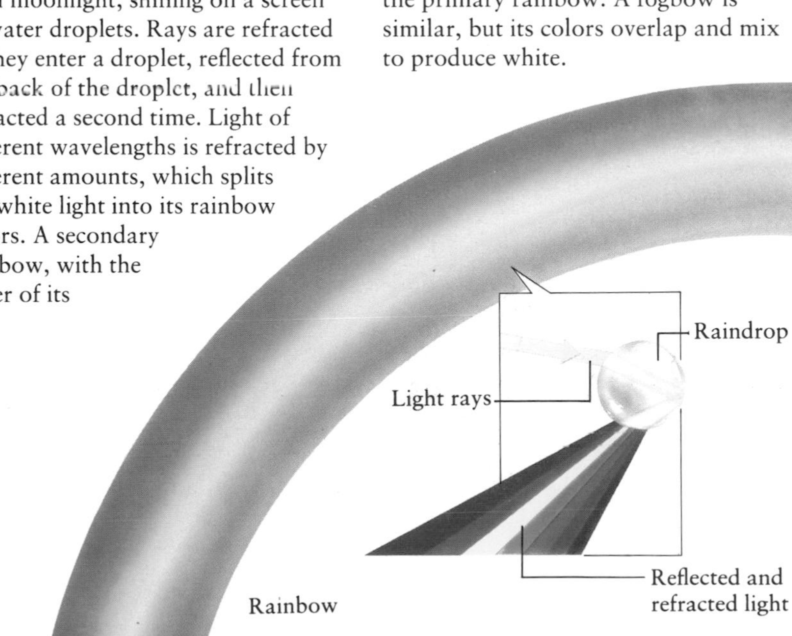

Weather Forecasting

We all like to know what the weather will be like over the next few hours or days. Indeed, it is very important for some people to know. Farmers must know when to plow or harvest their crops. Fishermen must know whether it is safe for them to leave port. Aircraft pilots must know what weather they will encounter during a flight, so that they can avoid dangers such as large thunderstorms. Scientists who study the weather are called meteorologists, and much of their work involves preparing weather forecasts.

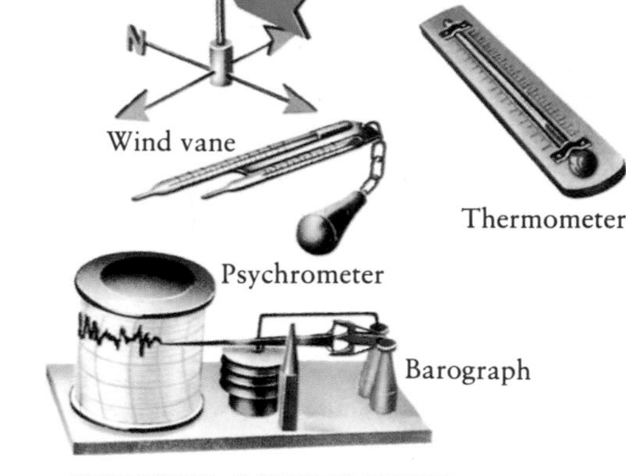

WEATHER INSTRUMENTS

Meteorologists use a barograph, a barometer linked to a pen and paper drum, to record atmospheric pressure. A thermometer records temperature, and a wind vane and anemometer record wind direction and speed. Wet- and dry-bulb thermometers (psychrometers) measure humidity.

WEATHER SYMBOLS

Weather maps summarize conditions at a particular time. They are called "synoptic charts" and use standard symbols. The most prominent of these symbols are "isobars," the lines connecting places where air pressure is the same. Winds flow roughly parallel to the isobars. The closer together the isobars are, the stronger the wind. The chart also shows warm and cold fronts (*see page 51*).

▲ *A weather map may show the pressure (in millibars) on isobars and at centers of low and high pressure.*

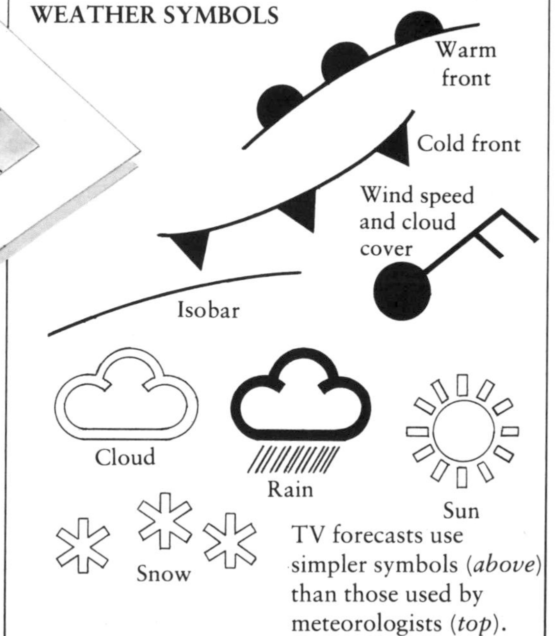

TV forecasts use simpler symbols (*above*) than those used by meteorologists (*top*).

COLORS IN THE SKY

There is a scientific explanation for much weather lore. If the wind is from the west, a red sky at sunset means the air to the west is dry and the next day is likely to be fine. If it is dull red with cloud, there may soon be rain. A red sky in the morning means clouds are arriving and the day may be rainy.

WEATHER TRACKING

Forecasters know the kind of weather associated with different cloud types. Cloud patterns often indicate areas of high and low pressure. By tracking weather systems, from surface reports and satellite images, meteorologists can predict their movements and speeds and the ways they will change. It is essential that the path of a hurricane is predicted accurately, so that people in threatened areas can be evacuated. Unfortunately, hurricane systems are so complicated that this is not always possible.

▲ *Weather planes monitor conditions in the upper atmosphere, using instruments attached to the aircraft's long nose. Conditions at sea are reported by specially equipped ocean weather ships* (left). *These ships are towed to positions far from shipping lanes, where they are anchored; they send reports up to eight times a day.*

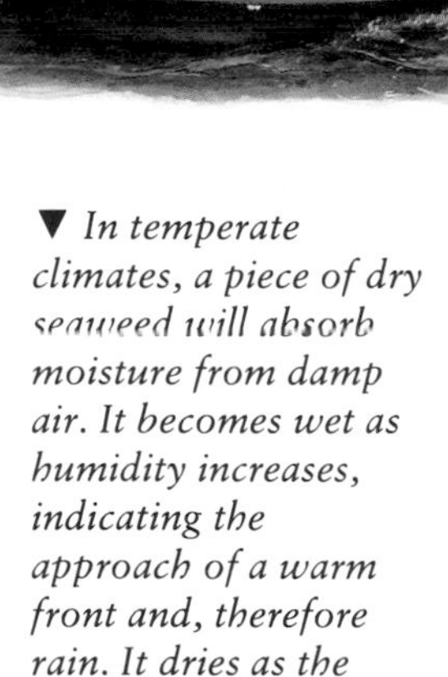

Meteosat

▲ *Weather satellites* (left) *transmit photographs of cloud patterns back to Earth, allowing scientists to study their type and movement. Satellites traveling in geostationary orbit remain above one point on the surface. Others orbit from pole to pole.*

▶ *Balloons called "radiosondes" carry instruments that are able to measure the temperature, pressure, and humidity of the upper atmosphere. The readings are sent by radio from the balloon to ground weather stations. The flight of some balloons, called "rawinsondes," is tracked. These balloons are filled with helium, and they expand as they rise into less dense air. Their path reveals the speed and direction of winds at high altitudes.*

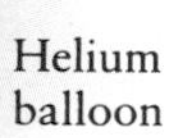

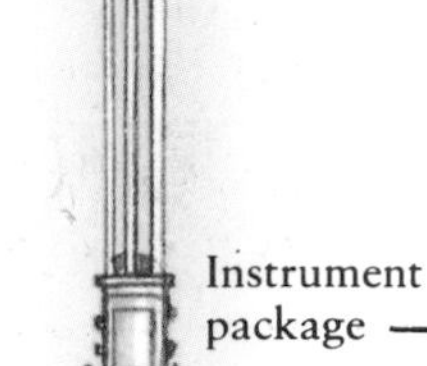

▼ *In temperate climates, a piece of dry seaweed will absorb moisture from damp air. It becomes wet as humidity increases, indicating the approach of a warm front and, therefore rain. It dries as the warm air passes.*

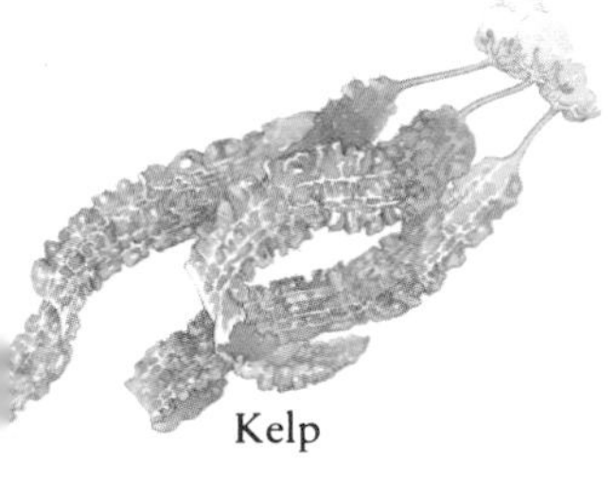

Kelp

CHAOS THEORY

Meteorologists are able to prepare a weather forecast for only a few days ahead. Long-range forecasts proved so unreliable that meteorologists no longer do them. The problem is that local differences in conditions, which are too small to record, can greatly alter the way a weather system moves and develops. So, for example, a small change in the air over the Arctic could cause a hurricane in the tropics. Scientists use a theory known as the "chaos" theory to describe this unpredictable behavior.

LANDSCAPES

The Changing Scene

The world's different landscapes have been made mainly by the action of the weather on rocks. Over thousands of years, mountains are worn away by wind, ice, and rain, until they become gently rolling hills and, eventually, level plains. As the rocks are broken into tiny fragments, living organisms can obtain the minerals they need from them, providing that they also have water. These organisms convert the mineral particles into soil. Plants grow in the soil, and animals can feed on the plants. Farmers may clear away the natural vegetation to plow fields and grow crops, where the soil and climate are suitable. Climate variations are recorded in the landscapes the farmers have formed. The hills, valleys, and soils of a desert are different from those of a forest, but plants may grow in soils that formed millions of years ago in a desert.

WORLD VEGETATION
The Earth can be divided into regions that have roughly the same climate. On land, it is mainly the climate that determines the kinds of plants that grow naturally in an area. Similar types of vegetation cover vast areas—these are called "biomes."

The tropical rain forest biome forms a belt on either side of the equator. Subtropical grasslands give way to scrub and semi-arid grassland, then to hot deserts, just outside the tropics. Beyond the tropics, temperate grasslands give way to temperate forests, then conifers, tundra, and eventually permanent ice.

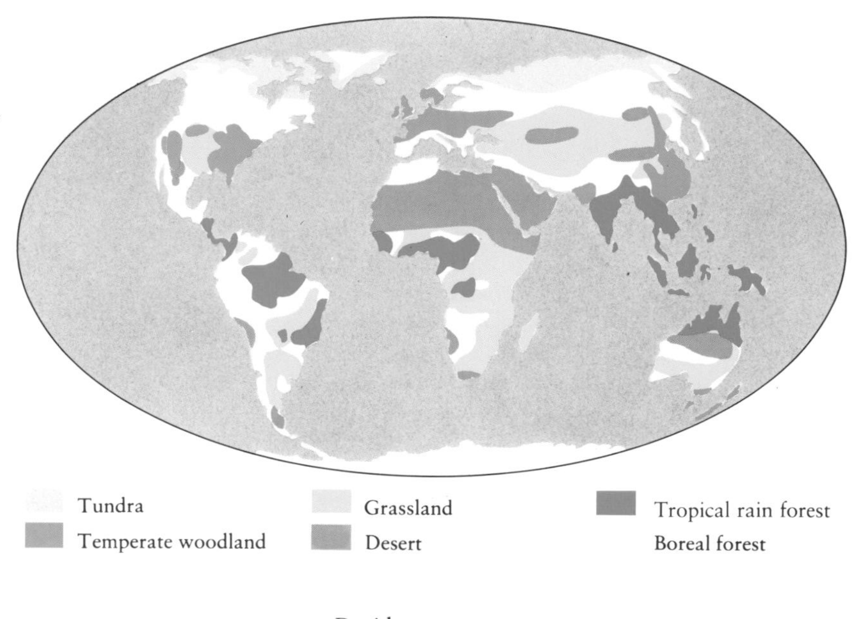

Tundra
Temperate woodland
Grassland
Desert
Tropical rain forest
Boreal forest

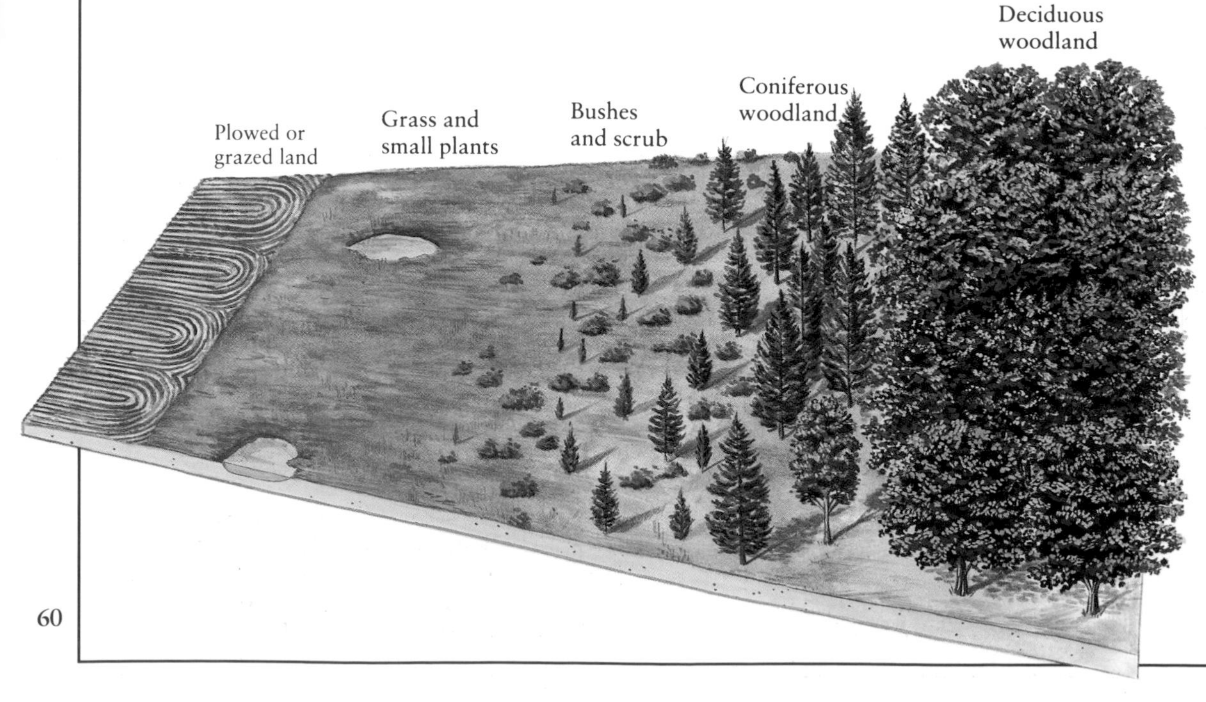

◀ *In a temperate climate bare land will not remain so for any length of time. The first plants to colonize are small herbs that grow quickly. These herbs are soon followed by grasses, which grow tall enough to shelter the seedlings of woody plants, such as blackberry bushes, and small trees, such as hawthorn and hazel. Then larger trees appear, such as beech and oak. This sequence is called a "succession."*

LAND USE

In Europe, the first villages were built near rivers, which supplied fish and fresh water. Later, forests were cleared to make fields for crops and grazing animals. The early farms grew until most of the valleys were cultivated. By the 11th century most of the original forest had disappeared, and many riverside settlements became large towns. Rock and metal ores were mined before Roman times.

Prehistoric

Medieval

◀ *Until people settled down and began to clear the land for fields and cut trees for houses, they had little impact on the landscape.*

► *In the 18th to 19th centuries as industry grew, so did the towns and the demand for raw materials and coal. The countryside became dirty and ugly. Today development continues, but we are aware of the damage we can do.*

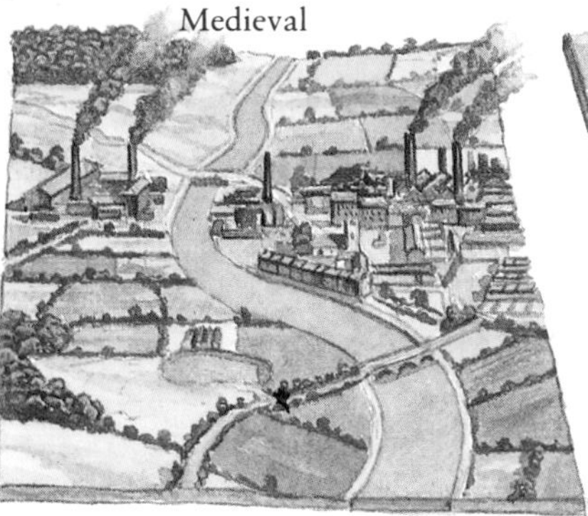

18th–19th Century

Today

Tundra soils	Chernozems (Mollisols)
Podzols (Spodosols)	Alfisols
Podzols and Brown Earths (Spodosols and Inceptisols)	Grumusols (Vertisols)
Podzols (Spodosols)	Desert soils (Aridisols)
Ferralsols (Oxisols)	Montane soils

Zonal soil types (shown above) were based on climatic factors. New names given in parentheses.

SOIL

The rate at which rock is changed into soil depends mainly on the climate. In places where the ground is frozen for most of the time, soil forms very slowly. The soils in the far north of America and Asia are said to be "young," because their formation has barely begun. Near the equator, where the climate is warm and wet, soils form rapidly and are said to be "ancient."

Scientists group different soil types into 10 orders. The very young ones are called Inceptisols, the ancient ones are Oxisols or Ultisols. Desert soils, also called Aridisols, are poorly developed because of the shortage of water and the lack of decomposed remains of plants (humus), which help make a soil fertile.

The best farm soils are the Mollisols of the prairies and steppes, and the Spodosols of temperate forests, found in northeastern North America and Britain.

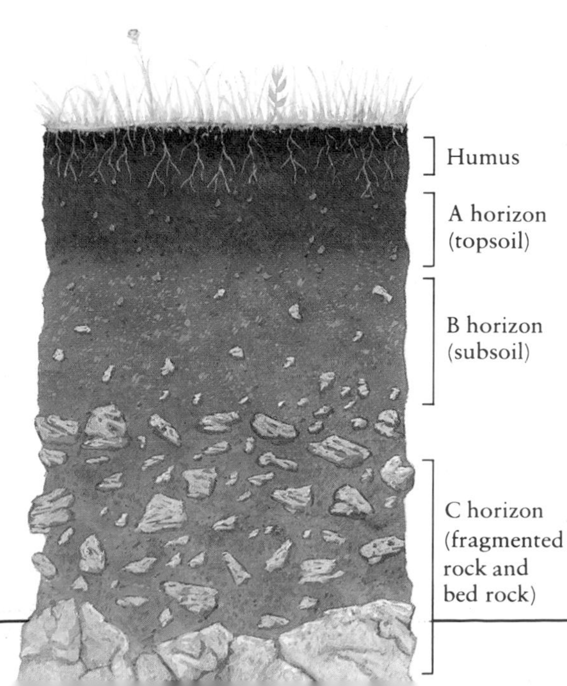

SOIL PROFILES

Once a soil has developed it forms layers, called "horizons." Beneath a surface layer of plant remains, the A horizon is rich in decomposed organic plant and animal matter. The B horizon is mainly mineral particles, with much less organic material. The C horizon is primarily small stones, and beneath them all lies the bedrock.

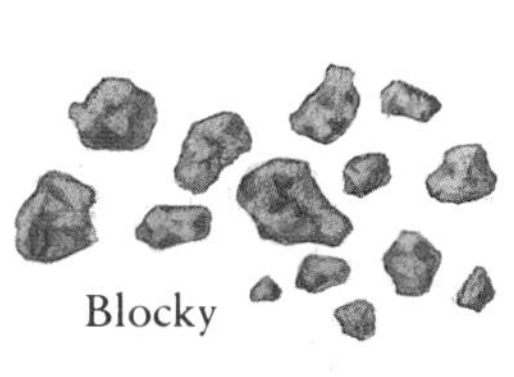
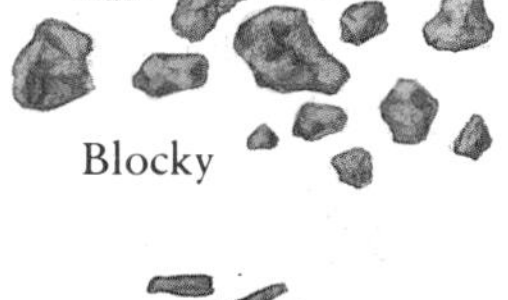

Blocky

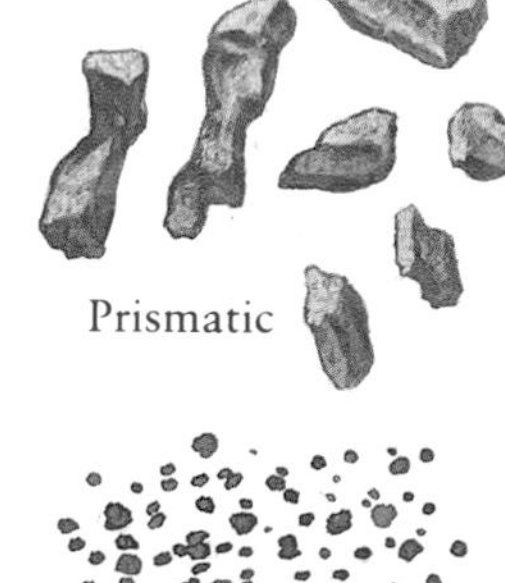

Prismatic

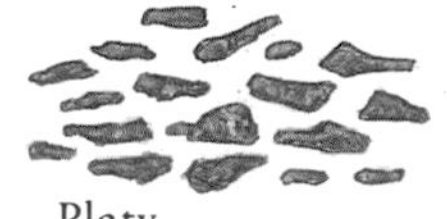

Platy

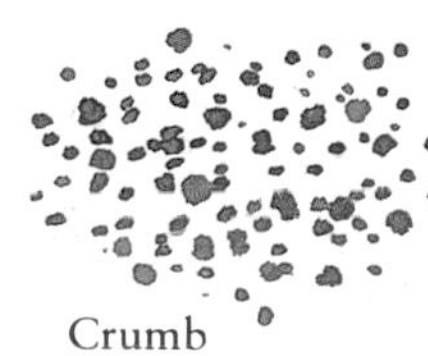

Crumb

▲ *The amount of air and water found in a soil depends on the particles from which it is made. A platy structure packs into watertight layers. A blocky structure drains well, a prismatic one less well. A crumb structure is best of all.*

Polar Regions and the Tundra

Within the Arctic and Antarctic circles there is at least one day a year when the Sun does not rise, and at least one when it does not set. The Arctic and Antarctic are lands of midnight Sun in summer, and noon darkness in winter. The polar regions are the coldest on Earth, and among the driest because there is little liquid water. Most of Greenland lies beneath ice 5,000 ft. (1,500 m) thick, that fills valleys and buries hills. The average thickness of the Antarctic ice sheet is more than 6,500 ft. (2,000 m).

ICEBERGS

An iceberg is a large block of floating ice. It is much larger than it looks because some nine-tenths of the ice floats below the surface. This can be dangerous to ships. Some Antarctic icebergs are more than 60 mi. (100 km) long.

THE ARCTIC AND THE ANTARCTIC

Most of Greenland and the northern parts of Alaska, Canada, Scandinavia, and Siberia lie within the Arctic Circle, but there is no land close to the North Pole itself. Antarctica is the world's fifth largest continent, divided into two parts by the Transantarctic Mountains. Beneath the ice, the land of East Antarctica is mostly rugged, in places rising to more than 13,000 ft. (4,000 m) above sea level. West Antarctica is lower. Much of it is made up of a peninsula and island archipelago. In places, the land around the South Pole is up to 8,200 ft. (2,500 m) below sea level.

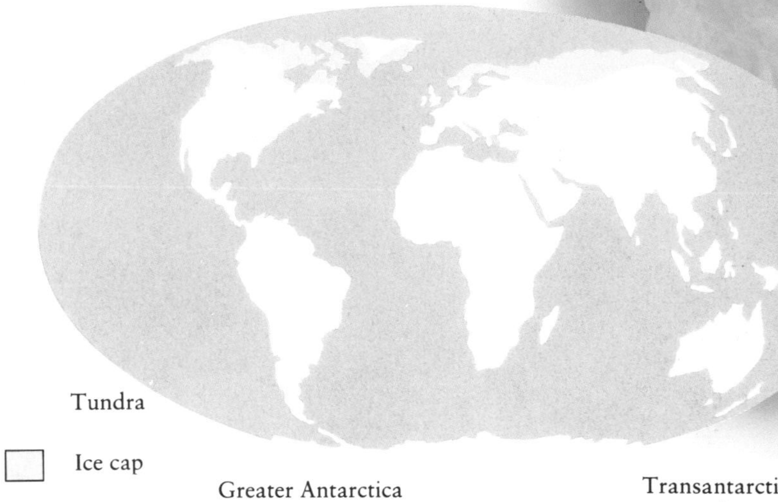

THE TUNDRA

Around the Arctic Circle, between the conifer forests farther south and the region of permanent ice to the north, the tundra extends as a vast treeless plain across all the northern continents. In summer the ground thaws for just a few weeks, triggering frantic activity for the region's animals and plants.

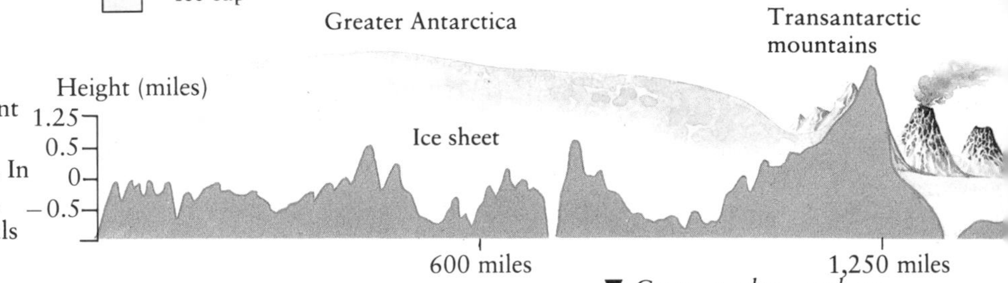

▼ *Geese, waders, and sea birds live on the tundra. Tundra mammals include polar and grizzly bears, musk ox, caribou, voles, and shrews.*

◀ *Tundra plants are small, as their roots can only grow to a depth of 12 in. (30 cm) before they reach frozen ground. There are heaths, dwarf birch trees, sedges and rushes, mosses and lichens. Many plants flower in the brief summer.*

Sea level

FACTS ABOUT POLAR LANDS

- During the dark nights, plant nutrients accumulate in the sea. As the light returns, marine plants multiply rapidly, providing food for small and larger animals, such as fish, sea birds, seals, and whales.
- Each year, more than 7,000 icebergs are carried south from Greenland in the Labrador Current.

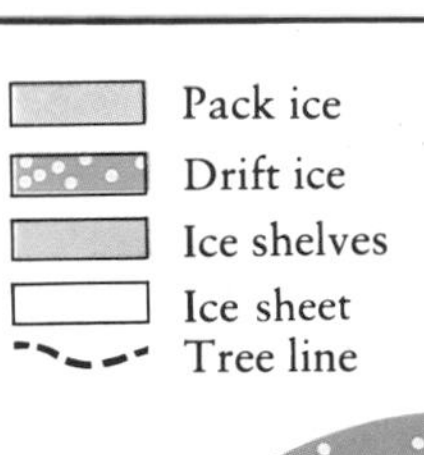

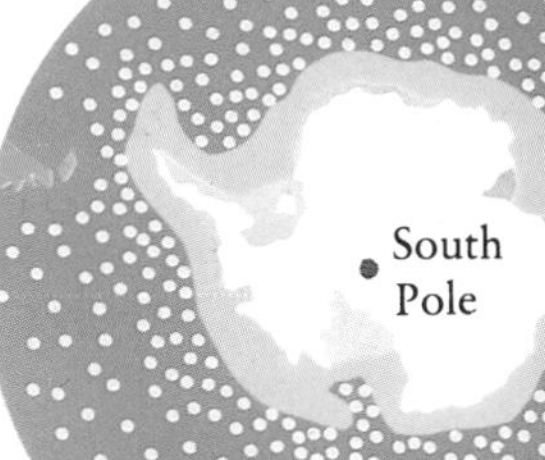

Glacier or ice sheet

▲ *Where a glacier enters the sea, the ice floats on the water. The end of the glacier snaps off to form an iceberg. Ice shelfs also break, forming much larger icebergs.*

▼ *Under the Antarctic ice sheet, unlike the Arctic, there is land. Near the coasts some glaciers have retreated, leaving dry, rocky valleys, called "oases." Inland, high mountain peaks project above the ice, as "nunataks."*

▲ *The extent of the Antarctic ice sheet varies with the changing seasons. In winter the drift ice extends out to the southern tip of South America. The Antarctic is home to a few plants and some insects. In summer, penguins, sea birds, and seals visit it.*

▲ *The Arctic is a mass of pack ice which also changes with the seasons. In winter, its ice covers all of Greenland and its drift ice reaches as far south as Iceland and northern Russia. The presence of the ice and tundra lands limits the growth of trees to areas south of the line through northern Canada, Norway, Sweden, and Russia.*

Lesser Antarctica

Ross Ice Shelf

Ice sheet

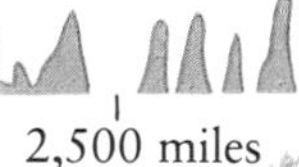

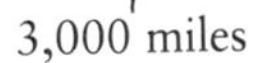

2,000 miles
2,500 miles
3,000 miles

PERMAFROST

In winter, in the Arctic and Antarctic, all the moisture in the soil freezes, but in some areas the top few inches of the soil thaw in summer. During the summer thaw, the ground turns to mud, with pools in the hollows. The subsoil and deeper layers remain permanently frozen. They are called "permafrost." If the permafrost thaws, for example, because of the heating effect of a house or oil pipeline, then the land will sink.

Soil thaws in summer

Permafrost

POLAR RESOURCES

Long ago, the polar regions lay in lower latitudes and had warmer climates. In Antarctica, there are deposits of coal up to 20 ft. (6 m) thick, formed 250 million years ago. Alaska also has vast coal reserves, and, in 1968, one of the world's largest oil fields was discovered at Prudhoe Bay.

▼ *Oil travels south from Alaska to ports by the Trans-Alaska Pipeline. The pipeline was built on supports above the ground, to prevent it thawing the permafrost.*

Temperate Woodland

During the last Ice Age, most of the northern latitudes, higher than 50°, lay beneath ice. When the ice retreated, it left bare rock and debris. As the climate warmed up, groups of plants spread north, until most of the landscape was covered by forest. Trees in far northern regions have to survive the equivalent of a dry season, when water is frozen. Deciduous trees save water by shedding leaves, conifers have needle-like leaves from which little water evaporates. Both types form large areas of forest.

SOIL

Soils of conifer forests have a light-colored, rather acid upper layer. Some soil minerals dissolve and drain into the subsoil. Broadleaved forests develop very fertile soil with an even, brown surface layer.

WOODLAND AREAS

Broadleaved evergreen forests grow around the Mediterranean and in those regions of the southern United States, China, South America, South Africa, and southern Australia, where temperatures rarely fall below freezing. Broadleaved deciduous forests grow farther north. Southern Chile is the only place they are found in the Southern Hemisphere. Most of Canada, northern Europe, and Asia is covered by coniferous or boreal forest. There is no boreal forest in the Southern Hemisphere.

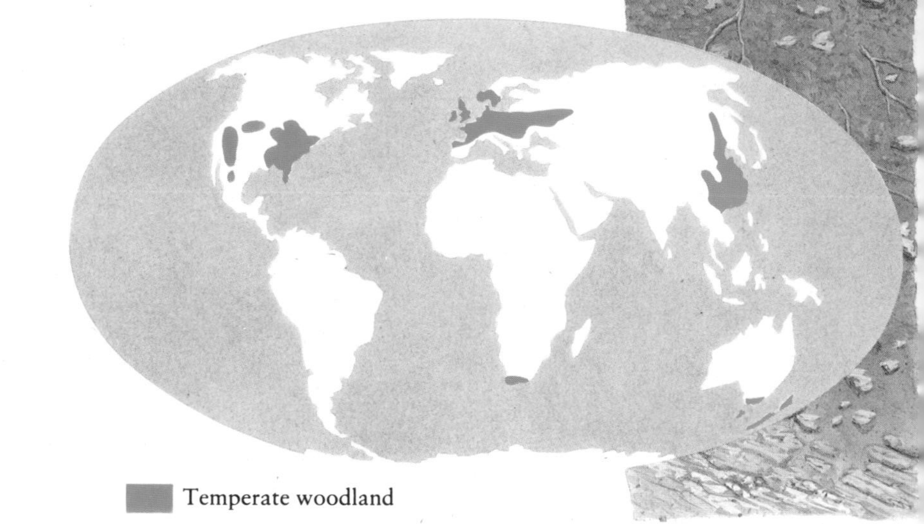

Temperate woodland

THE FOREST ECOSYSTEM

The trees of the broadleaved evergreen forest include holm oak, cork oak, and a few species, such as holly, that also grow in mild, moist regions farther north. These species are usually mixed with pines and cedars. The deciduous forests have a wider variety of species. They are often dominated by oak, beech, and maple, with different species in Europe and North America. Coniferous forest is made up entirely of pines, spruces, firs, and larches. Each type of forest has its own type of wildlife. The evergreen forests have fewer animal species than the others. Species found in the conifer forests include moose, bears, and wolves.

Mixed woodland

◄ *Near the edges of the temperate deciduous forest area, conifers grow side by side with broadleaved trees. This mixed forest of aspens and larch is in the Nevada Rockies. North American woodlands contain more tree species than European woodlands. The richest deciduous forests are in the Appalachians, in the east. The species include tulip trees and oaks, and basswood and buckeye.*

WOODLANDS

- Approximately 29 percent of the United States is forested, compared to 24 percent of Western Europe. About 850 native and naturalized species of trees are grown in the U.S.
- All the paper we use comes from conifers grown in temperate regions.

DEFORESTATION IN EUROPE

Most of the original forest that once covered Europe was cleared long ago to provide farmland. In Britain, this clearance was well advanced by the end of the Roman occupation. By the 11th century, trees covered a smaller area than they do today. The woodlands that remained were made up of species that had established themselves naturally. Some of these survive as "ancient woodlands." These are woods where trees have grown since before the forest plantations were begun in the 18th century. Scientists identify them from historical records, and also by the type of plant species they contain.

Natural extent of forest

Present extent of forest

▲ *The New England forests are famous for their autumn colors. As leaves die, they lose their green chlorophyll, revealing many shades of red and yellow.*

THE NITROGEN CYCLE

Nitrogen is constantly being recycled. All proteins contain atoms of nitrogen. The nitrogen is taken from the air by soil bacteria and made into soluble nitrate ("fixed"), which is absorbed by plants. Other bacteria decompose organic matter, releasing nitrate and returning some nitrogen to the air.

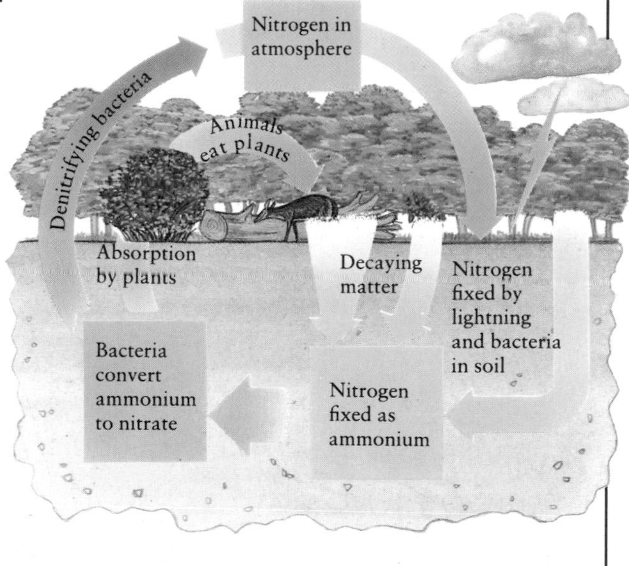

▲ *Nitrogen passes from air to soil, to plants, to animals which eat plants, and eventually back into the air. The energy of lightning also fixes some nitrogen, by making the gas react with oxygen to form nitrogen oxides. These dissolve in rain. Some nitrate drains from the soil into rivers and then into the sea. This supplies nitrogen for freshwater and marine plants, and the animals that feed on them.*

Deciduous woodland

▲ *A European broadleaved forest typically contains oak, ash, beech, and chestnut. The trees provide shelter and food for many species of birds, insects, and small mammals. For example, the common European oak (*Quercus robur*) is said to support more than 300 animal species.*

Grasslands

There are regions where, for most of the year, the rate at which water can evaporate is greater than the amount of rainfall. Such regions would be deserts, were it not for the rain that falls during one season of the year. Just beyond the edges of the deserts, rain falls in the summer. In the dry interior of continents it falls in winter. The rain allows plants to grow, turning the dry, brown landscape green. Few trees can survive in these conditions, but flowering herbs and grasses abound. These regions are the grasslands.

SOIL

Grassland soils have a deep, dark-colored, surface layer rich in humus. They are among the most fertile of all soils and are often farmed. The lower layers vary from place to place.

GRASSLANDS OF THE WORLD

The middle of continents are dry, because of their distance from the sea. The temperate, continental grasslands are called "steppes" in Europe and Asia, "prairies" in North America, and "pampas" in South America. Temperate grassland is also found in eastern Australia. The subtropical grasslands of South America, Africa, India, and northern Australia are called "savanna."

Grassland vegetation is mainly made up of drought-resistant grasses. In parts of the steppe these grasses are short, but they can be up to 7 ft. (2 m) tall. Grass leaves grow from the base and can survive and grow even if grazed.

THE LANDSCAPES

In the rainy season, grasses and herbs grow rapidly. The land turns green or is blanketed by a mass of brightly colored flowers. The flowers set seed quickly and then die, with nutrients stored in their roots. As the rains stop, the plants turn brown. The dry vegetation burns easily and fires are common. The fires nourish the soil with ash, which encourages new growth next time it rains. The plants have deep roots.

In Europe, the climate and geography have led to forests as the natural landscape, and almost all grassland has been reclaimed from forest.

▼ *Much of the South American pampas is covered by tussocks, or humps of feather grass. Elsewhere there is scrub. Underneath the plants the ground is hard.*

◀ *The prairies were once home to large herds of bison (buffalo). Today, most of the prairie is used to grow wheat and corn.*

◀ *Grasslands differ mainly by the species in each location. In Australia, the grasslands cover a wide range of soils and feature tough grasses and scattered acacia and gum trees. These grasslands are home to kangaroos, koalas, emus, and kookaburras.*

SLASH AND BURN

People have extended grasslands by setting fire to vegetation during the dry season. This encourages new growth, but destroys woody plants. Livestock also destroy tree seedlings by trampling them.

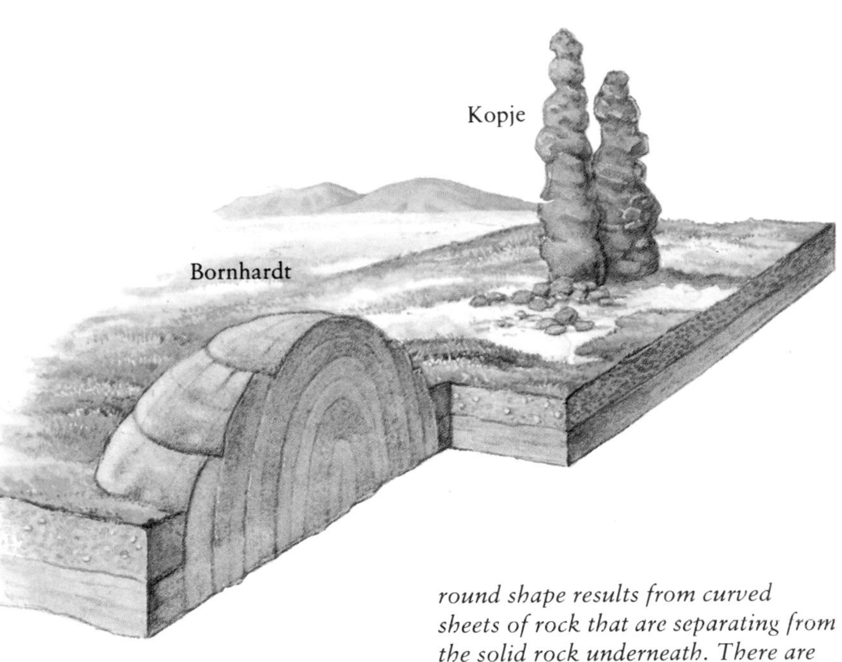

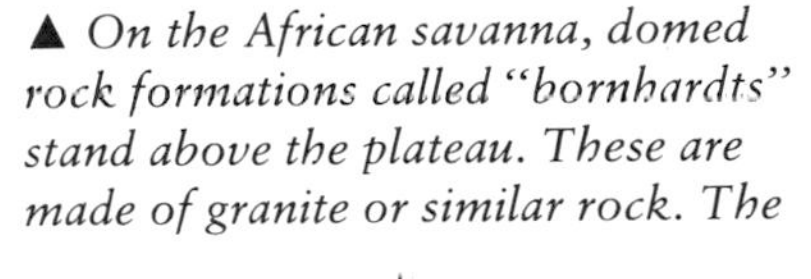

▲ *On the African savanna, domed rock formations called "bornhardts" stand above the plateau. These are made of granite or similar rock. The round shape results from curved sheets of rock that are separating from the solid rock underneath. There are other small, isolated hills about 30 ft. (10 m) tall formed from exposed rock. They are called castle "kopjes" (pronounced koppies).*

THE DUST BOWL

The grassland climate is dry and sometimes there are long droughts. One in the prairies of the central United States lasted from 1933 to 1939. Severe dust storms in 1934 and 1935 turned the area into what came to be known as the "Dust Bowl." The situation had been made worse by years of overgrazing and poor farming.

▲ *African savanna is found on both sides of the equator. It is home to many species of grazing animals, such as wildebeest which move in large herds. Each species feeds differently, so they do not compete with each other. The grazers provide food for lions, cheetahs, hyenas, dogs, and other carnivores. Not all African grasslands are tropical savanna, the South African "veld" is temperate grassland.*

FACTS ABOUT GRASSLANDS

- Grasslands are cultivated in North America and to a limited extent in Europe, but farming is restricted by the dry climate.
- In South America and Australia the grasslands are used for cattle ranching.
- The southern part of the African savanna is farmed, but over most of the area people live mainly by herding cattle. In the dry season people and cattle move around in search of pasture, as do all grazing animals.
- Wheat is grown on the Indian savanna.
- There are nearly 8,000 species of grasses. Some have roots that can reach water 16 ft. (5 m) below ground.

Deserts

When rain falls on the ground some of it evaporates. If the amount of water that evaporates is greater than the rainfall, a desert will form. Any region where the annual rainfall is less than 10 in. (250 mm) is a desert. Deserts are usually windy, but not necessarily hot. The polar regions are deserts, and temperatures in the Gobi Desert are below freezing for 6 months of every year. Most deserts are rocky, not sandy. Sand covers about 2 percent of the North American desert and only 11 percent of the Sahara.

► *Many plants survive in the desert by having a very fast life cycle. Their seeds can lie dormant for many years. On the rare occasions of rainfall, these plants, such as this African grass, germinate, flower, set seed, and die, all in a matter of days.*

ARID LANDS
Deserts are also known as arid or dry lands. Desert soils contain almost no plant or animal (organic) matter. They are made up of dry sand grains and stones, often with gravel, because wind blows away smaller particles. Some desert soils can be cultivated, if water is provided. Most of the world's hot deserts are spreading, mostly due to a change in climate, bringing drier weather to bordering lands. Overgrazing may make this worse by removing vegetation, causing soil erosion.

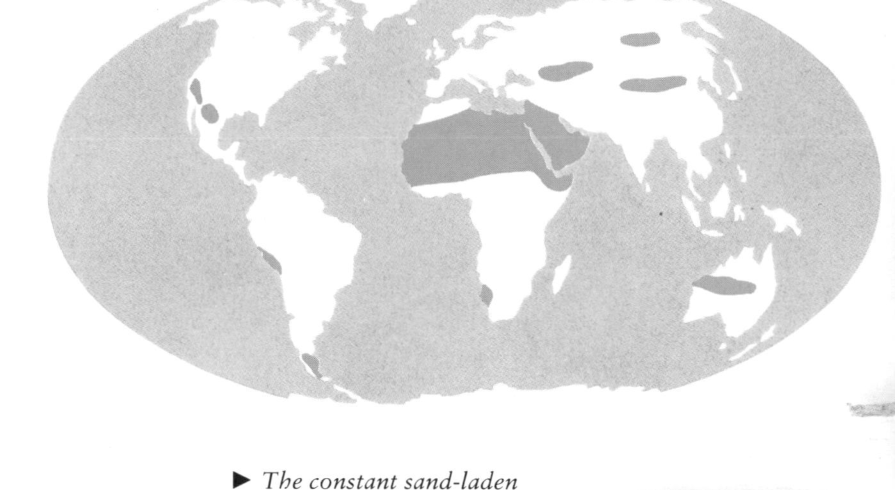

DESERT ECOSYSTEMS
Few plants can grow in shifting sand. There are two main types of desert plants. Shrubs and small trees—such as the Joshua trees, saguaro cactus trees, and sagebrushes of North America and the acacias and tamarisks of the Sahara—store water. Others lie dormant waiting for rain.

► *The constant sand-laden wind sculpts rocks into strange shapes. The rocks are also heated by the Sun—they expand, and then cool and shrink. This strain splits them.*

SAND DUNES

Many different types of sand dunes are formed in sandy deserts. They are shaped by the wind. A "barchan" has a crescent-shaped front and a long tail made from sand blown by the wind. They form long series. "Linear" dunes are created in strong steady winds, which cut troughs in the desert floor. The sand is piled up into long, rounded dunes. A "seif" dune is a long, sharp ridge lying parallel to the wind direction. A ridge that is formed at right angles to the wind is an "aklé" dune. It is formed where there are cross currents. "Star" dunes, with several sharp ridges, occur where the wind direction is constantly changing.

Barchan

Linear dune

FACTS ABOUT DESERTS

- In some hot deserts, a "sand sea" may form, called an "erg." The biggest is the Grand Erg Oriental, covering 76,000 sq. mi. (196,000 sq. km) in Algeria and Tunisia. Some dunes in the Grand Erg are more than 1,000 ft. (300 m) high.
- Death Valley, in the Mojave Desert of California, is the hottest place in North America. Summer temperatures often exceed 120°F (50°C), and 135°F (57°C) has been recorded. Temperatures are similar in the Libyan Sahara, where 136°F (58°C) has been recorded. Desert nights are cool.
- The coldest place in the world is in Antarctica, which is also a desert. The winter temperature can fall to −130°F (−90°C).

▼ *High, rocky Saharan plateaus* (hamada) *are cut through by deep canyons.*

Hamada

Sand dunes

Salt pan

Oasis

Sandy desert

▼ *Oases are natural desert features, but people can make them. In places where water lies close to the surface wells are dug, and the underground water from distant wetter regions is released under pressure from the aquifer.*

OASES

An oasis is a fertile place in a desert where the water table reaches the surface. Sometimes water will fill an aquifer (a rock that holds water). If there is a fault above the aquifer the water will be forced up it and an oasis will form.

◀ *A stony desert, or the surface of small, rounded pebbles, is called a "reg."*

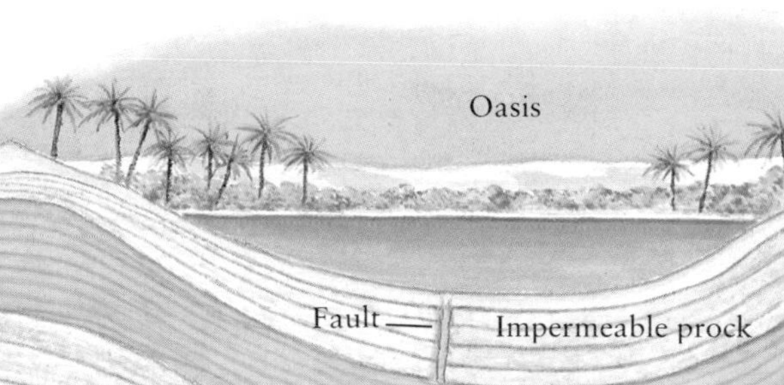

Tropical Rain Forests

The ice sheets have advanced and retreated many times over land near the poles. But in a belt of land around the equator the temperature has never fallen below freezing. As the climate changed near the poles, so did the forests. But in the tropics, forest of one kind or another has grown in some places for millions of years. In this time, species have evolved to fill every corner and use every source of food. This is why tropical rain forests contain a greater variety of plant and animal species than any other forests.

SOILS
Tropical soils are red or yellow in color and up to 33 ft. (10 m) deep. They lie on top of clay. Most nutrients have been lost from the surface layers. Plants feed on the rapidly recycled organic matter.

HUMID HABITATS
The equatorial climate is warm, with heavy rainfall. Plants grow rapidly, and in order to expose their leaves to sunlight, trees grow very tall. The tops of the trees form a continuous canopy, at a height of about 130 ft. (40 m), shading the ground. Most of the trees have shallow roots. They obtain their nutrients from the uppermost layer of soil. Many support themselves with roots that grow outward as stilts or props. Smaller trees and seedlings form a lower layer of forest, and shrubs grow near ground level.

Tropical rain forest

THE ECOSYSTEM
Rain forest grows in lowland areas, including shallow swamps. But much of the equatorial region is mountainous. As you go higher, the lowland forest changes into forest with smaller trees. There is more abundant undergrowth, often with palms, and many more plants growing at ground level. This is called "montane" forest. Higher still, the forest becomes more open. The trees are shorter and covered in epiphytes and climbers. Mosses, ferns, and herbs blanket the ground. This is called "cloud" forest because it is often shrouded in low cloud, from which it obtains moisture.

FACTS ABOUT RAIN FORESTS

- When tropical forest is cleared, new growth often forms a dense "jungle."
- Because of the shade in a rain forest, the air temperature may not be high. But the lack of wind and high humidity make it feel hotter.
- Despite the high rainfall, in most places the ground dries out quickly.
- Tropical forests are being cleared mainly to provide land for farming and plantation forestry. This is often successful on richer soils in valleys, but elsewhere crops soon fail as surface nutrients are removed.

EROSION

When trees are cleared from a hillside the soil may be left bare. Rain can then wash away topsoil, which is carried down the slope. Sometimes it falls into a river, causing pollution.

▲ *In Madagascar, most of the original rain forest has been cut down to provide land for growing crops. Removing the protective forest cover has caused severe soil erosion.*

◄ *In Colombia, as elsewhere in the tropics, isolated hills with steep, smooth sides rise 1,300 ft. (400 m) or more above the plains. These "inselbergs," or "sugar-loaf mountains," are made from layers of rock that have separated and are peeling away.*

◄ *The crowns, or tops, of trees in the rain forest merge to form an interlocking canopy of leafy vegetation about 145 ft. (44 m) from the ground. Food is more abundant here than on the ground, and many of the snakes, lizards, frogs, birds, mammals, and insects living here never visit the ground.*

MANGROVE SWAMPS

Mangrove forests form dense thickets in coastal swamps. The trees produce stilt roots which then develop more roots of their own. Some of these roots stick out above the mud in which they grow. The roots trap shifting sediment, gradually extending the land seaward. Snails and other small animals live among the roots.

PAST, PRESENT, FUTURE

Natural Resources: Energy

Substances we use to make the things we need are called "resources." To prepare a meal, for example, we need food, water, pots and pans, cutlery, an oven, and a source of heat. The food and water, the metals from which the oven and utensils are made, and the fuel that is burned to produce heat are all resources. They are called natural resources because we obtain them from the world around us. If using a resource does not reduce the amount of it available to us, it is called a "renewable" resource. River water is a renewable resource, provided we take no more of it than is replaced by the rain. Food is a renewable resource because farmers can grow more. Most of our energy resources are "nonrenewable," either because the amount of the resource is fixed and cannot be replaced, or because it is replaced, but only very slowly.

FOSSIL FUELS
The word "fossil" is from a Latin word meaning "dug from the ground." Coal, oil, and natural gas are called fossil fuels because they are made from the remains of ancient plants or animals. They were formed very slowly over millions of years and are nonrenewable.

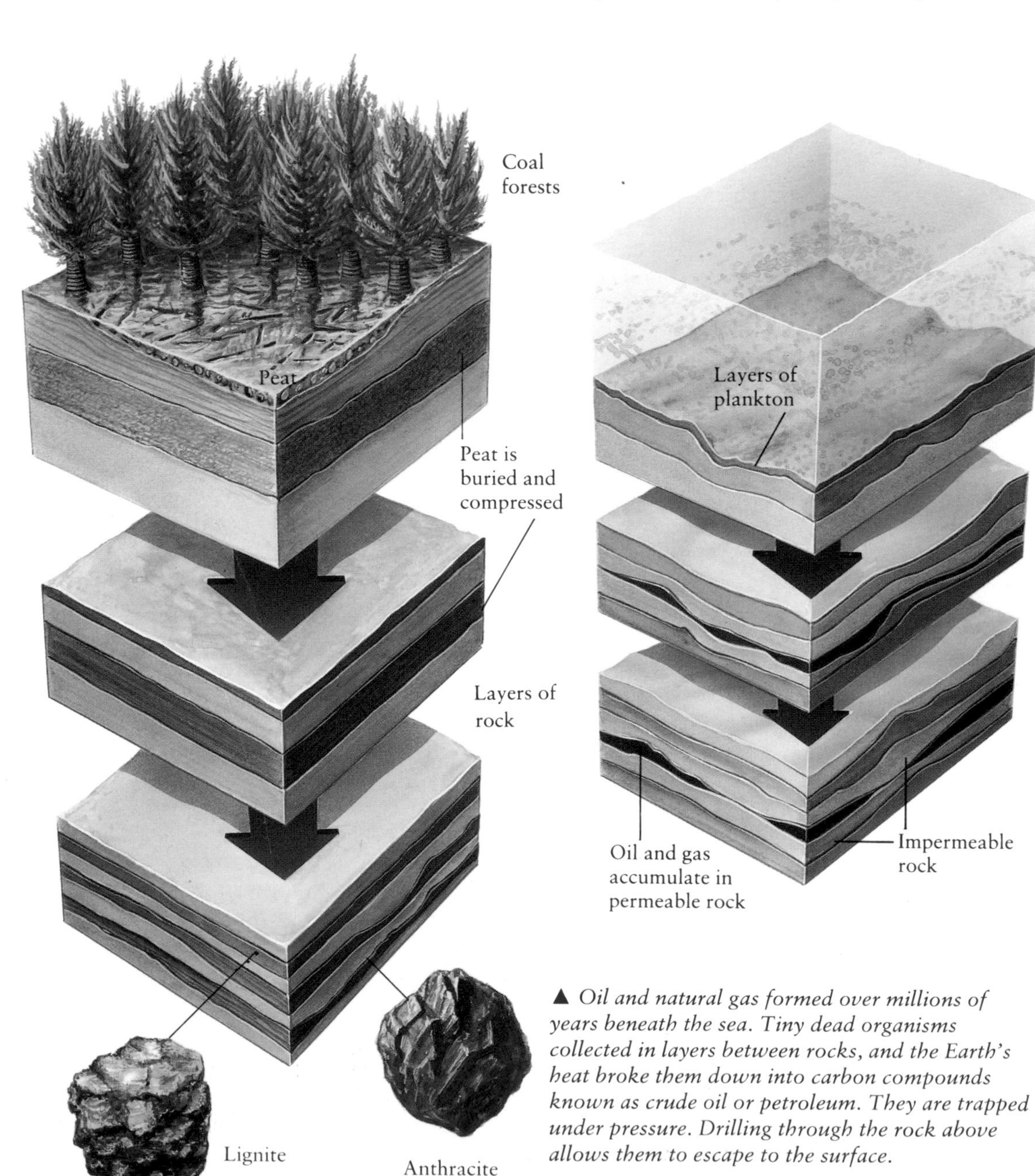

► *Coal formed from trees and other plants that grew beside water. When the trees died, they could not rot away fully because the ground was waterlogged. They accumulated as peat and were eventually buried. The peat was squashed by its own weight and the weight of the rocks above it, making it harder, and turning into "lignite," or "brown coal."*

Movements of the rocks underneath then squashed some of the coal even more and heated it, forming a hard black coal called anthracite. Anthracite burns better than lignite as it contains more carbon.

▲ *Oil and natural gas formed over millions of years beneath the sea. Tiny dead organisms collected in layers between rocks, and the Earth's heat broke them down into carbon compounds known as crude oil or petroleum. They are trapped under pressure. Drilling through the rock above allows them to escape to the surface.*

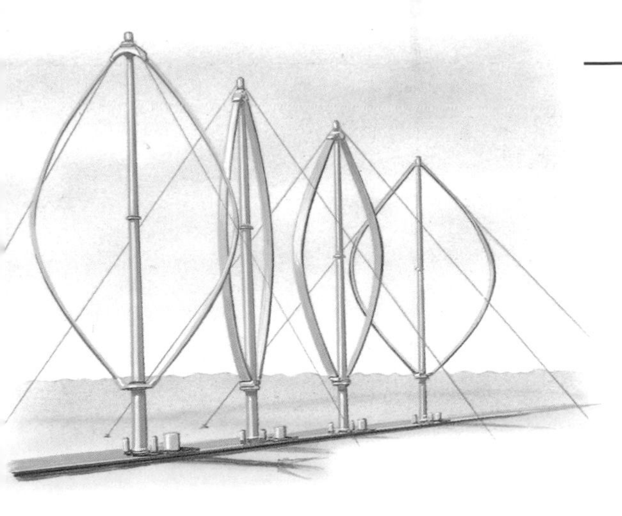

WIND POWER

Wind generators convert the motion of wind into electrical power. Wind is a renewable energy source, but it takes hundreds of very large, costly machines to obtain useful amounts.

GEOTHERMAL POWER

If a mass of rock or water below ground is hotter than its surroundings, the heat can be recovered as "geothermal energy." Drilling into the rock allows the hot water to flow to the surface. If the rock is dry, water is pumped down one hole, heated, and recovered from another. This resource is nonrenewable, because the rock is cooled, or the hot water is removed.

▲ *In some places, the pressure of water heated below ground forces it to the surface as a geyser* (see page 19). *The geyser can be capped and the steam is used to drive a turbine to produce electrical power.*

HYDROELECTRIC POWER

Hydroelectric power is generated by turbines driven by falling water. A dam is built across a river to form a lake. Gates in the dam wall allow the water to fall to the level of the river below, flowing past turbines inside the wall as it does so.

◄ *On the La Rance estuary, France, and in Fundy Bay, Canada, the ebb and flow of the tides turns turbines in a tidal barrier.*

NUCLEAR POWER

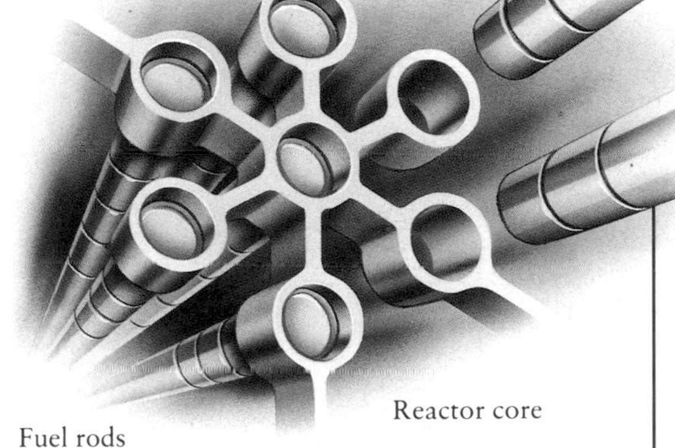

In the core of a nuclear reactor, uranium atoms are split to produce heat. The heat is used to boil water for steam to drive turbines. Uranium is mined from rocks.

HOW LONG WILL IT LAST?

No one knows how much of the nonrenewable resources remain. Uranium and coal will last several centuries, but oil and gas are less abundant and may soon run out.

Oil 45 years

Gas 76 years

Coal 521 years

SOLAR POWER

Solar panels absorb heat from the Sun, which warms water flowing through pipes beneath the surface of the panels. The pipes pass inside the building to the hot water tank, and heat up the water. Solar cells convert sunlight into electricity. They work in warm or cold weather.

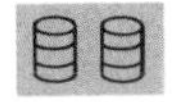

Natural Resources: Metals, Land, and Water

Metals are extremely valuable resources. Many of the everyday articles in our homes are made from metal. A few metals, such as gold and copper, occur in the Earth's crust in pure form, as "native elements." But most are found as minerals called ores, which are chemical compounds containing a high proportion of the metal. Some of our mineral resources are nonmetals. Paper is made whiter by adding kaolin, a clay mineral. Land and water are also valuable resources.

◀ *Round lumps of metal form on the bed of most oceans. In some places there are nearly 52,000 tons in a square mile. They are called "manganese nodules," but also contain other metals.*

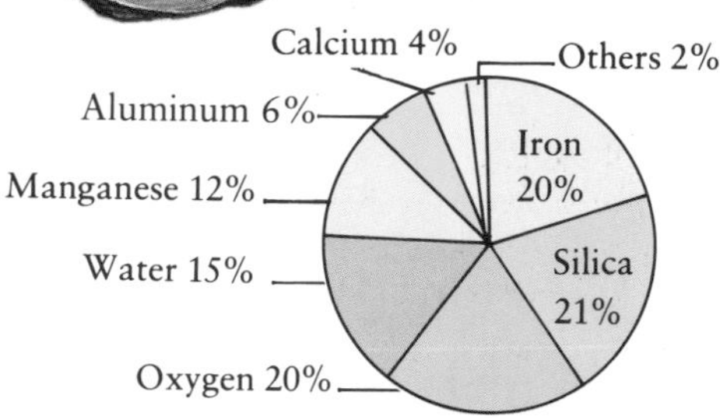

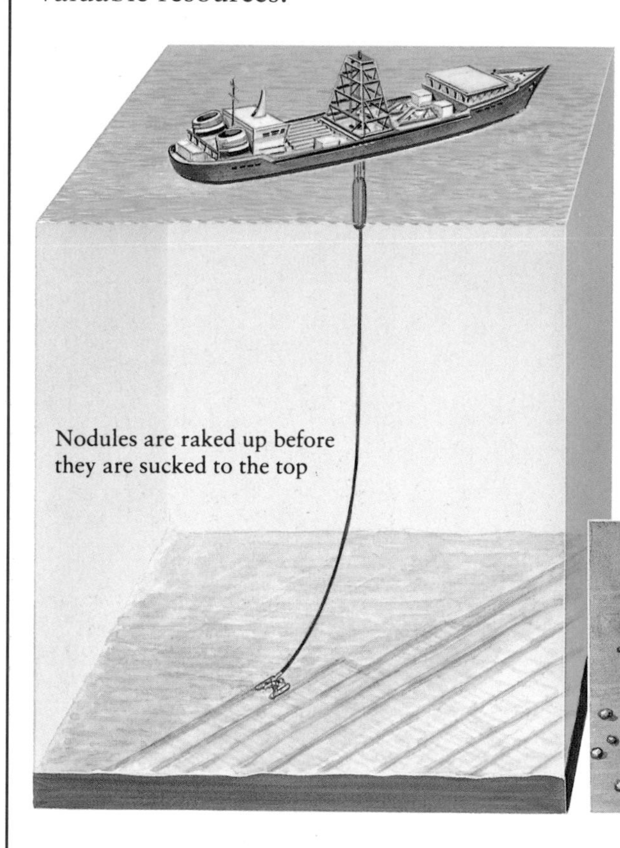

UNDERWATER RICHES
Mineral resources are nonrenewable, although many can be recycled and used again. We obtain many minerals from quarries and mines, but the seabed and seawater itself are rich in minerals.

Manganese nodules grow very slowly on deep seabeds. They contain enough of some metals to supply us for centuries. The nodules are dredged from the seabed, but it is more costly than mining the metals on land. Phosphorus also forms nodules, which are mined off the coast of California.

DATAFILE

Amount of metal produced each year worldwide (tons):
- **Iron:** 815 million
- **Bauxite (aluminum ore):** 85 million
- **Manganese:** 24 million
- **Zinc:** 7 million
- **Lead:** 3 million
- **Nickel:** 760,000
- **Copper:** 8,800
- **Gold:** 1,650
- About 15 percent by weight of the uppermost 10 mi. (16 km) of the Earth's crust is aluminum oxide.
- Sixty percent of the world's gold is mined in South Africa. The mines are up to 12,000 ft. (3,700 m) deep.

METALS
We use metals in widely differing amounts. Some, such as aluminum, iron, and magnesium, are abundant, but tin, silver, and platinum are already scarce. New deposits may be discovered, or new technology might allow existing resources to be mined more efficiently, but the costs will rise, and substitutes for some metals will be needed. The graph shows how long the known stocks of some metals may last, given the present rate of consumption.

Gold 28 years
Zinc 35 years
Nickel 75 years
Lead 40 years
Copper 60 years
Manganese 180 years
Iron ore 400 years
Bauxite (aluminum ore) 255 years

Years' supply
0 100 200 300

Open pit mine

MINING ORES

Metal ores are cut or blasted from the surrounding rock. The ore is crushed, and the worthless rock removed. Many metal ores contain oxygen or sulfur. The pure metal can be separated by heating.

RARE METALS

Gold is usually found as small grains or nuggets of the pure metal. Where there is gold there may also be platinum, either pure or mixed with ores of copper, nickel, lead, or other metals. Silver occurs as a pure metal, or as silver sulfide, with the sulfides of other metals.

DRINKING WATER

In regions where rainfall is low, drinking water can be obtained by purifying seawater. The process is called "desalination," and there are two methods. The most common is distillation. Seawater is boiled and the vapor, which contains very little salt, is condensed and collected. The process is repeated until the water is fit to drink. The other method is to force water through a membrane that allows water molecules to pass through, but traps the salts.

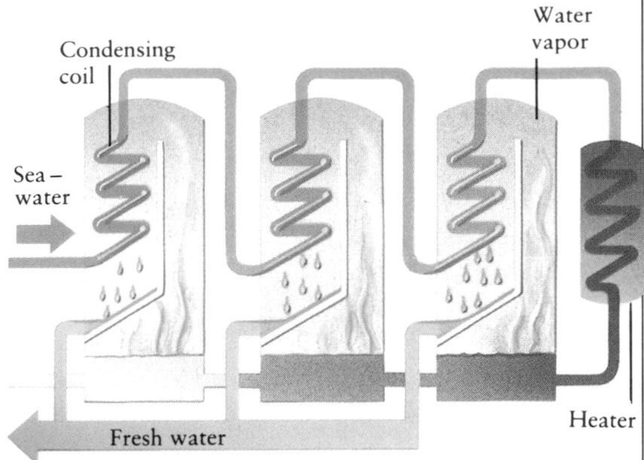

▲ *Some metals can be obtained from sands. Titanium is extracted from the rich sands on the Australian coasts.*

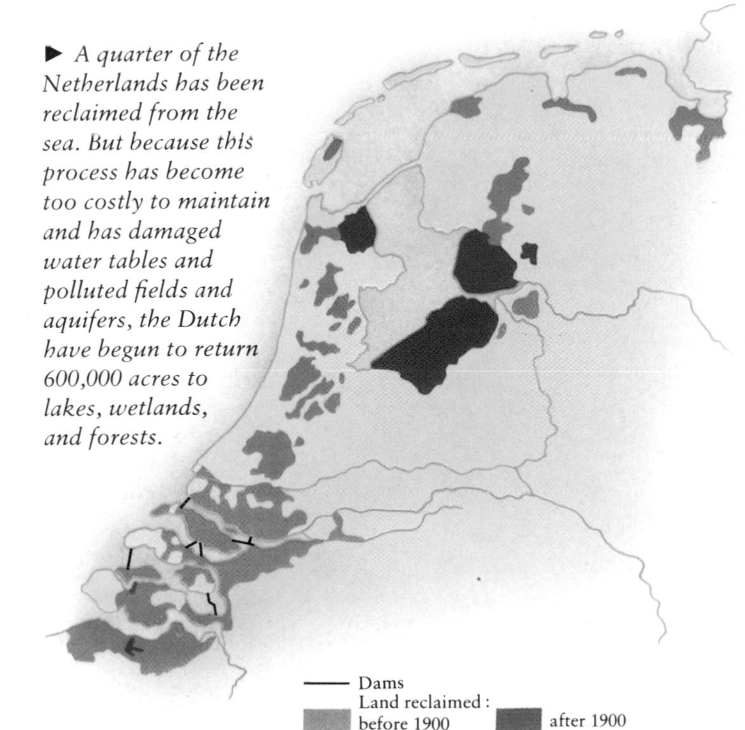

► *A quarter of the Netherlands has been reclaimed from the sea. But because this process has become too costly to maintain and has damaged water tables and polluted fields and aquifers, the Dutch have begun to return 600,000 acres to lakes, wetlands, and forests.*

RECLAIMING LAND FROM THE SEA

Since medieval times, earth banks or dikes were built to protect the reclaimed Dutch "polders" from flooding. In the 1920s, a large part of the Zuider Zee in the Netherlands was reclaimed by enclosing it with a dam 18 mi. (29 km) long. The fertile polders are valuable farmland. Parts of England, Italy, and Japan have also been reclaimed.

▲ *The windmills, for which the Netherlands is famous, were used to pump water from the polders into drainage channels. Continous pumping is needed as the water seeps back in.*

Air Pollution

Technology makes our lives easier, but factories, cars, and power stations also pollute the air we breathe. Incinerating waste and burning fuel to produce power releases millions of tons of gases such as carbon dioxide, sulfur dioxide, and nitrogen oxides into the air every year, together with ash, dust, and soot particles. Air pollution damages human health and harms wildlife; it can also alter the finely balanced atmospheric processes of the Earth, with potentially serious consequences.

BURNING

Although some air pollution is caused by natural sources, such as volcanic eruptions which release sulfur dioxide, most is caused by waste gases released by burning fuels and incinerating waste from homes and factories. Some of the waste contains toxic (poisonous) chemicals such as mercury which are then released into the air. Tiny particles of solids and liquids are also given off, which can cause breathing problems.

THE OZONE LAYER

Ozone is a form of oxygen in which the molecule is made up of three atoms (O_3), rather than the usual two (O_2). It forms in the stratosphere. Ultraviolet (UV) radiation from the Sun splits oxygen molecules into free oxygen atoms. Each oxygen atom joins an oxygen molecule to form ozone. UV radiation also splits ozone molecules. So ozone is constantly forming, splitting, and reforming. The UV radiation absorbed by this process cannot reach Earth. UV radiation causes sunburn, skin cancer, and eye problems; it also affects plant growth.

THE HOLE OVER THE ANTARCTIC

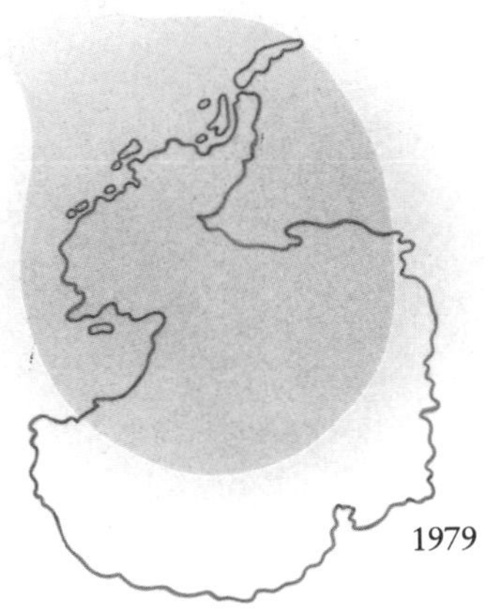

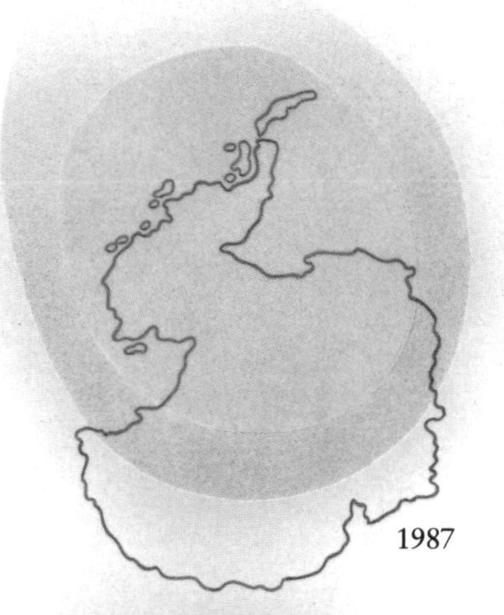

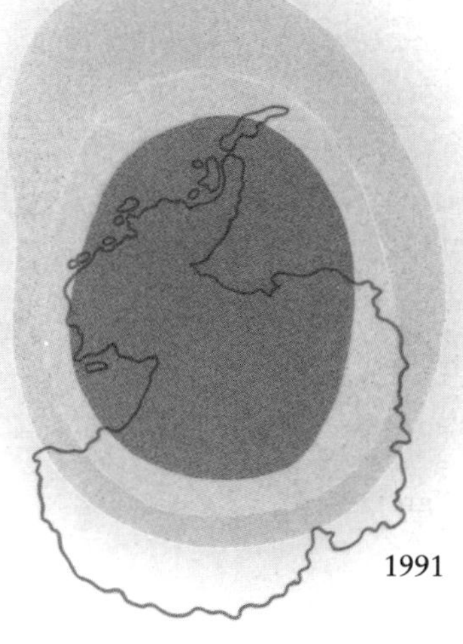

► *Every spring (October) the ozone layer over Antarctica thins by up to 50 percent. This "hole" disappears in summer, but reappears every year. In 1987 it covered an area the size of the United States. No "hole" has been detected over the Arctic, but the ozone decreases slightly in January and February.*

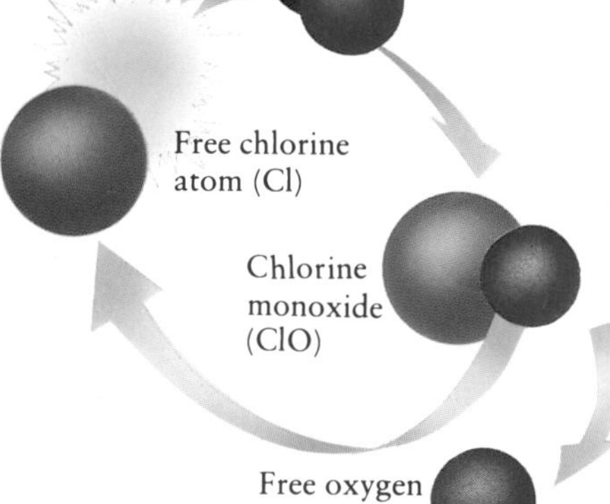

◄ *Chlorine (Cl) reacts with ozone to form chlorine monoxide (ClO) and oxygen (O_2). The chlorine is then released to go around the cycle again.*

CFCs

Ozone is destroyed by compounds such as chlorofluorocarbons (CFCs). CFCs are used in some refrigerators and packaging materials. They drift up into the stratosphere, where they break down and release chlorine. Each chlorine atom can destroy hundreds of thousands of ozone molecules. Many scientists are concerned that as the ozone layer is damaged, a greater amount of harmful UV radiation will reach the Earth's surface.

▲ *Photochemical smog is a health hazard in many major cities. It is the result of chemical reactions caused by the action of sunlight on nitrogen oxides and unburned fuel from car exhaust fumes.*

SMOG

Earlier this century, in London and other European cities, a mixture of fog and smoke caused choking smogs, known as "pea-soupers." Today, in places where there is a lot of sunshine, such as Los Angeles, traffic fumes cause "photochemical smog."

ACID RAIN

Cloud droplets are naturally acidic, because the carbon dioxide in air dissolves to form a weak acid. But sulfur dioxide and nitrogen oxides produced by burning fossil fuels form stronger acids. The moisture reaches the ground as acid mist, snow, or rain. Acid rain damages forests and acidifies lakes, harming aquatic animals.

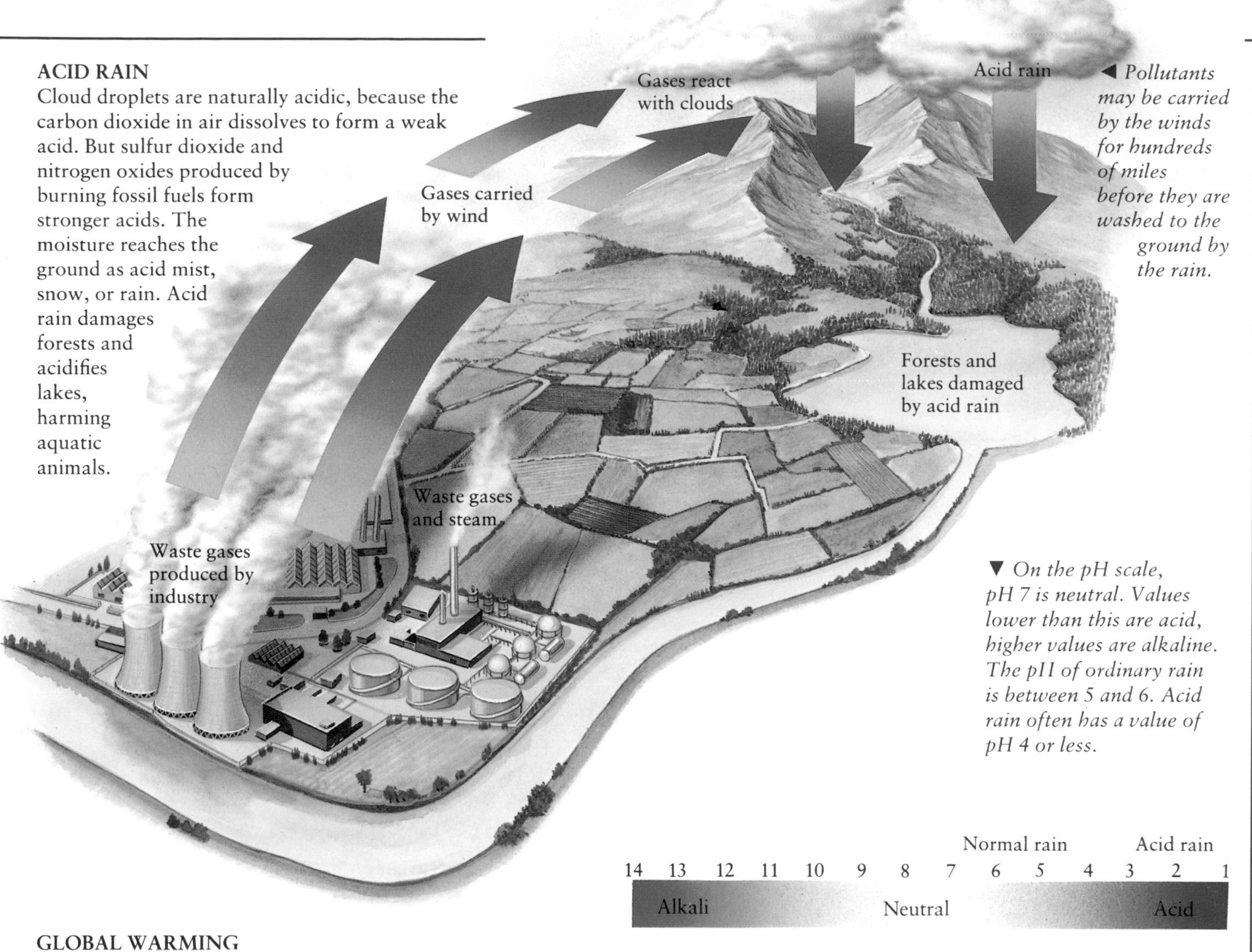

◀ *Pollutants may be carried by the winds for hundreds of miles before they are washed to the ground by the rain.*

▼ *On the pH scale, pH 7 is neutral. Values lower than this are acid, higher values are alkaline. The pH of ordinary rain is between 5 and 6. Acid rain often has a value of pH 4 or less.*

GLOBAL WARMING

The Earth's surface is warmed by the Sun and radiates heat back into space. Gases such as carbon dioxide, nitrogen oxides, methane, and CFCs in the atmosphere trap some of this heat and warm the lower atmosphere. The atmosphere radiates heat back to Earth. This is called the "greenhouse effect," and without it the Earth would be so cold that life could not exist. But many scientists fear that the huge amounts of these "greenhouse gases" released into the atmosphere by industrial processes and burning fossil fuels are warming the Earth so much that they will eventually upset the world's climate, and cause sea levels to rise.

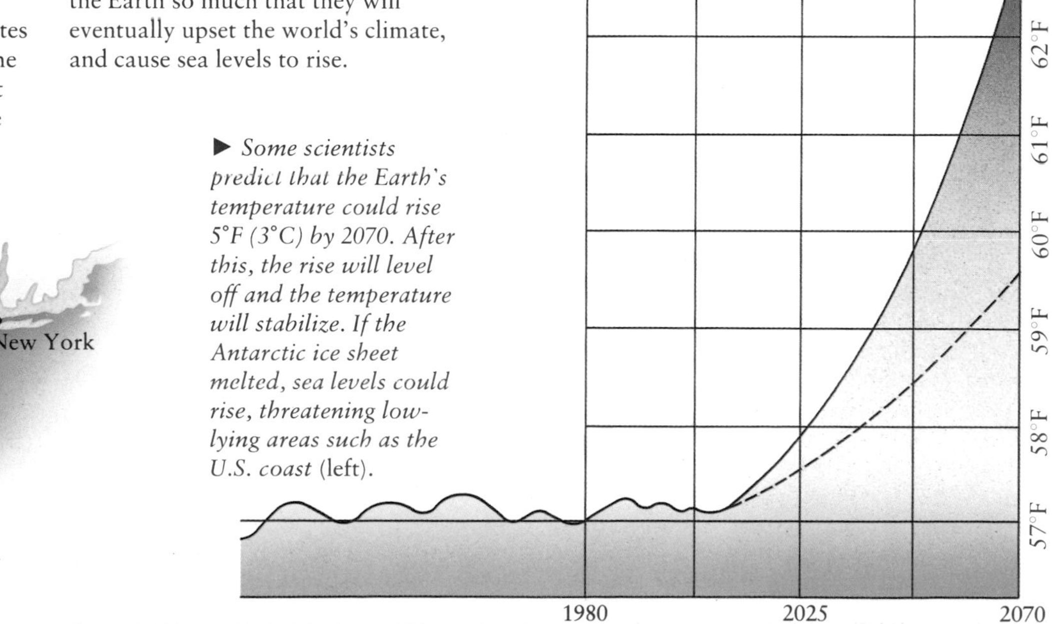

▶ *Some scientists predict that the Earth's temperature could rise 5°F (3°C) by 2070. After this, the rise will level off and the temperature will stabilize. If the Antarctic ice sheet melted, sea levels could rise, threatening low-lying areas such as the U.S. coast* (left).

Environmental Problems

Every living organism changes the world around it, including humans. We clear massive areas of land to grow food and to build homes, cities, and roads. We quarry and mine for building materials, minerals, and fuels, and use these to manufacture the things that improve our lives. When we are careless we harm the environment in many ways: poor people are forced to farm using methods that damage the land; we dirty water with our waste; and we harm wildlife by destroying its food and shelter.

LAND AT RISK
Land bordering all deserts is at risk. As the deserts expand, people are forced onto smaller grazing areas, increasing the risk of soil erosion. The problem is most severe in the Sahel, south of the Sahara Desert.

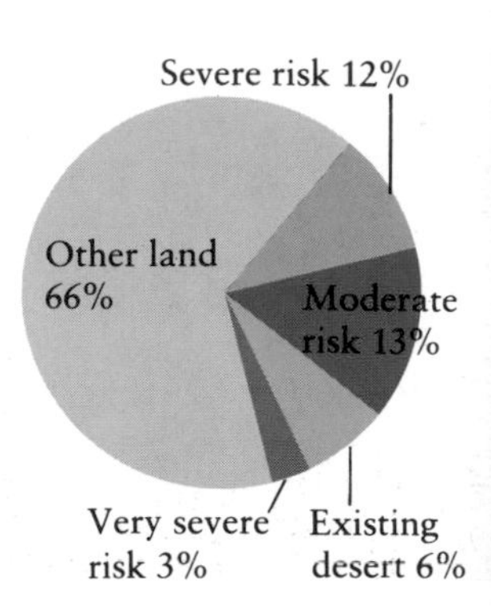

MARCHING SAND
Much of the world's grazing land is in semiarid areas. From time to time there are severe droughts lasting several years. During these droughts the ground dries, plants disappear, and wind-blown sand and dust may bury more fertile soil and destroy crops. Many people in the semiarid areas are nomadic, moving with the seasons in search of pasture for their animals. When the pasture fails, they are crowded into the small areas that remain. This leads to overgrazing, which causes soil erosion.

▲ *Irrigated land grows crops, but the equipment is expensive. Unless surplus water is drained away, the ground may become waterlogged, and salts accumulate, so that crops cannot grow.*

DEFORESTATION
In the tropics, forests are cleared to provide timber. Some of the land is replanted as plantation forest and some is converted into farms and cattle ranches. Forest valleys are also flooded to make lakes for hydroelectric schemes. Plant and animal species die out; those found nowhere else become extinct.

▼ *Soil is washed into rivers, causing them to silt up.*

◀ *Clearing plants from high ground may cause flooding on low ground, as rain water quickly runs off the surface. In the dry season, exposed, infertile soils may be baked hard and crack.*

▶ *The soil in many parts of the rain forests is poor. The land can only support a few harvests before farmers have to move to a new area.*

RIVERS IN PERIL

Small quantities of waste do not usually cause serious harm if they are dumped into a river, because the river quickly purifies itself. But if many factories dump their waste into the same river, it cannot cope, and the water becomes very polluted.

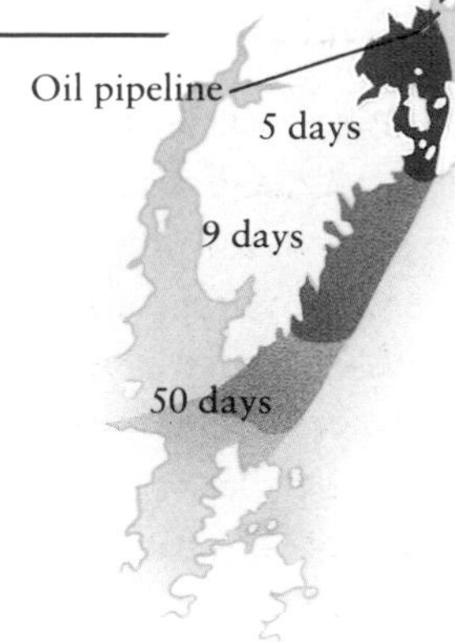

Spread of oil slick

EXXON VALDEZ

On March 24, 1989, the tanker *Exxon Valdez* ran aground in Alaska, releasing about 8.5 million gallons of crude oil. The oil formed a slick covering more than 1,400 sq. mi. (3,600 sq. km) and about 1,100 mi. (1,700 km) of shoreline was badly polluted.

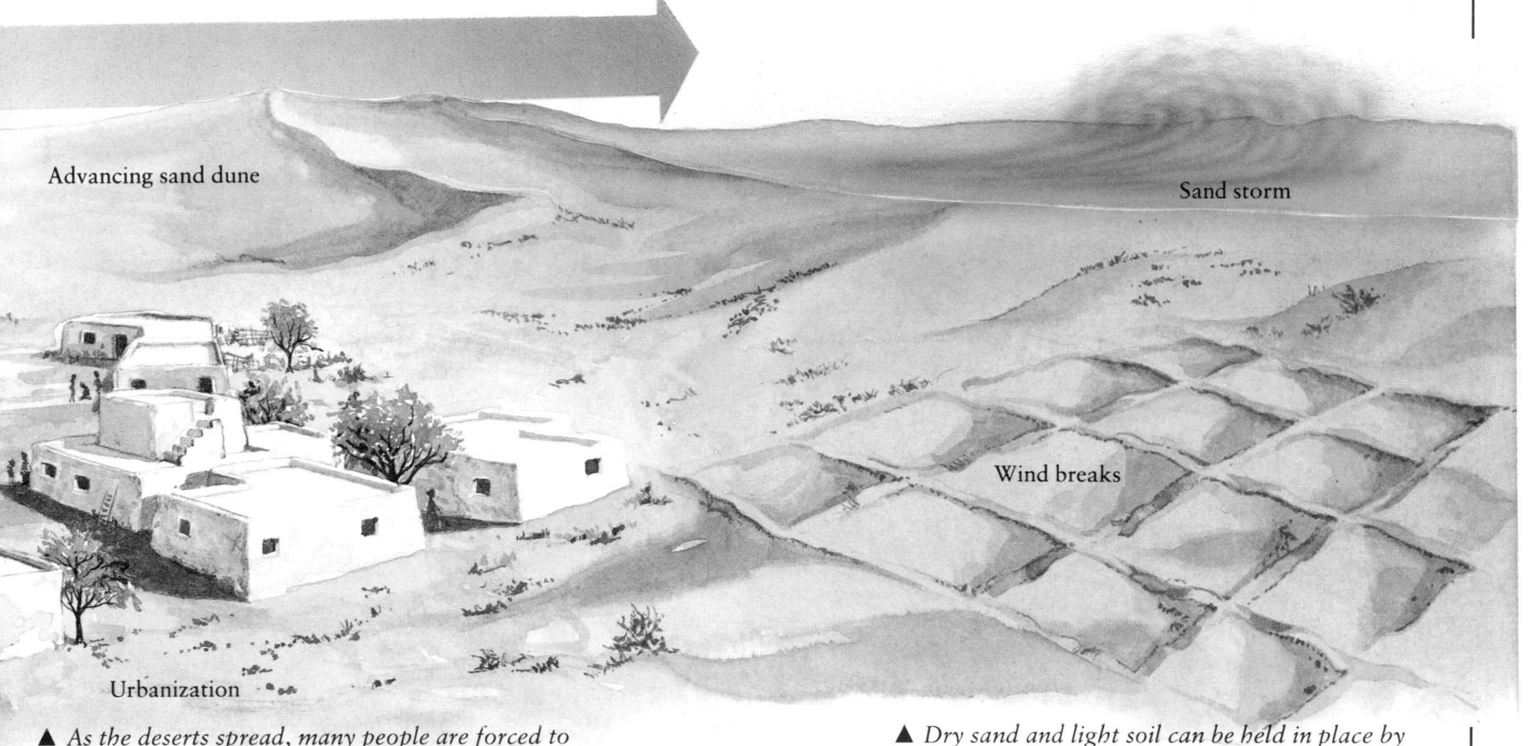

▲ *As the deserts spread, many people are forced to leave their homes and try to make a living elsewhere. Settlements grow larger, increasing pressure on the land.*

▲ *Dry sand and light soil can be held in place by nets or by spraying a protective film over it, to stop the processes of erosion.*

WASTE DISPOSAL

We produce millions of tons of waste every year from mines, factories, and homes. Most of the waste is buried in landfill sites. Although new sites are strictly controlled to ensure they are safe, toxic chemicals have leached from some old sites, contaminating the land and water supplies.

◄ *The base of a landfill site is lined to prevent dirty liquids from leaking into nearby water. At the end of each day, the rubbish is leveled and covered with topsoil.*

Pollutants leach into water table

► *Nuclear waste remains dangerous for several centuries. It must be stored deep underground.*

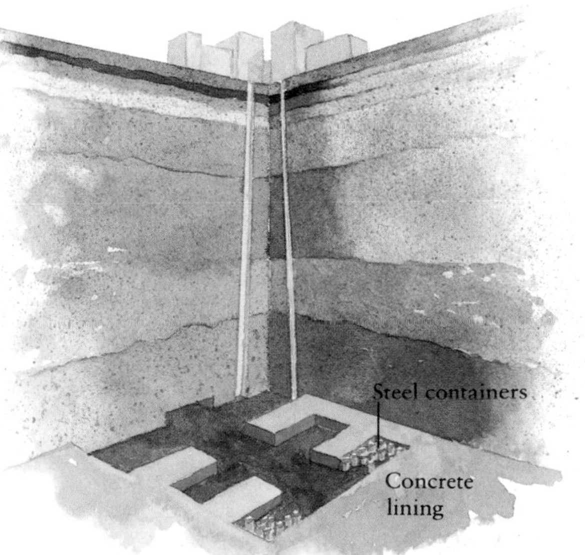

Conservation

Over the last 30 years, we have become aware of the damage we do to the environment. People have begun to find ways to reduce the amount of waste released into the air and water, and strict laws are applied by many governments. Ecologists and conservationists have found ways to help wildlife. Recycling has encouraged new uses for things that we used to throw away as garbage. Many problems remain, but progress has been made, and there have been many important improvements.

RECYCLED PAPER

About half of our domestic waste is paper which could be recycled. Recycling paper causes less pollution and protects the natural habitats cleared to plant the softwoods used to make new paper.

Symbol for recycled paper

GLASS AND ALUMINUM RECYCLING

Processing raw materials uses energy and causes pollution. Recycling often saves energy and resources, and reduces domestic waste as well as pollution. Old glass jars and bottles can be crushed and melted to make new glass objects. Extracting aluminum from its ore uses a great deal of energy, so recycling aluminum cans saves energy. It used to be expensive to recycle the cans, because they contained steel which had to be separated from the aluminum. Cans are now made from aluminum only.

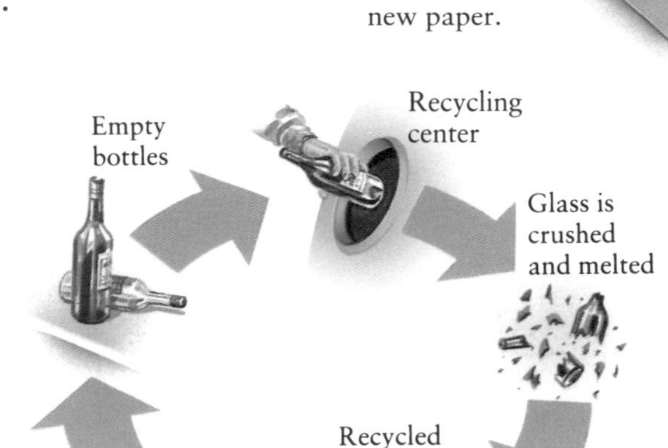

◄ *Many towns now have recycling points, with containers for different items. Separating glass bottles into different colors makes the process easier. Some plastic bottles can also be recycled.*

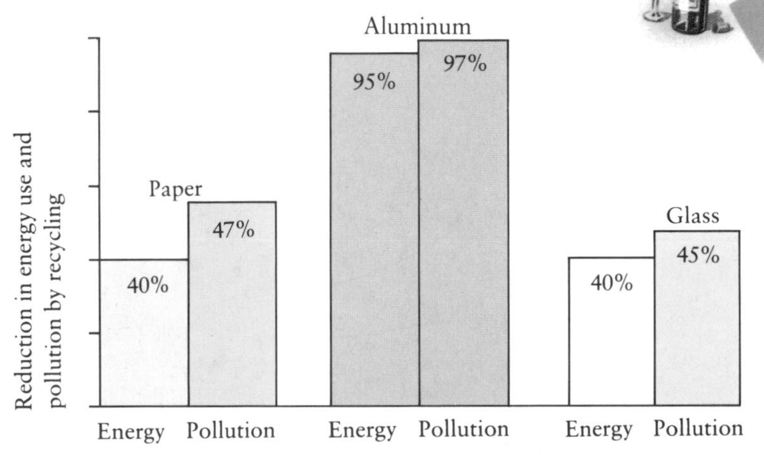

Core area
Inner buffer zone

BIOSPHERE RESERVES

"Biosphere reserves" are a network of areas that will include examples of all the world's major types of vegetation. The reserves contain undisturbed vegetation and farmland and are managed for conservation. About 60 have been set up so far by the United Nations.

LANDSCAPE PRESERVATION

Attempts are now being made to reclaim land that has been used for mining or industry. Open-pit mines can be flooded, and turned into lakes for watersports and wildlife. Old industrial sites and garbage dumps tips can be covered with a thick layer of soil and turned into recreational land.

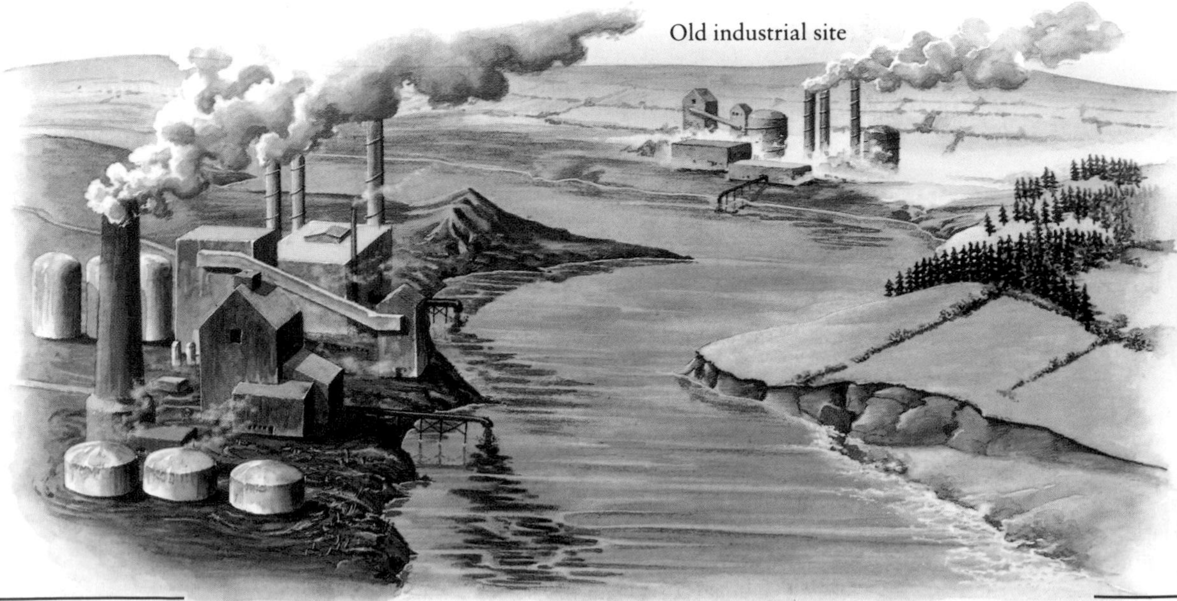

CONSERVING HABITATS

International agreements exist to protect endangered species and to preserve their habitats. The habitats of migratory water birds are protected by the Ramsar Convention on Wetlands of International Importance. Other places are recognized by the United Nations as World Heritage Sites.

COUNTRIES WITH AGREEMENTS TO PROTECT WORLD HERITAGE SITES AND WETLANDS OF INTERNATIONAL IMPORTANCE

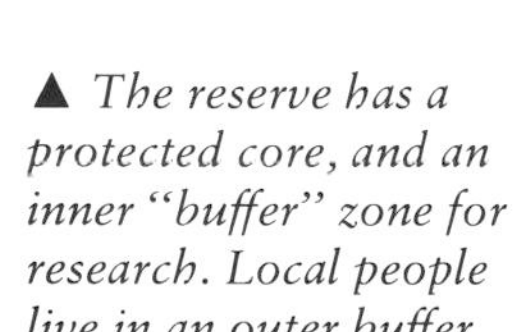

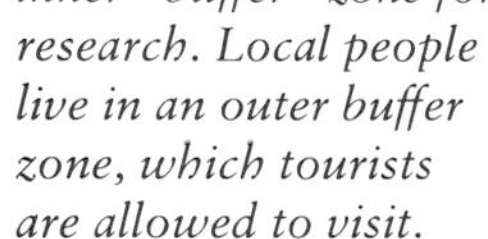

▲ *The reserve has a protected core, and an inner "buffer" zone for research. Local people live in an outer buffer zone, which tourists are allowed to visit.*

BIOLOGICAL CONTROL

Pesticides are expensive and can cause pollution. "Biological control" uses natural enemies to control pests and weeds. For example, geese eat the weeds in fruit orchards, at the same time fertilizing the ground. Ladybugs are used to control aphids. Ducks are used by farmers in China to eat the insects that damage crops.

THE EARTH SUMMIT

In June 1992, the UN Conference on Environment and Development (the "Earth Summit") took place in Rio de Janeiro. Agreements were signed by many countries to limit climate change, and to protect species and habitats ("biodiversity"), sustainable development, and forest management. A program to achieve these aims was outlined.

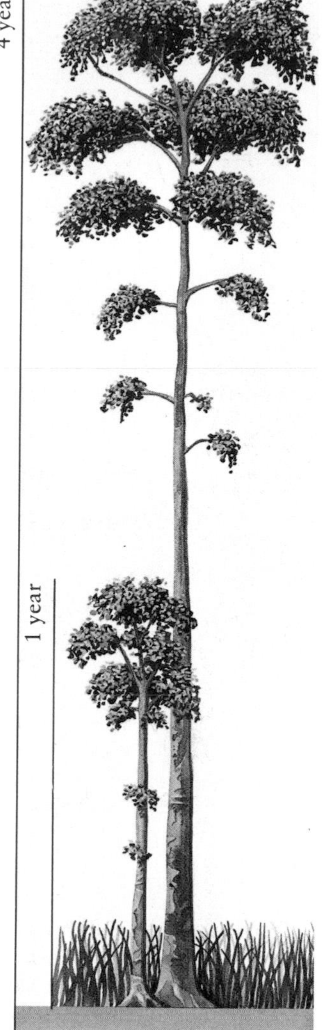

FAST-GROWING TIMBER

Tropical timber can be produced in plantations of fast-growing native species, to keep rain forests from being destroyed. One such tree, *Kadam*, is grown in Indonesia. It reaches 10 ft. (3 m) in its first year, then adds 6–10 ft. (2–3 m) a year for up to 8 years. Another Asian species, *Erima*, can grow to 80 ft. (25 m) in 4 years and 160 ft. (50 m) in 60 years.

◄ *Industrial wasteland can become a park with lakes and trees or playing fields.*

Emperor penguins are remarkable parents, providing food and warmth for their chicks through the Antarctic winter.

THE

LIVING WORLD

Life on Earth **84** Animal and Plant Classification **86** Animal and Plant Habitats **88**
The World of Plants **90** Bacteria, Algae, Lichens, and Fungi **92**
Liverworts, Mosses, Horsetails, and Ferns **94** Ginkgos, Cycads, and Conifers **96**
Monocotyledons and Dicotyledons **98** Fruits or Vegetables? **102**
Trees **104** Plants and People **106** The World of Animals **108**
Marine Invertebrates, Worms, Snails, and Slugs **110**
Millipedes, Crabs, and Spiders **112** Insects **114** Fish **116** Amphibians **118**
Reptiles **120** Birds **122** Bird Behavior **124** Mammals **126**
Mammal Senses **128** Animal Homes **130** Animal Movement **132**
Animals and their Young **134** Animals and People **136**
Endangered Animals **138** Prehistoric Animals **140**
Systems of the Body **142** Skeleton and Muscles **144**
The Nervous System **146**
Heart, Blood, and Skin **148**
Digestion and Respiration **150**
Reproduction **152**
Growing and Aging **154**
Looking After Your Body **156**

An incredible number of changes have taken place for life on Earth to get where it is today. Some plants and animals have appeared and died out; others have undergone countless mutations to evolve into the forms we recognize; a few have cheated evolution and have remained virtually as living fossils.

A staggering number of life forms have been cataloged by researchers. Each of these holds an important place in the organization of life on Earth. The web of relationships among the living things of our planet is intricate. Much like the pieces of a puzzle, each life touches many others — and should one go missing, the whole is inevitably affected.

THE LIVING EARTH

Life on Earth

The story of life on Earth begins many millions of years before the appearance of the first human beings. From dating the rocks, we can estimate the age of the Earth at around 4.5 billion years. How life began is uncertain. It may have been due to a chemical reaction, a haphazard coming together of lifeless molecules to form a tiny organism able to reproduce itself. The oldest known forms of life are the fossils of simple bacteria and algae, over 3.5 billion years old. Today there are more than 2 million living things on our planet. Many are so microscopically tiny that they are invisible to the naked human eye. Others are giants, such as the redwood tree and the blue whale. All the kinds, or species, of plants and animals have evolved as the result of gradual adaptation to the widely differing environments the Earth offers its inhabitants.

EVOLUTION

We mark the prehistory of the Earth by eras lasting many millions of years: the Precambrian, Paleozoic, Mesozoic, and Cenozoic. Life began in the oceans over 3.5 billion years ago. The first living things were simple, single-celled organisms. Scientists are still working out the relationship between these early groups. The Paleozoic Era brought an enormous expansion of life with some animals coming out of the warm, shallow seas onto the land. The evolution of species has shaped the "family tree" of life. Many plant and animal species died out. Other species developed new forms to create the diversity of animals and plants of today.

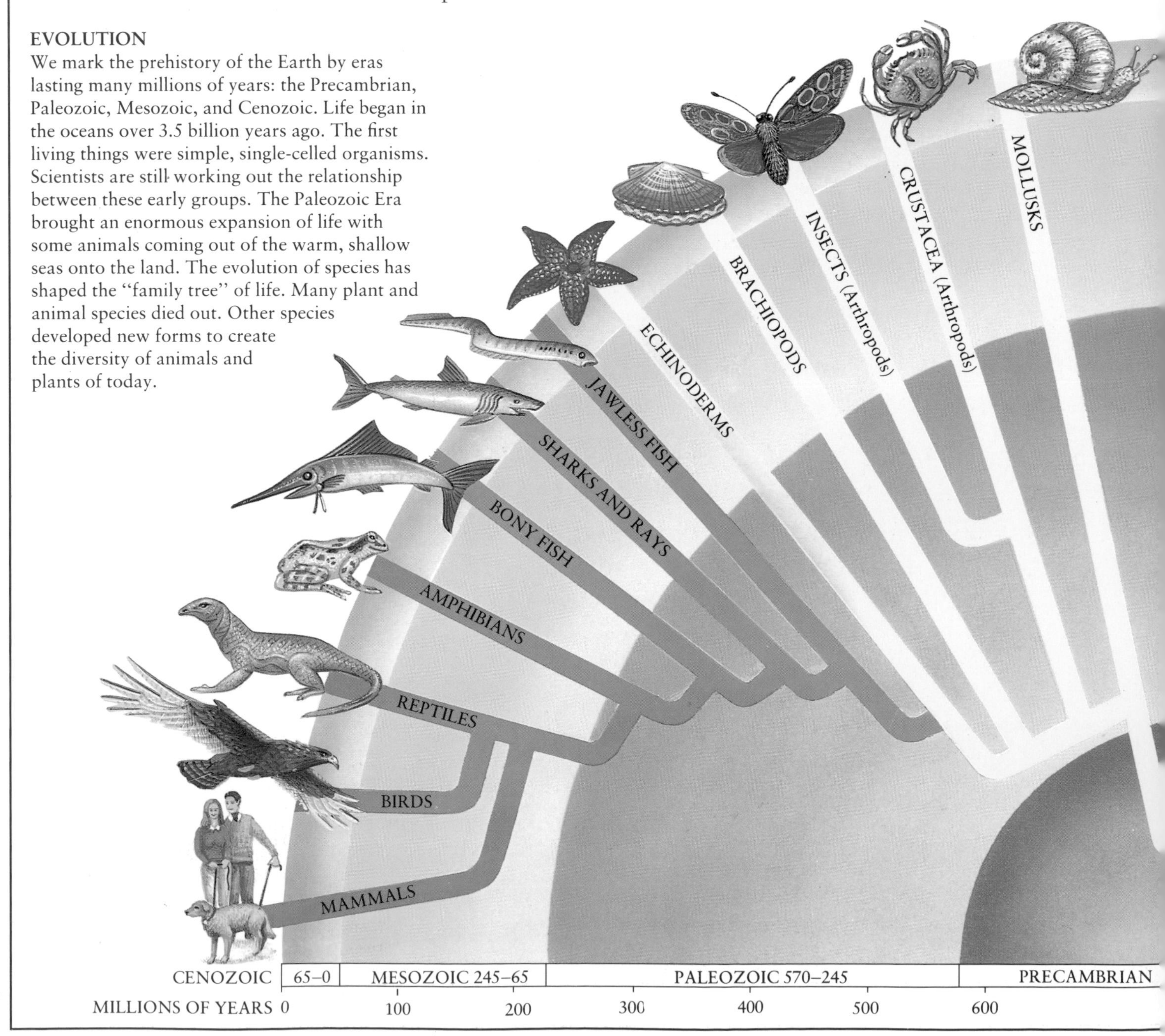

FOSSILS: THE EVIDENCE
Fossils like this trilobite *(left)* are the hardened remains of dead animals and plants. Fossils are found in rocks millions of years old. The earliest soft-bodied animals left no fossils. By 600 million years ago, larger and more advanced animals and plants lived in the warm seas *(above)*. These animals, which had hard shells and skeletons, left fossil remains. Dating the rocks where fossils are found fills in the calendar of life.

Animal and Plant Classification

Classification, grouping living things together by similarities, shows how one group is related to another and how modern organisms may have evolved from earlier forms. The science of classifying plants and animals is called taxonomy, and Greek and Latin scientific names are used to identify each species, or kind, of living thing. Each species can be classed in levels: by kingdom (the largest group), then by phylum, class, order, family, genus, and lastly by species.

HOW MANY LIVING THINGS?
No one knows how many living things there are. About 2 million species have been named. But perhaps four times as many species remain unknown to science. Of the species we know about, 75 percent are animals (mostly insects), 18 percent are plants, and 7 percent are "in-betweens"—things that do not fit easily into either animal or plant groups.

ANIMAL CLASSIFICATION

KINGDOM
All animals belong to the kingdom Animalia. The other four kingdoms are Plants, Protoctists, Bacteria, and Fungi.

PHYLUM
Within the animal kingdom are 20 or more phyla. All animals with backbones belong to the phylum Chordata.

CLASS
Animals with hair on their bodies that feed their young with milk are mammals, members of the class Mammalia.

ORDER
Mammals that eat meat such as bears, dogs (including foxes), and cats, belong to the order Carnivora.

FAMILY
Dogs, foxes, and wolves look similar. These animals all belong to the same family, the Canidae.

GENUS
Animals of the same genus may not interbreed. Several foxes belong to the genus *Vulpes*.

SPECIES
Members of a species can interbreed. All fennec foxes belong to the fox species *Vulpes zerda*.

▲ *The fennec is a small fox of North Africa and Arabia. Its "family tree," from kingdom to species, is illustrated here.*

FIVE KINGDOMS
Three groups of living things are classed separately from animals and plants. Some of the simple cells are claimed to be plants and some of them are claimed to be animals. Bacteria and blue-green algae-like cells form the kingdom Bacteria. These organisms are tiny single cells. Fungi, (mushrooms and toadstools, for example), are like plants in some ways but have no chlorophyll and so cannot make their own food. Protoctists are the third "outsiders"; they contain species claimed by both botanists and zoologists and some groups with no clear relationship to any species. Some are single-celled, (such as diatoms and amoebas), and some are groups of cells, such as red and brown seaweeds.

PLANT CLASSIFICATION

KINGDOM
Every multicellular green plant, from the tiniest to the tallest, belongs to the plant kingdom.

PHYLUM
All seed plants that reproduce themselves by flowers making covered seeds are Angiosperms.

CLASS
The Angiosperms are divided into two classes, Monocotyledons and Dicotyledons *(right)*.

ORDER
Oak trees, along with their close relatives beeches and chestnuts, belong to the order Fagales.

FAMILY
Some 900 species of trees including beeches, chestnuts, and oaks, belong to the family Fagaceae.

GENUS
All oaks belong to the genus *Quercus*. There are more than 600 species: some are tall; others are shrubby.

SPECIES
The evergreen holm oak is *Quercus ilex*. The English oak is *Quercus robur*; the American white oak is *Quercus alba*.

Oaks vary in size and in the way they grow, and each species of oak has a distinctive leaf, flower, and fruit.

Animal and Plant Habitats

Animals and plants live in places, or habitats, that provide the food and shelter they need. For example, giraffes (Africa), kangaroos (Australia), and prairie dogs (North America) are animals of the grasslands. Nature has equipped them to survive in this particular habitat. The Earth's regions offer many habitats, from freezing polar wastes to hot, tropical rain forests. Animals and plants live together in biological communities. Ecology is the study of how living things interact within such communities.

▲ *Oceans and seas form the marine habitat. The seashore, continental shelf, coral reefs, and deep, cold, ocean depths all have their own communities of plants and animals.*

► *Rivers and wetlands (marshes and swamps) are usually rich in plant and animal life. Animals can include fish, amphibians, reptiles, and birds such as cranes.*

WORLDWIDE NATURAL REGIONS

► *Plant and animal communities are grouped worldwide into "biomes"—natural regions with similar climates and vegetation that provide similar habitats. The map shows the main land biomes. The oceans form a vast biome of their own.*

◄ *In tropical rain forest, plants thrive and animals (such as monkeys, birds, snakes, and insects) live in the different layers of the forest canopy.*

▲ *Desert plants and animals must conserve water and keep cool. Reptiles such as lizards seek shade in the midday heat. Many desert animals are small and nocturnal.*

► *In hot climates, savanna grasslands support herds of grazing animals, as well as the carnivores (such as African lions) that prey on these grass-eaters.*

◀ Polar bears and seals live in the cold Arctic. On the tundra plains, plants grow quickly in the brief summer warmth.

▶ Some plants are adapted to life on alpine slopes. Mountain animals include sheep, goats, and soaring birds.

▼ The cold northern forests of evergreen conifer trees are home to animals as varied as owls, beavers, porcupines, moose, bears, and wolves.

▼ Temperate regions have warm summers and cool winters. In areas of mixed forest and farmland small mammals are found, as well as larger animals such as deer.

THE PLANT KINGDOM

The World of Plants

Without plants, our planet would be a lifeless world. Plants give off the oxygen all animals need to breathe; they provide much of our food, materials such as timber and cotton, as well as many health-restoring drugs. Scientists have named more than 375,000 kinds of plants, ranging from simple algae to trees. There could be the same number of undiscovered plants growing in remote forests and on mountains. Even so, there are far fewer plants than there are animals. Some plants are widespread; others grow only in one place.

Plants form the largest and longest-living things on Earth. All true plants are made up of many cells containing a material called cellulose. They develop from embryos (tiny forms of the adult plant). Most plants make their own food from water and carbon dioxide by a chemical process called photosynthesis which requires sunlight.

PLANT CLASSIFICATION
The system for naming plants and animals was drawn up by the Swedish naturalist Carl von Linné (Linnaeus) in 1758. The different groups of plants have been arranged in various ways since this time. Modern classification allows for many different phyla, of which the main ones are the ones shown here.

NEITHER ANIMALS, OR PLANTS
The Protoctists, the Bacteria, and the Fungi are not considered plants because they do not make their own food and their cells are different from those of animals and plants.

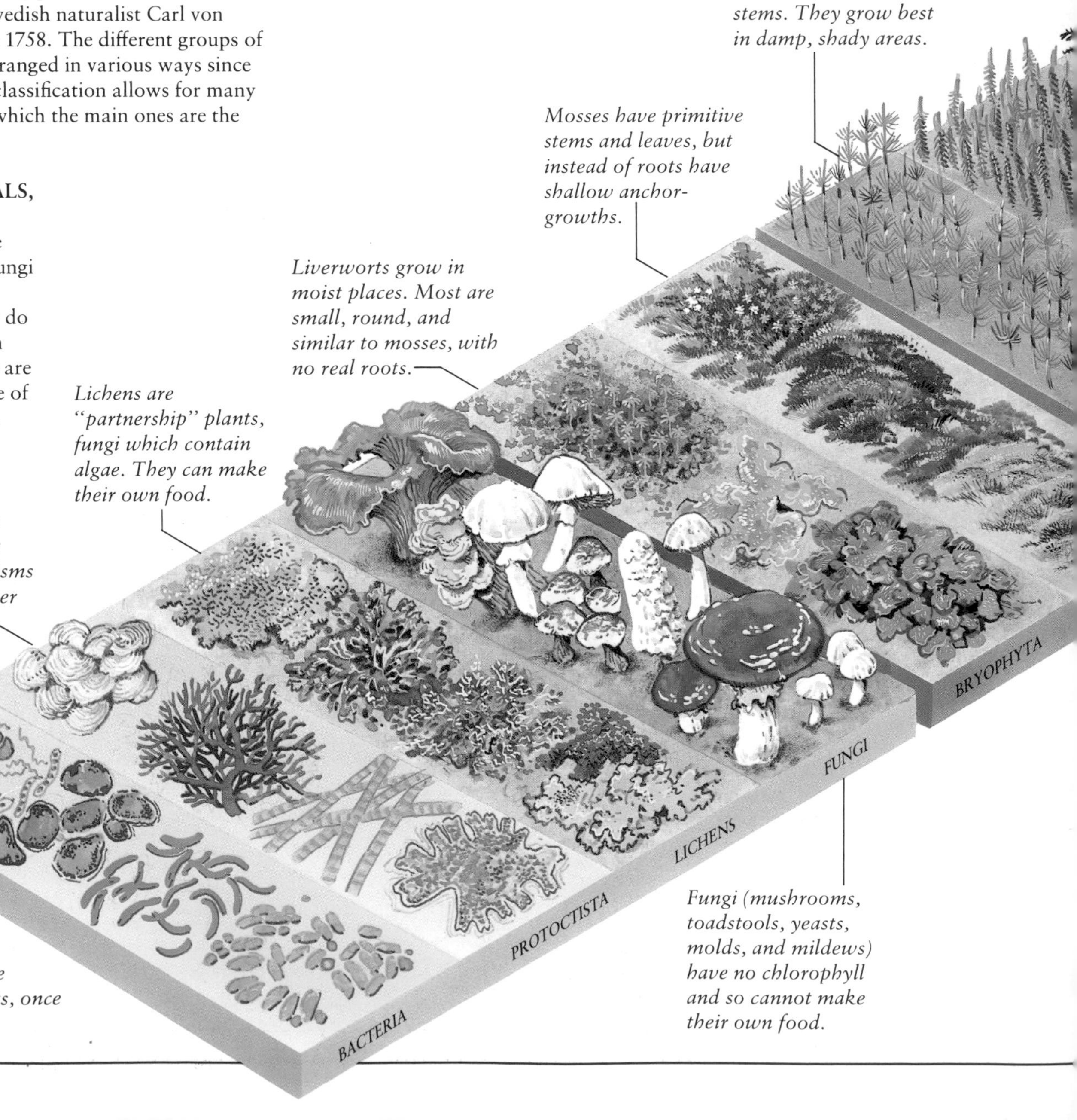

Horsetails have small leaves and hollow stems. They grow best in damp, shady areas.

Mosses have primitive stems and leaves, but instead of roots have shallow anchor-growths.

Liverworts grow in moist places. Most are small, round, and similar to mosses, with no real roots.

Lichens are "partnership" plants, fungi which contain algae. They can make their own food.

Algae include large seaweeds as well as tiny floating organisms that can live in either fresh or salt water.

Bacteria are tiny, simple single-celled organisms, classified within the kingdom Protoctists, once known as Monera.

Fungi (mushrooms, toadstools, yeasts, molds, and mildews) have no chlorophyll and so cannot make their own food.

BACTERIA

PROTOCTISTA

LICHENS

FUNGI

BRYOPHYTA

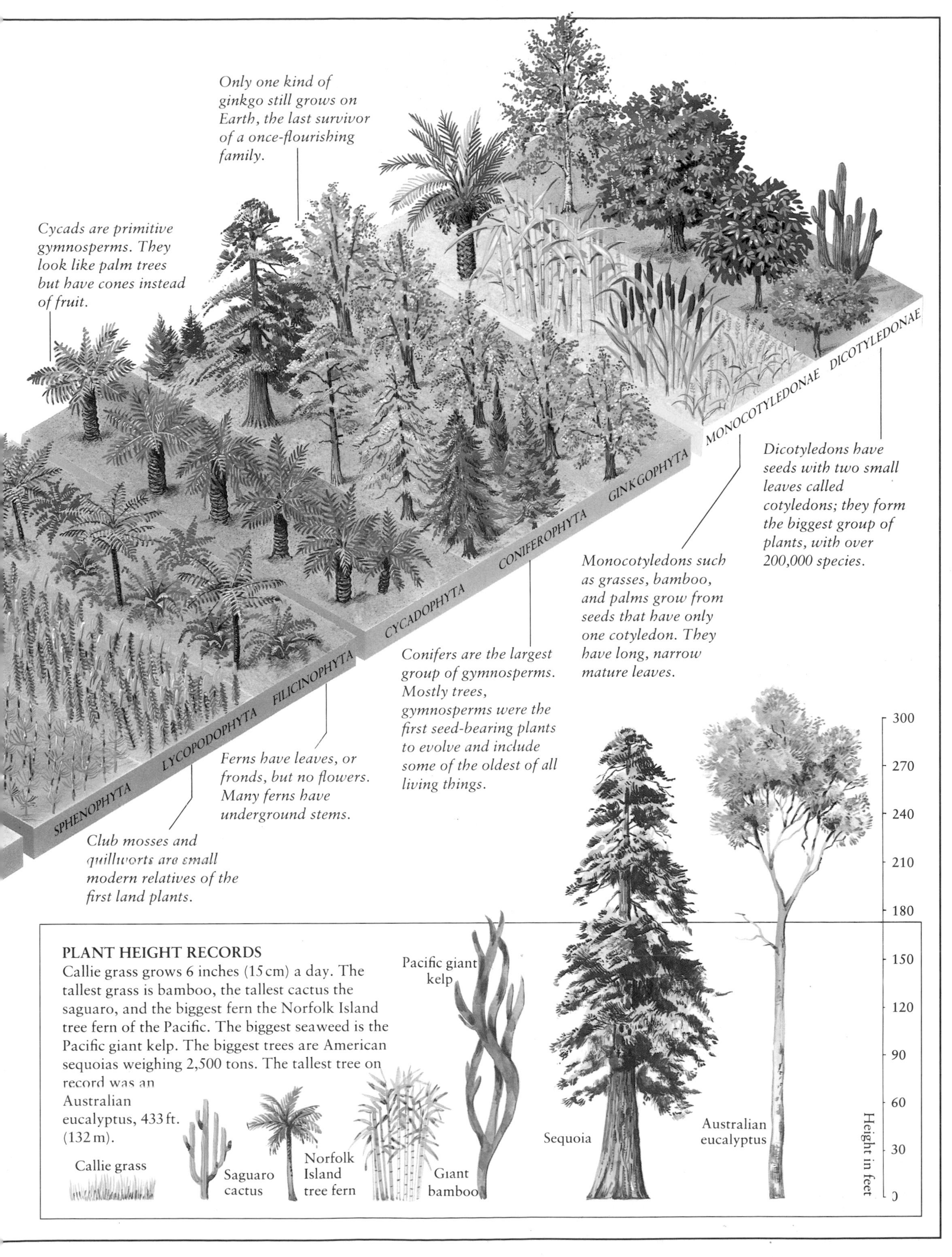

PLANT HEIGHT RECORDS

Callie grass grows 6 inches (15 cm) a day. The tallest grass is bamboo, the tallest cactus the saguaro, and the biggest fern the Norfolk Island tree fern of the Pacific. The biggest seaweed is the Pacific giant kelp. The biggest trees are American sequoias weighing 2,500 tons. The tallest tree on record was an Australian eucalyptus, 433 ft. (132 m).

Bacteria, Algae, Lichens, and Fungi

These organisms are no longer classified as plants. Bacteria are tiny and single-celled, and were probably the first living things on Earth. They, and the microscopic blue-green algae, are widespread on land and in water. Other algae, classed separately as Protoctists, include the seaweeds, many of which more closely resemble plants. Fungi (mushrooms and toadstools, mildews, yeasts, and molds) are simple nongreen organisms without leaves, roots or stems; they are grouped on their own.

See pages 90–91

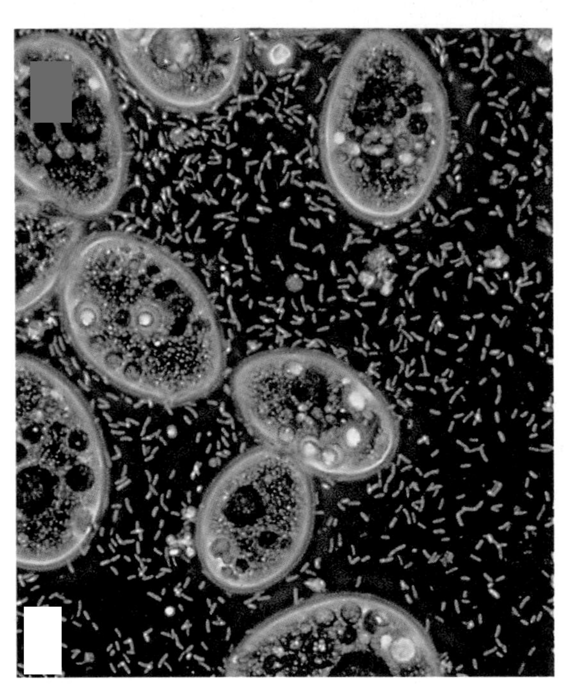

BACTERIA
Most bacteria can be seen only through a microscope. They have a simple structure, usually with a cell wall that stops them drying out. Huge numbers of bacteria live in the soil. They help to break down dead matter.

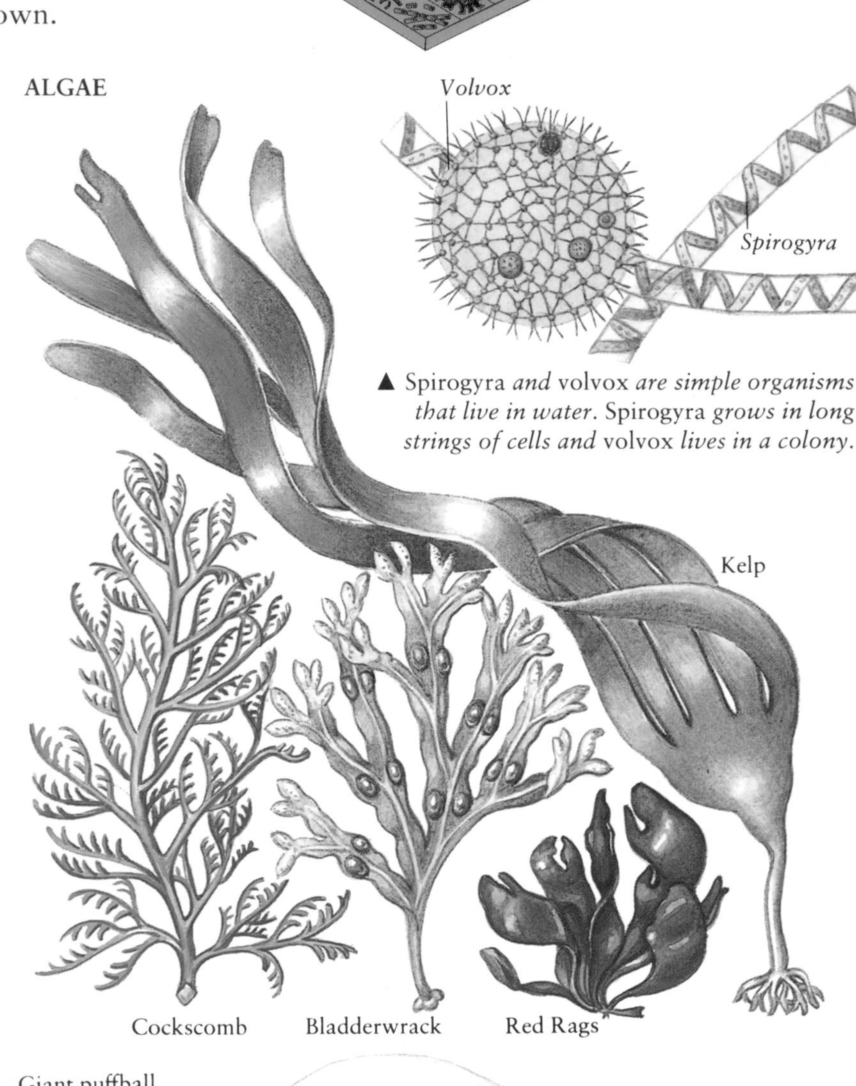

▲ *Spirogyra and volvox are simple organisms that live in water. Spirogyra grows in long strings of cells and volvox lives in a colony.*

► *Algae include the diatoms and the seaweeds. Classed by color (green, brown, and red), the 7,000 kinds of seaweeds are plants of ocean and shore. Some have air bladders to help them float. Others cling to the seabed.*

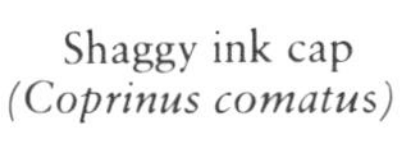

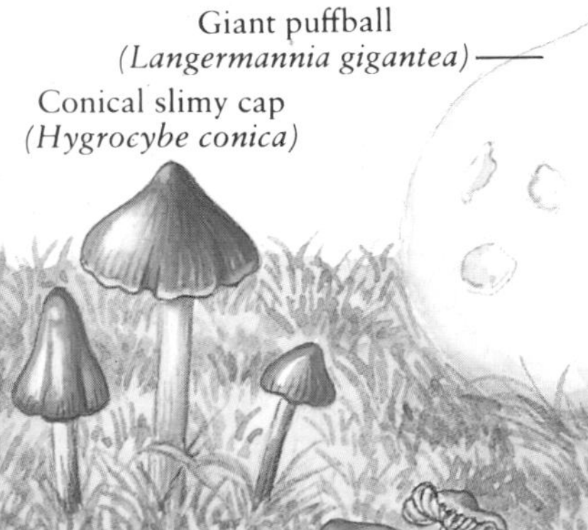

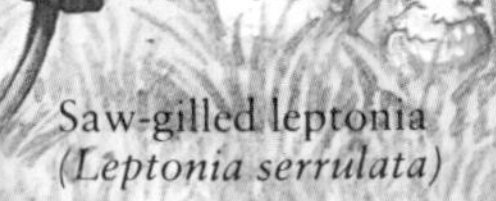

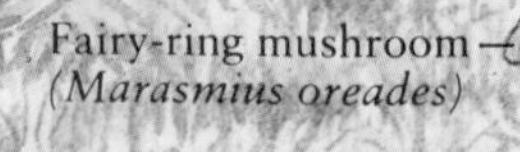

LICHENS

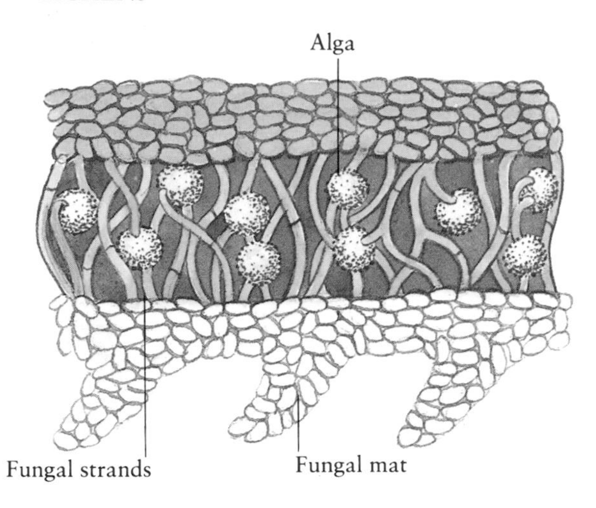

◀ *A lichen is made up of two living things in symbiosis, or partnership. Within the lichen is a single-celled alga enclosed in a fungus. Using photosynthesis, the alga makes food both for itself and for the fungus, which cannot survive on its own.*

▶ *Some lichens grow in soil, but most grow on rocks, walls, or tree bark. Lichens are low-growing, but can live for over 4,000 years, enduring extremes of heat and cold.*

THE LIFE CYCLE OF A FUNGUS

The mushrooms and toadstools we see are the "fruiting bodies" of fungi. The hidden part of the fungus, growing under the soil or in the wood of trees, consists of thousands of threadlike cells that form a tangled mass, the mycelium. The fruiting body appears when the fungus is ready to produce spores, which develop into new plants.

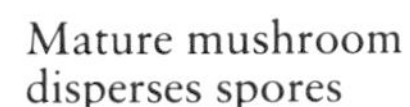

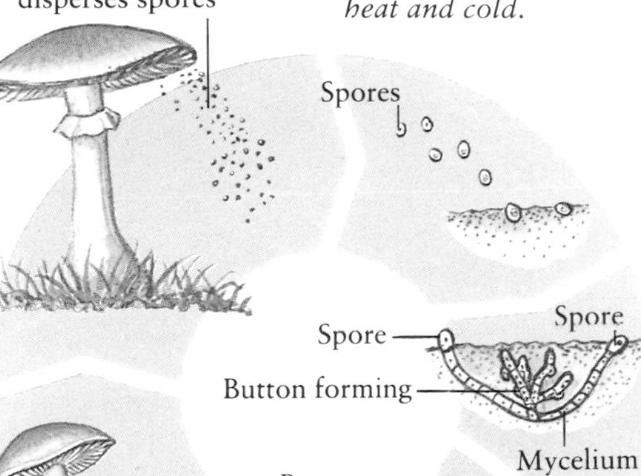

FACTS ABOUT FUNGI

- A field mushroom produces 16 billion spores in just under a week.
- Raindrops help to disperse the spores of puffballs. The paper-thin outer wall of the ball encloses the powdery spores. When a raindrop hits the wall of the ball, it bulges inward and puffs out a cloud of spores.
- The most deadly fungus is the yellowish-green death cap *Amanita phalloides*, which is commonly found with beech and oak trees. If eaten, it can kill in 6 to 15 hours.

MUSHROOMS AND TOADSTOOLS

Most fungi produce fruiting bodies in autumn—a good time to spot colorful mushrooms and toadstools. Some are good to eat, but others are poisonous. Never pick or eat a mushroom until you are certain it is not poisonous.

Blood-stained bracket (*Daedaleopsis confragosa*)

Many-zoned bracket fungus [also known as Varicolored bracket] (*Coriolus versicolor*)

Mealy tubaria (*Simocybe centuncula*)

Dryad's saddle (*Polyporus squamosus*)

Yellow brain fungus (*Tremella mesenterica*)

Devil's boletus (*Boletus Satanas*)

Collared earthstar (*Geastrum triplex*)

Coral spot fungus (*Nectria cinnabarina*)

Death cap (*Amanita phalloides*)

Common morel (*Morchella esculenta*)

Common stinkhorn (*Phallus impudicus*)

Wood blewit (*Lepista nuda*)

Cystolepiota aspera (*Lepiota friessii*)

Chanterelle (*Cantharellus cibarius*)

Liverworts, Mosses, Horsetails, and Ferns

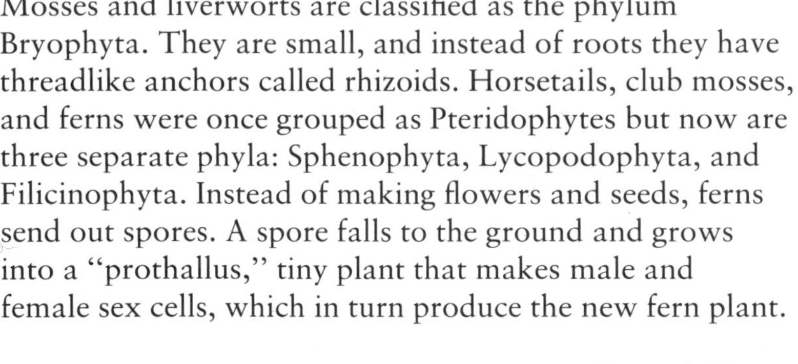

Mosses and liverworts are classified as the phylum Bryophyta. They are small, and instead of roots they have threadlike anchors called rhizoids. Horsetails, club mosses, and ferns were once grouped as Pteridophytes but now are three separate phyla: Sphenophyta, Lycopodophyta, and Filicinophyta. Instead of making flowers and seeds, ferns send out spores. A spore falls to the ground and grows into a "prothallus," tiny plant that makes male and female sex cells, which in turn produce the new fern plant.

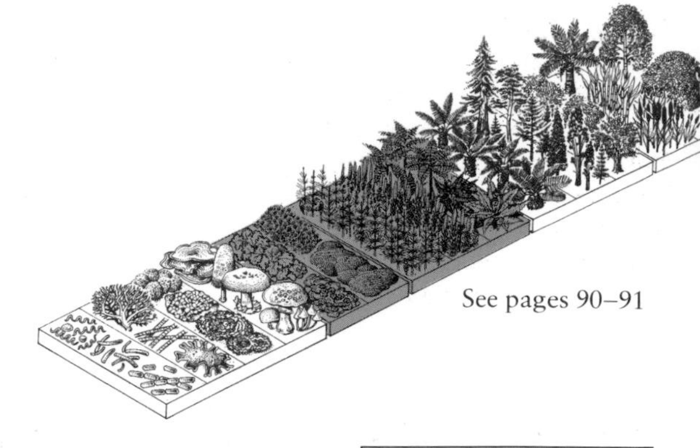

See pages 90–91

Marchantia polymorpha

Pellia epiphylla

◀ *Liverworts grow in damp places. Some have a flat body, or "thallus." Others look more leafy, with rows of leaves growing on a stem. There are about 8,000 species of liverworts, found in both hot and cold climates.*

MOSS FACTS

- Millions of years ago, ferns, horsetails, and club mosses forested the Earth.
- Peat moss is so absorbent it can be used to dress wounds.

MOSSES

Mosses are small and usually grow in clumps or dense mats, often clinging to a rock or a stone wall. As part of their complicated reproductive cycle, many mosses send out stalks with a pod at the tip. The pod releases thousands of spores to form new plants.

Leucobryum glaucum

Sphagnum papillosum

Racomitrium lanuginosum

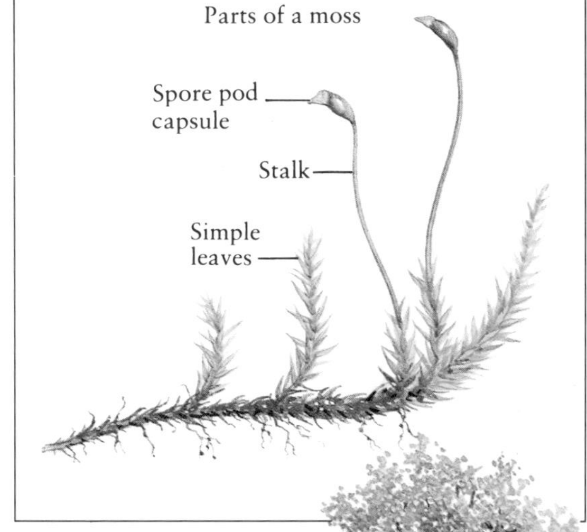

FORMATION OF A PEAT BOG

Peat mosses are often called sphagnum mosses. They may float on top of a lake, forming a thick green carpet. This is how a peat bog starts.

The submerged parts of the moss plants die. Dead and decaying matter sinks to the bottom of the lake, forming a mat below water level.

In time, the mat of decaying vegetation builds up into a dense mass. The plant matter absorbs water and gradually turns the lake into a bog.

The lower layers are squashed by the matter above, and slowly turn into mud-like peat. The bog dries out and new plants colonize the surface.

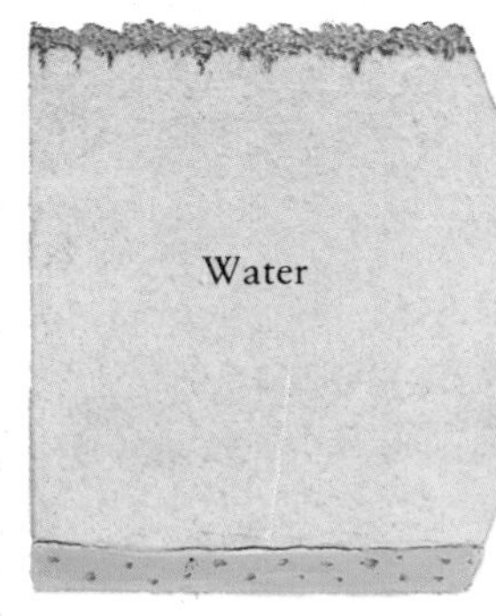

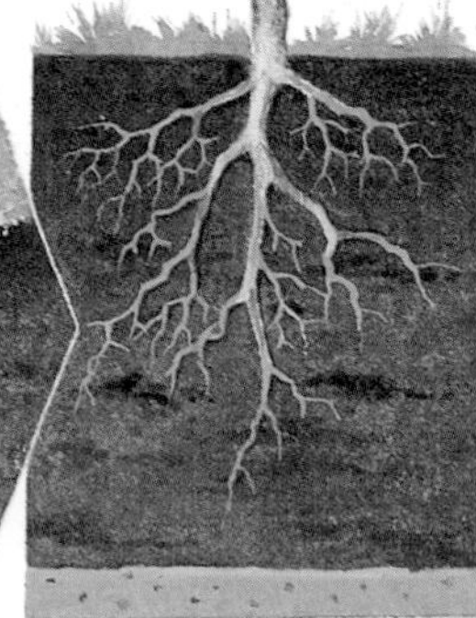

CLUB MOSSES

Club mosses are not true mosses. They are related to ferns. Club mosses have an underground root from which grows a stem with branches and small leaves. Club-like cones, which contain the spores from which new plants grow, form at the tips of these branches.

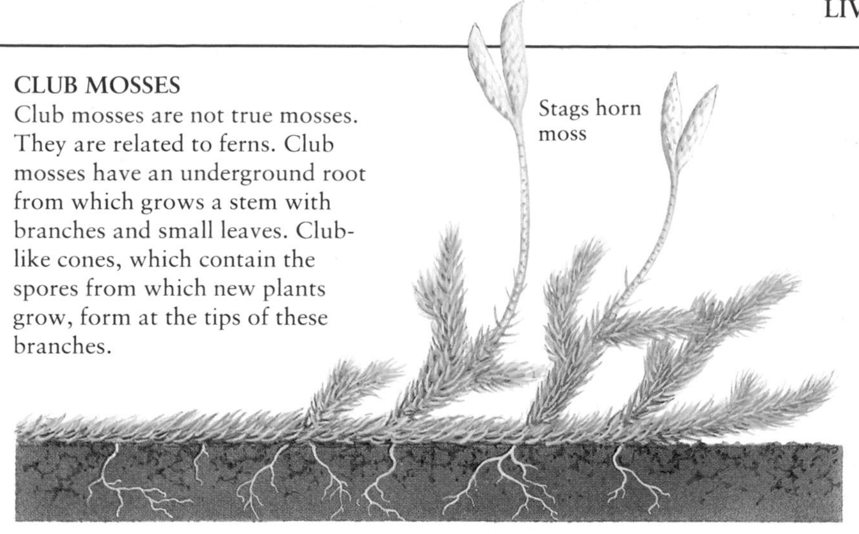

HORSETAILS

Horsetails are small plants with hollow, jointed stems and stalks that often look like miniature trees. They have no flowers and can be found in damp places.

HOW A NEW FERN GROWS

Ferns have fronds. Under each frond are spore cases, or "sporangia," lined with hundreds of spores.

The sporangia burst and spores are blown away. Fern spores survive best in shaded, moist soil. In suitable ground, a spore grows into a prothallus.

The prothallus has male and female cells from which a young fern develops; it feeds on the prothallus until it roots and can live on its own.

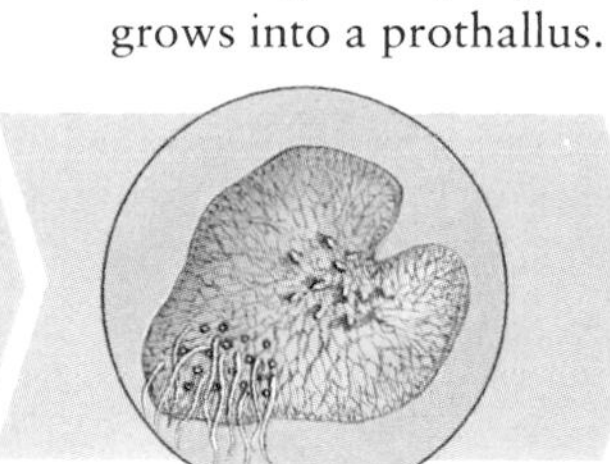

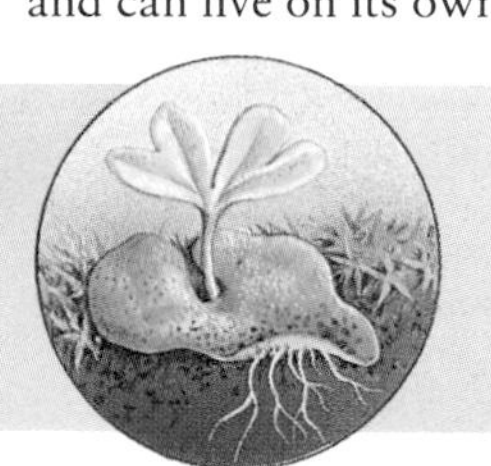

TYPES OF FERN

There are about 10,000 kinds of ferns on the Earth today. Tree ferns grow in the tropics. The leaves or fronds of many ferns are long and lacy. Other ferns have simple oval or round leaves.

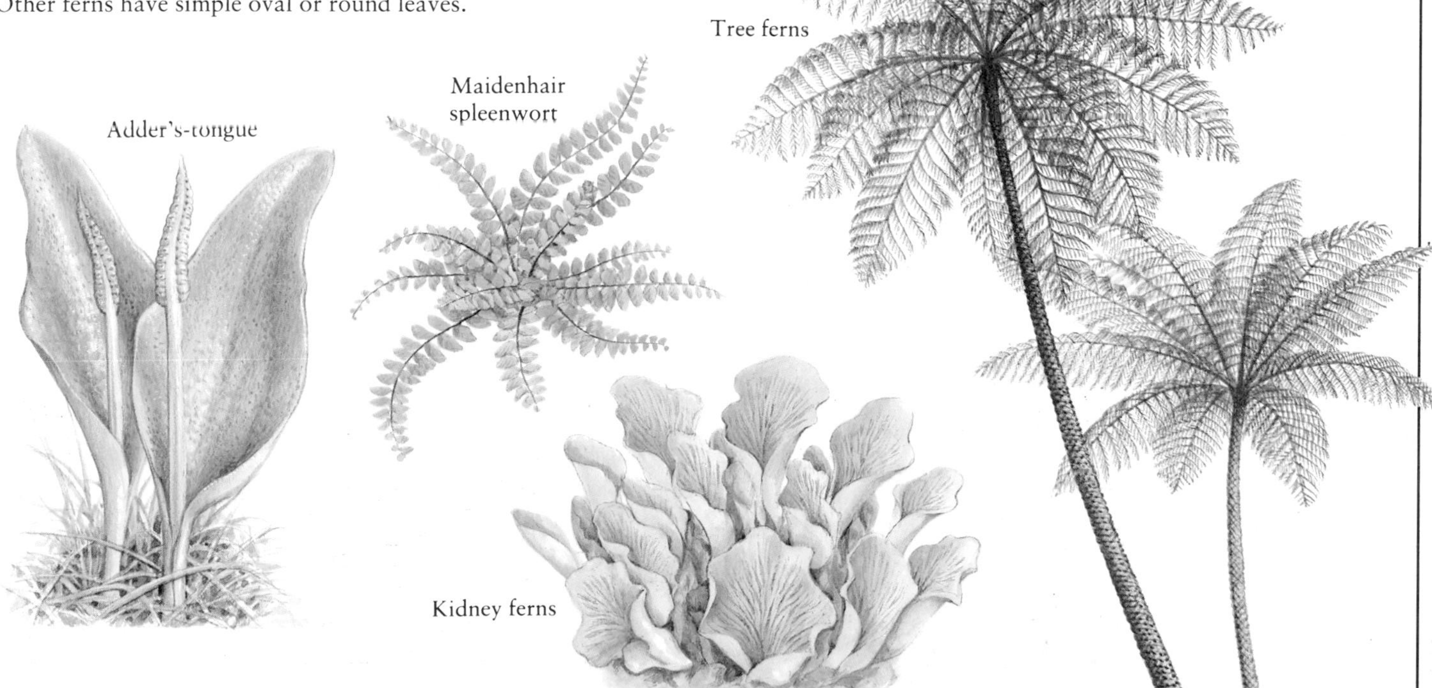

Ginkgos, Cycads, and Conifers

Ginkgos and cycads are the survivors of a group of plants that were growing over 300 million years ago when the first amphibians crawled onto the land. The sole surviving ginkgo is the maidenhair tree. Only nine kinds of cycads remain today. Both these plants are gymnosperms—plants that bear seeds in cones. The most successful gymnosperms are the conifers (Coniferophyta), which include the pines, spruces, larches, cedars, firs, and cypresses. All except the larch and swamp cypress are evergreen trees.

See pages 90–91

GINKGOS

◀ *The maidenhair tree* (Ginkgo biloba) *from China has fan-shaped leaves. The seed has a hard, nutlike center.*

◀ *The maidenhair tree is a "living fossil," the only survivor of an ancient family of trees. Fossil leaves of ginkgos show how little this plant has changed over millions of years.*

▼ *The* Welwitschia *is a gymnosperm, found in Africa, that lives for over 100 years. Two large woody leaves with a cone in the middle grow from its short stem.*

CYCADS

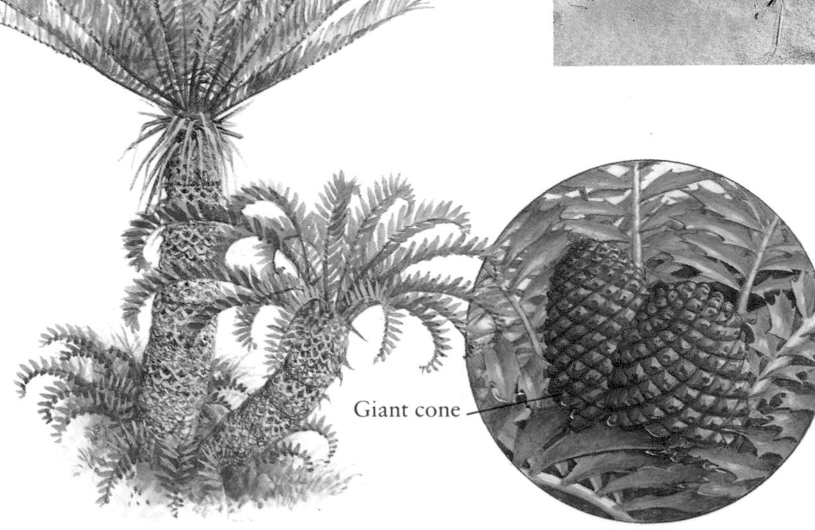

◀ *Cycads first grew on Earth in the Triassic period (from 225 million years ago). Cycads resemble palm trees and some are very long-lived (up to 1,000 years). The fern-like leaves sprout from the top of the stem. The seeds are inside a large cone that forms in the middle of the leaf cluster.*

CONIFERS

Silver fir

▼ Conifers grow mostly in the Northern Hemisphere. Only a few species, such as the South American monkey puzzle, grow south of the equator. Conifers have long, needle-shaped leaves. The leaf shape helps to keep in water, so that conifers can grow in very dry soils and can also tolerate extremes of cold.

CONIFER REPRODUCTION
Male cones produce pollen. Female cones produce ovules which are sticky and attract pollen grains during fertilization. Seeds form in the scales of the female cone and are released in spring when they are dispersed by the wind.

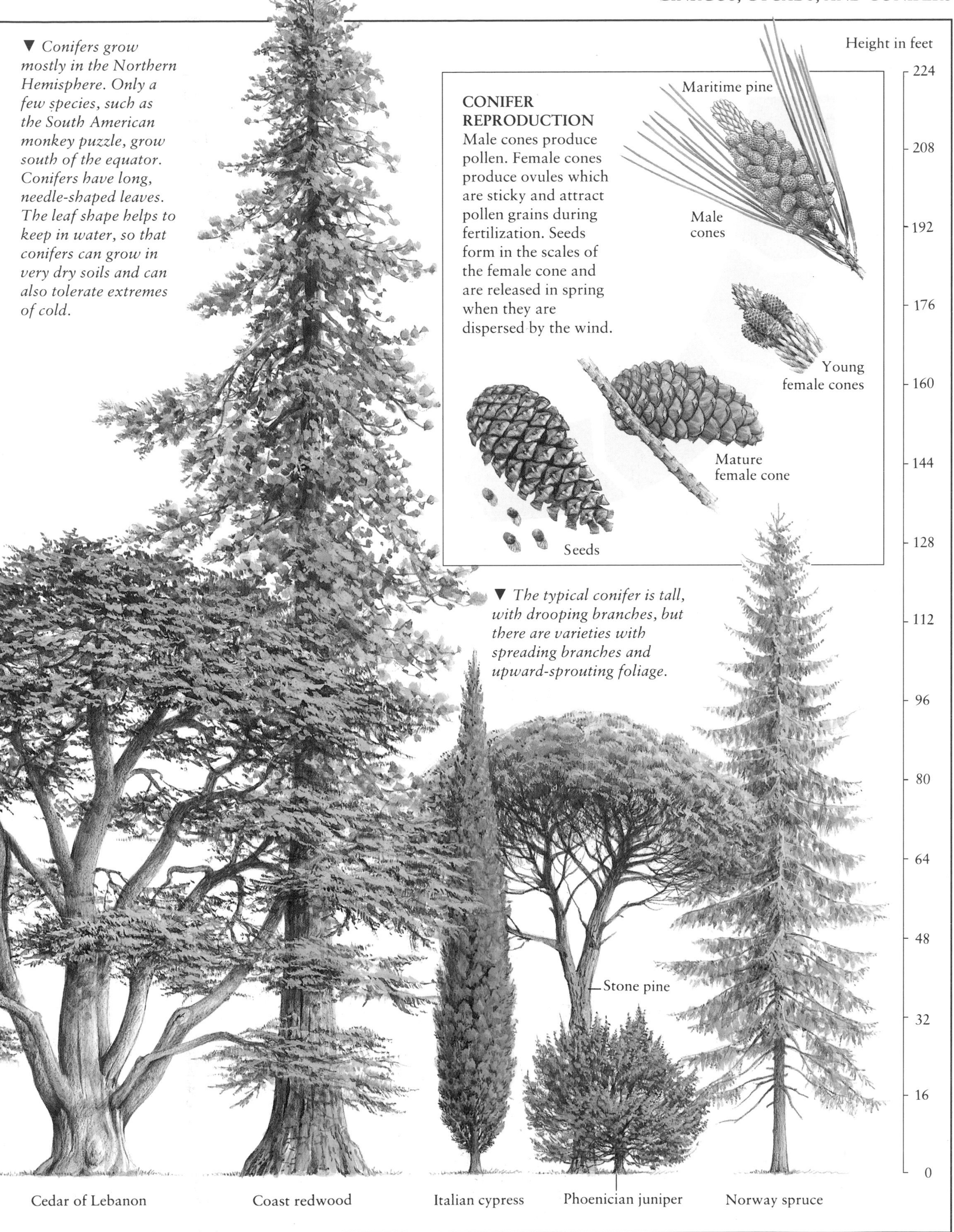

▼ The typical conifer is tall, with drooping branches, but there are varieties with spreading branches and upward-sprouting foliage.

Monocotyledons and Dicotvledons

Monocotyledons and dicotyledons are the two classes of flowering plants. They are the most diverse of all plants, with the most efficient reproductive system in the plant world. The basic difference, which gives the two classes their names, is in the number of cotyledons, or seed leaves. Monocots have one, dicots have two. There are also differences in the mature leaves. Monocots usually have long, narrow leaves. Grass is a good example. Dicot leaves are usually broad, with smooth, rounded, or toothed edges.

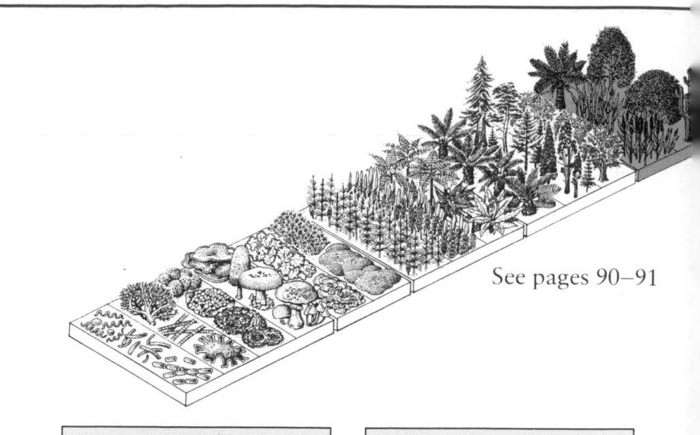

See pages 90–91

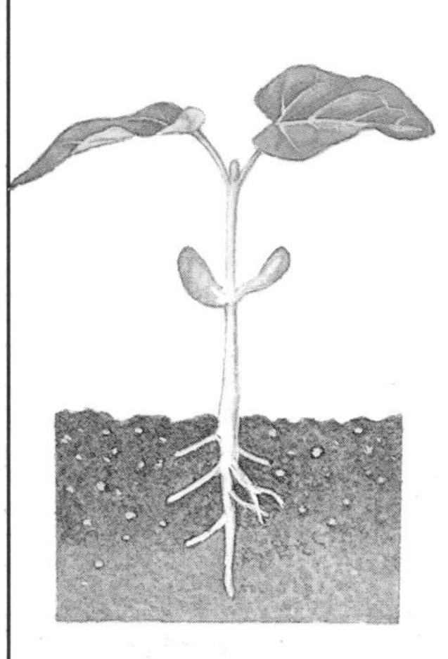

◄ *The cotyledon is the leaf part of a plant embryo, from which a new plant grows. Dicot seeds have two cotyledons.*

► *Monocot leaves are smooth-edged with parallel veins. The leaves grow from the base so, for example, grass keeps growing even when mowed or nibbled by animals.*

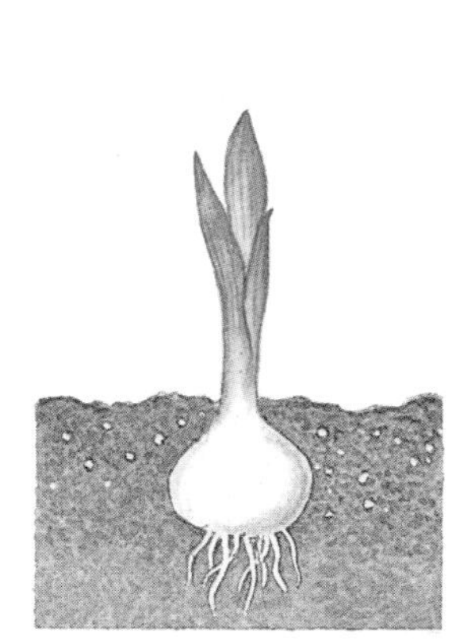

MONOCOTS
The 40,000 species of monocots include: **Grasses** **Cereals** such as rice, wheat, and corn **Bulb plants** such as tulips and lilies **Orchids** **Bananas** **Bulrushes and reeds** **Palm trees**

DICOTYLEDONS
Most flowering plants are dicots. Typical examples include: **Foxgloves** **Rhododendrons** **Deciduous trees** such as oak, beech, maple **Roses, grapes** **Carrots, cucumbers** **Potatoes, beans, and peas**

► *Typical plants of the cool temperate forest include deciduous trees (ones that shed their leaves in autumn) such as oak, maple, and beech, and woodland flowers, such as bluebells, which bloom in spring.*

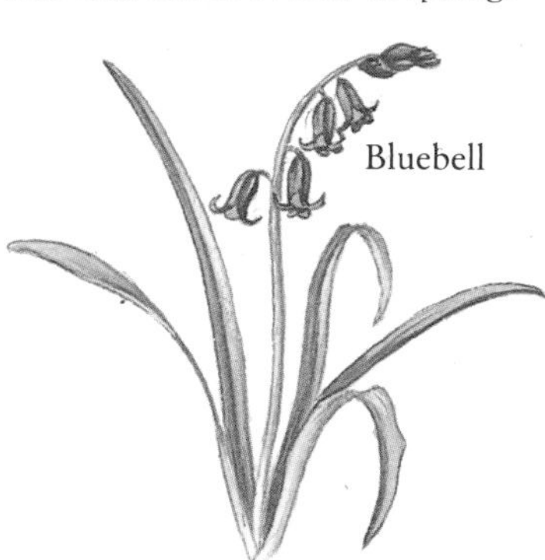

◄ *Grasslands are divided into three types: prairies, steppes, or savannas. Prairies have longer grasses than steppes, while savannas have trees such as palm and acacia as well as grass.*

◄ *Plants such as mosses, shrubs, and flowers survive the wind and cold on high mountains by growing near the ground and having long, clinging roots. Conifers are the trees best adapted to alpine conditions.*

Alpine forget-me-not

► *Tropical forests contain half the known plant species. Most tropical forest trees are evergreen. Plants requiring little light, such as ferns, grow at ground level, while vines and orchids grow high in the trees.*

Rafflesia

◄ *Wetland plants include water plants such as lilies, reeds, and willow and mangrove trees. Some wetland plants live completely under water; others have air spaces in their stems, and leaves that carry air to their roots and so keep them afloat in the water.*

Giant waterlily

► *Plants can survive in hot deserts, although some deserts have only sand dunes. Many desert plants—cactuses, palm trees, yuccas—have spiny leaves and fleshy stems for storing water. They flower quickly after rain.*

Prickly pear

POLLINATION

A flower is the reproductive part of a seed plant. Pollen from the anther reaches the stigma. It then unites with an ovule to make a seed. Self-pollination occurs when pollen reaches a stigma on the same plant. Cross-pollination is when pollen from one plant transfers to another.

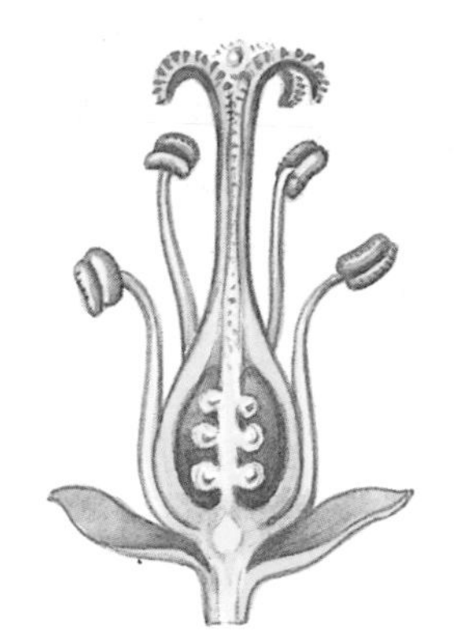

FLOWER TYPES

Flowers vary in form. Many plants produce clusters of flower heads. Some flower heads, for example, the daisy, are made up of many tiny flowers. These are called composite flowers. Three typical flower types are shown here.

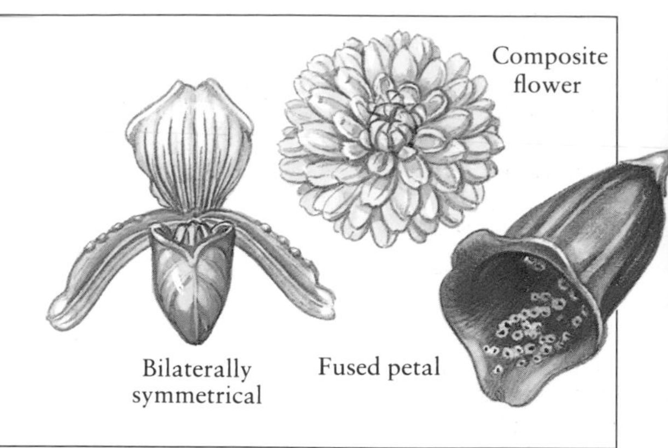

THE DEVELOPMENT OF A SEED

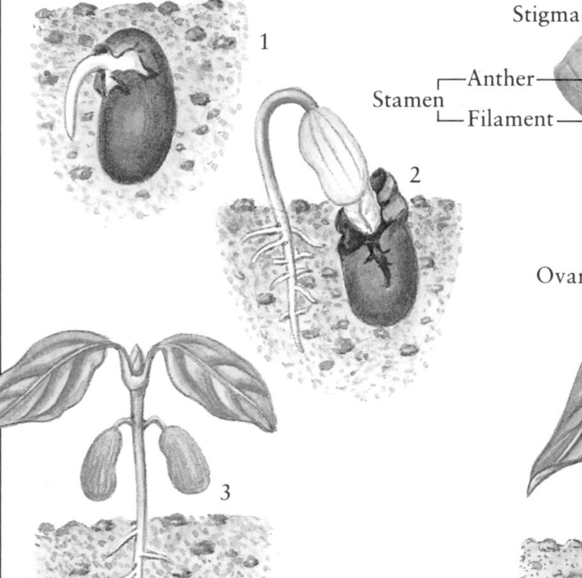

PARTS OF A FLOWER

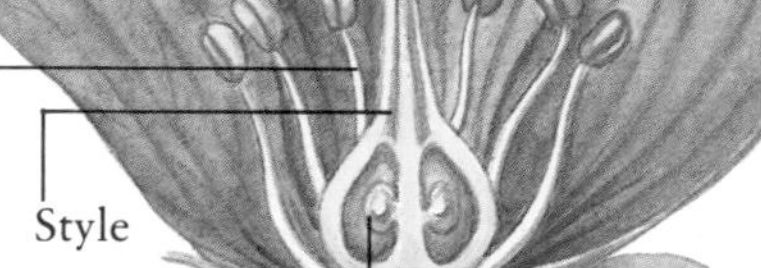

THE LEAF

Green leaves contain chlorophyll, a substance that absorbs energy from sunlight to make food from carbon dioxide gas in the air and water in the soil. This process, unique to plants, is called photosynthesis.

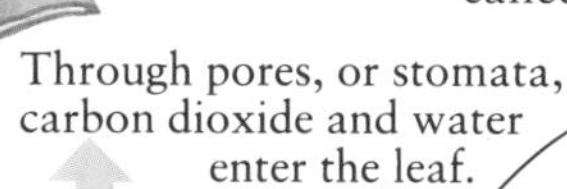

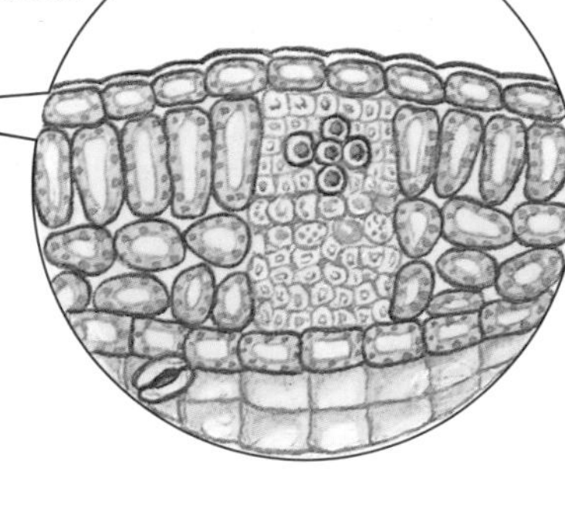

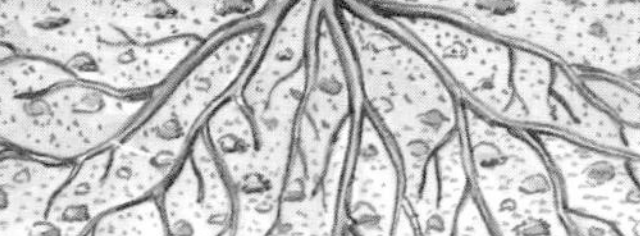

▲ *First the seed absorbs water, swells and splits. It sends out an embryo root, or radicle, which pushes downward into the soil. The shoot then pushes upward, bending toward the sunlight. Finally, the first leaves sprout.*

ROOTS

A plant's roots anchor it in the soil, and absorb water and minerals. Plants such as grass have a fibrous root system, with slender spreading roots; plants such as carrots have a taproot system, where one root is much larger than the rest.

▼ *Plants have organs other than roots under the ground. Bulbs, tubers, corms, and rhizomes store food to help the plant survive, and produce whole new plants without sexual reproduction, as runners also do on the surface.*

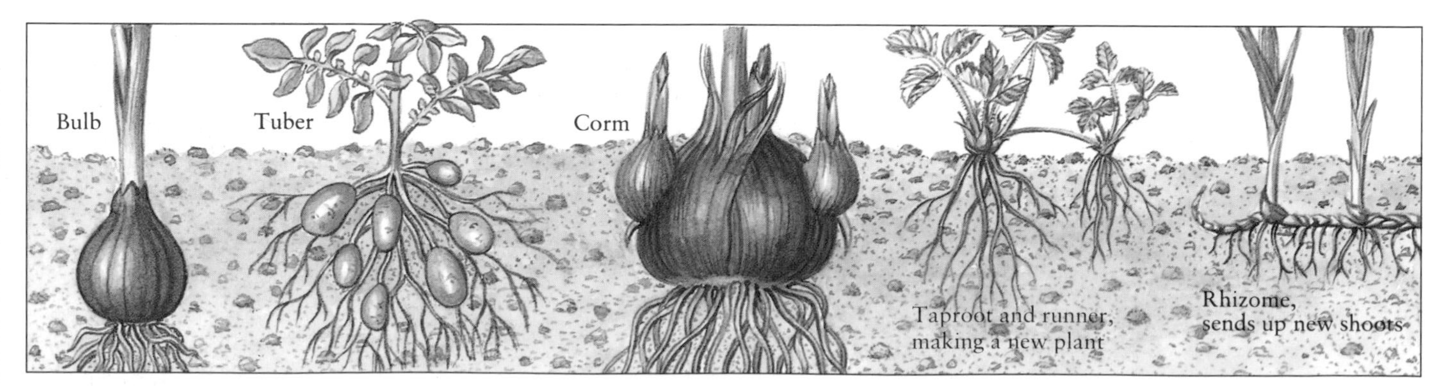

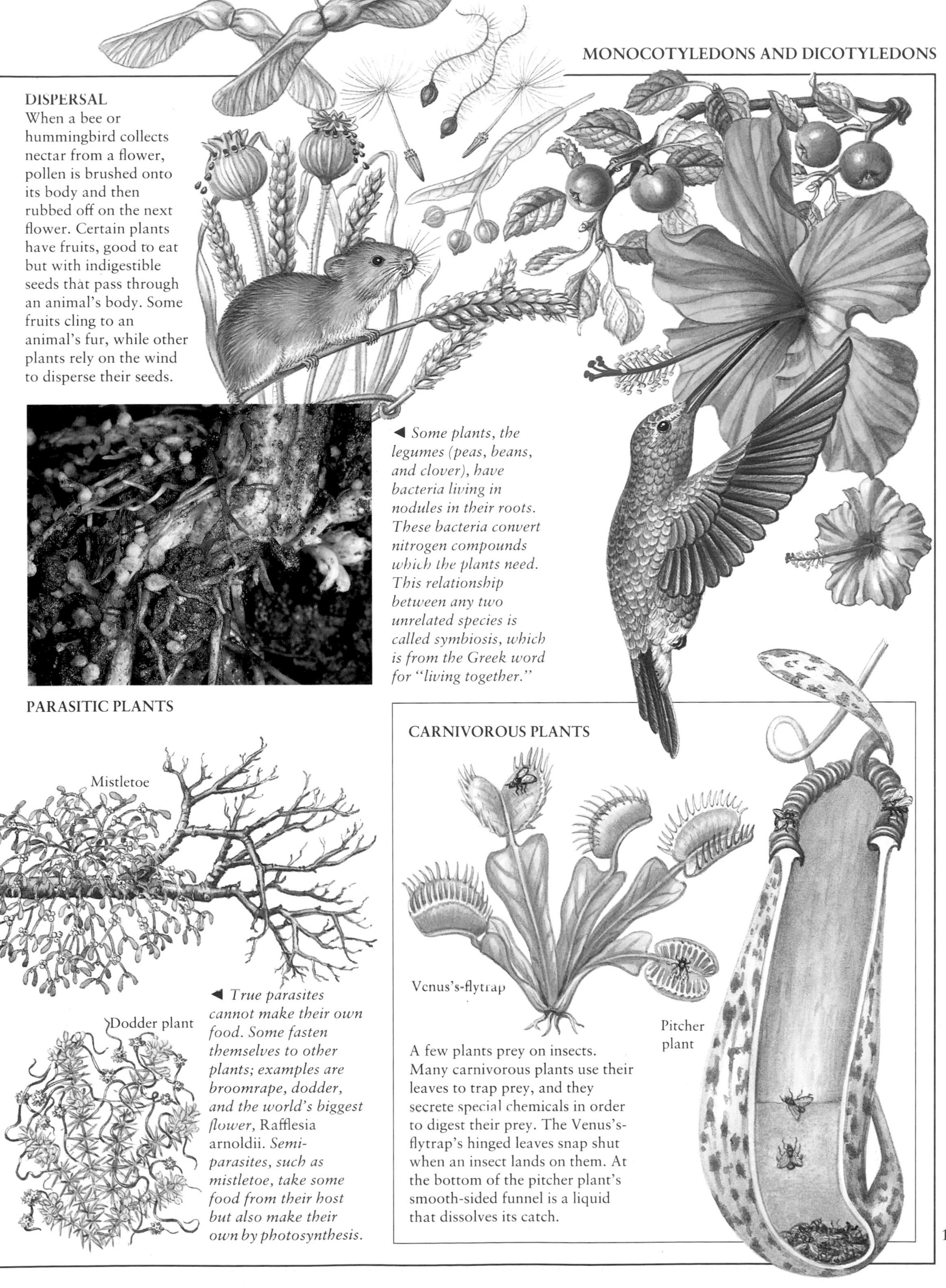

DISPERSAL

When a bee or hummingbird collects nectar from a flower, pollen is brushed onto its body and then rubbed off on the next flower. Certain plants have fruits, good to eat but with indigestible seeds that pass through an animal's body. Some fruits cling to an animal's fur, while other plants rely on the wind to disperse their seeds.

◄ *Some plants, the legumes (peas, beans, and clover), have bacteria living in nodules in their roots. These bacteria convert nitrogen compounds which the plants need. This relationship between any two unrelated species is called symbiosis, which is from the Greek word for "living together."*

PARASITIC PLANTS

◄ *True parasites cannot make their own food. Some fasten themselves to other plants; examples are broomrape, dodder, and the world's biggest flower,* Rafflesia arnoldii. *Semi-parasites, such as mistletoe, take some food from their host but also make their own by photosynthesis.*

CARNIVOROUS PLANTS

A few plants prey on insects. Many carnivorous plants use their leaves to trap prey, and they secrete special chemicals in order to digest their prey. The Venus's-flytrap's hinged leaves snap shut when an insect lands on them. At the bottom of the pitcher plant's smooth-sided funnel is a liquid that dissolves its catch.

Fruits or Vegetables?

People have many uses for plants that are most valuable as a source of food. Prehistoric people first gathered seeds, berries, and roots. About 10,000 years ago people began to grow cereals (such as wheat) and other crops. To the modern shopper, "fruit" means a juicy food, such as apples, oranges, or raspberries, grown on a bush or tree. These foods provide minerals, sugar, and vitamins. As a rule, vegetables are less sweet-tasting. The part of the plant that we eat may be its leaf, stem, root, seed, or fruit.

WHAT PARTS DO WE EAT?

BULB	Onion, garlic
TUBER	Potato, Jerusalem artichoke, yam, cassava
ROOT	Carrot, parsnip, beet, radish, turnip, Swedish turnip, sweet potato
LEAF	Brussels sprouts, cabbage, chard, Chinese cabbage, watercress, endive, kale, lettuce, spinach
FLOWER	Broccoli, cauliflower
FRUIT	Cucumber, zucchini, eggplant, apple, pepper, pumpkin, tomato, watermelon
NUT	Coconut, almond, chestnut, filbert, pistachio, pine nut, cashew
SEED	Brazil nut, peanut, bean, pea, lentil, corn, rice, oats, wheat, sunflower seed
STEM	Asparagus, kohlrabi, bamboo shoots, green onions. Celery and rhubarb are leafstalks

INFINITE VARIETY
The fruits and vegetables we enjoy are the result of cross-breeding from wild plants. There are many varieties, from all over the world. The familiar foods on the table come from a fascinating variety of plants.

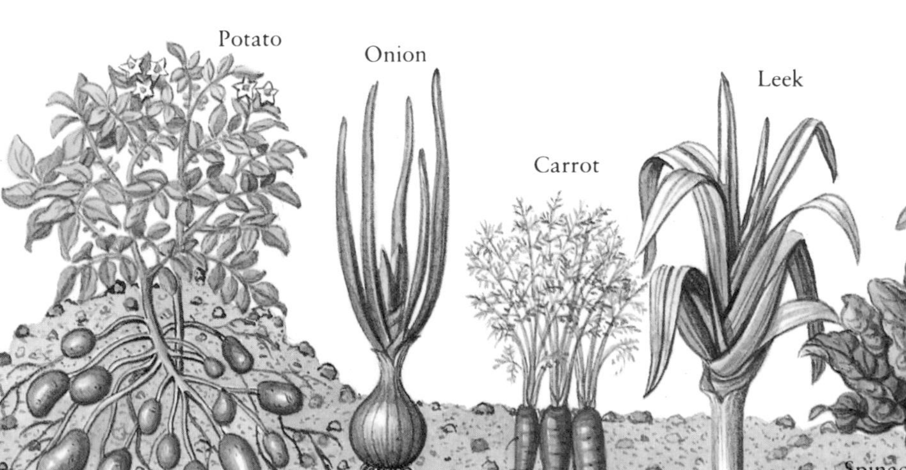

TYPES OF FRUIT

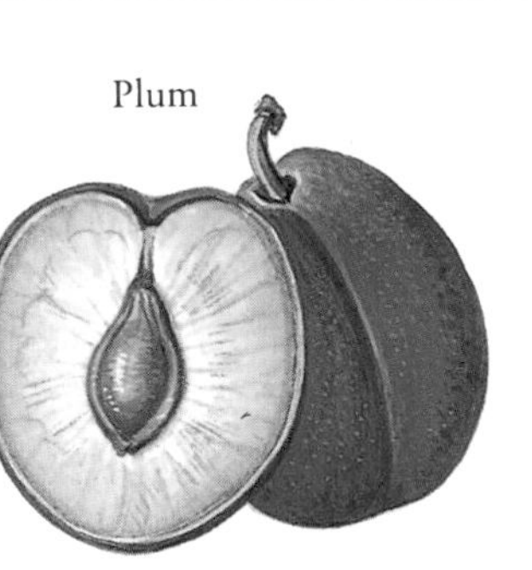

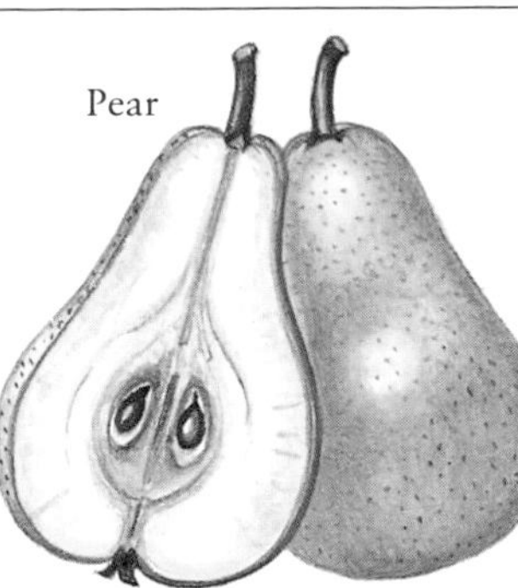

BERRIES

Seeds are enclosed in soft pulp: e.g. oranges. A blackberry and a pineapple are a cluster of drupes.

DRUPES

Fruits with pits inside: e.g. peach or plum. The seed is the pit in the middle of the fleshy fruit.

POMES

Apples, quinces, and pears are pomes. The seeds are held in a core (formed from the carpel of the flower).

DRY FRUITS

Can look as different as a chestnut, a corncob, and a pea pod: they are all seed-bearing parts of plants.

WHAT FRUITS ARE FOR

The fruit protects the plant seeds and aids their dispersal. It develops from the flower, forming a cover around the seeds. When ripe, the fruit splits. Fruits may be eaten, blown by the wind, or stick to animals to disperse the seeds inside them.

Orange

Corn

Raspberry

Pepper

Tomato

HERBS AND SPICES

A herb is any soft-stemmed plant, but in the kitchen, herbs are scented plants used for flavoring: they include sage, parsley, thyme, mint, and basil. Some herbs (e.g. camomile and feverfew) have medicinal uses. Spices, sweet or hot-tasting herbs, include vanilla, chillies, ginger, and cloves.

Celery

Squash

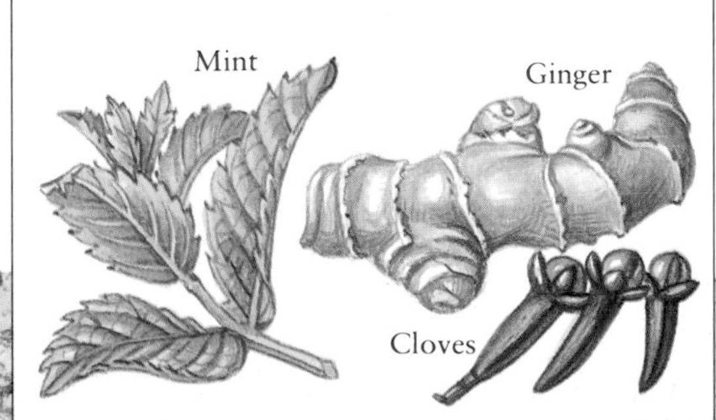

Trees

There are two main groups of trees, conifers (softwoods) and broadleaf trees (hardwoods). Conifers carry their leaves all year round, as do many tropical trees. Broadleaf trees in cooler climates are deciduous: they shed their leaves in the fall. Trees need more internal stiffening than smaller plants. In trees the tubes called xylem, which carry water through the stem (or trunk) from the roots, are thick and stiffened. Thinner-walled tubes called phloem carry food made in the leaves to other parts of the plant.

▼ *Most deciduous trees have broad, flat leaves. They may be oval, with smooth or toothed edges. Others are narrow (peach), pinnate—compound —(acacia or ash), and forked (horse chestnut or sycamore).*

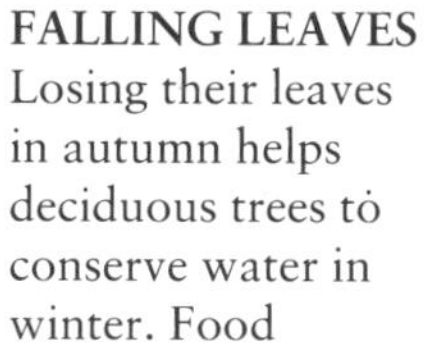

FALLING LEAVES
Losing their leaves in autumn helps deciduous trees to conserve water in winter. Food "pipes" to the stem are sealed: food is stored for next year's bud. The leaf is sealed off from the stem; the chlorophyll that makes the leaf green breaks down and hidden colors are seen.

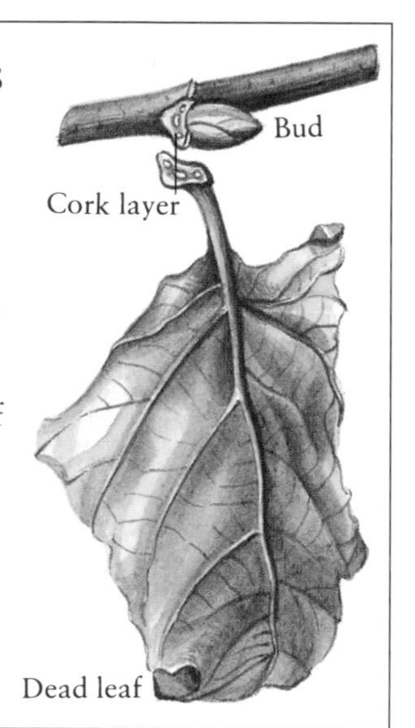

DEEP ROOTS

Roots take in water and minerals. Some trees have long roots, with as much growth below the ground as above it. Others have massive trunks, but shallow roots.

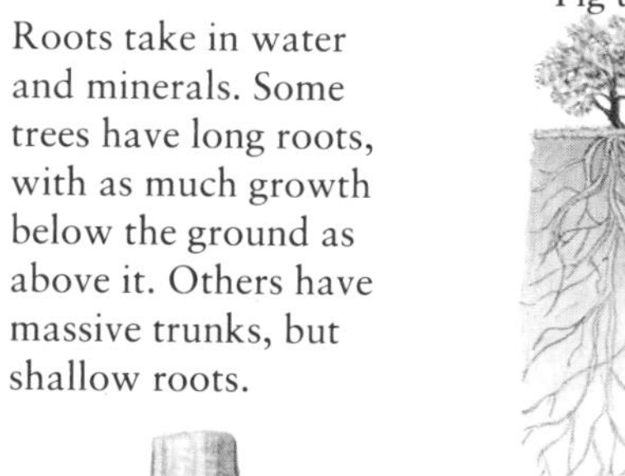

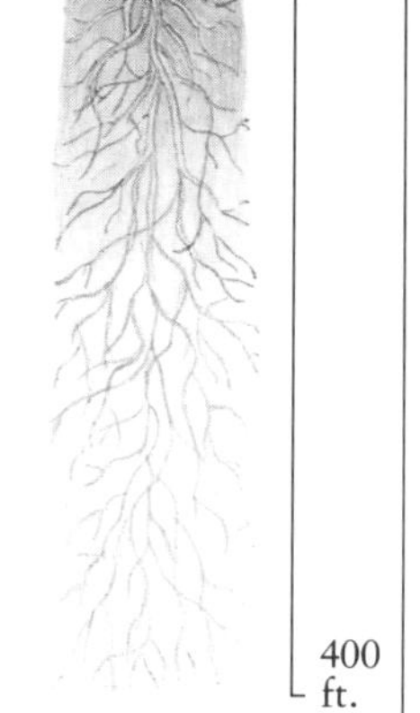

DECIDUOUS TREES
Most broadleaf trees are deciduous, although some tropical broadleaves are evergreen. A typical broadleaf tree has spring flowers, which develop into fruits, and a spreading crown and roots.

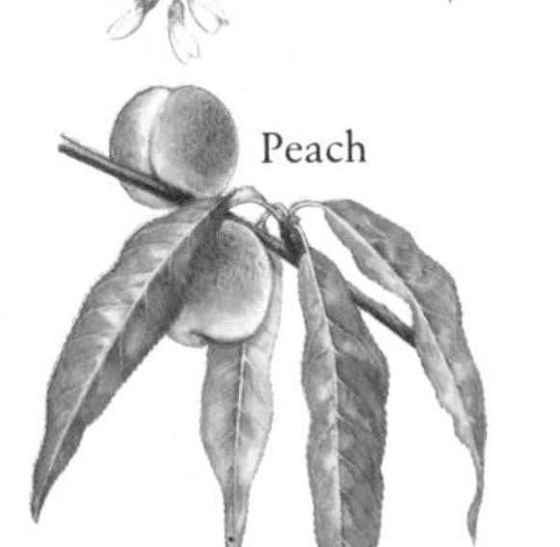

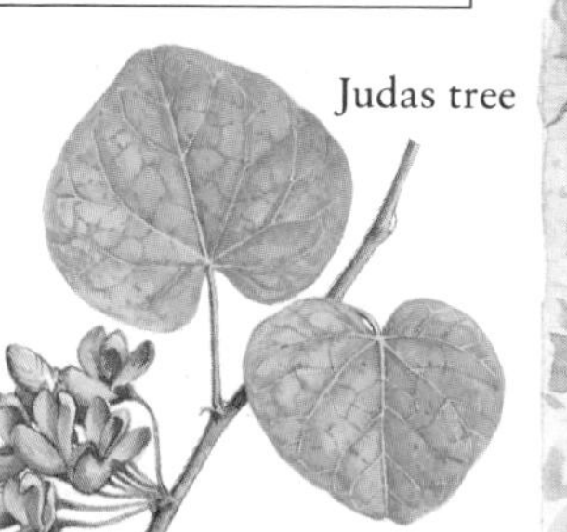

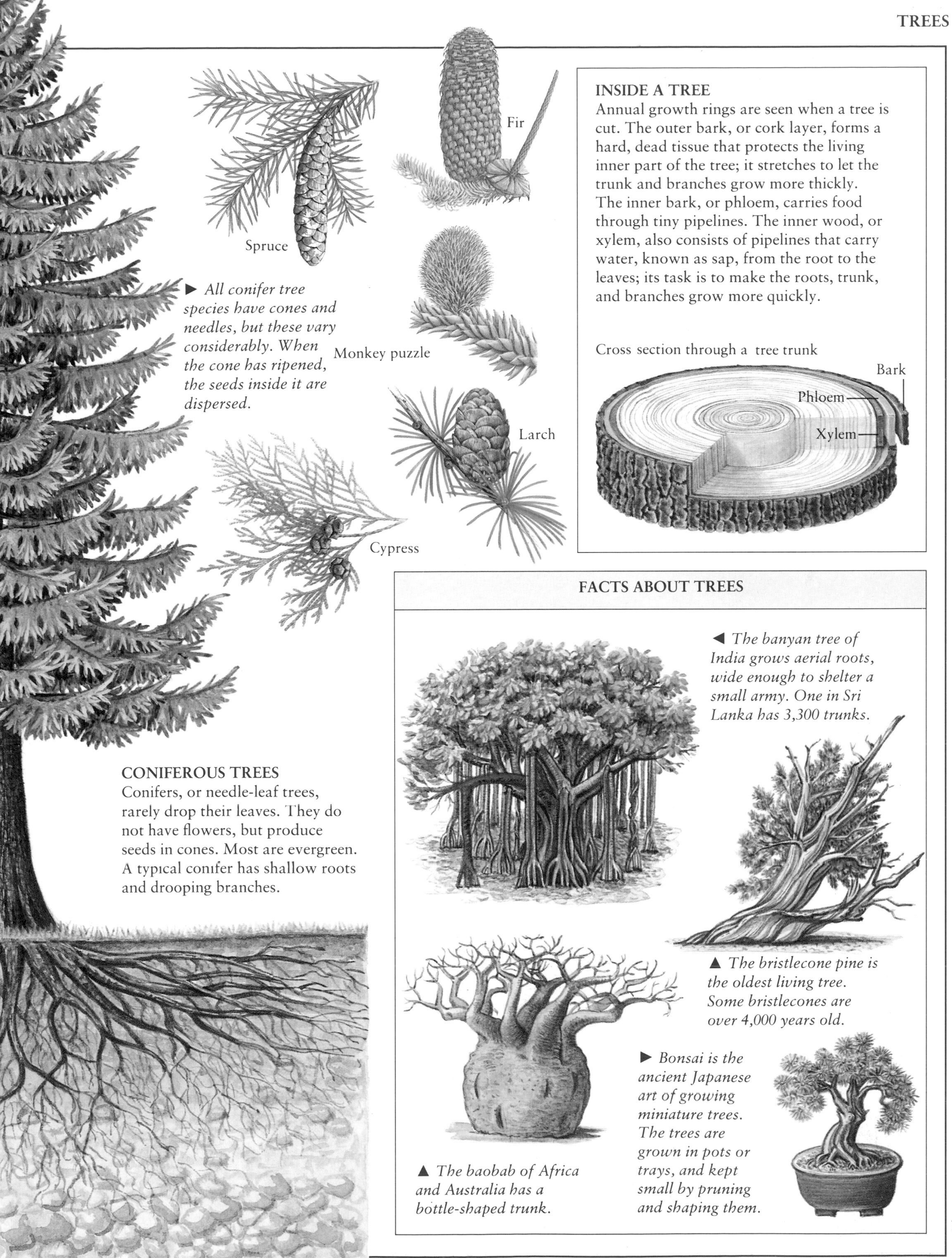

► *All conifer tree species have cones and needles, but these vary considerably. When the cone has ripened, the seeds inside it are dispersed.*

INSIDE A TREE

Annual growth rings are seen when a tree is cut. The outer bark, or cork layer, forms a hard, dead tissue that protects the living inner part of the tree; it stretches to let the trunk and branches grow more thickly. The inner bark, or phloem, carries food through tiny pipelines. The inner wood, or xylem, also consists of pipelines that carry water, known as sap, from the root to the leaves; its task is to make the roots, trunk, and branches grow more quickly.

CONIFEROUS TREES

Conifers, or needle-leaf trees, rarely drop their leaves. They do not have flowers, but produce seeds in cones. Most are evergreen. A typical conifer has shallow roots and drooping branches.

FACTS ABOUT TREES

◄ *The banyan tree of India grows aerial roots, wide enough to shelter a small army. One in Sri Lanka has 3,300 trunks.*

▲ *The bristlecone pine is the oldest living tree. Some bristlecones are over 4,000 years old.*

▲ *The baobab of Africa and Australia has a bottle-shaped trunk.*

► *Bonsai is the ancient Japanese art of growing miniature trees. The trees are grown in pots or trays, and kept small by pruning and shaping them.*

Plants and People

Plants are important to us, both as sources of food and as raw materials. About 10,000 years ago people began to grow plants, rather than simply collecting them from the wild. The basic food crops, such as cereals, were developed in this way by selective breeding from wild plants. Today, cultivated plants may look very different from their wild ancestors, and genetic engineering is making it possible to develop plants that yield large crops, resist pests, and grow in unfavorable conditions.

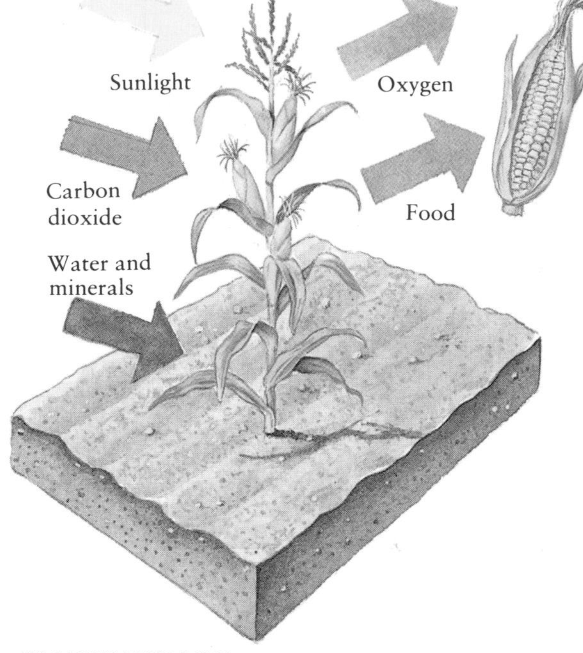

THE USES OF PLANTS

Many different parts of plants can be used. The sap of the rubber tree is tapped to give the latex from which natural rubber is made. Cotton comes from the ripe fruit, or "bolls," of the cotton plant. Inside the boll is a mass of fibers enclosing the seeds. Cork comes from the bark of the Mediterranean cork oak. The cork is stripped from the tree once every nine or ten years.

Rubber tree being tapped

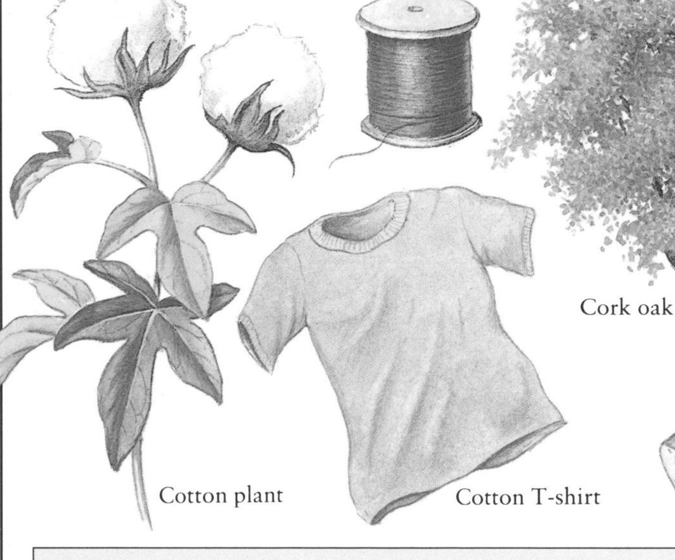

Cork oak

Cork

PLANTS FOR LIFE

Plants are essential to life on Earth. The chlorophyll in the leaves of green plants absorbs light energy from the Sun. Water is drawn up through the roots and carbon dioxide gas taken in from the air by the leaves, combining to make glucose (sugar) and oxygen. The plants use the sugar for food, releasing the oxygen into the air.

► *All animals depend on plants for their food in some way. The most important food crops grown by people include wheat, corn, rice, potatoes, beans, cassava, fruits, and vegetables.*

FACTS ABOUT PLANTS AND PEOPLE

- The painkilling drug morphine is made from the opium poppy.
- Quinine, once used for treating malaria, comes from the cinchona tree.
- The drug digitalis, a treatment for heart disease, comes from the leaves of the foxglove.
- The first antibiotic drug effective against infections was penicillin. It was developed in the 1940s from a mold found in 1928 growing in a dish that contained bacteria. The mold killed the bacteria.

Cinchona tree (flower of)

HOW PEOPLE CHANGE PLANTS

There are thousands of varieties of apples, developed over the past 2,000 years. Most apple trees are grown from a bud of one variety grafted onto the roots of another. Plant breeders have also cross-bred flowers such as roses to give them better colors or perfumes.

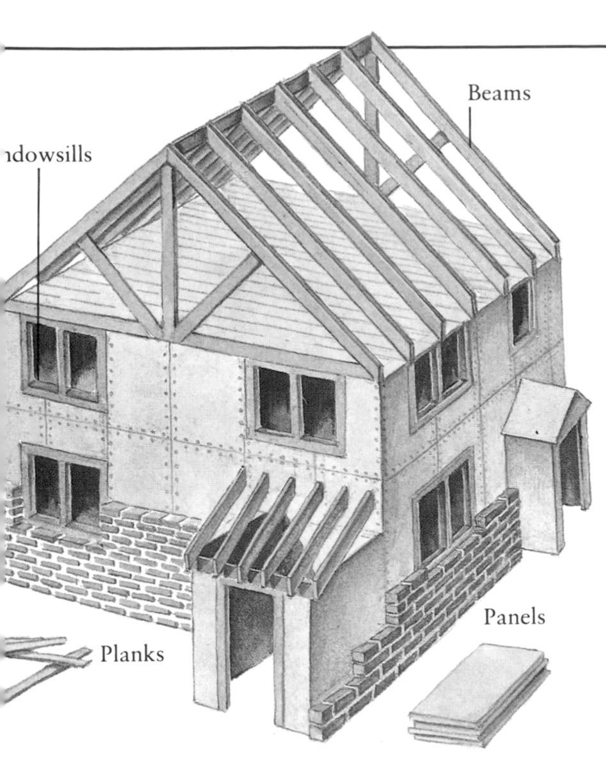

▲ *Timber for building use is mostly sawn softwood planking or factory-made laminate (such as plywood) and chipboard. Softwoods are far easier to saw, plane, and bore than hardwoods, making them ideal for the frame of a house. However, hardwoods such as oak or maple are often used for a house's interior paneling and for finished floors.*

HARDWOODS AND SOFTWOODS

These terms refer to the trees from which timber comes, not to its hardness. Softwood comes from conifer trees such as pine and cedar. Hardwood comes from broadleaf trees growing either in cool regions (trees such as oak and ash) or in the tropics (trees such as mahogany and ebony).

THE USES OF HARDWOOD	THE USES OF SOFTWOOD
Hardwoods can be very strong. Oak, for example, was used to make sailing ships. Beech is hard-wearing and elm is water-resistant. Furniture made from tropical hardwoods has an attractive color.	Softwoods grow quickly, and are easy to cut and shape by hand or on machines. They are used for making boxes, furniture, toys, and for building materials such as planks, frames, doors, posts, and beams.

▲ *Most paper is made from the pulp of trees such as beech, fir, pine, and oak, although other plant fibers can also be pulped to make paper.*

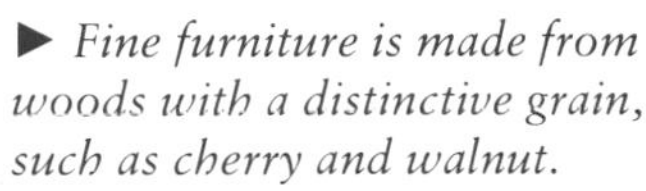

► *Fine furniture is made from woods with a distinctive grain, such as cherry and walnut.*

► *Reeds and grasses have been used for thousands of years as materials for house-building and in basket-making.*

PLANTS AND LANDSCAPE

People have changed complete landscapes by cutting down forests *(above)*, plowing prairies, and by introducing new kinds of plants. Plants are a vital ingredient of our landscape. Planting trees makes cities more pleasant to live in, and screens busy roads or factories.

DANGEROUS PLANTS

Some food plants have parts that are dangerous to eat—rhubarb and potato leaves, for example. Certain mushrooms are harmful if eaten. Poison ivy contains an irritant oil. Every part of the azalea, deadly nightshade, foxglove, oleander, and rhododendron is poisonous. Yew and laburnum seeds are poisonous, as are the berries of mistletoe and the bulb of a hyacinth.

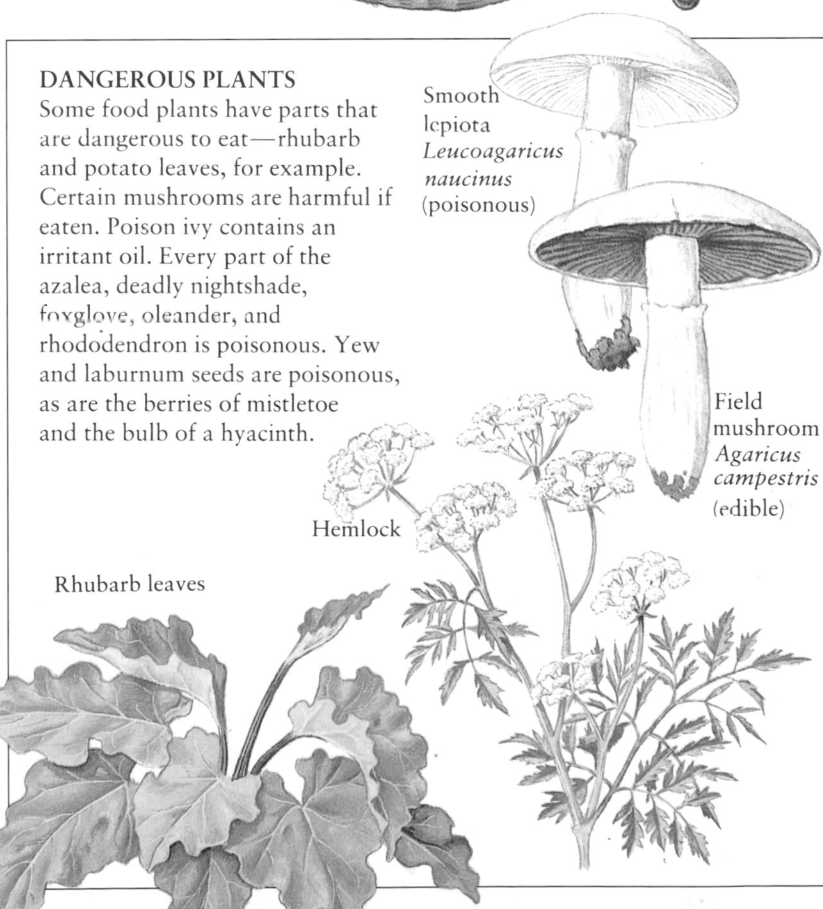

THE ANIMAL KINGDOM

The World of Animals

What makes an animal? A general rule is that animals move (plants are anchored by their roots). Unlike most plants, which make their own food, animals must eat either plants or other animals. Some live on dry land, others in water. Some have two legs, others have four, six, eight, or hundreds. Some are constantly warm-blooded; others have body temperatures that vary with their surroundings. Animals with similar body characteristics are grouped together. Scientists do this to identify each distinct species, and also to show how species are related within larger groups. Common features reveal how animal species have evolved over millions of years. An elephant looks very different from a dog, yet both are mammals and they share with birds, fish, amphibians, and reptiles an important body feature—a backbone of vertebrae.

THE ANIMAL KINGDOM
There are over one million animal species, classified into 20 or more phyla. For example, all animals with backbones (vertebrates) belong to the phylum Chordata. This includes all reptiles, birds, and mammals, but even so, the chordates make up only a small part of the vast animal kingdom. Only major phyla are shown here.

Protozoans are single-celled organisms. They move by floating or waving hairlike organs on their bodies. Protozoans are now usually classed in the separate Protoctist kingdom.

Sponges are the most primitive of multi-cellular animals; the 5,000 species make up the phylum Porifera. Sponges live in either fresh water or oceans; like most animals, they eat their food but they cannot move from place to place.

Flatworms, flukes, and tapeworms belong to the phylum Platyhelminthes. These animals have soft, thin, flat bodies. Most flatworms live as parasites in other animals.

Nematodes are thin, round worms. Some are too small to be seen without a microscope. There are more than 15,000 species, living in soil and water. Many, such as hookworms, are parasites.

Worms with long bodies made up of segments are annelids. They are all soft-bodied, without hard skeletons. This phylum includes earthworms, leeches, and lugworms.

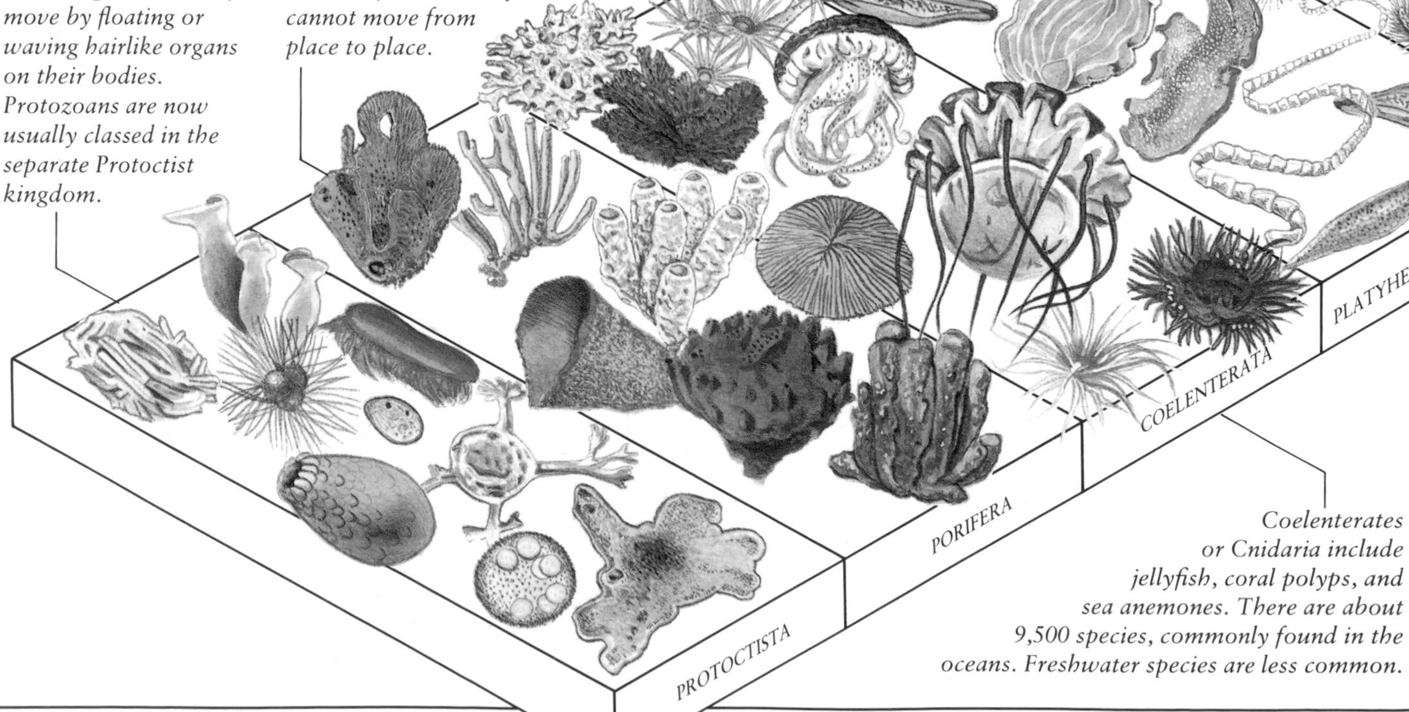

Coelenterates or Cnidaria include jellyfish, coral polyps, and sea anemones. There are about 9,500 species, commonly found in the oceans. Freshwater species are less common.

FACTS ABOUT ANIMALS

- The two main groups of animals are the vertebrates (animals with backbones) and the invertebrates (animals without backbones). About 96 percent of all animals are invertebrates (including insects).
- There are at least one million insect species—compared with about 4,000 mammals, 8,800 birds, 6,500 reptiles, 4,000 amphibians, and around 21,000 fish.

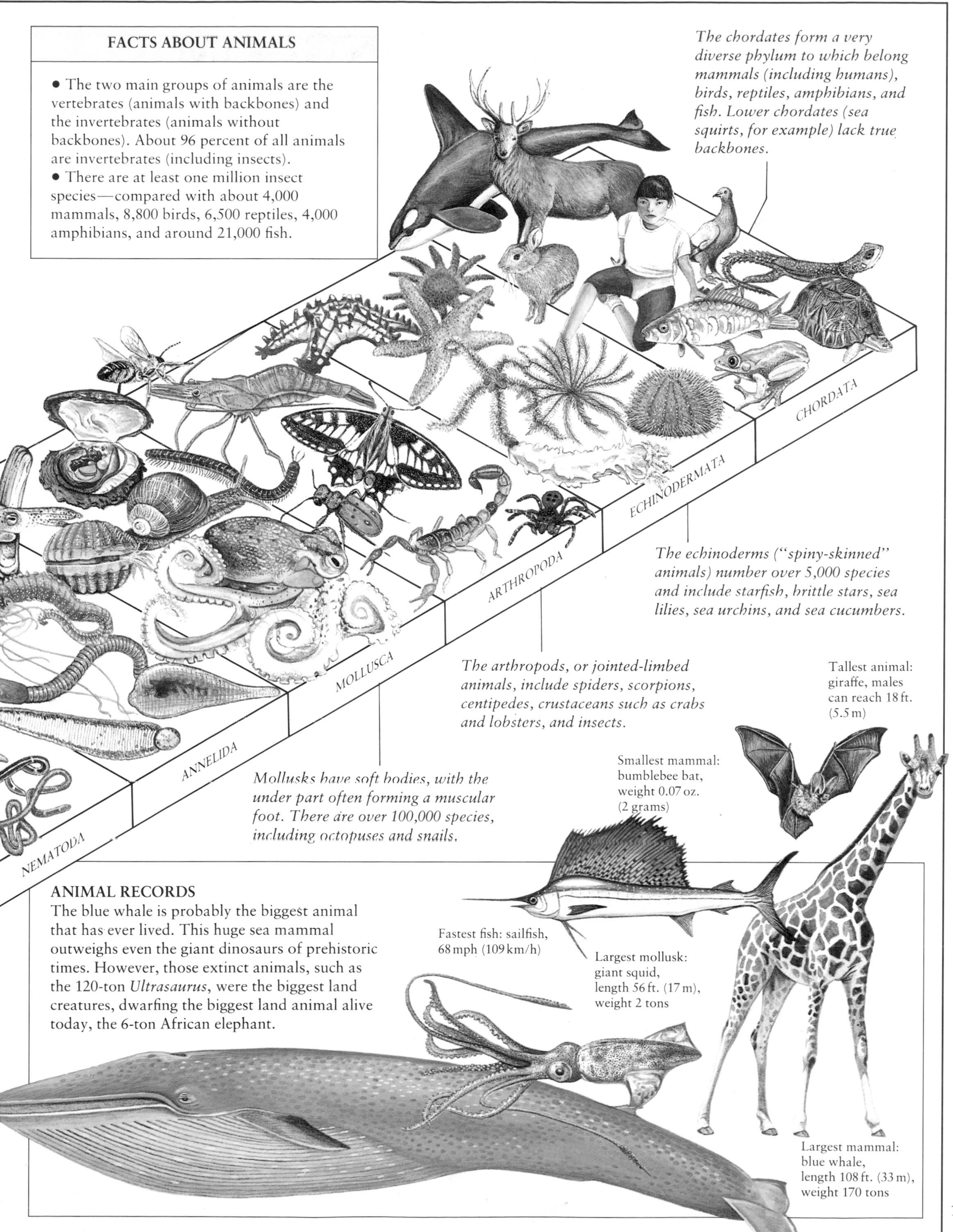

ANIMAL RECORDS

The blue whale is probably the biggest animal that has ever lived. This huge sea mammal outweighs even the giant dinosaurs of prehistoric times. However, those extinct animals, such as the 120-ton *Ultrasaurus*, were the biggest land creatures, dwarfing the biggest land animal alive today, the 6-ton African elephant.

Marine Invertebrates, Worms, Snails, and Slugs

The first multi-celled animals were sea-living invertebrates, creatures without backbones that swam and crawled in the oceans millions of years before the first backboned animals (fish). Their modern descendants include worms, corals, clams, snails, starfish, and squid. Even without the arthropods (for example, insects, spiders, and crabs), these "lower" animals are enormously successful: there are more than 100,000 mollusks, ranging from tiny snails to giant squid, and several thousand kinds of worms.

See pages 108–109

FACTS ABOUT MARINE INVERTEBRATES

- The bootlace worm of the North Sea can grow up to 180 ft. (55 m) long.
- Sea cucumbers (echinoderms) force out their own insides to confuse enemies, while they crawl to safety.
- The venom of a sea wasp jellyfish can kill a person in 1 to 3 minutes.
- The biggest snail is the African giant snail, weighing 2 lb. (900 grams).
- The quahog clam of the Atlantic Ocean can live for 220 years.
- Squids, cuttlefish, and octopuses are the most active mollusks. They are carnivores, and swim rapidly by squirting out jets of water behind them.

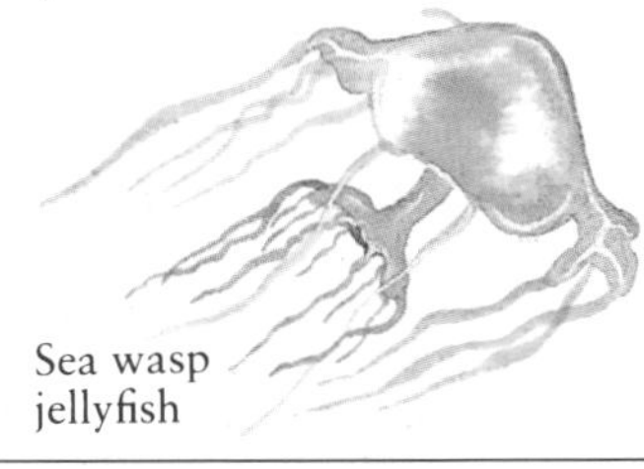
Sea wasp jellyfish

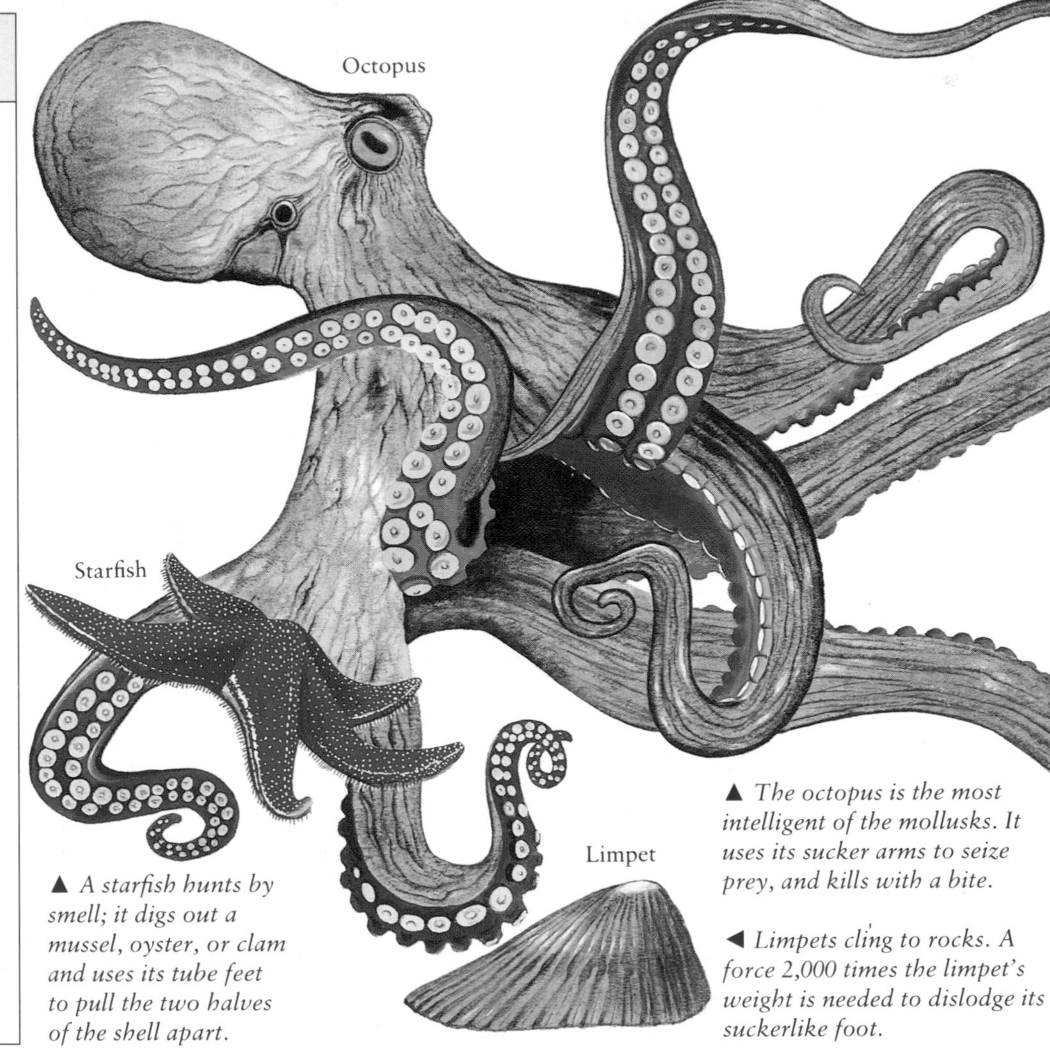

▲ *A starfish hunts by smell; it digs out a mussel, oyster, or clam and uses its tube feet to pull the two halves of the shell apart.*

▲ *The octopus is the most intelligent of the mollusks. It uses its sucker arms to seize prey, and kills with a bite.*

◄ *Limpets cling to rocks. A force 2,000 times the limpet's weight is needed to dislodge its suckerlike foot.*

► *Coral reefs are only found in tropical seas, as the corals that form the reefs cannot live in cold water. Coral is made of limestone formed by millions of tiny marine animals.*

◄ *Small fish feed and shelter among the colorful branches and fronds of coral.*

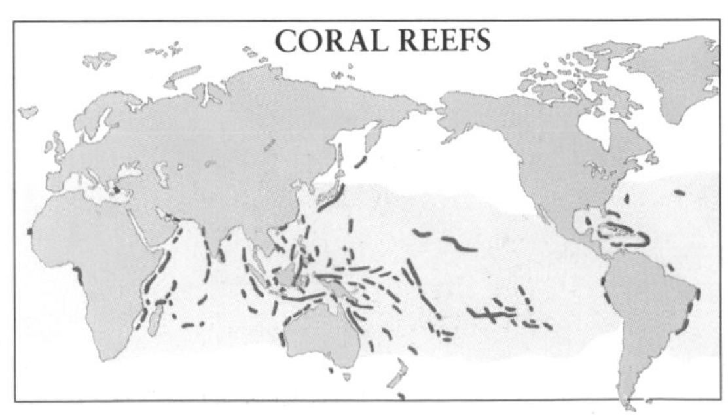

Coral reefs

70°F (coral cannot survive below this temperature)

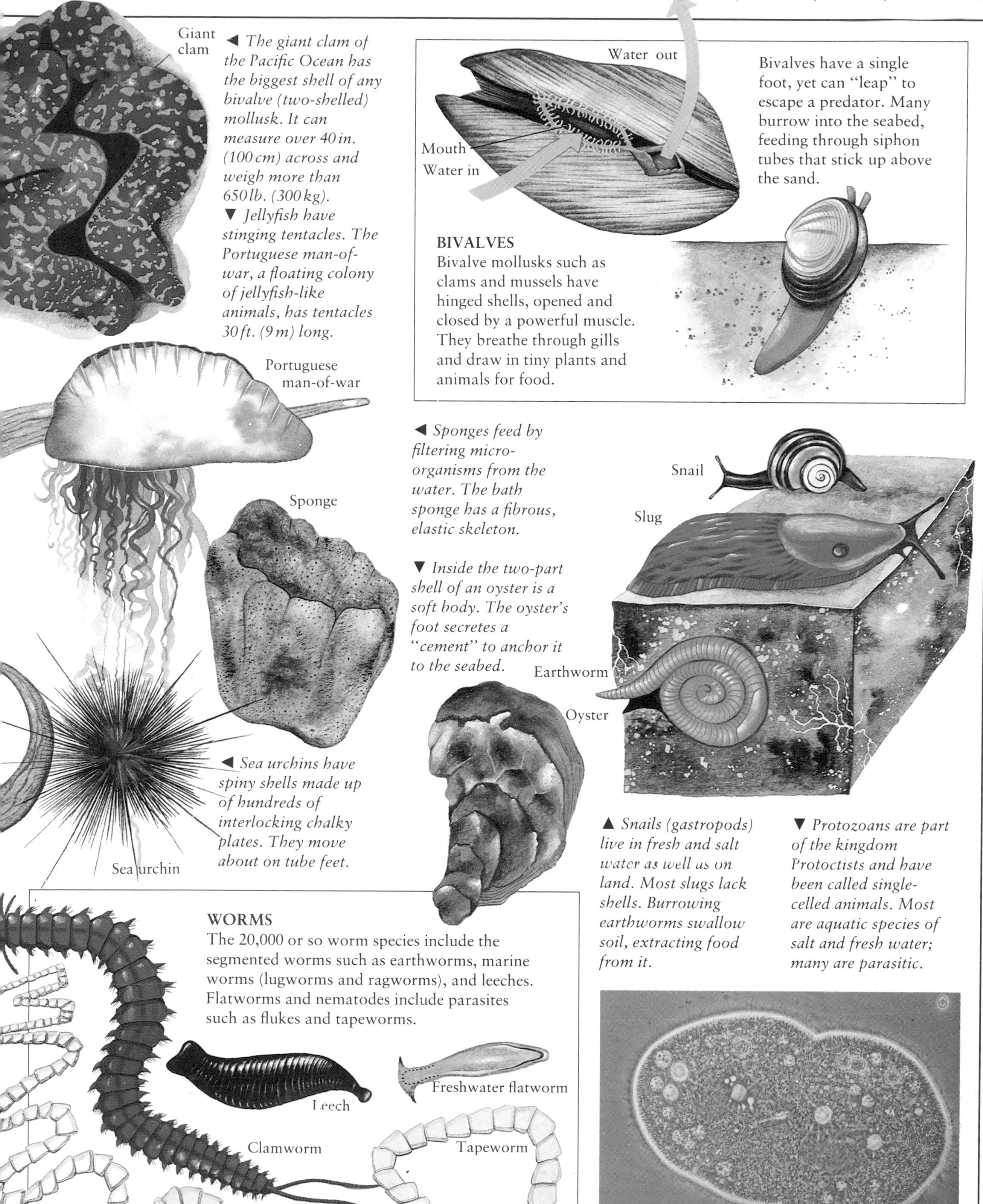

◄ *The giant clam of the Pacific Ocean has the biggest shell of any bivalve (two-shelled) mollusk. It can measure over 40 in. (100 cm) across and weigh more than 650 lb. (300 kg).*

▼ *Jellyfish have stinging tentacles. The Portuguese man-of-war, a floating colony of jellyfish-like animals, has tentacles 30 ft. (9 m) long.*

BIVALVES
Bivalve mollusks such as clams and mussels have hinged shells, opened and closed by a powerful muscle. They breathe through gills and draw in tiny plants and animals for food.

Bivalves have a single foot, yet can "leap" to escape a predator. Many burrow into the seabed, feeding through siphon tubes that stick up above the sand.

◄ *Sponges feed by filtering micro-organisms from the water. The bath sponge has a fibrous, elastic skeleton.*

▼ *Inside the two-part shell of an oyster is a soft body. The oyster's foot secretes a "cement" to anchor it to the seabed.*

◄ *Sea urchins have spiny shells made up of hundreds of interlocking chalky plates. They move about on tube feet.*

▲ *Snails (gastropods) live in fresh and salt water as well as on land. Most slugs lack shells. Burrowing earthworms swallow soil, extracting food from it.*

▼ *Protozoans are part of the kingdom Protoctists and have been called single-celled animals. Most are aquatic species of salt and fresh water; many are parasitic.*

WORMS
The 20,000 or so worm species include the segmented worms such as earthworms, marine worms (lugworms and ragworms), and leeches. Flatworms and nematodes include parasites such as flukes and tapeworms.

Millipedes, Crabs, and Spiders

Like insects, these animals are arthropods, members of the largest animal phylum. All arthropods have jointed limbs and most have a plated body-covering, which the animal molts, or sheds, as it grows. Centipedes and millipedes are wormlike, with a pair of limbs on almost every segment of their bodies. Crustaceans (for example, crabs, lobsters, and woodlice) number more than 30,000 species. The biggest group, with over 50,000 species, is the arachnids: spiders, scorpions, and mites.

See pages 108–109

FACTS ABOUT ARTHROPODS

- The first true arthropods lived in the sea. They were the trilobites, now extinct.
- Millipedes use chemical defenses. Stink glands in their bodies secrete a venom capable of killing or repelling insects.
- Most crabs live in the oceans and seas, but robber crabs are so adapted to life on land that they drown if kept under water.

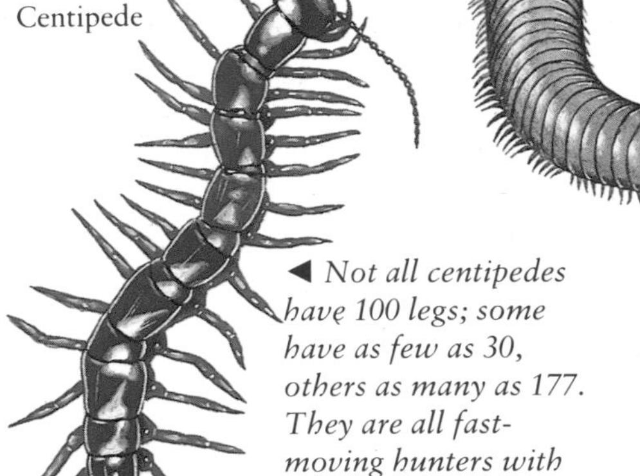

◀ *Millipedes have even more legs than centipedes: as many as 375 pairs. They live in soil and leaf litter, feeding on decaying vegetation.*

◀ *Not all centipedes have 100 legs; some have as few as 30, others as many as 177. They are all fast-moving hunters with poisonous claws.*

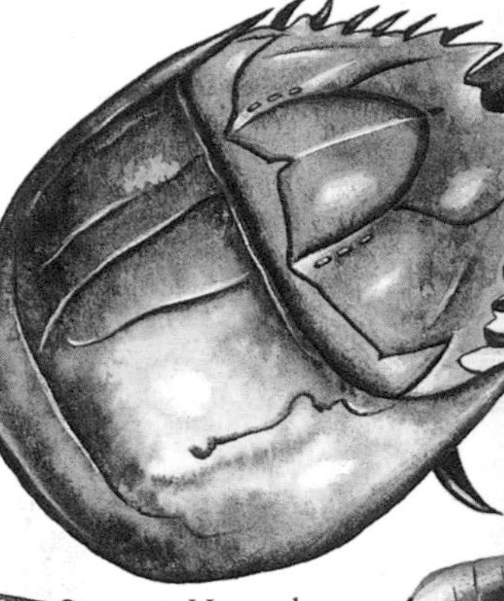

▼ *The horseshoe, or king, crab is closer to spiders than to true crabs. Up to 2 ft. (60 cm) long, it has a horny shield.*

▼ *Lobsters are the heaviest crustaceans, weighing up to 44 pounds (20 kg). Spiny lobsters migrate in long columns.*

Lobster

LIFE CYCLE OF A SHRIMP

Marine crustaceans such as shrimps lay eggs, often carried by the female until they hatch. The tiny larvae look very different from the adults. The larvae drift through the water, gradually changing body shape until fully developed.

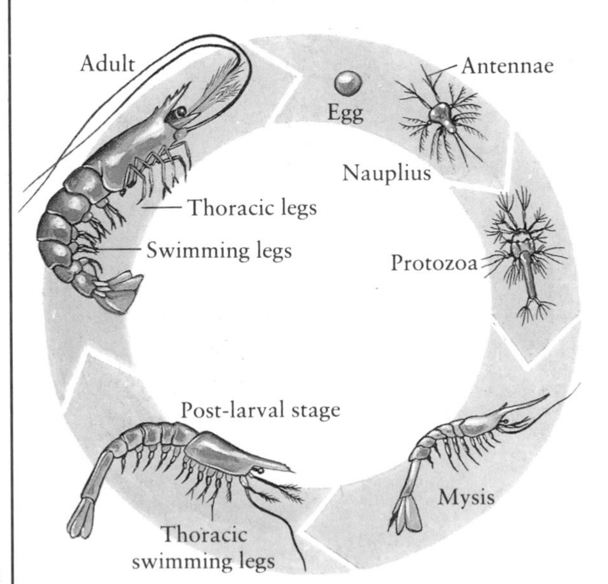

CARING PARENTS

Few arthropods are caring parents. The arthropods that show most concern for their young are those with the fiercest reputations as killers: spiders and scorpions. A female scorpion gives birth to live young. She carries the babies around on her back until after their first molt.

Female scorpion

DEFENSE

The wood louse, or pill bug, is one of the few crustaceans that live on land. A small, flat creature with seven pairs of legs, it has the ability, when alarmed, to roll itself up tightly so that its segmented body becomes an armored ball.

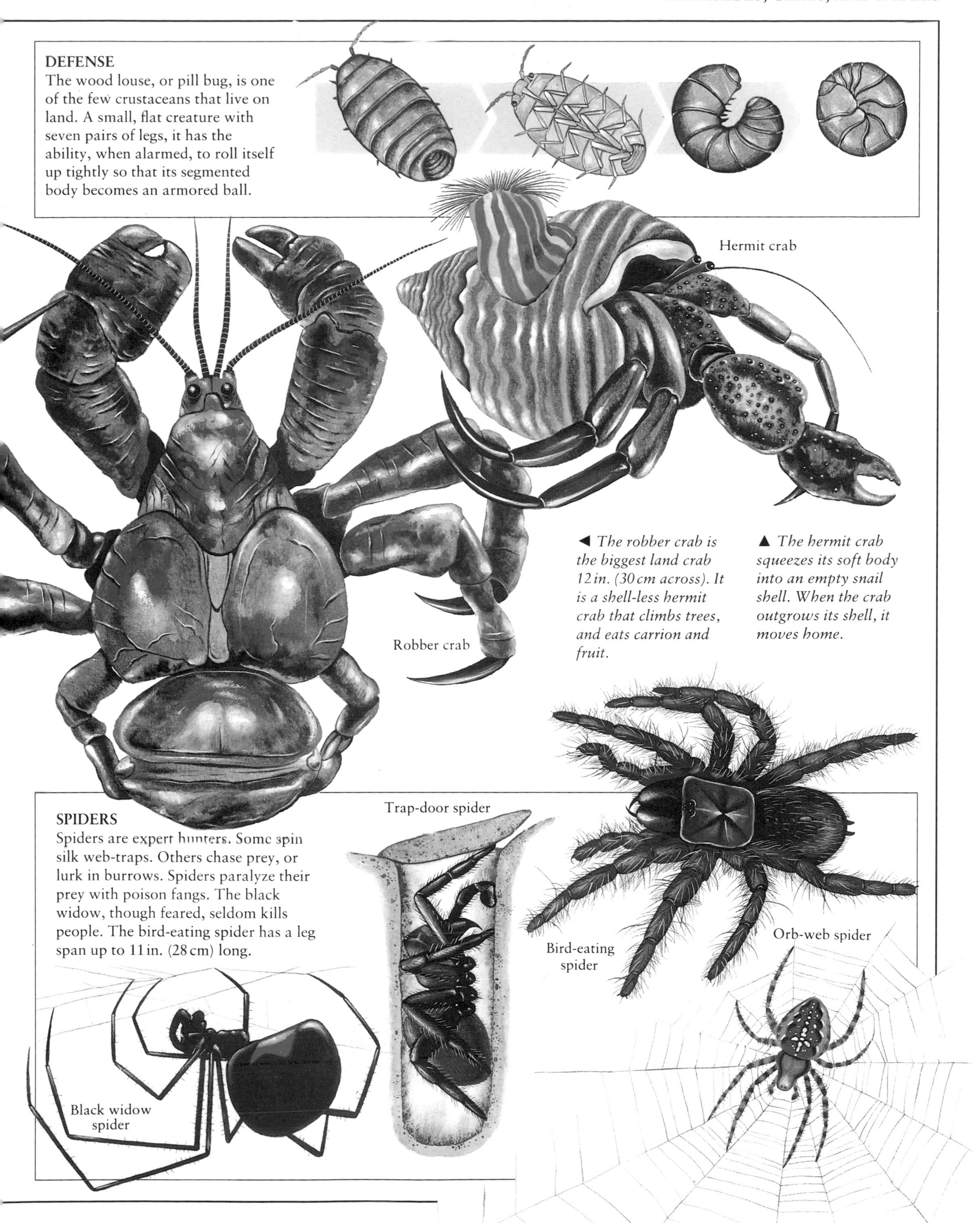

◀ *The robber crab is the biggest land crab 12 in. (30 cm across). It is a shell-less hermit crab that climbs trees, and eats carrion and fruit.*

▲ *The hermit crab squeezes its soft body into an empty snail shell. When the crab outgrows its shell, it moves home.*

SPIDERS

Spiders are expert hunters. Some spin silk web-traps. Others chase prey, or lurk in burrows. Spiders paralyze their prey with poison fangs. The black widow, though feared, seldom kills people. The bird-eating spider has a leg span up to 11 in. (28 cm) long.

Insects

There are about one million known species of insects, and millions more are probably still to be identified; about 80 percent of all known animals are insects. Some scientists believe there could be as many as 10 million insect species. The secret of the insects' success is their adaptability. Although they are limited in size by their body design, they have conquered all environments, from the hottest to the coldest places. Evolution has also equipped them to eat an astonishing variety of foods.

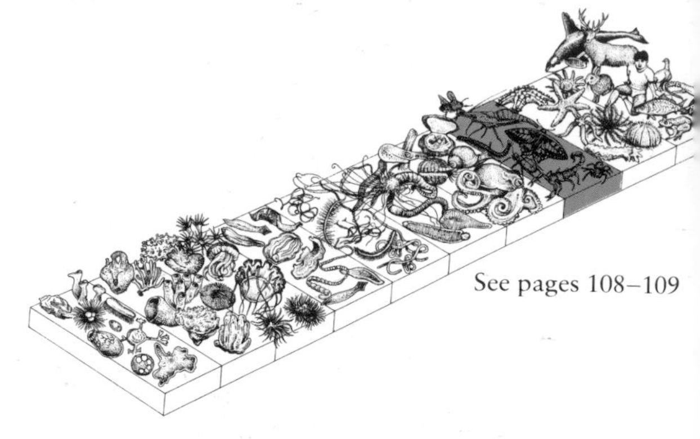
See pages 108–109

FACTS ABOUT INSECTS

- Insects breathe through tiny holes called "spiracles" along the sides of their bodies. Each hole allows air to pass into a system of tubes branching all around the insect's body. These tubes carry oxygen to the cells and carry away carbon dioxide.
- Many insects can lift or drag an object 20 times their own weight. A caterpillar has from 2,000 to 4,000 muscles—six times as many as a human.
- Botflies and horseflies can fly at 24 mph (39 km/h). A tiny midge holds the record for fast wing beats: more than 62,000 times a minute.

THE BODIES OF INSECTS

All insects, like this honeybee, have three pairs of legs. An insect's body has three parts: a head, a middle, or thorax, where the legs are attached, and an abdomen. Most adult insects have wings and a pair of antennae. But while the majority of insects (like the bee) have four wings, flies only have two.

▼ *Hornets are large social wasps. They build nests of paper made from chewed-up plant matter. Hornets sting if disturbed, and hunt flies and caterpillars.*

► *Dragonflies are the fastest insects, with a top speed of over 34 mph (55 km/h).*

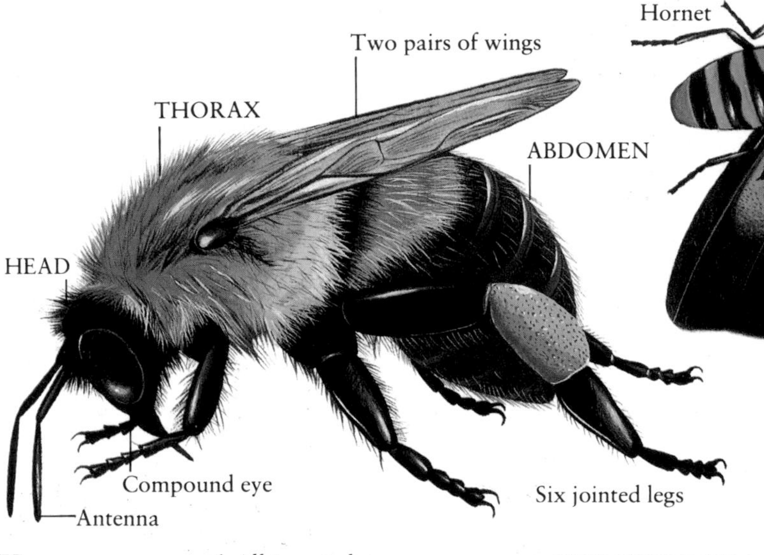

▲ *The largest insect is the Queen Alexandra birdwing butterfly of Papua New Guinea – 11 in. (28 cm) from wingtip to wingtip.*

LIFE CYCLE OF A MOTH

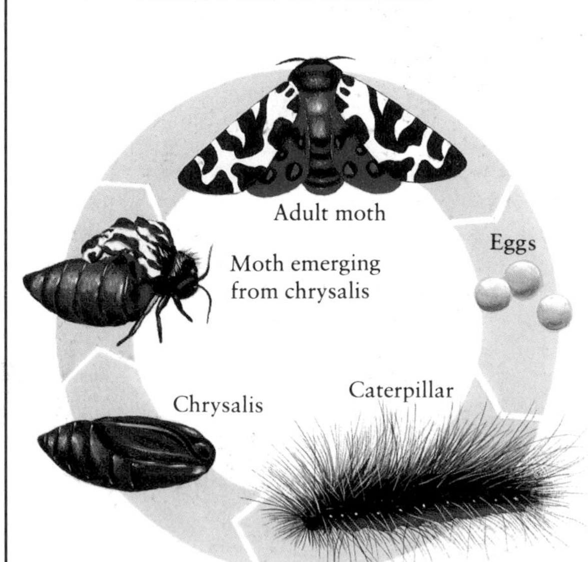

◄ *All insects lay eggs. The young of most insects go through four stages of growth and development. A moth develops from egg to caterpillar (larva) to chrysalis before emerging as an adult.*

► *Grasshoppers and insects such as cockroaches, earwigs, and aphids, go through three stages of growth. After hatching from the egg, the young look like miniature adults, though at first they lack wings.*

LIFE CYCLE OF A GRASSHOPPER

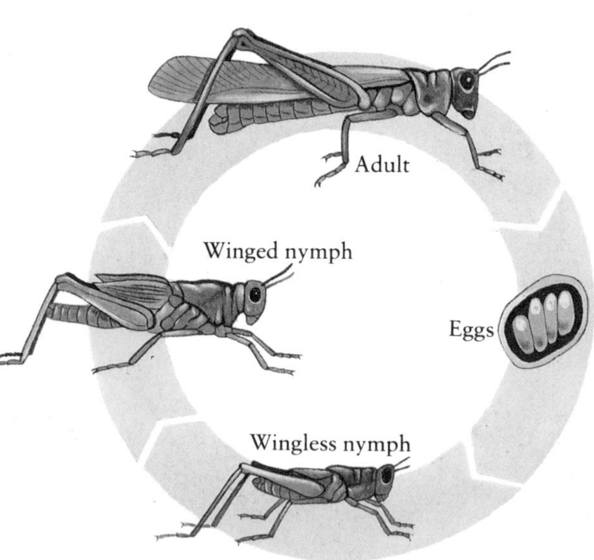

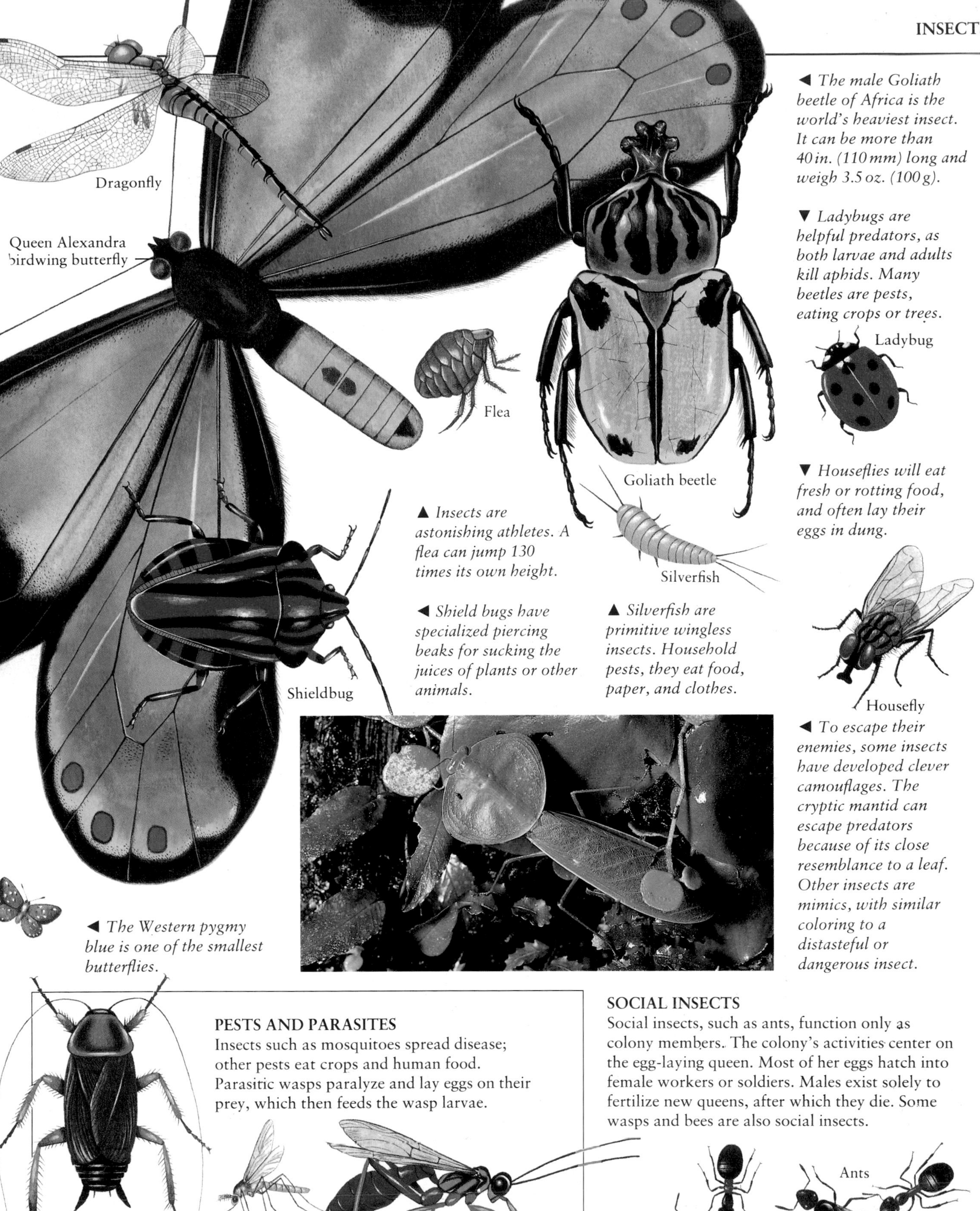

◀ *The male Goliath beetle of Africa is the world's heaviest insect. It can be more than 40 in. (110 mm) long and weigh 3.5 oz. (100 g).*

▼ *Ladybugs are helpful predators, as both larvae and adults kill aphids. Many beetles are pests, eating crops or trees.*

▼ *Houseflies will eat fresh or rotting food, and often lay their eggs in dung.*

▲ *Insects are astonishing athletes. A flea can jump 130 times its own height.*

◀ *Shield bugs have specialized piercing beaks for sucking the juices of plants or other animals.*

▲ *Silverfish are primitive wingless insects. Household pests, they eat food, paper, and clothes.*

◀ *To escape their enemies, some insects have developed clever camouflages. The cryptic mantid can escape predators because of its close resemblance to a leaf. Other insects are mimics, with similar coloring to a distasteful or dangerous insect.*

◀ *The Western pygmy blue is one of the smallest butterflies.*

PESTS AND PARASITES

Insects such as mosquitoes spread disease; other pests eat crops and human food. Parasitic wasps paralyze and lay eggs on their prey, which then feeds the wasp larvae.

SOCIAL INSECTS

Social insects, such as ants, function only as colony members. The colony's activities center on the egg-laying queen. Most of her eggs hatch into female workers or soldiers. Males exist solely to fertilize new queens, after which they die. Some wasps and bees are also social insects.

Fish

The first fish appeared in the oceans some 540 million years ago. By breathing through gills, fish have always been fully adapted to life in water. About 60 percent of all fish species live in salt water; a few kinds can live in either salt or fresh water. There are three main groups of fish: Agnatha or jawless fish (lampreys and hagfish), Chondrichthyes or fish with a skeleton of cartilage (sharks, chimaeras, and rays), and Osteichthyes, a group that includes some 20,000 species of fish with a bony skeleton.

See pages 108–109

▼ *Stingrays have whiplike tails armed with poisonous spines. They are ocean bottom-dwellers with flattened bodies.*

FISH RECORDS

- The biggest freshwater fish is the giant catfish, at 16 ft. (5 m) long (a record set in the 1800s).
- The longest bony fish is the oarfish, up to 50 ft. (15 m) long.
- The smallest fish is the dwarf goby of the Pacific Ocean. Few grow more than 0.3 in. (9 mm) long.
- Largest of all fish is the whale shark: 60 ft. (18 m) long and weighing 15 tons.

Whale shark

- Coelacanths live deep in the Indian Ocean. Scientists believed they had died out 70 million years ago—until one was caught in 1938.
- Fish known to attack people include some sharks (whites, blues, hammerheads), barracudas, and moray eels. Venomous fish include stonefish, with poisonous spines.

Coelacanth

HOW FISH BREATHE

Fish breathe oxygen from water by means of gills; water contains amounts of dissolved oxygen. The fish gulps in water through its mouth. The water passes across the gills and out again. Blood flows through the gills in tiny filters that take oxygen in, and release waste carbon dioxide.

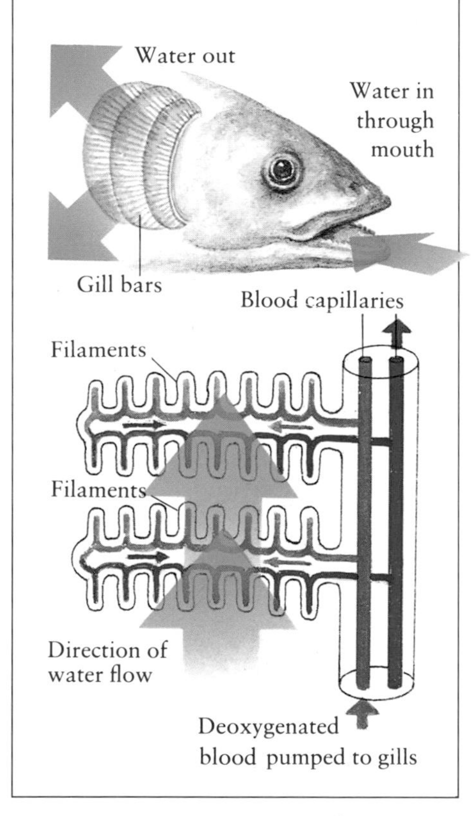

► *A mako shark, is about 11 ft. (3.5 m) long. Makos prey on fish but sometimes attack human bathers and small boats.*

Mako shark

▼ *The alligator gar is a large 10 ft. (3 m) freshwater hunter of North America. It has an alligatorlike snout.*

▼ *The saltwater herring is an important North Atlantic food fish. Members of the herring family include shad and sardine.*

► *Flying fish glide rather than fly. To escape predators they take off, using their long pectoral fins as wings.*

Flying fish

JAWLESS FISH

Lampreys and hagfish lack true jaws, but have sucking mouths with horny teeth. They clamp onto a victim and tear its flesh with file-toothed tongues.

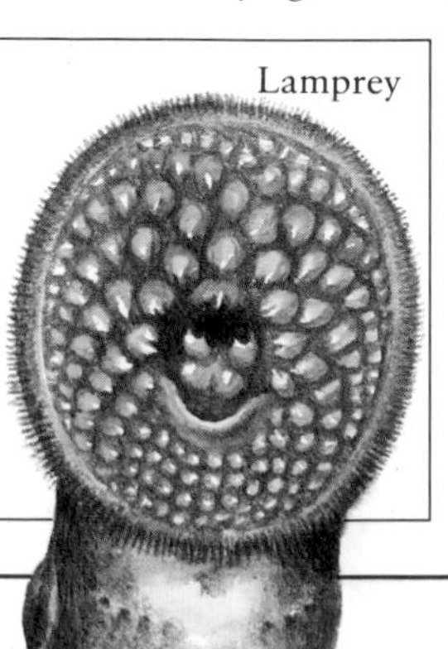
Lamprey

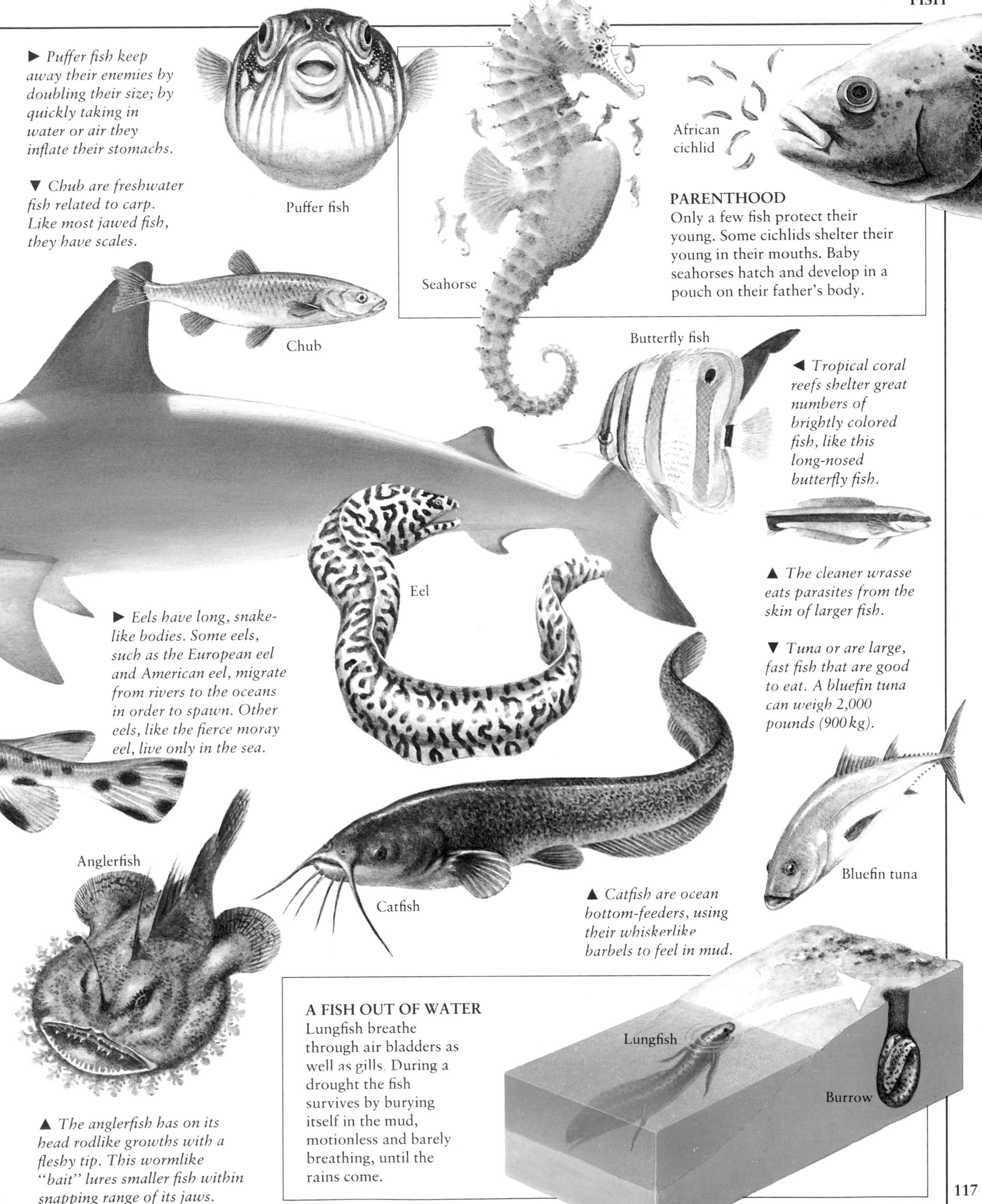

► *Puffer fish keep away their enemies by doubling their size; by quickly taking in water or air they inflate their stomachs.*

▼ *Chub are freshwater fish related to carp. Like most jawed fish, they have scales.*

PARENTHOOD
Only a few fish protect their young. Some cichlids shelter their young in their mouths. Baby seahorses hatch and develop in a pouch on their father's body.

◄ *Tropical coral reefs shelter great numbers of brightly colored fish, like this long-nosed butterfly fish.*

▲ *The cleaner wrasse eats parasites from the skin of larger fish.*

▼ *Tuna or are large, fast fish that are good to eat. A bluefin tuna can weigh 2,000 pounds (900 kg).*

► *Eels have long, snake-like bodies. Some eels, such as the European eel and American eel, migrate from rivers to the oceans in order to spawn. Other eels, like the fierce moray eel, live only in the sea.*

▲ *Catfish are ocean bottom-feeders, using their whiskerlike barbels to feel in mud.*

▲ *The anglerfish has on its head rodlike growths with a fleshy tip. This wormlike "bait" lures smaller fish within snapping range of its jaws.*

A FISH OUT OF WATER
Lungfish breathe through air bladders as well as gills. During a drought the fish survives by burying itself in the mud, motionless and barely breathing, until the rains come.

Amphibians

Amphibians are a relatively small group of coldblooded vertebrate animals: about 3,000 species. Many are water creatures. Others live on land, in trees, and even in deserts. Most amphibians need water (a river, pond, or even a droplet on a leaf) to lay their eggs. There are three orders: wormlike caecilians (Apoda); newts and salamanders (Urodela), with long tails and usually four legs; and toads and frogs (Anura), tailless and four-legged. Amphibians are most common in warm climates.

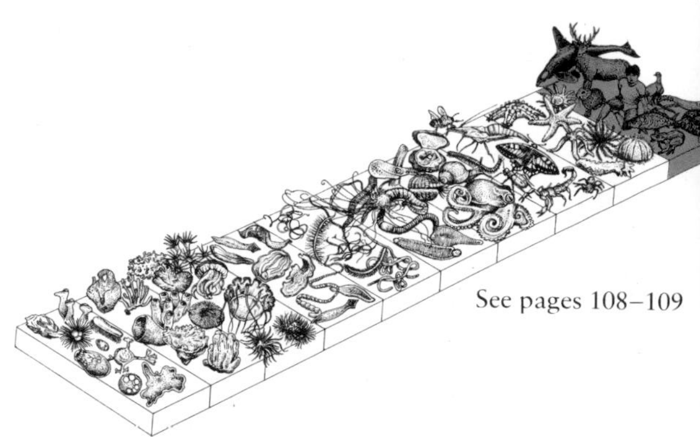
See pages 108–109

LIFE CYCLE OF A FROG

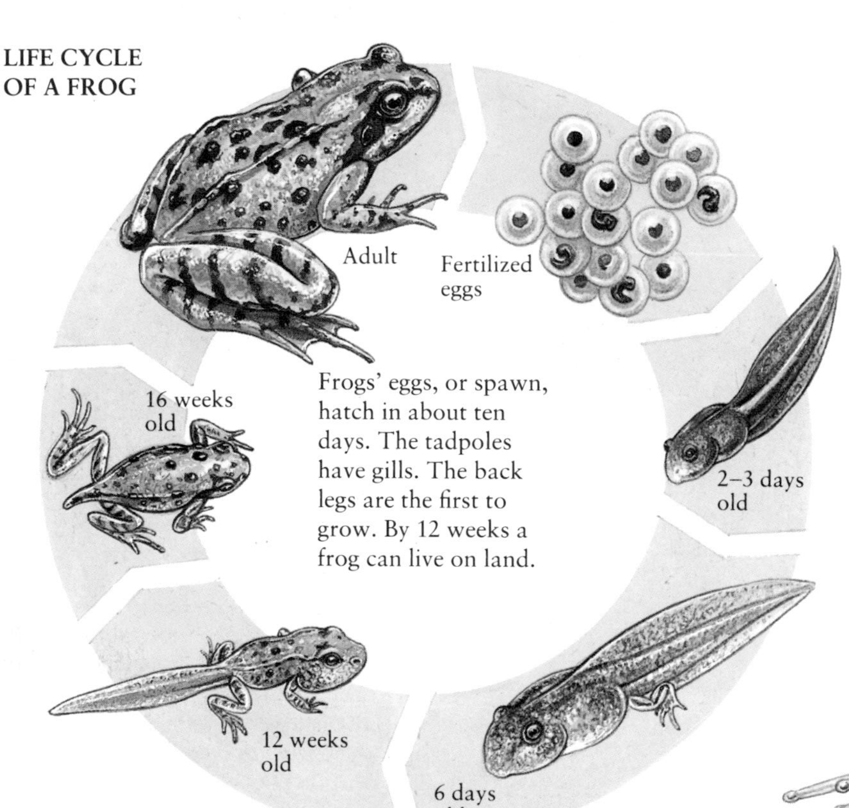

Frogs' eggs, or spawn, hatch in about ten days. The tadpoles have gills. The back legs are the first to grow. By 12 weeks a frog can live on land.

HOW DO FROGS AND TOADS DIFFER?

Frogs spend more time in water than toads, which can live in drier places.

Frog skin is smooth, toad skin is lumpy.

Frogs use their long legs for jumping. Toads crawl.

Frogs have moist skin, toads have dry skin.

► *Most amphibians simply deposit their spawn and leave the young to fend for themselves. But some make nests in leaves or burrows, or even out of a special foam. A few carry their offspring around with them, or even hide them in their mouths. Baby Surinam toads emerge from eggs encased in pockets of their mother's skin. The male midwife toad attaches its eggs to its hind legs and carries them for about three weeks until they hatch.*

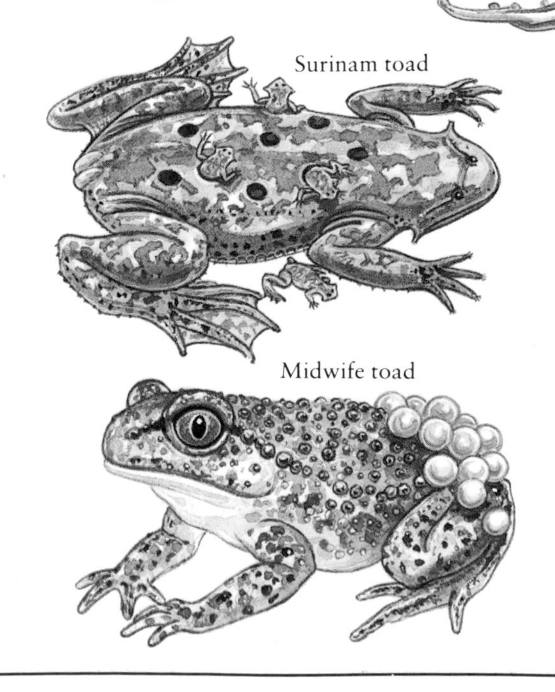

▲ *Frogs and toads are the largest order of amphibians, with some 2,700 species. They are very diverse in their habitats. Some, like the spadefoots, burrow in the ground; others spend almost all their time in water. Some have disklike pads on their toes for climbing. There is even a species (Rattray's frog) whose tadpoles drown in water.*

WHY DO FROGS AND TOADS CROAK?

Male frogs croak to call females during the mating season by forcing air over their vocal chords. In some species, females also call, but less loudly. The loudest croakers are species with an expanding vocal sac.

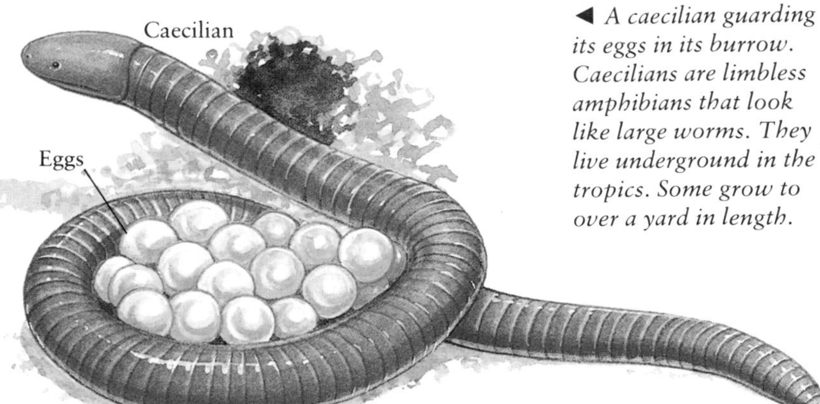

◀ *A caecilian guarding its eggs in its burrow. Caecilians are limbless amphibians that look like large worms. They live underground in the tropics. Some grow to over a yard in length.*

◀ *A toad capturing a meal with its long, sticky-tipped tongue. Most frogs and toads eat insects and other small animals. They usually hunt by staying still and aiming precisely at their target. Some frogs have teeth.*

▲ *The Mexican axolotl is an amphibian that never grows up. This newtlike creature spends its life in water, breathing by means of gills. Only in unusual conditions does it develop lungs. The axolotl usually breeds in its gilled state.*

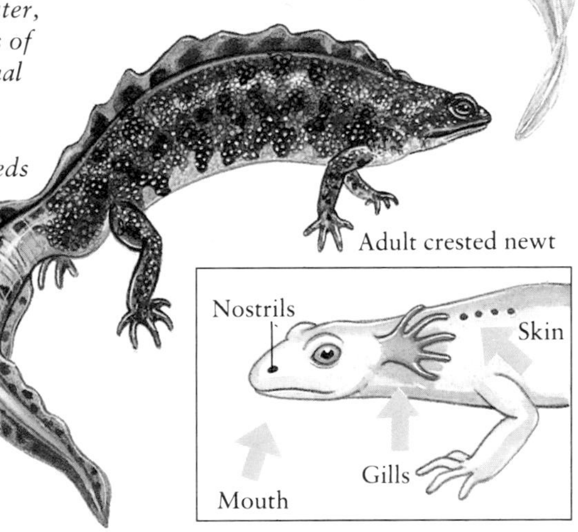

Common frog

FACTS ABOUT AMPHIBIANS

- The largest amphibian is the giant salamander of China, 3 ft. (1 m) long and weighing up to 66 pounds (30 kg).
- The smallest amphibian is the Cuban frog, *Sminthillus limbatus*, less than half an inch (12 mm) long.
- One Colombian golden poison-dart frog has enough toxin in its body to kill 1,000 people.
- Mudpuppies, waterdogs, and olms never leave the water.
- The biggest frog is the African goliath frog: it is 3 ft. (88 cm) long and weighs nearly 9 pounds (4 kg).
- There is no truth in the legend that a salamander can pass through fire unhurt.

COURTSHIP

Not all amphibians mate in water. Courtship takes place in spring, and is often a frenzied affair. Males and females make their way to suitable breeding locations and large numbers may congregate to mate. The male long-tailed salamander indulges in an energetic display; dancing and grappling with the female to persuade her to accept a small packet of his sperm. The eggs are fertilized inside the female's body.

HOW AMPHIBIANS BREATHE

Amphibians breathe by taking air into their lungs, but they also breathe in air through the skin, mouth, and throat. A young newt or frog starts its life as a tadpole with gills, but develops lungs as it matures and is able to leave the water. Newts, also called efts (see above), are very like salamanders, but are more aquatic.

Reptiles

There are more than 6,500 species of reptiles: some 250 kinds of tortoises and turtles (order Testudinidae); 25 species of crocodiles and alligators (order Crocodilia); about 2,800 species of snakes (order Squamata); 3,700 species of lizards (order Squamata); and the unique tuatara (order Rhynchocephalia). Like amphibians, reptiles are cold blooded. Most live in the tropics, though a few snakes and lizards live in cooler climates. Reptiles have scaly skins and most lay leathery-shelled eggs.

See pages 108–109

SNAKES

► *Poisonous snakes bite with grooved fangs which inject venom from saclike glands in the head. Many venomous snakes are brightly colored as a warning.*

Venom sac

Fang

◄ *Male rattlesnakes wrestle for mates, but do not use their poison fangs. The rattle is formed by horny plates at the tip of the tail. Rattlesnakes detect prey with heat-sensing organs.*

▲ *Vipers, which include copperheads and rattlesnakes, have long fangs that unfold from the roof of the mouth as the snake strikes. Cobras and sea snakes have short, fixed fangs. Poisonous snakes can bite as soon as they have hatched.*

◄ *Snakes can swallow objects larger than their heads. The egg-eating snake swallows the egg, then crushes the shell.*

LIZARDS

► *The little gecko has suckerlike pads on its feet. It can run across the ceiling of a room with ease when hunting insects.*

▲ *The Gila monster is a poisonous lizard from the southwestern United States. Like all coldblooded animals, this desert-dweller is most active when warmed by the sun; its body is as warm or as cool as its surroundings.*

◄ *The chameleon's color changes are the result of hormone activity triggered by a change of light or temperature, or by fear or anger. Chameleons catch insects by shooting out their long, sticky tongues.*

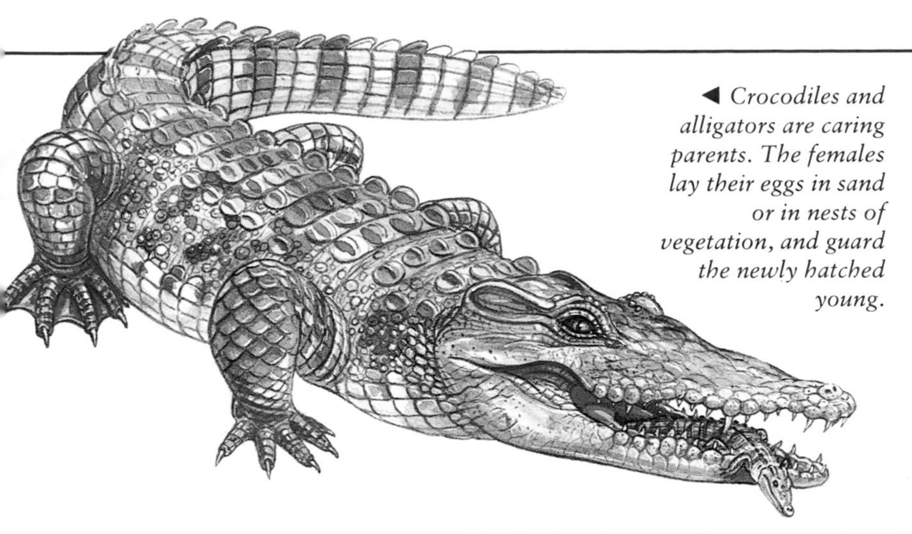

◀ *Crocodiles and alligators are caring parents. The females lay their eggs in sand or in nests of vegetation, and guard the newly hatched young.*

FACTS ABOUT REPTILES

Aldabra giant tortoise

- Largest lizard: Komodo dragon (a monitor lizard) of Indonesia, 10 ft. (3 m) long.
- Largest snakes: anaconda (South America) and reticulated python (Asia): up to 30 ft. (9 m).
- Largest crocodile: estuarine crocodile of Southeast Asia: up to 23 ft. (7 m) long.
- Heaviest turtle: leatherback, about 6.5 ft. (2 m) long and wide, and weighing over 1,000 pounds (450 kg).
- There are two species of giant tortoise, one living on the Seychelles and Aldabra islands, the other on the Galápagos Islands.
- The slowworm looks like a snake, but is actually a lizard that has lost its limbs.
- There are other legless lizards, such as the glass snake and amphisbaena.

CROCODILES AND THEIR RELATIVES

Crocodiles, alligators, caimans, and gavials are large carnivores with strong jaws and powerful tails. They either bask on river banks or lie almost submerged in the water with only their eyes, nostrils, and ears showing. The crocodile's fourth tooth sticks out when its jaws are shut: in the broader-headed alligator this tooth is hidden. Gavials, or gharials, have long, thin snouts.

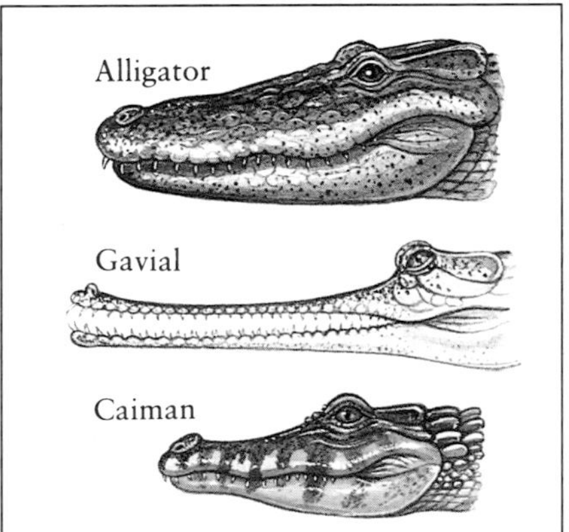

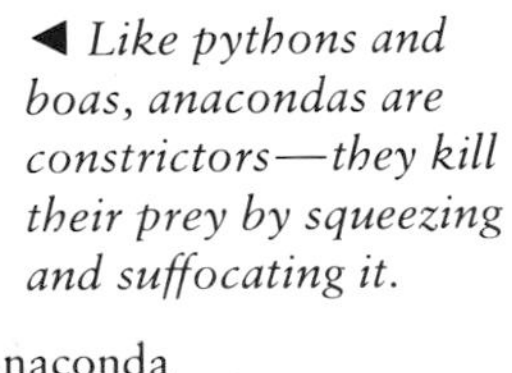

◀ *Like pythons and boas, anacondas are constrictors—they kill their prey by squeezing and suffocating it.*

Anaconda

TURTLES AND TORTOISES

Turtles and tortoises are reptiles with shells. Only the head, legs, and tail stick out. The name tortoise is often used for land-living Testudinidae. Marine turtles are excellent swimmers. So are freshwater terrapins. Turtles and tortoises prefer warm climates. Some are plant-eaters, while others are carnivorous.

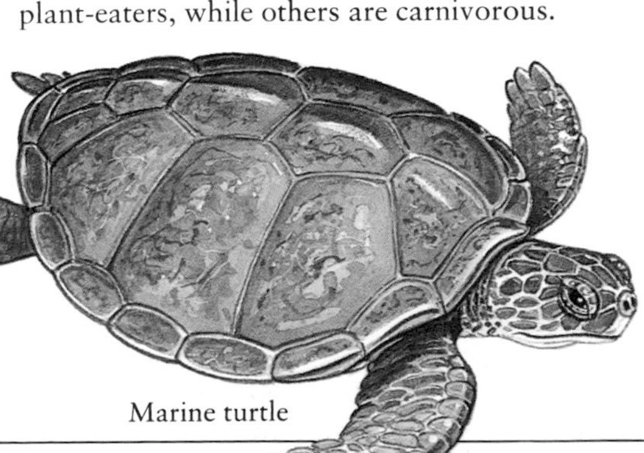

Marine turtle

▶ *Sea turtles come ashore to lay their eggs in sand. The hatchlings scramble out and race for the safety of the water. Only a few escape the waiting predators.*

Birds

Some prehistoric reptiles jumped or glided from tree to tree and, over millions of years, reptile scales became feathers. Birds are the only animals with feathers—they keep a bird warm, even in sub-zero temperatures, and they make flight possible. There are 28 orders and about 8,600 species of birds. These include water birds (wildfowl, waders, and seabirds), birds of prey (eagles, hawks, and owls), and "passerine" or perching birds, most of which live among trees or other high places when not flying.

See pages 108–109

THE 28 ORDERS OF THE CLASS AVES (BIRDS)

1 **Struthioniformes**: large, flightless birds, one species survives; ostrich.

2 **Rheiformes**: two species of flightless birds of South America; rhea.

3 **Casuariiformes**: large flightless birds of Australia, New Guinea; cassowary, emu.

4 **Apterygiformes**: nocturnal, flightless birds of New Zealand; kiwi.

5 **Tinamiformes**: weak-flying ground birds of S. and C. America; tinamou.

6 **Sphenisciformes**: flightless swimming birds with erect posture; penguin.

7 **Gaviiformes**: web-footed diving birds of Asia, America, Europe; diver, loon.

8 **Podicipediformes**: diving birds with long toes with flaplike lobes; grebe.

9 **Procellariiformes**: ocean birds with tubelike noses; petrel, shearwater, albatross.

10 **Pelecaniformes**: fully webbed feet, beak, pouch; pelican, cormorant, gannet.

11 **Ciconiiformes**: long-legged waders; heron, egret, stork, spoonbill, ibis.

12 **Anseriformes**: water birds; duck, goose, swan, screamer.

17 Pintailed sandgrouse

13 **Falconiformes**: birds of prey; eagle, buzzard, hawk, falcon, vulture.

14 **Galliformes**: fowl-like birds; quail, chicken, peacock, pheasant, turkey.

15 **Gruiformes**: marsh and land birds; rail, crane, coot, moorhen, bustard, trumpeter.

16 **Charadriiformes**: waders and water-birds; gull, sandpiper, plover, curlew, tern, oystercatcher, auk.

17 **Pteroclidiformes**: medium-sized birds with long, pointed wings; sandgrouse.

18 **Columbiformes**: medium-sized short-legged birds; pigeons, doves.

19 **Psittaciformes**: seed- and fruit-eaters with grasping claws; parrots, lories, cockatoos.

20 **Cuculiformes**: tree and ground-dwelling; cuckoo, touraco, roadrunner.

21 **Strigiformes**: mostly nighttime silent birds of prey with large heads; owl.

22 **Caprimulgiformes**: nighttime insect-eaters; nightjar, oilbird, frogmouth.

23 **Apodiformes**: weak-footed, strong wings, spend most of their lives flying; swift, hummingbird.

24 **Trogoniformes**: long-tailed forest birds with small, weak feet; trogon.

25 **Coliiformes**: small, long-tailed African fruit-eaters with four toes; coly, mousebird.

26 **Coraciiformes**: long bills, short legs; kingfisher, bee-eater, roller, hoopoe.

27 **Piciformes**: woodland birds that nest in holes; toucan, woodpecker, barbet.

28 **Passeriformes**: 60 families, over 5,000 species; broadbills, all songbirds (crow, lark, finch, thrush, etc.).

Bird Behavior

To call someone "bird-brained" should be a compliment, for bird behavior is amazingly complex, a mixture of learned skills—such as a pigeon getting food from a bird feeder—and instinct, as in the territorial aggression of a European robin. Flight enables birds to be extraordinary travelers and some species migrate across oceans and continents. Finding food in all kinds of habitats, mating, and nest-building, birds around the world demonstrate a remarkable range of adaptations and techniques.

BIRDS' FEET
Most birds have four, clawed toes, adapted to suit different ways of life. Perching birds (for example, the corn bunting) have three forward-pointing toes and one backward-pointing one. Ducks have webbed feet, for swimming. The osprey's talons seize and crush prey.

► *Depending on its species, a bird has between 940 and 25,000 feathers. In most species, the male has more colorful feathers than the female.*

FEATHERS
Feathers are replaced (molted) once or twice a year. A flight feather (*right*) has a central rod or quill. In close-up, the threadlike barbs can be seen; these are held together by smaller hooked fibers called barbules.

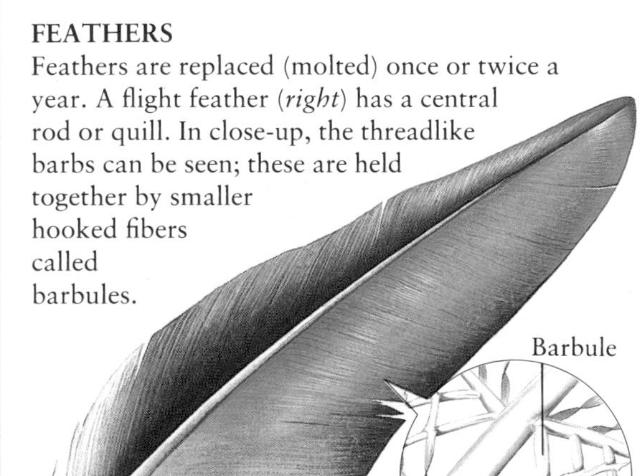

WING SHAPES
Long, slender wings like those of the albatross are best for effortless, gliding flight. Fast fliers, such as hawks, have narrow, pointed wings. Game birds like the partridge have stubby wings; good for quick takeoffs and brief dashes.

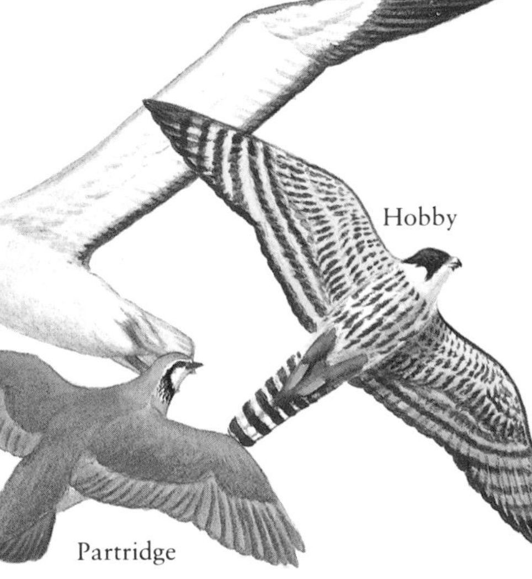

BIRDS' BILLS
Many water birds use their bills as probes or sieves. Woodpeckers' bills are wood drills. Seed- and nut-eaters have bills for cracking hard outer shells. Birds of prey have hooked bills for tearing flesh.

NESTS

Some birds, such as the plover, lay eggs in a scrape on the ground among sand or stones. Water birds, such as grebes, nest on or beside the water. Swallows are mud-builders, often nesting against walls of buildings. Many songbirds, such as the thrush, nest in trees or bushes, building a nest from twigs, leaves, and grass in which to lay their eggs.

Swallow's nest

Robin's nest

FACTS ABOUT BIRDS

- The ostrich is the biggest bird, 9 ft. (2.7 m) tall, weighing 340 pounds (156 kg), and lays the biggest egg, weighing about 4 pounds (1.7 kg).
- The smallest bird is the bee hummingbird of Cuba, less than 2 inches (5 cm) long and weighing only 0.05 oz. (1.6 grams).
- The wandering albatross has the biggest wingspan: over 10 ft. (3 m).
- Bird song is a signal, usually telling other birds to stay off the singer's territory. Parents and chicks can recognize each other's voices.
- The peregrine falcon is credited with a top diving speed of more than 180 mph (300 km/h).

▼ *Eider ducks are valued for the soft aown on their breasts, used in bedding; they are also one of the world's fastest flying birds.*

COURTSHIP

Many birds have elaborate courtship behavior, in which males dance or display colorful plumage to attract females. The Australian lyrebird has long tail feathers which it displays during its courtship.

MIGRATION

Many birds make amazingly long migrations. Different species may be seen in flight along favored routes, often in large flocks. The Arctic tern makes the longest migratory journey of any animal. It flies up to 22,000 miles (36,000 km) in a year, journeying south from its Arctic breeding grounds to the Antarctic summer, and back again.

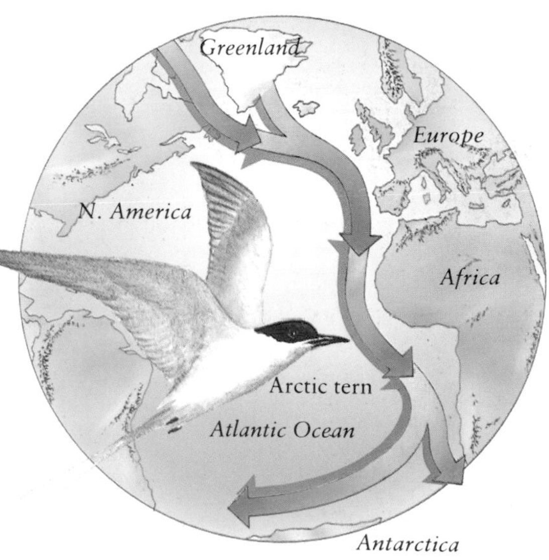

Mammals

Mammals are one of the eight classes of vertebrate (backboned) animals. There are far fewer species of mammals than other groups of animals—but mammals have adapted to a wider range of habitats. Mammals live on land, in hot or cold climates, in the sea, and have even taken to the air. Mammals feed their young on the mother's milk, they protect their young, they have hair, they maintain a constant body temperature, and they have relatively larger brains than other animals.

THE 18 ORDERS OF MAMMALIA (MAMMALS)

1 **Monotremata**: primitive mammals that lay eggs; echidna, platypus.

2 **Marsupialia**: young born tiny, develop in mother's pouch; kangaroo, koala.

3 **Insectivora**: small insect-eaters with snouts; shrew, hedgehog, mole.

4 **Dermoptera**: tree-living Asian mammals capable of gliding; flying lemur.

5 **Chiroptera**: true fliers, with skin-covered forelimbs adapted as wings; bat.

6 **Primates**: uniquely grasping hands and feet; lemur, ape, monkey, human.

7 **Edentata**: no or few teeth, diggers and climbers; anteater, armadillo, sloth.

8 **Pholidota**: insect-eaters with no teeth and scaly bodies; pangolin.

9 **Lagomorpha:** small plant-eaters with nibbling teeth; pika, hare, rabbit.

10 **Rodentia**: gnawers, with chisel-like upper incisors; beaver, rat, squirrel, porcupine.

11 **Cetacea**: Aquatic, flat tail, paddlelike front limbs; whale, dolphin, porpoise.

12 **Carnivora:** meat-eaters, with claws; bear, raccoon, seal, cat, dog, wolf, weasel.

13 **Tubulidentata**: burrowing insect-eater with long, sticky tongue; aardvark.

14 **Proboscidea**: large, thick-skinned plant-eaters with trunks and tusks; elephant.

15 **Hyracoidea**: small and rodentlike with hooflike claws and short tails; hyrax.

16 **Sirenia**: aquatic, with paddlelike front limbs, flat noses: dugong, manatee.

17 **Perissodactyla**: hoofs (one or three toes); horse, tapir, rhinoceros.

18 **Artiodactyla**: hoofs (two or four toes): pig, antelope, deer, camel, giraffe, cattle, sheep.

Mammal Senses

Mammals are constantly receiving messages from their senses, on which their next meal or their lives may depend. Many mammals have senses far more acute than our own: keener eyesight and hearing, for example, as well as others (like the bat's sonar or the mole's sensitive whiskers) which we simply do not need. The variations in mammal body design are the result of millions of years of evolution and adaptation. So too is mammal behavior, either as individuals or in cooperating groups.

Loris

SIGHT

In most mammals the two eyes are set either side of the head. So each eye gives a different image. A few have binocular vision, with the eyes set in the front of the head and able to work together to focus on one image. This enables the mammal to judge distance more accurately—an important aid for tree climbers, like the loris (a primate), and for hunters, such as cats.

ANATOMY OF A MAMMAL

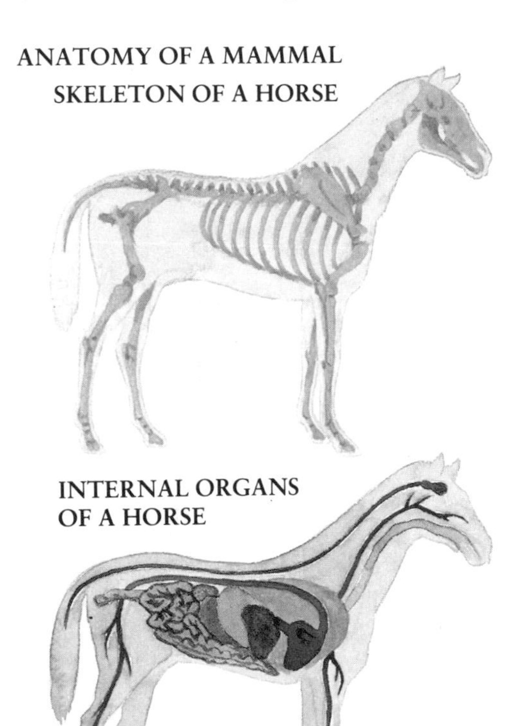

SKELETON OF A HORSE

◀ *The skeleton of a mammal acts as a framework for its body and protects vital organs such as the heart and stomach; also, the muscles that enable it to move are attached to its skeleton. An adult mammal—whether a mouse or an elephant—has over 200 bones.*

INTERNAL ORGANS OF A HORSE

◀ *The main internal body systems of a mammal, such as a horse, are concerned with digestion and waste disposal, and reproduction. The skull protects the brain and houses important sense organs such as the eyes, ears, nose, and mouth, linked to the nervous system.*

Wild dogs

HIBERNATION

Some species of mammals hibernate: they sleep through all or part of the cold winter when food is scarce. Before hibernating, the animal stores fat in its body. It becomes chilled and its heartbeat slows. A hibernating dormouse will not wake up even if it is touched. Other animals, such as squirrels and badgers, emerge in mild weather to seek food.

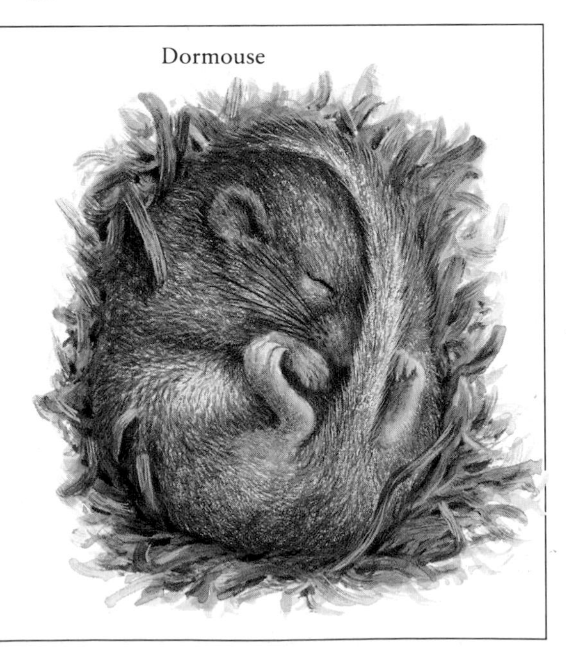

Dormouse

TOUCH

In many mammals the sense of touch is highly developed. They use sensitive hairs or whiskers and inquisitive snouts to investigate their surroundings when either burrowing underground or moving about in darkness. Moles have weak eyes, but rely on touch and smell to find their way through their tunnels.

Mole

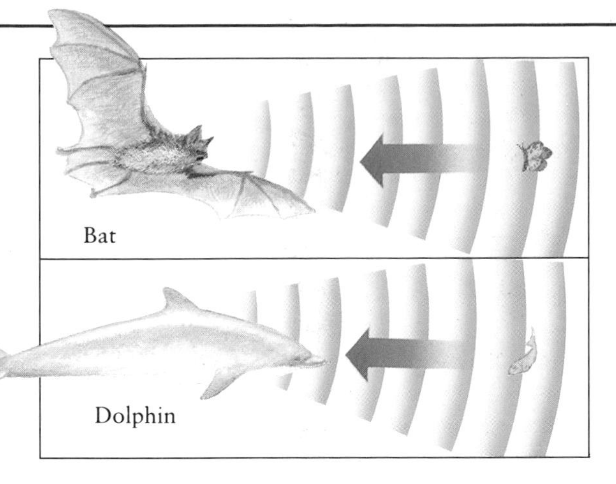

TASTE
Taste is situated in the tastebuds of the tongue. A dog has about 8,000, a cow four times as many. Anteaters use their long tongues to raid ants' nests.

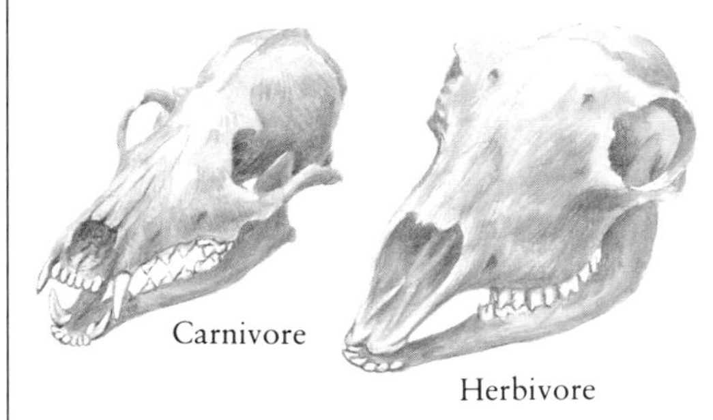

TEETH
Almost all mammals have teeth. Carnivores have sharp incisors and canines for tearing flesh. Herbivores nibble with their front teeth and use grinding molars to crush fibers.

HEARING
Some animals can hear much better than people. Bats in flight and dolphins swimming send out sound waves to detect prey by echolocation. The bat's sonar makes it an aerobatic flier even in the dark, though its eyesight is poor.

▼ *Carnivores kill for food, but some also feed on dead animals, that they have seen or scented.*

SMELL AND SCENT
Smell is an important sense for many mammals. Some deposit scent-messages to mark territory. Hunting animals, especially dogs (wild dogs, wolves, and foxes), track prey by smell. Hunting lions approach a herd of zebras from downwind, so that their smell is not carried toward the prey, alerting them to the danger.

ANIMAL INTELLIGENCE
Chimpanzees are the most intelligent apes. Inquisitive and persevering, they will imitate human actions and can solve simple problems. Only dolphins rival chimps in intelligence. Rats, dogs, and pigs also perform well in animal intelligence tests.

Animal Homes

Most animals need homes only to shelter their young. Birds build nests, a female bear seeks a den, a vixen (a female fox) takes over a burrow. Some social animals live in large colonies, used for many generations. The colony's home may be a structure of remarkable size—like a prairie dog town or a termite nest. Most hunting and grazing animals have no fixed homes, but wander their territories in search of food. Each individual or group may fiercely defend its territory against rivals of the same species.

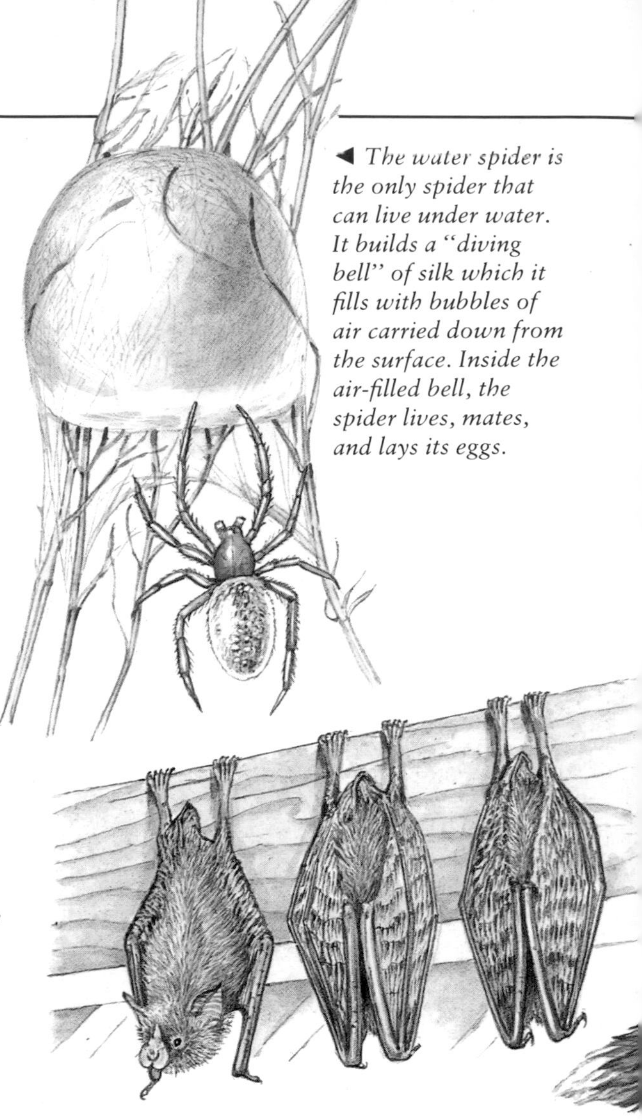

◄ *The water spider is the only spider that can live under water. It builds a "diving bell" of silk which it fills with bubbles of air carried down from the surface. Inside the air-filled bell, the spider lives, mates, and lays its eggs.*

◄ *Carmine bee-eaters, African birds related to kingfishers, nest in holes in river banks. The nest protects the young until they are old enough to fly. Birds' nests vary from complex woven or mud structures to holes or simple scrapes in the ground.*

► *Most bats are active at night. By day they shelter in caves, trees, or the roofs or cellars of buildings. Large numbers may roost together, hanging upside down by their feet and huddled close for warmth.*

A BEAVER'S LODGE
Beavers build island homes in rivers. They use chisel-like teeth and strong claws to build a log dam. The dam creates a pond in which they build a lodge, a mound of sticks and mud with dry inner chambers and underwater entrances.

PRAIRIE DOG TOWN

Prairie dogs are burrowing rodents of the grasslands of North America. Hundreds may live together in a colony or town. Family groups dig territorial burrows as deep as 16 ft. (5 m). Sentries keep watch above ground for predators.

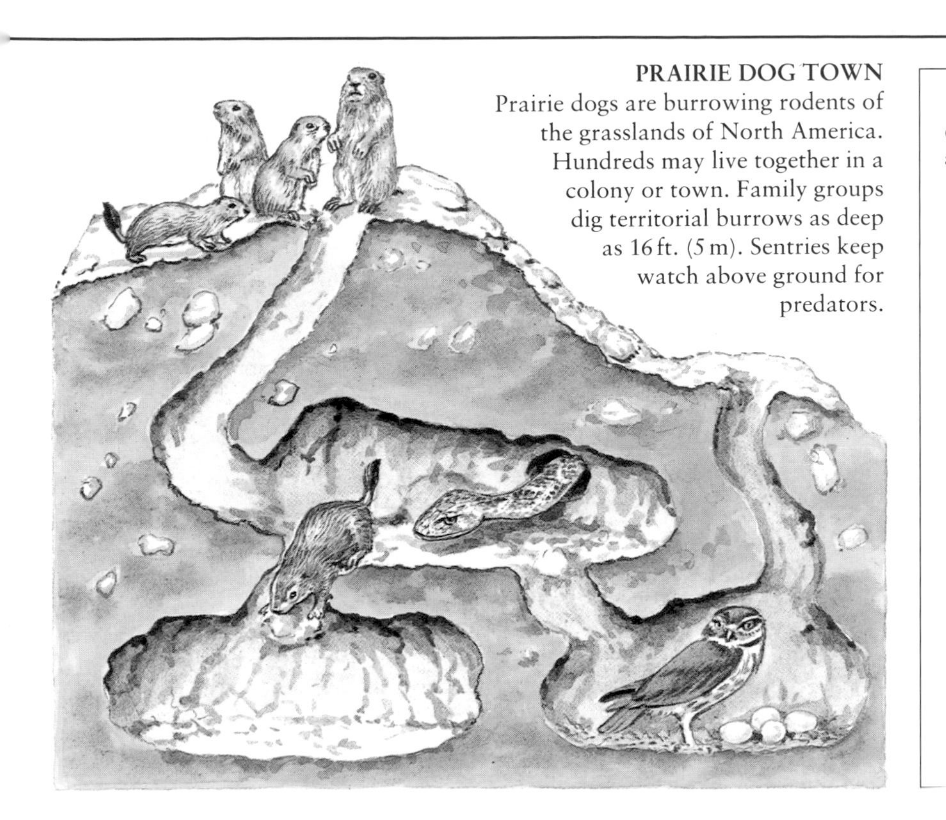

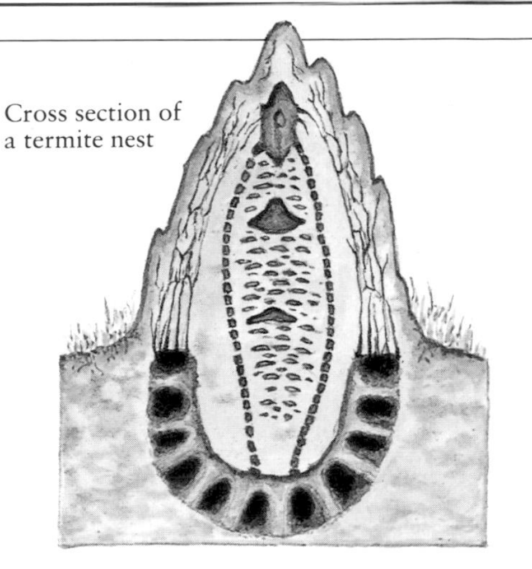

Cross section of a termite nest

TERMITE NEST

Termites have the most amazing homes; these social insects build mud mounds 30 ft. (9 m) high. Some termite nests have sloping roofs to deflect rain. Australian compass termites align their narrow nests north-south to escape too much hot sun.

TERRITORY

Many animals are territorial. They will fight rivals for an area that is big enough to provide its occupants with food, breeding space, and shelter. Territorial limits are respected. A rabbit will retreat, even when being chased, if it reaches a territorial "no-go" zone.

SAFETY IN NUMBERS

Many grazers and browsers (cattle, sheep, deer, antelope, horses) live in herds, sometimes hundreds strong. Deer form groups of females and young, each led by a strong male. Herd living has advantages. Every animal is a lookout, alert for danger. The strongest animals will often defend the herd. Females may help one another with young. By being one of a crowd, an individual has a smaller chance of being singled out for attack by predators and more chance of escaping in the confusion as the herd runs away.

Animal Movement

All animals move at some time in their lives—even the limpet clinging to its rock began life as a free-swimming juvenile. Fast movement is essential for many animals, to hunt and to escape when hunted. To conserve energy, fast-moving animals usually sprint only when they have to, in bursts. Some have no need of speed, relying on other strategies such as camouflage or armor for protection. Other animals are marathon athletes, traveling immense distances during seasonal migrations.

FAST- AND SLOW-MOVERS

A lion, 50 mph (80 km/h), can outsprint a racehorse, 40 mph (65 km/h). The snail crawls about 30 inches (80 cm) in a minute. This is about half the speed of a sloth on the ground (it moves a little faster when it is hanging upside-down from the branches of trees).

ANIMALS ON LAND
Legs act like props and levers. In motion, the legs push against the ground, propeling the animals forward.

◄ *Gazelles are among the fastest four-legged animals; at top speed they only have one foot on the ground, and may even lift all four legs in the air during each stride.*

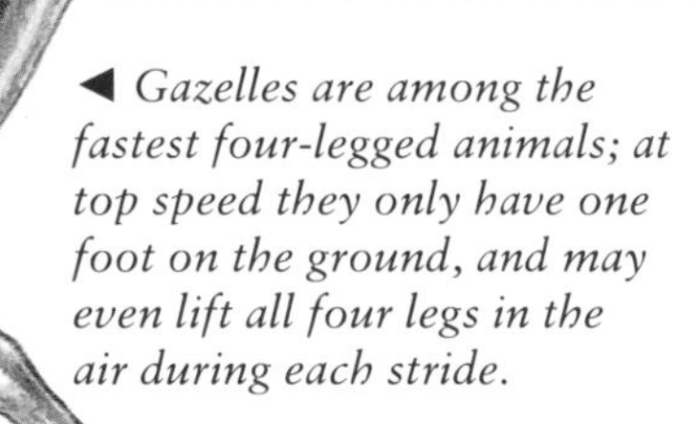

◄ *Arachnids (spiders and scorpions) have eight legs that move rather like the oars in a rowboat.*

▼ *Squids and octopuses swim by taking water in and then squirting it out through a tube; the expelled water causes them to shoot backward.*

WITHOUT LEGS
Not all animals need legs to move quickly. The fastest snake, the black mamba, can reach over 18 mph (30 km/h). Many snakes move with a wriggling motion *(top)*. Burrowing snakes move in accordianlike contractions *(below)*.

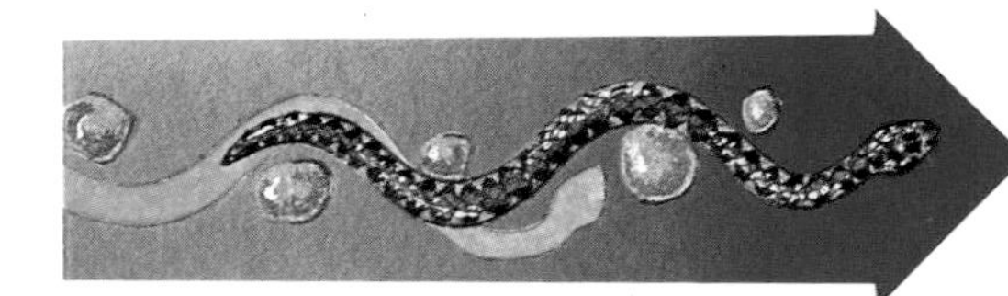

▼ *Snails and slugs have a single foot used for clinging and motion. The animal moves in rhythmic muscular waves by extending and withdrawing its foot. Slime helps the snail move, and leaves the familiar glistening trail.*

ANIMAL SPEEDS
Animal speeds are difficult to measure accurately. The fastest fish, the sailfin, narrowly outsprints the fastest land animal, the cheetah. But in flight, an eider duck in level flight would be overtaken by a diving peregrine falcon.

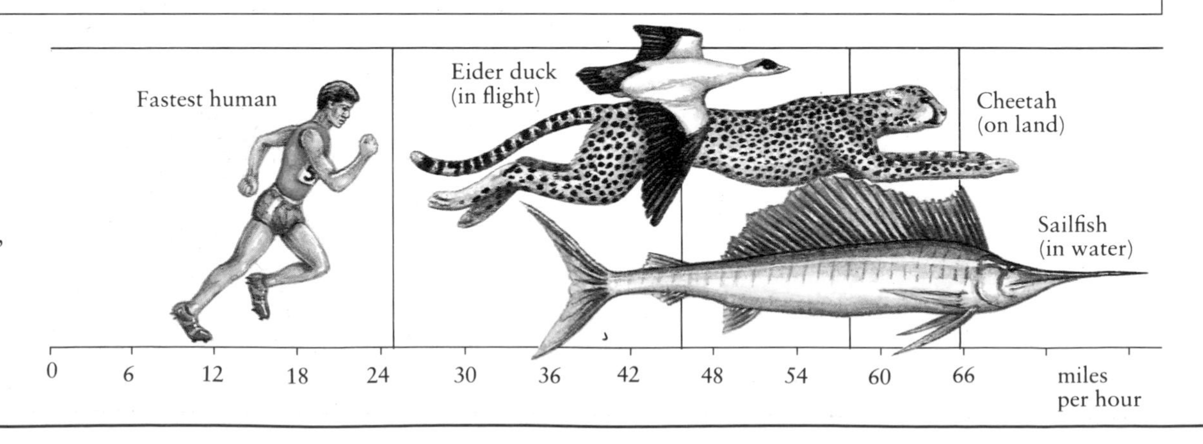

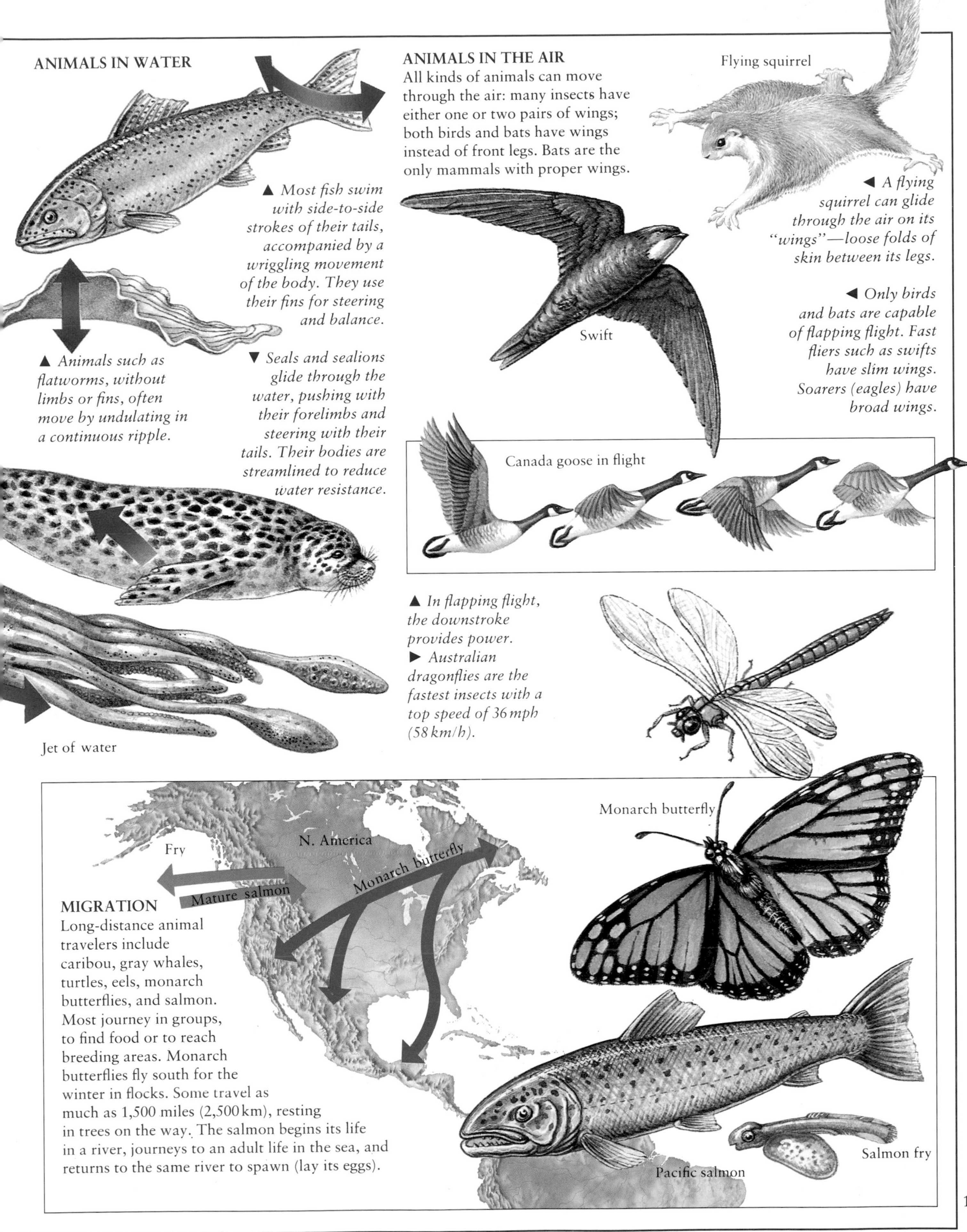

ANIMALS IN WATER

▲ *Most fish swim with side-to-side strokes of their tails, accompanied by a wriggling movement of the body. They use their fins for steering and balance.*

▲ *Animals such as flatworms, without limbs or fins, often move by undulating in a continuous ripple.*

▼ *Seals and sealions glide through the water, pushing with their forelimbs and steering with their tails. Their bodies are streamlined to reduce water resistance.*

ANIMALS IN THE AIR

All kinds of animals can move through the air: many insects have either one or two pairs of wings; both birds and bats have wings instead of front legs. Bats are the only mammals with proper wings.

◀ *A flying squirrel can glide through the air on its "wings"—loose folds of skin between its legs.*

◀ *Only birds and bats are capable of flapping flight. Fast fliers such as swifts have slim wings. Soarers (eagles) have broad wings.*

▲ *In flapping flight, the downstroke provides power.*
▶ *Australian dragonflies are the fastest insects with a top speed of 36 mph (58 km/h).*

MIGRATION

Long-distance animal travelers include caribou, gray whales, turtles, eels, monarch butterflies, and salmon. Most journey in groups, to find food or to reach breeding areas. Monarch butterflies fly south for the winter in flocks. Some travel as much as 1,500 miles (2,500 km), resting in trees on the way. The salmon begins its life in a river, journeys to an adult life in the sea, and returns to the same river to spawn (lay its eggs).

Animals and their Young

An animal's life span is determined chiefly by the time it needs to reproduce. Wild animals face many hazards and few survive to extreme old age. Most records for longevity have been set by captive animals. Reproduction in animals takes two forms: asexual, when only one parent produces the young (such as budding in sponges or corals), and sexual (when male and female sex cells combine to form a new animal). Some animals can regenerate parts of their bodies; for example, a crab can grow a new claw.

LIFESPANS

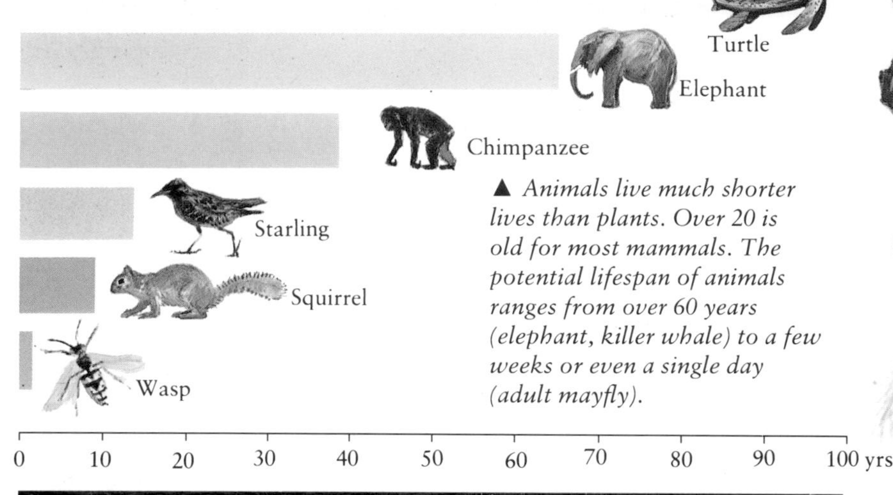

▲ *Animals live much shorter lives than plants. Over 20 is old for most mammals. The potential lifespan of animals ranges from over 60 years (elephant, killer whale) to a few weeks or even a single day (adult mayfly).*

COURTSHIP
Mating involves pairing of males and females. Courtship rituals often involve elaborate behavior, especially in birds. Egrets grow long plumes during the mating season and display these feathers as part of their courtship dance. Some animals pair for life, others mate and then part.

LIVING TOGETHER
Mammal babies take months or even years to develop. A female bear, by nature a solitary hunter, guards her young with care and teaches the cubs to catch fish. Bears are carnivores, but they also eat other kinds of food including grubs, birds' eggs, and berries. The cubs are energetic and playful: through play they learn the skills necessary to survive. Usually the cubs will stay with their mother for one or two years.

► *Elephants are sociable animals. They move in herds, feeding as they go. The herd is led by a dominant older female. A female giving birth is watched over by other female "midwives." If one elephant is trapped or wounded, other herd members will come to its aid.*

▲ *Male elephants (bulls) usually stay behind the herd. Rogues are aggressive males that live apart from a herd.*

MAMMAL REPRODUCTION

Placental mammals

Most mammal species have a placenta, a two-way filter that joins the unborn baby to its mother's body. Through it the baby gets food and oxygen from the mother's blood. After birth, the mother suckles it on her milk.

Red kangaroo

Zebra with foal

Marsupials

Pouched mammals, or marsupials, give birth to live young, but the young are born only partially developed. The tiny baby crawls into a pouch on its mother's body. Inside the pouch, it feeds on milk from her body, and will return to the pouch for shelter even when big enough to emerge.

Monotremes

These are the most primitive mammals, found only in Australia and New Guinea. The female lays eggs, but when the young hatch they are fed on milk from her body.

Duckbilled platypus

LOOKING AFTER THE YOUNG

King penguin

Trout

◄ *A male king penguin of the Antarctic keeps its egg warm beneath a flap of skin on its feet. The chick is sheltered in the same place.*

▲ *Fish, with few exceptions, simply deposit their eggs and swim away. The young that hatch must look after themselves.*

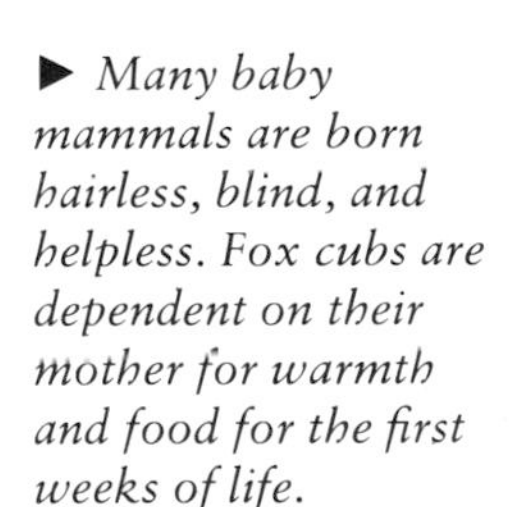

► *Many baby mammals are born hairless, blind, and helpless. Fox cubs are dependent on their mother for warmth and food for the first weeks of life.*

Fox with cubs

Macaque

◄ *Parents teach by example. Many animals have complex behavior patterns which the young inherit. Monkeys, for example, exhibit learned skills—such as specialized food-gathering techniques. By watching its mother, this baby macaque learns to copy her behavior.*

GESTATION PERIODS

Gestation is the time between fertilization and birth. Incubation is the time between fertilization and the hatching of an egg. A female elephant's pregnancy lasts 20 to 22 months. A fruit fly takes less than a day to change from egg to larva.

0 0.25 0.5 0.75 1.0 1.25 1.5 1.75 2.0 years

Elephant

Giraffe

Gibbon

Lion

Dog

Rabbit

FACTS ABOUT ANIMAL LIVES

- The longest-lived sea creature is the quahog clam (150 years).
- A queen ant may live for 18 years, some spiders for as long as 25 years.
- A mayfly emerges from the larval stage, breeds, and dies in a few hours.
- Sturgeon (80) and carp (50) are among the longest-lived fish.
- Record litters are 19 (cat), 23 (dog), 34 (mouse). These records were all set by pet animals.

Mayfly

Animals and People

People first hunted animals, then domesticated some species for food or wool, or as beasts of burden. Today domestic animals still carry goods and provide us with food, textiles, and other materials. Through selective breeding people have changed animals. The growth of population has destroyed many animal habitats. Wild animals live alongside people in town and country. Some thrive (pigeons, cockroaches, rats, fleas). Many others face an uncertain future, possibly extinction.

LOAD CARRIERS
The dog was probably the first domestic animal. Native Americans used dogs as pack animals. Over 5,000 years ago horses, asses, and camels were tamed for riding and for carrying loads. Oxen pulled heavy plows and carts.

ANIMALS THAT ARE USEFUL TO PEOPLE

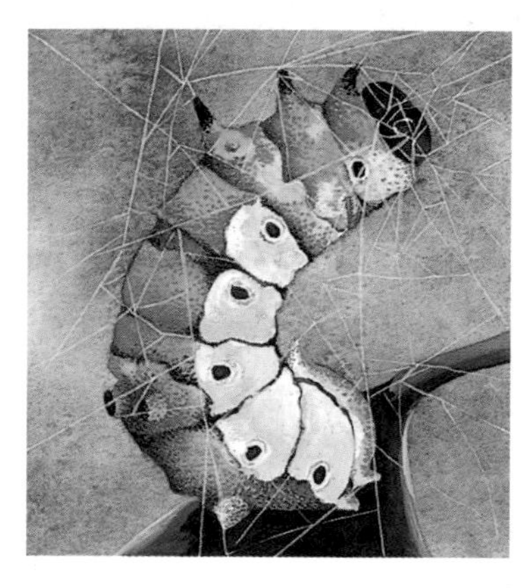

◀ *The animals that are most useful to people provide fur, skin, or wool as well as meat and milk; they include sheep, cattle, llamas, and camels. Other animals are useful because they can help people to hunt or get around, while some animals can carry loads or messages.*

▲ *One of the most unusual animal providers is the silkmoth larva, which produces a cocoon from which natural silk is obtained. One type of silkworm is raised on silk farms; wild silk comes from silkworms living wild in China and India.*

DAIRY FOODS
Milk, butter, yogurt, cheese: from cows, sheep, horses, goats, reindeer, and camels.

MEAT AND FISH
Beef (cattle), pork and bacon (pigs), lamb and mutton (sheep), goat, poultry, fish.

LEATHER
Hides from cattle, sheep, goats, even farmed alligators, for making leather.

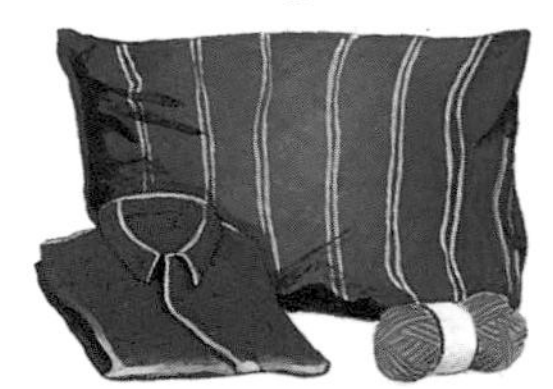

TEXTILES
Wool from sheep, goats, camels, llamas. Silk from silkworm. Down from ducks.

ANIMAL COMPANIONS

The first pets were probably baby animals (wolf cubs, goat kids, birds) brought back by prehistoric hunters for children to play with. For thousands of years, and in every society, people have valued animals as companions. To the lonely and elderly a dog or cat can be both a friend and a comfort; pets can sometimes have a beneficial effect on people who are ill.

◀ *People have kept dogs as pets and helpers since prehistoric times. No dog is more valued than the guide dog, trained as the "eyes" of its blind or partially sighted human owner.*

▶ *Bees not only provide us with honey, they also help to pollinate many plants, including fruit trees and garden flowers. People have kept bees for centuries as honey used to be the only sweetener for foods.*

DANGEROUS ANIMALS

Few animals attack people unless provoked and genuine man-eaters are rare. Venomous insects and spiders, and disease-carrying flies are more likely to cause people harm than are sharks, tigers, alligators, or snakes.

Brazilian wandering spider

▲ *The most venomous spider is a species of Brazilian wandering spider. It can hide in clothing or shoes, and give a fatal bite.*

Great white shark

▲ *Sharks have a worse reputation than they deserve, but great whites can be very dangerous.*

Crocodile

▲ *Crocodiles and alligators occasionally attack people, dragging them under the water.*

Tsetse fly

◄ *The tsetse fly feeds on blood and spreads the sleeping sickness, a serious disease.*

Snapping turtle

◄ *Snapping turtles protect themselves with their strong jaws; these American freshwater turtles can give a painful bite to unwary swimmers.*

FACTS ABOUT PETS

- The best "talkers" among pet birds are African gray parrots, parakeets, and mynahs.
- Cats usually live longer than dogs. The oldest cat on record lived to be 36.
- Cats kept ancient Egypt's granaries rat-free. People worshiped the cat-goddess Bastet (Bubastis), mourned cats' deaths, and often mummified their bodies.
- Guinea pigs are descendants of wild South American rodents called cavies. Hamsters come from Syria; all pet hamsters are descended from a pair brought to England in 1930.

ANIMALS AS PETS

Favorite pets include hamsters, gerbils, goldfish, birds such as parakeets, parrots, and canaries, rabbits, guinea pigs, cats, and dogs. Only animals bred in captivity should be kept as pets. As a rule, wild animals do not make good pets.

SELECTIVE BREEDING

The many breeds of dog, from Great Danes to Chihuahuas, share a common wolflike ancestor. Domestic cats are thought to be descended from African wildcats that were originally tamed by the ancient Egyptians. Since the 1800s, the increasing popularity of cats has resulted in much specialist breeding.

▲ *Breeding changed wild horses into strong war horses, the ancestors of the heavy cart and farm horses.*

▼ *A sheepdog obeys the calls of the shepherd as it drives sheep. The dog is carefully trained so that it chases but does not attack the sheep.*

Endangered Animals

Animal species become extinct or die out usually because they cannot adapt to changing conditions. The problem for wildlife today is lack of living space. People compete with wild animals for space, and win. Even prehistoric hunters were efficient enough to wipe out animal populations. The rate of extinctions has accelerated since the 1600s. Many species are in danger; some are being hunted, some are losing their habitat, some are being overrun by other animals, introduced by people.

EXTINCT ANIMALS
Species with no natural enemies are defenseless. The dodo, a flightless pigeon from Mauritius, was extinct by 1680, victim of sailors, cats, and rats who had landed on the island. The great auk was slaughtered for its feathers. The last two were killed in 1844.

VANISHING ANIMALS
Animals close to extinction include the Javan rhino and South China tiger (about 50 left), kakapo of New Zealand (40 or so) and Southeast Asian kouprey or wild ox (about 300). The European bison or wisent, nearly extinct by 1920, survives in Polish reserves.

▲ *Habitat destruction can cause rapid extinction. Forest animals of the tropics are endangered as forests shrink before the chainsaw and bulldozer. Forest monkeys like South America's bearded sakis have only slim chances of survival without protection.*

THE SKIN TRADE
Fashion and vanity has brought about the decline of many animal species. Birds such as the egret were hunted for their feathers. Snakes and alligators are killed and skinned to make bags and shoes. The fur trade, though declining, still takes its toll, especially of spotted cats such as the margay.

▲ *Even when protected in game reserves, rare animals are not safe from poachers. In Africa, poachers have killed most of the rhinos for their horns. In Indian nature reserves, tigers are poisoned for their skins and bones, which are made into a medicine drunk by Chinese and Koreans.*

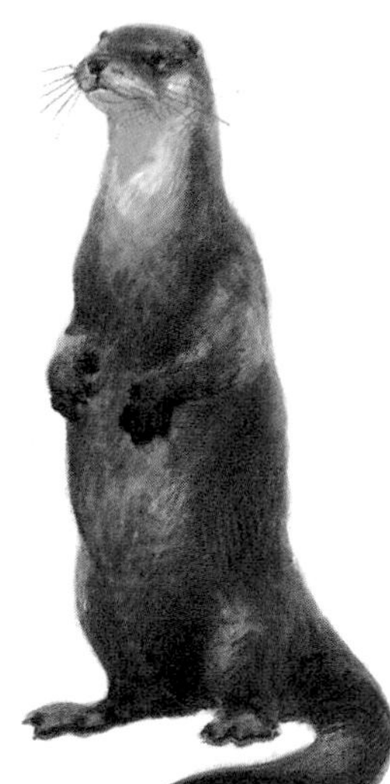

RIVER POLLUTION

Animals, like people, need clean water. River and lake animals are sensitive to any change in their environment. In the past 50 years farm pesticides, fertilizers, and chemical waste from factories have steadily poisoned many rivers. The European otter *(left)* is no longer seen in most of the rivers where it was once commonly found.

◀ *In the 19th century collectors took large numbers of insects, such as butterflies, and birds' eggs for display in their homes. Big game hunters shot animals for trophies, to be similarly displayed. Such activities are frowned on today.*

ZOOS

Since the 1600s the number of human beings has risen from 450 million to over 5 billion. For many people, seeing a bear at the zoo or a lion in a game reserve is the nearest they get to seeing an animal in the wild. Zoos have a role to play in conservation, through education and schemes to save endangered species. However, many people no longer want to stare at lonely animals penned in unsuitable cages.

WOLVES AND PEOPLE

Wolves have survived centuries of persecution, often unjust, by people. These intelligent and adaptable predators were once widespread, as the map shows *(right)*, but now their distribution is greatly reduced. There are now only small numbers of wild wolves in the United States and a few in Europe.

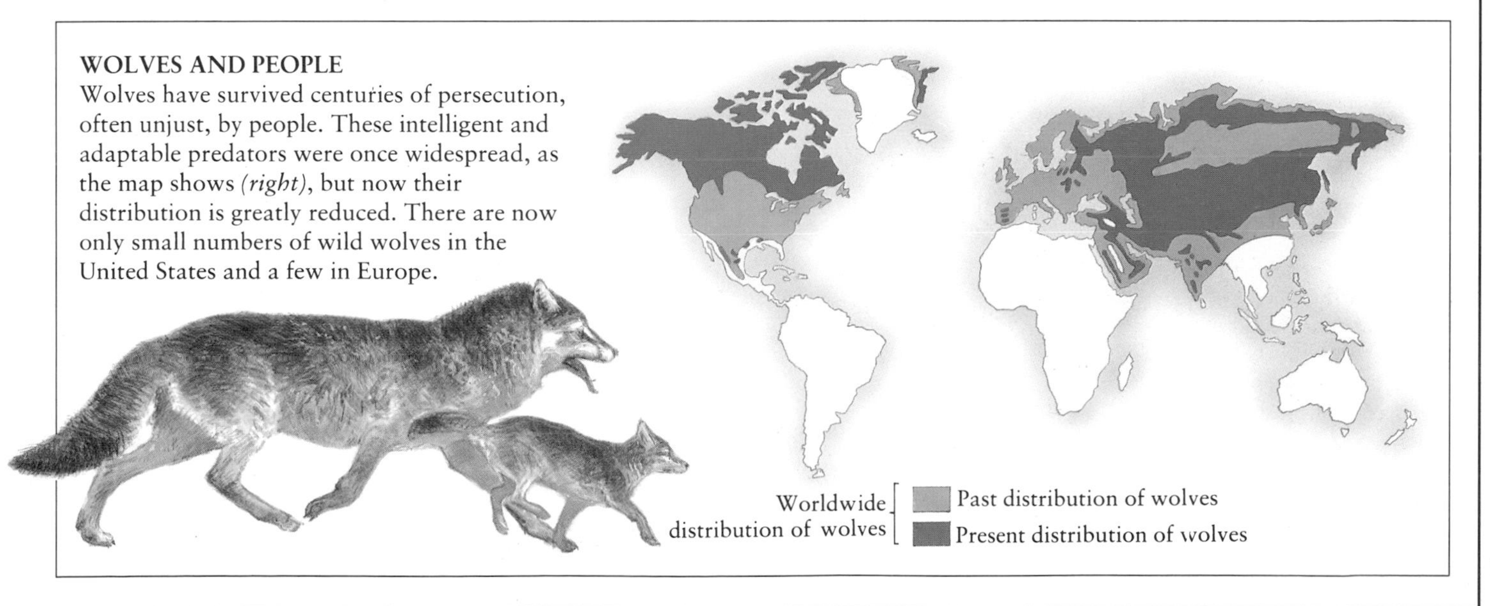

Prehistoric Animals

Our knowledge of most prehistoric animals comes from their fossil remains, mostly bones and shells. No humans observed the mighty dinosaurs, as the last of these prehistoric reptiles died out 65 million years ago. Prehistoric mammals then became the dominant animals, and from about 4 million years ago mammals such as saber-toothed cats and woolly mammoths shared the Earth with prehistoric humans. By about 10,000 years ago these early mammals had died out or evolved into new species.

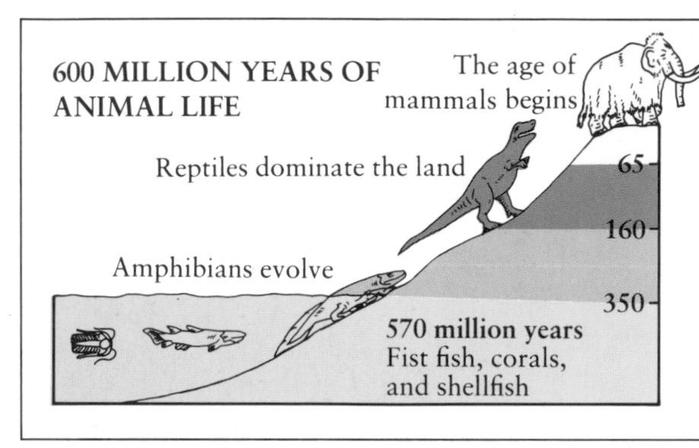

Evolution is a natural process of gradual change. Some species are better fitted to life in changing surroundings. They survive. Other, less adaptable species, die out.

BEFORE DINOSAURS
By 350 million years ago, when *Ichthyostega* became the first four-legged animal to invade the land, many types of animals had evolved in the seas; they included early true fish and crustaceans.

DINOSAUR RECORDS
Tallest and heaviest dinosaur: (complete skeleton) *Brachiosaurus*, 72 ft. (22 m), 70 tons; an incomplete skeleton of a *Brachiosaurus*, named *Ultrasaurus*, estimated at 82 ft. (25 m) long and 130 tons.
Longest: *Diplodocus*, almost 88 ft. (27 m) long.
Smallest: chicken-sized *Compsognathus*.
Fiercest carnivore: *Tyrannosaurus rex*, 46 ft. (14 m) long, 12 tons.
Most intelligent?: *Stenonychosaurus* was dog-sized, large-eyed and large-brained.
Most stupid?: *Stegosaurus* had a walnut-size brain in an elephant-size body.

OTHER ANIMALS
There were other animals just as remarkable as the dinosaurs—flying lobe-finned fish like *Osteolepsis*, and a reptilian ancestor of birds, *Archaeopteryx*.

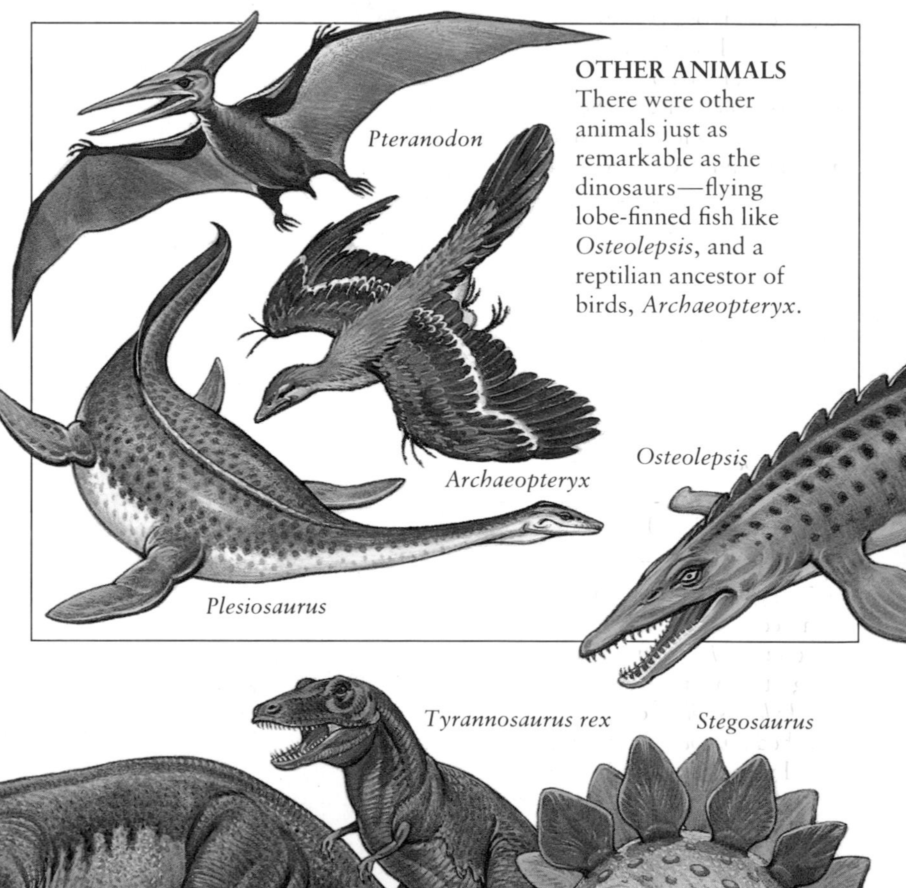

Pteranodon

Archaeopteryx

Plesiosaurus

Osteolepsis

Tyrannosaurus rex

Stegosaurus

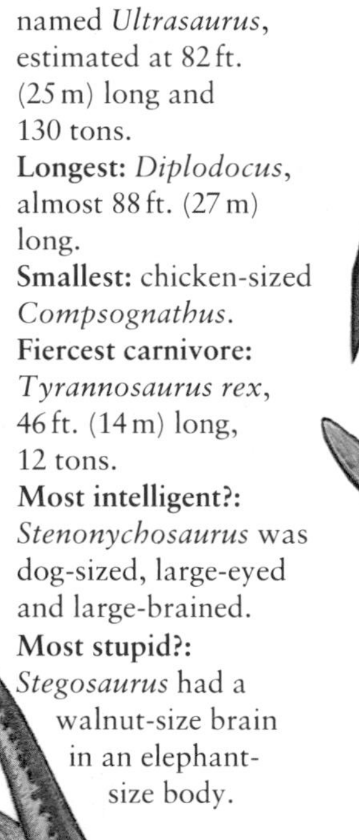

Brachiosaurus

Diplodocus

Compsognathus

Stenonychosaurus

◀ *Like all reptiles, dinosaurs laid eggs. Some skeletons have been found with complete nests. Females may have incubated eggs with their bodies, like some snakes, or buried them as turtles do. The young were born as miniature versions of their parents.*

AFTER THE DINOSAURS

When the dinosaurs vanished, mammals and birds took over the land. *Diatryma* was a large flightless bird, *Smilodon* a saber-toothed big cat. The woolly mammoth was closely related to the modern elephant. All three are extinct.

HUNTERS AND HUNTED

The largest dinosaurs were plant-eaters. Some species lived in herds for safety. Others, like *Stegosaurus*, relied on armor for defense against predators. The agile, scythe-clawed *Deinonychus* was among the most efficient of dinosaur killers. It may have hunted in groups to kill larger prey.

FACTS ABOUT EXTINCT ANIMALS

- The giant Steppe mammoth that once roamed central Europe, (*Mammuthus trogontherii*), 148 ft. (45 m) tall, was the biggest elephant that has ever lived.
- *Thylacosmilus* looked like the saber-toothed *Smilodon* but was not a cat at all.
- Why dinosaurs vanished is still being debated; climatic changes or an asteroid hitting the Earth are possible causes.
- The earliest ancestor of the horse was the dog-sized *Hyracotherium*, a forest animal of 50 million years ago.
- The Cretaceous pterosaur *Quetzalcoatlus* had wings as long as a bus; these flying reptiles probably had hair, rather than feathers, on the skin that formed their wings.

REPTILE SURVIVORS

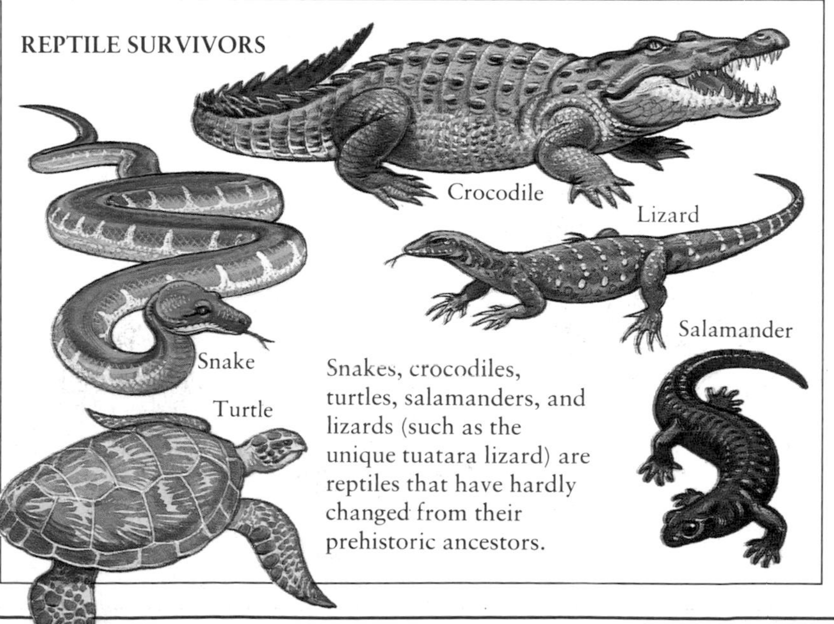

Snakes, crocodiles, turtles, salamanders, and lizards (such as the unique tuatara lizard) are reptiles that have hardly changed from their prehistoric ancestors.

THE HUMAN BODY

Systems of the Body

Human beings have more advanced brains than any other living thing. Human brainpower has given us abilities beyond those of any other animal—such as language and the transfer of knowledge from generation to generation. Human beings are primates, members of the species *Homo sapiens sapiens*. We share many characteristics with apes but unlike apes we walk erect on two legs. The body has parts and systems, like a machine, yet it can do things beyond the ability of any machine. It can grow, rebuild, and fight off diseases. The brain is the control center of our bodies; it receives information from our senses and then sends out commands that affect our development, movements, and sensations as well as the involuntary actions of our internal organs. The brain also stores information and is the source of all our feelings, speech, and thoughts.

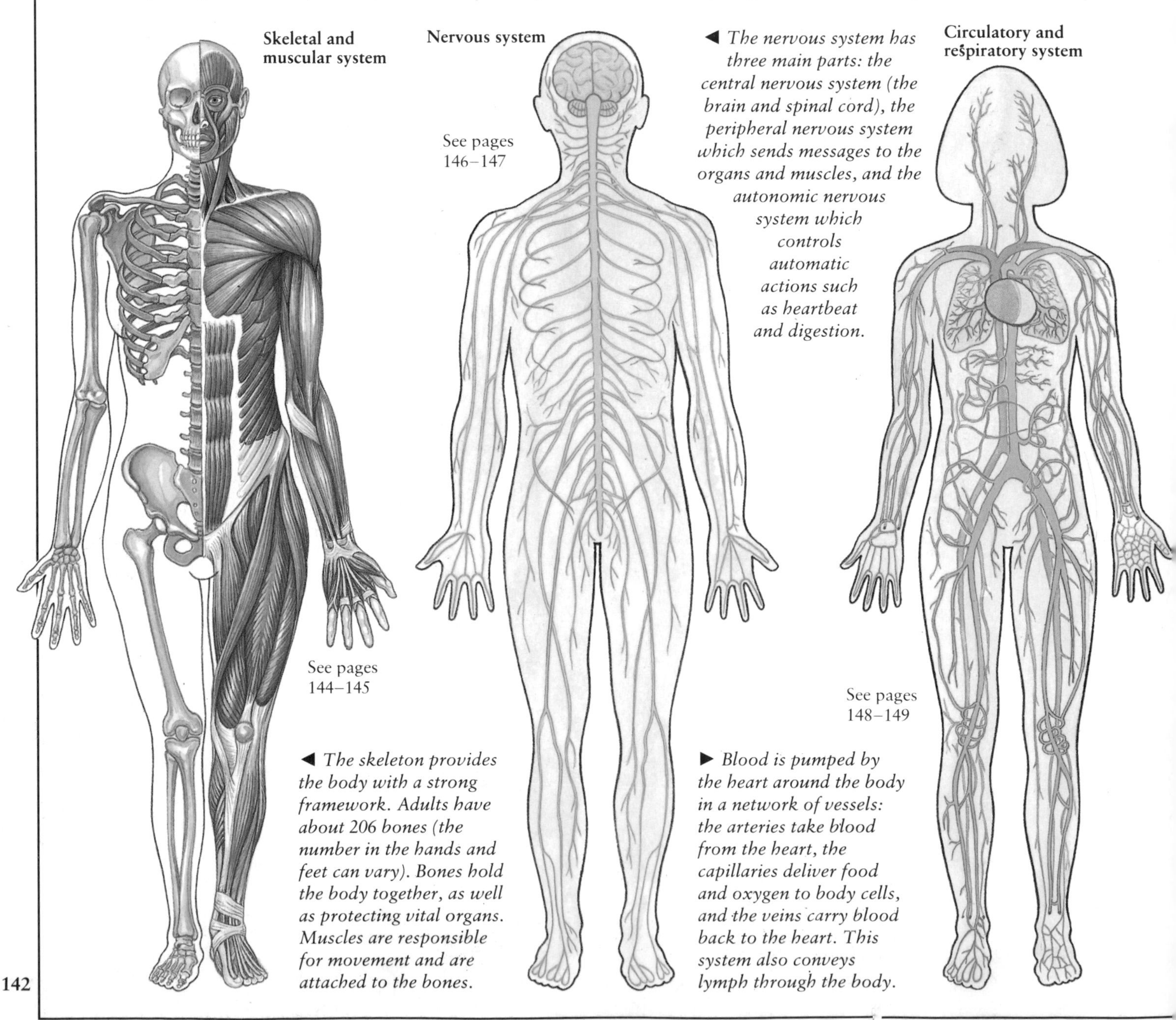

See pages 146–147

◀ *The nervous system has three main parts: the central nervous system (the brain and spinal cord), the peripheral nervous system which sends messages to the organs and muscles, and the autonomic nervous system which controls automatic actions such as heartbeat and digestion.*

See pages 144–145

See pages 148–149

◀ *The skeleton provides the body with a strong framework. Adults have about 206 bones (the number in the hands and feet can vary). Bones hold the body together, as well as protecting vital organs. Muscles are responsible for movement and are attached to the bones.*

▶ *Blood is pumped by the heart around the body in a network of vessels: the arteries take blood from the heart, the capillaries deliver food and oxygen to body cells, and the veins carry blood back to the heart. This system also conveys lymph through the body.*

FACTS ABOUT THE BODY

- The tallest recorded human was Robert Wadlow of the U.S. (1918–1940) who was 8ft. 11.1 in. (2.72 m).
- The oldest human (with an authenticated birth-date) was a Japanese, Shigechiyo Izumi, who died in 1986 aged 120 years 237 days.
- The strongest muscles are the masseters on each side of the mouth, which are used for biting; the most active muscles move the eye.
- An adult's body contains about 10 pints (5 liters) of blood. To pump blood around the body, the heart beats about 70 times a minute.

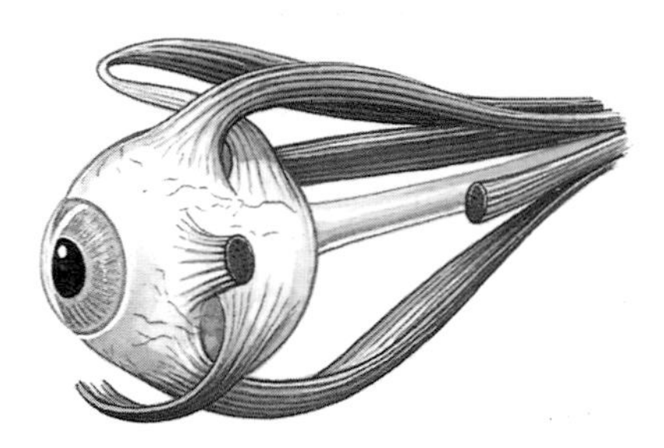

▲ *About 80 percent of the eyeball is made up of a jellylike substance. Six muscles move the eye about in its socket.*

- The fastest nerve signals travel at 250 mph (400 km/h).
- A person takes about 23,000 breaths each day.
- Children have more bones than adults —about 300. As a child grows, some bones fuse together.
- Each of a woman's ovaries contains about 400,000 eggs. Only about 400 mature during her childbearing years.
- The eyeball measures about an inch (25 mm) across.
- There are about 50 million cells in the body and 60,000 miles (100,000 km) of blood vessels.

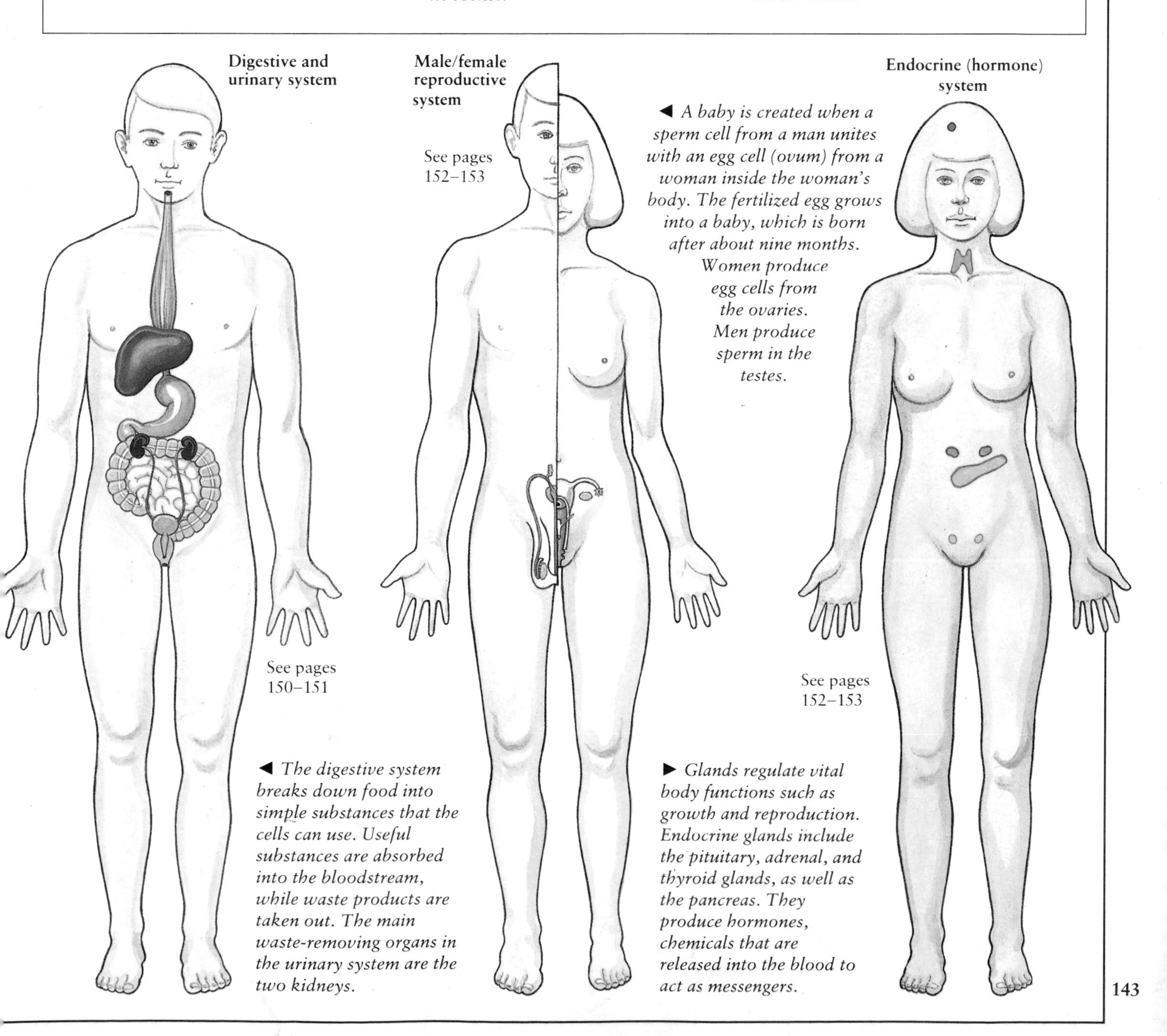

See pages 150–151

See pages 152–153

See pages 152–153

◄ *A baby is created when a sperm cell from a man unites with an egg cell (ovum) from a woman inside the woman's body. The fertilized egg grows into a baby, which is born after about nine months. Women produce egg cells from the ovaries. Men produce sperm in the testes.*

◄ *The digestive system breaks down food into simple substances that the cells can use. Useful substances are absorbed into the bloodstream, while waste products are taken out. The main waste-removing organs in the urinary system are the two kidneys.*

► *Glands regulate vital body functions such as growth and reproduction. Endocrine glands include the pituitary, adrenal, and thyroid glands, as well as the pancreas. They produce hormones, chemicals that are released into the blood to act as messengers.*

Skeleton and Muscles

Bones are made of living cells. The largest bone in the body is the femur or thighbone. The smallest is a bone in the ear, the stirrup. The ribs form a cage to protect the heart and lungs; the skull similarly encloses the soft brain. Where bones meet, they form a joint. Joints are held together by elastic ligaments and soft tissue called cartilage. Muscles are attached to the bones by tendons. When the brain orders muscles to contract, the muscles pull the bones—this is how we move.

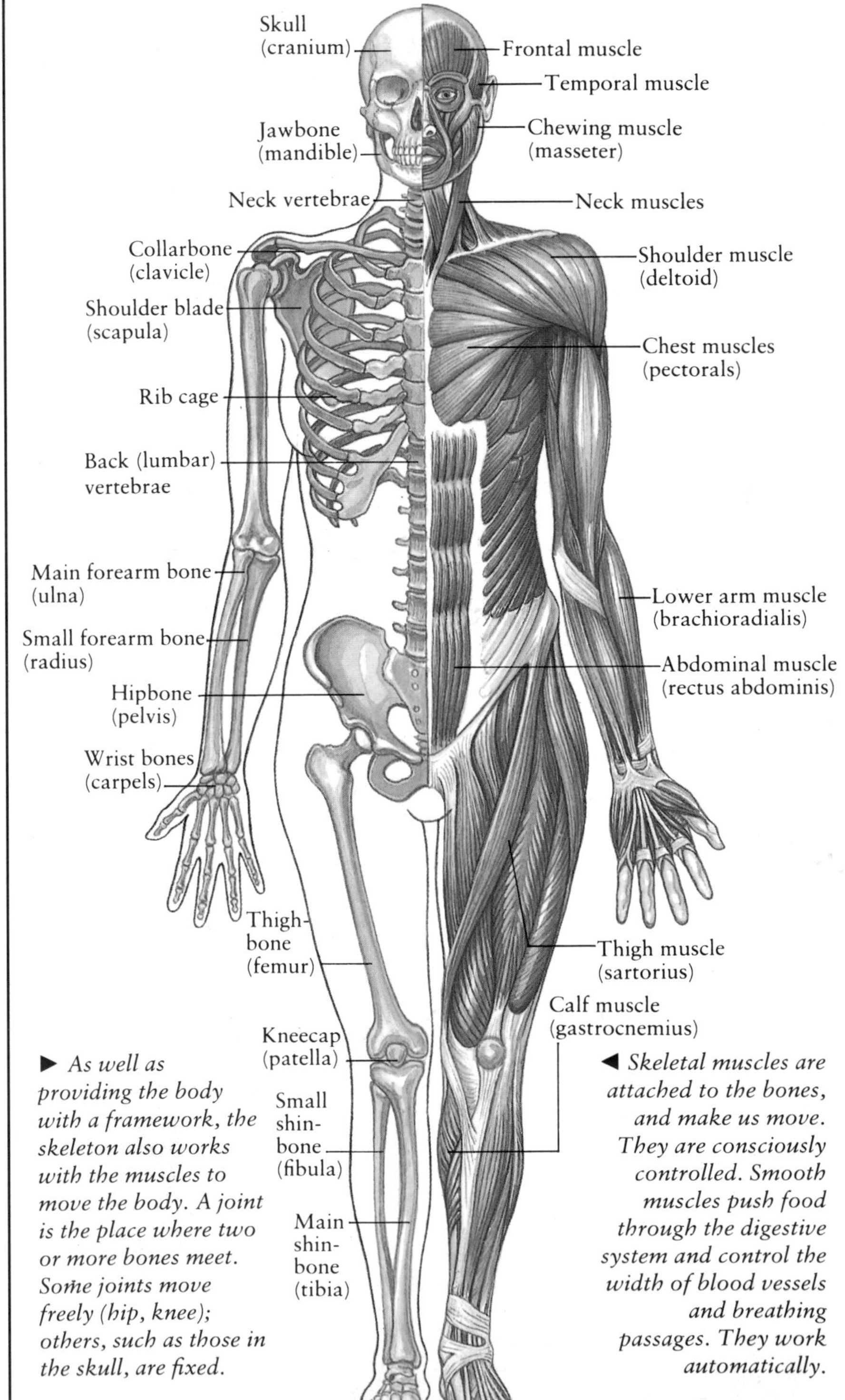

► *As well as providing the body with a framework, the skeleton also works with the muscles to move the body. A joint is the place where two or more bones meet. Some joints move freely (hip, knee); others, such as those in the skull, are fixed.*

◄ *Skeletal muscles are attached to the bones, and make us move. They are consciously controlled. Smooth muscles push food through the digestive system and control the width of blood vessels and breathing passages. They work automatically.*

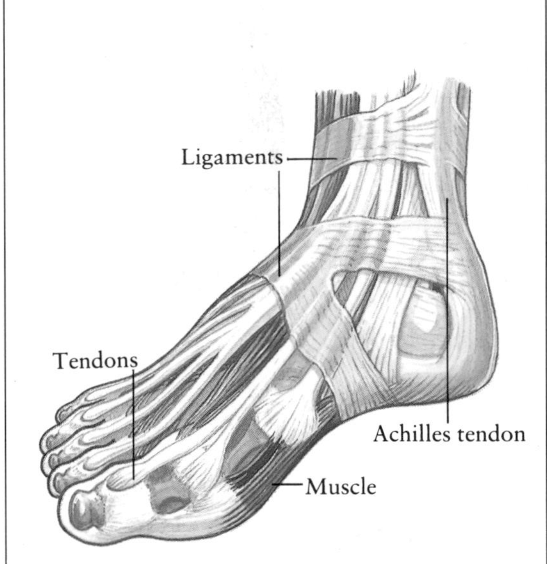

TENDONS AND LIGAMENTS
Tendons and ligaments are tough elastic tissues that hold joints together while allowing them to move. Ligaments connect one bone to another. Tendons connect a muscle to a bone. As the muscle contracts, the tendon acts like a cable, pulling the bone into the new position. In the foot, the Achilles tendon joins the calf muscle to the heel bone. We can consciously control such movements.

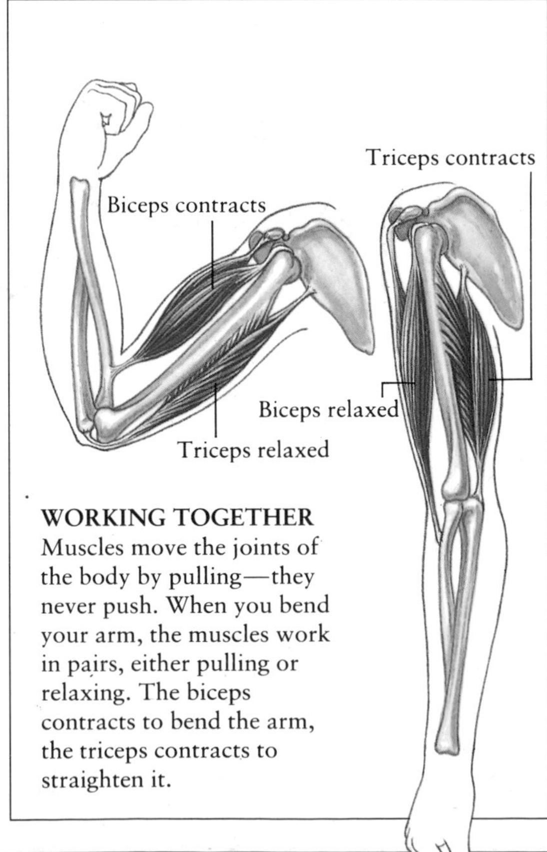

WORKING TOGETHER
Muscles move the joints of the body by pulling—they never push. When you bend your arm, the muscles work in pairs, either pulling or relaxing. The biceps contracts to bend the arm, the triceps contracts to straighten it.

MUSCLE

The interior of a muscle looks like bundles of cables (*far right*). Skeletal muscle is made up of long cells. Each cell has many nuclei. Smooth muscle and heart (cardiac) muscle both have shorter cells, each with only one nucleus.

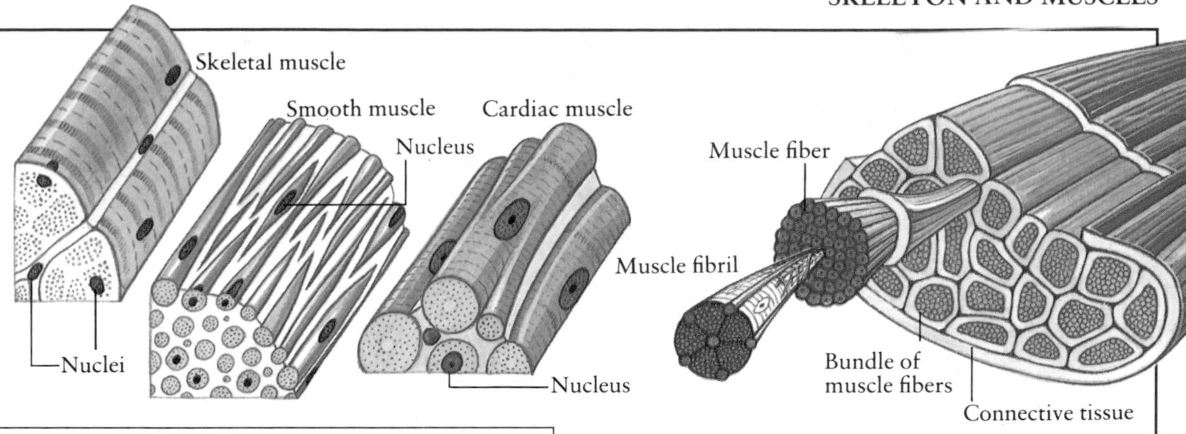

Hinge joint

Ball and socket joint

Plane joint

JOINTS

Hinge joints (elbow, knee) allow movement in one direction only. The hip and shoulder are swivelling ball and socket joints. Other joints allow a range of movement: the saddle joint at the base of the thumb, the pivot that allows the forearm to twist, or the plane that allows sideways movement.

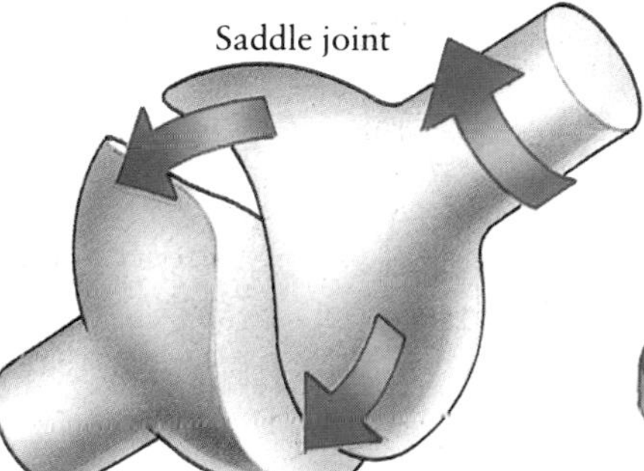

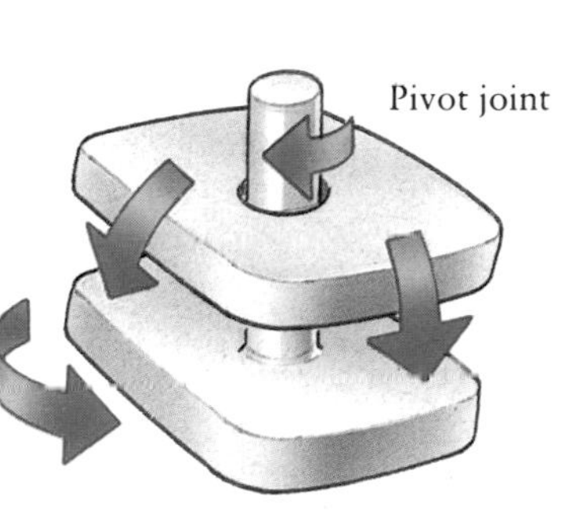

INSIDE A BONE

The outer layer of a bone is made up of hard compact bone that forms rings around the Haversian canals. Inside each canal are blood vessels carrying food and oxygen to the bone cells. The compact bone is covered with an even tougher layer, the periosteum. The inner part of a bone is often called the spongy bone, but it is very strong. Bone strength comes from a protein called collagen. The hardness comes from phosphorus and calcium. The soft, fatty core of many bones is called the marrow.

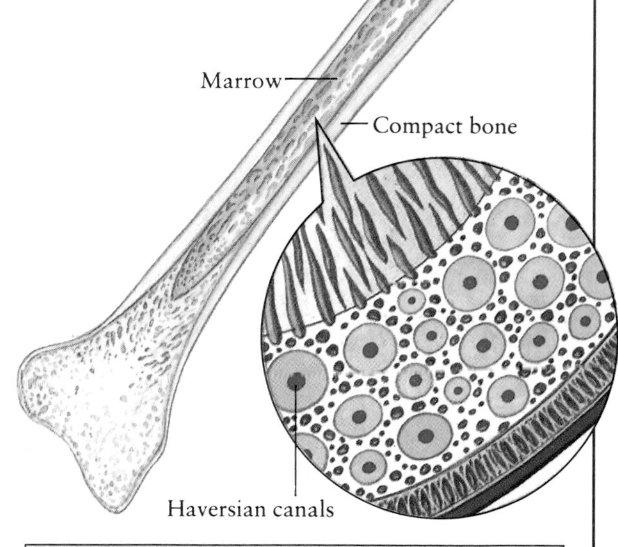

Fibula

Tibia

Cast

Bone knits

◀ *A broken (fractured) bone will heal itself. Doctors set a simple fracture by placing the broken ends together, allowing the repair cells (osteoblasts) to knit the bone together. Compound fractures (with tissue damage) are more serious and the broken bone may need pinning.*

FACTS ABOUT BONES AND MUSCLES

- Bones keep our bodies healthy. The cells of bone marrow produce new blood cells and release them into the bloodstream.
- Muscles make up about 40 percent of a person's body weight.
- When you walk, your body calls over 200 different muscles into use.
- Muscles produce heat when they use energy; this is why people become warm when they exercise.

The Nervous System

The nervous system is a complex network of nerves—bundles of long fibers made up of nerve cells. The nerves collect information from inside and outside the body and send messages to the brain. These messages are signals produced by sensory cells and passed to nerve fibers in the brain or spinal column; signals can also be sent from the brain to the body's organs. The part of the nervous system that controls such automatic body processes as breathing and digestion is called the autonomic nervous system.

► *The central nervous system—the brain and spinal cord—carries messages between the brain and body. The peripheral nervous system consists of sensory and motor nerve cells, linked with the central nervous system by special connector cells.*

Brain
Spinal cord
A nerve
Peripheral nerves

THE BRAIN
The cortex is in the cerebrum. It receives sense-messages and sends out nerve impulses to the muscles. It is also responsible for conscious feelings, thought, memory, and learning ability. The areas of the brain responsible for conscious thought and speech are at the front of the cortex. The left-hand side of the cortex controls activities on the right of the body; the right side controls the left of the body. The speech, reading, and writing of a right-hander is directed by the left side of the cortex; the right side controls the actions of a left-handed person.

Broca's area sends instructions to the motor cortex to give orders for the speech organ muscles to move.

The motor cortex sends out signals to the skeletal muscles. Each area controls a different movement.

Interpretations of touch from all areas of the body are received by zones within the sensory cortex.

The sounds we hear are interpreted by the sensory area of the cortex. Other zones receive impulses of taste and smell.

The images that we see through the eyes are interpreted by the visual cortex, a sensory area at the back of the brain.

CEREBRUM
BRAIN STEM
CEREBELLUM
MEDULLA

The cerebellum is concerned with balance and coordination. The medulla controls involuntary actions such as breathing.

The brain is the most important part of the nervous system. It uses large amounts of energy and needs a constant supply of blood. Brain cells die if starved of oxygen for as little as five minutes. The brain has three main parts: the cerebrum (about 85 percent of brain weight), the cerebellum, and the brain stem.

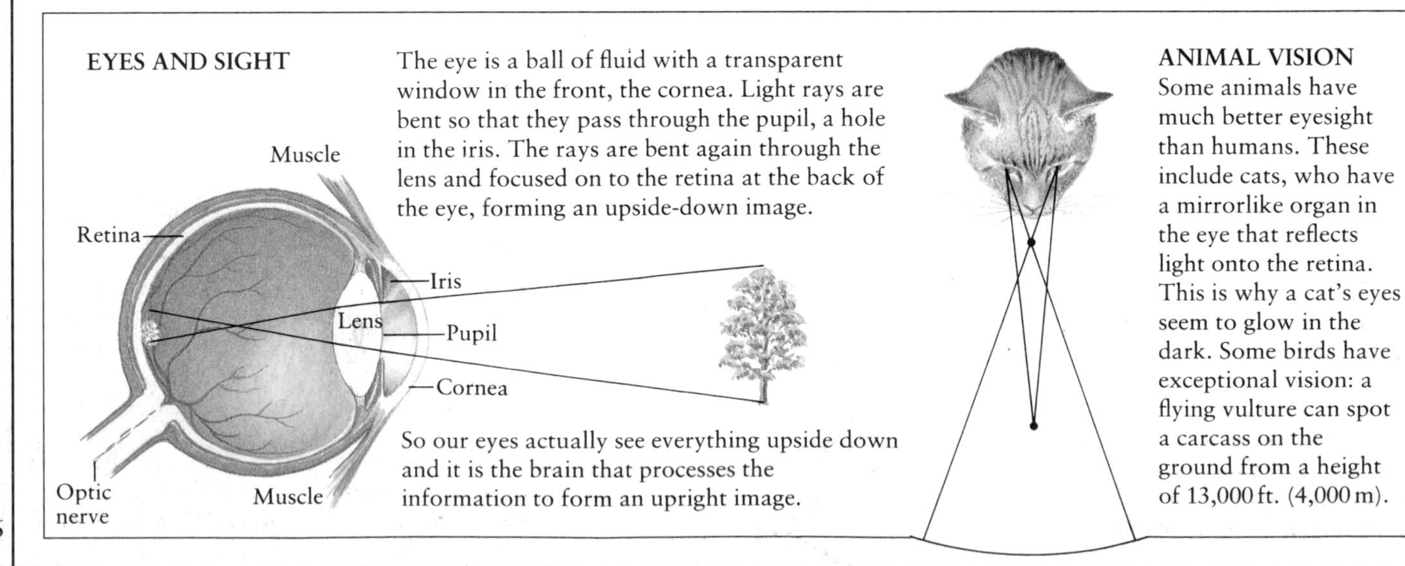

EYES AND SIGHT

The eye is a ball of fluid with a transparent window in the front, the cornea. Light rays are bent so that they pass through the pupil, a hole in the iris. The rays are bent again through the lens and focused on to the retina at the back of the eye, forming an upside-down image.

So our eyes actually see everything upside down and it is the brain that processes the information to form an upright image.

ANIMAL VISION
Some animals have much better eyesight than humans. These include cats, who have a mirrorlike organ in the eye that reflects light onto the retina. This is why a cat's eyes seem to glow in the dark. Some birds have exceptional vision: a flying vulture can spot a carcass on the ground from a height of 13,000 ft. (4,000 m).

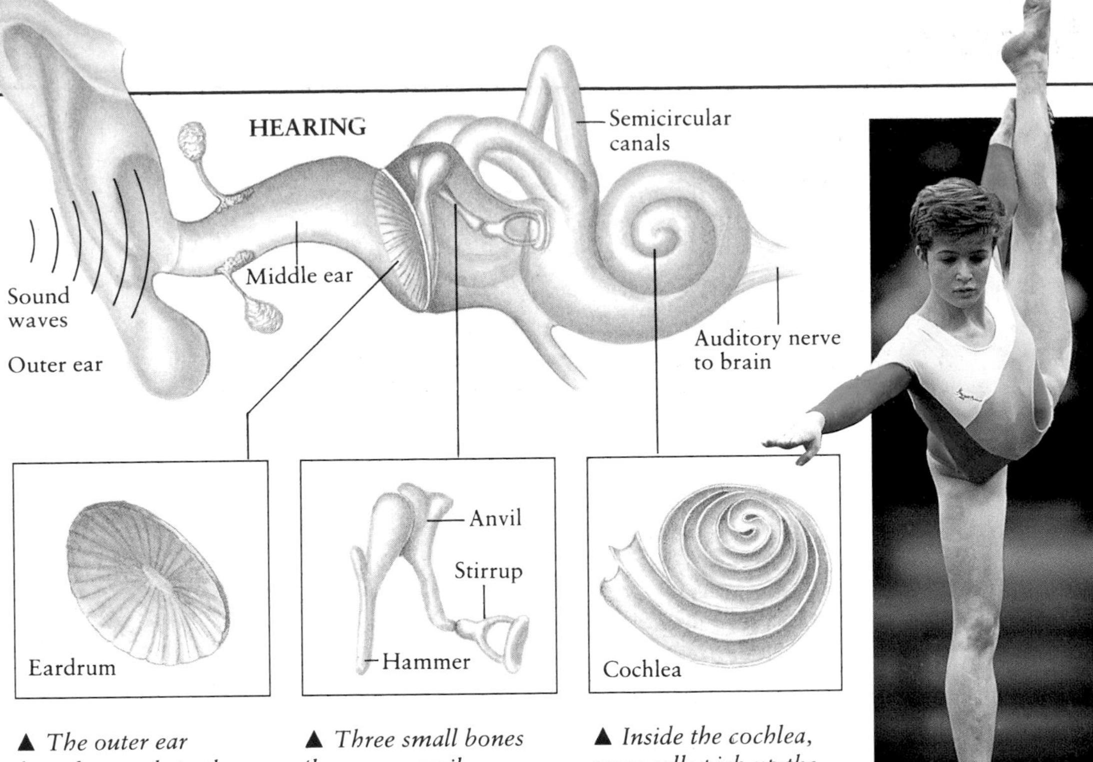

▲ *The outer ear funnels sounds to the eardrum, which vibrates when the sound waves hit it.*

▲ *Three small bones (hammer, anvil, stirrup) amplify the sounds and carry them through the middle ear.*

▲ *Inside the cochlea, sense cells pick up the vibrations, which are transmitted as impulses to the brain.*

◄ *The human ear has three main regions. The inner ear contains three semicircular canals filled with fluid which help us to keep our balance. As you move, the fluid moves. These canals, together with two other sense organs, the utricle and saccule, are called the vestibular organs. They send messages to the brain about the position of the head so that it can direct movements of the muscles that keep the body and the head steady. Any abnormal messages to the brain make a person feel dizzy. Gymnasts (left) must learn to keep their balance.*

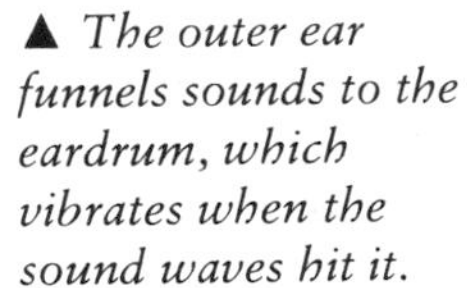

TOUCH

Touch is a vital sense, because it helps to protect the body from damage. It operates in five ways, sensing pressure, heat, cold, touch, and pain. Receptors are grouped in the dermis layer *(see p. 149)* of the skin and pass signals to the brain along nerves. The fingertips and lips are among the most sensitive parts of the human body.

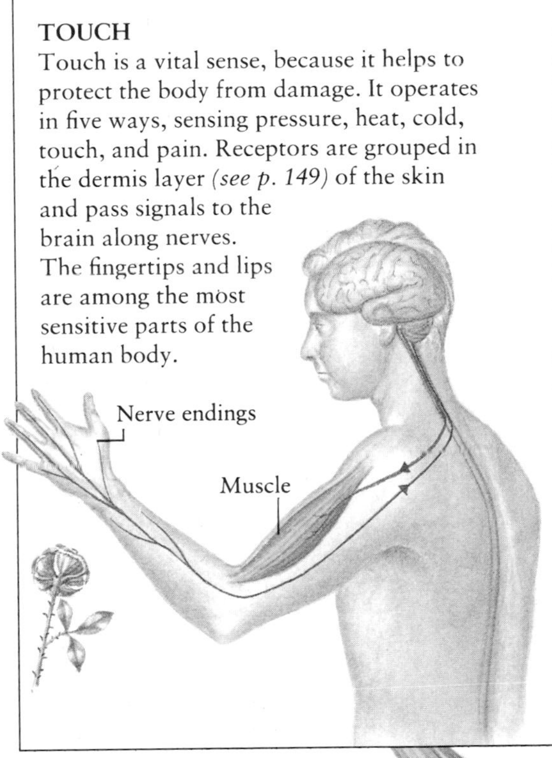

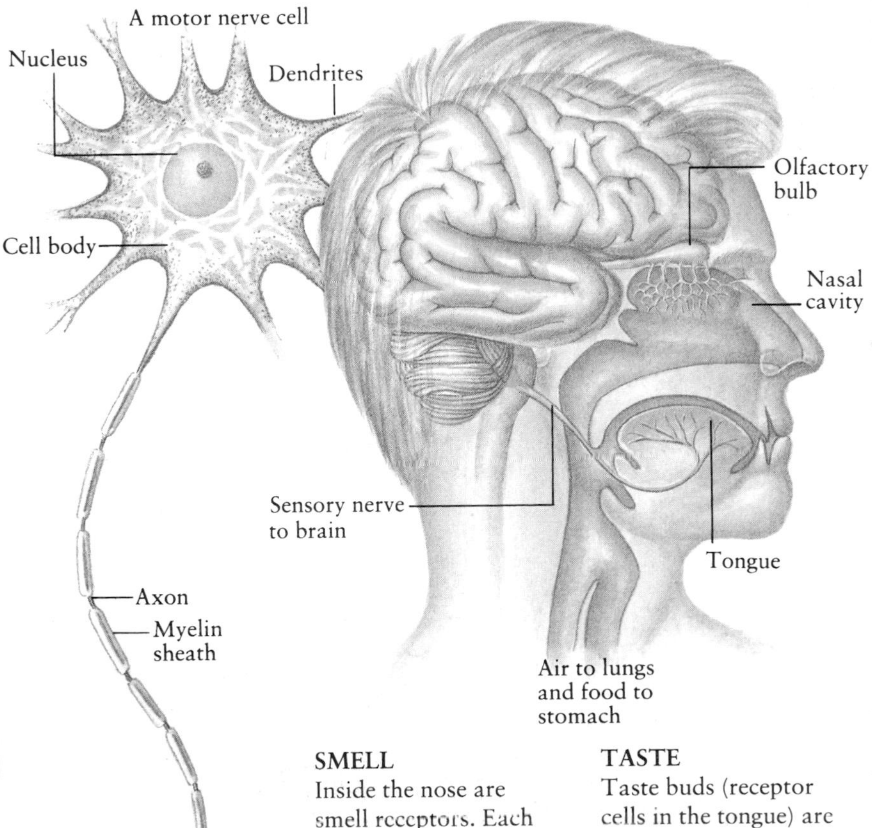

► *A nerve cell (neuron) has a cell body with fibers branching from it. Short dendrites carry signals to the cell body. A long fiber, or axon, carries messages away from the cell body to the muscle. Messages are passed chemically across the gap between dendrites.*

SMELL

Inside the nose are smell receptors. Each has minute hairs covered with sticky mucus. Scent particles dissolve in the mucus, and the receptors send messages to the brain to identify the smell.

TASTE

Taste buds (receptor cells in the tongue) are sensitive to four basic tastes: bitter, sweet, sour, and salt. Different areas of the tongue respond to different tastes. Taste and smell work closely together.

Heart, Blood, and Skin

The heart works continuously to pump blood around the body, through the arteries and veins. The blood carries oxygen from the lungs and food-energy from the food we eat through the arteries to the rest of the body. The veins carry away waste products and return "exhausted" blood from the body to the heart, for the cycle to begin again. The skin acts as a waterproof protective layer, shielding the body from infection and injury; it also keeps the body's internal temperature to a normal level.

FACTS ABOUT BLOOD

- There are four blood groups: A, B, AB, and O. Someone who is given a blood transfusion must be given blood that matches their own type.
- One microliter (millionth of a liter) of blood normally contains up to six million red blood cells, up to 10,000 white blood cells, and as many as 500,000 platelets.

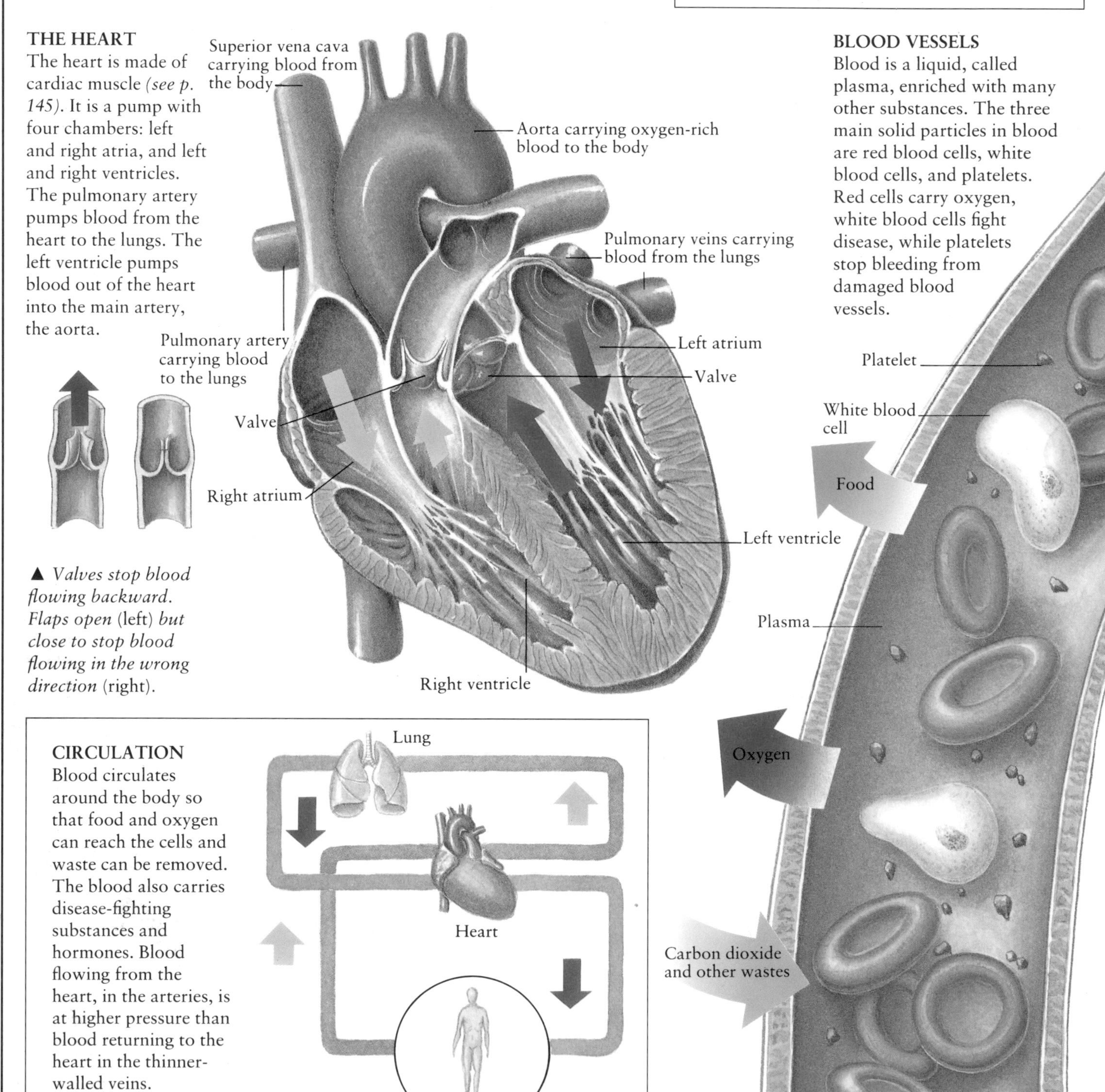

THE HEART
The heart is made of cardiac muscle *(see p. 145)*. It is a pump with four chambers: left and right atria, and left and right ventricles. The pulmonary artery pumps blood from the heart to the lungs. The left ventricle pumps blood out of the heart into the main artery, the aorta.

▲ *Valves stop blood flowing backward. Flaps open* (left) *but close to stop blood flowing in the wrong direction* (right).

BLOOD VESSELS
Blood is a liquid, called plasma, enriched with many other substances. The three main solid particles in blood are red blood cells, white blood cells, and platelets. Red cells carry oxygen, white blood cells fight disease, while platelets stop bleeding from damaged blood vessels.

CIRCULATION
Blood circulates around the body so that food and oxygen can reach the cells and waste can be removed. The blood also carries disease-fighting substances and hormones. Blood flowing from the heart, in the arteries, is at higher pressure than blood returning to the heart in the thinner-walled veins.

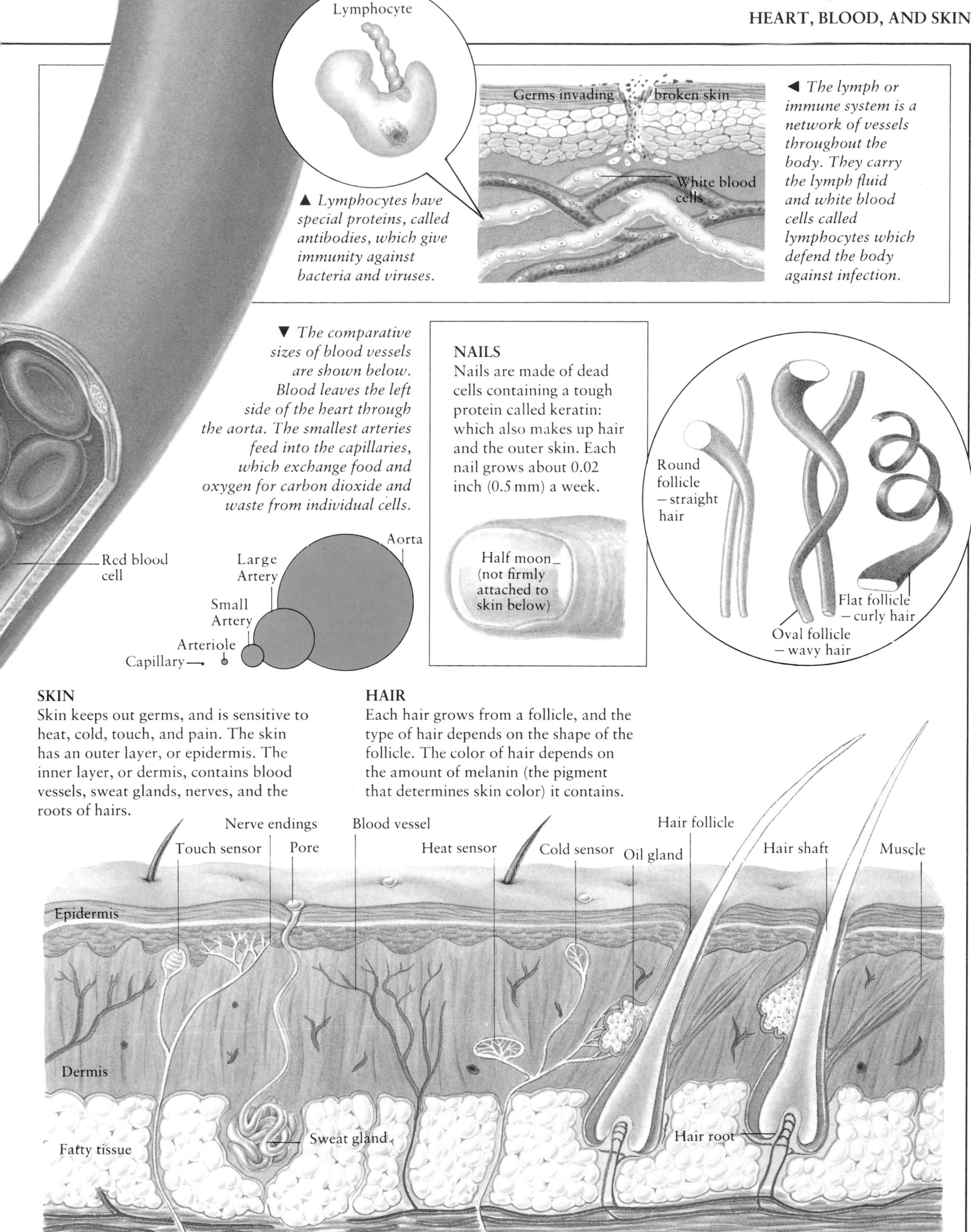

▲ *Lymphocytes have special proteins, called antibodies, which give immunity against bacteria and viruses.*

◀ *The lymph or immune system is a network of vessels throughout the body. They carry the lymph fluid and white blood cells called lymphocytes which defend the body against infection.*

▼ *The comparative sizes of blood vessels are shown below. Blood leaves the left side of the heart through the aorta. The smallest arteries feed into the capillaries, which exchange food and oxygen for carbon dioxide and waste from individual cells.*

NAILS
Nails are made of dead cells containing a tough protein called keratin: which also makes up hair and the outer skin. Each nail grows about 0.02 inch (0.5 mm) a week.

SKIN
Skin keeps out germs, and is sensitive to heat, cold, touch, and pain. The skin has an outer layer, or epidermis. The inner layer, or dermis, contains blood vessels, sweat glands, nerves, and the roots of hairs.

HAIR
Each hair grows from a follicle, and the type of hair depends on the shape of the follicle. The color of hair depends on the amount of melanin (the pigment that determines skin color) it contains.

Digestion and Respiration

The digestive system breaks down food into simple substances for the body cells to use. These substances are absorbed into the bloodstream and waste matter is passed out of the body as urine or feces. Cells need oxygen to break down and release the energy in food. The oxygen is taken into the body through the respiratory system—the nose, windpipe, or trachea, and two lungs. You take in oxygen from the air when you breathe in, and release waste carbon dioxide when you breathe out.

FACTS ABOUT BREATHING

- A baby is born with pink lungs. As we get older, our lungs darken from breathing polluted air.
- The voice box, or larynx, is at the upper end of the trachea. Sounds are made when air is forced through the vocal cords, two bands of cartilage stretched across the opening into the larynx.

BREATHING IN AND OUT

Breathing in: the diaphragm pushes down and the ribs move up and out to increase the chest space. Pressure is then greater outside the lungs than inside, and air moves into them. When breathing out, the process is reversed in order to expel air from the lungs.

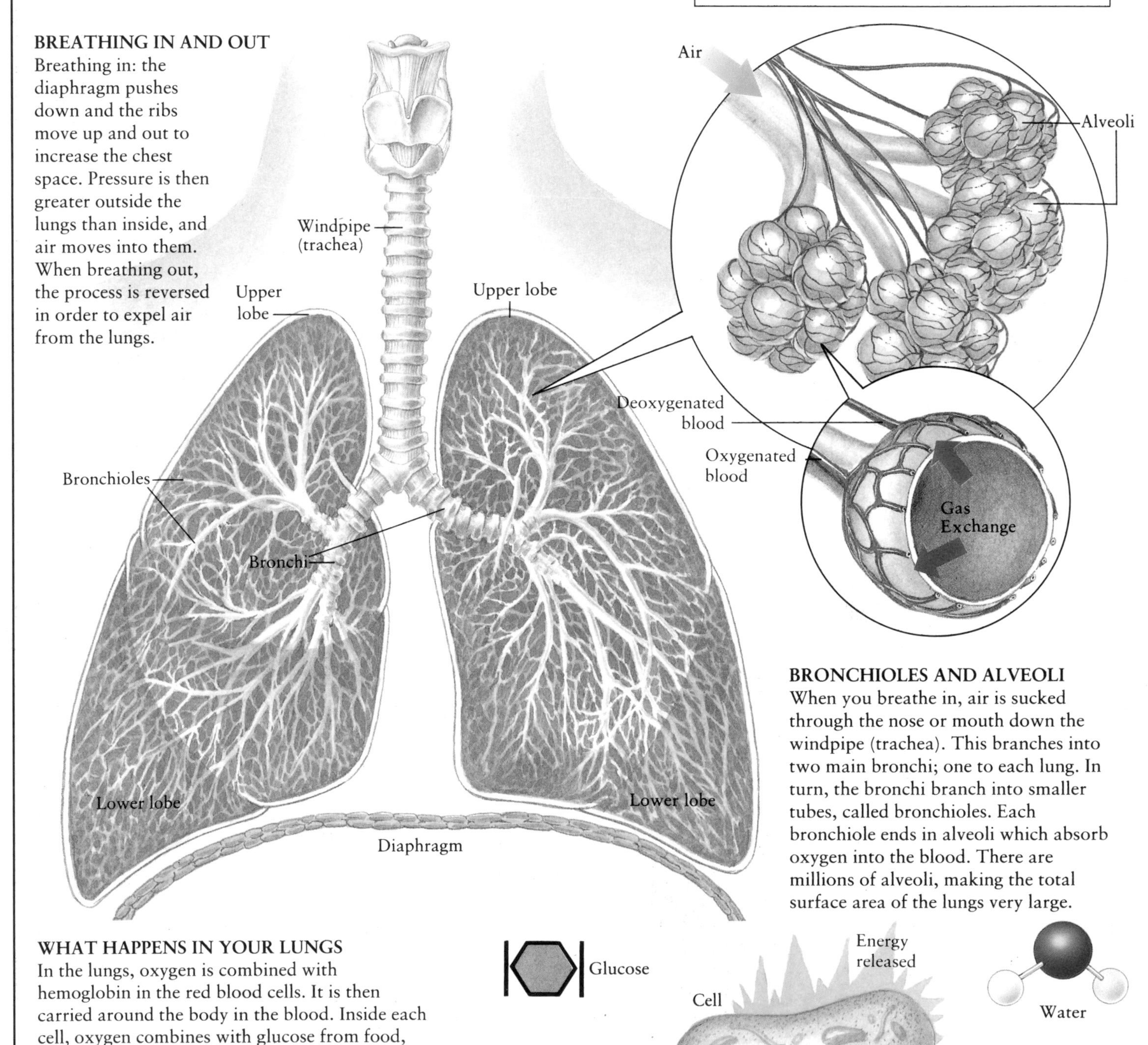

BRONCHIOLES AND ALVEOLI

When you breathe in, air is sucked through the nose or mouth down the windpipe (trachea). This branches into two main bronchi; one to each lung. In turn, the bronchi branch into smaller tubes, called bronchioles. Each bronchiole ends in alveoli which absorb oxygen into the blood. There are millions of alveoli, making the total surface area of the lungs very large.

WHAT HAPPENS IN YOUR LUNGS

In the lungs, oxygen is combined with hemoglobin in the red blood cells. It is then carried around the body in the blood. Inside each cell, oxygen combines with glucose from food, and energy is released. The carbon dioxide produced during respiration is passed from the blood to the alveoli and released as waste when we breathe out.

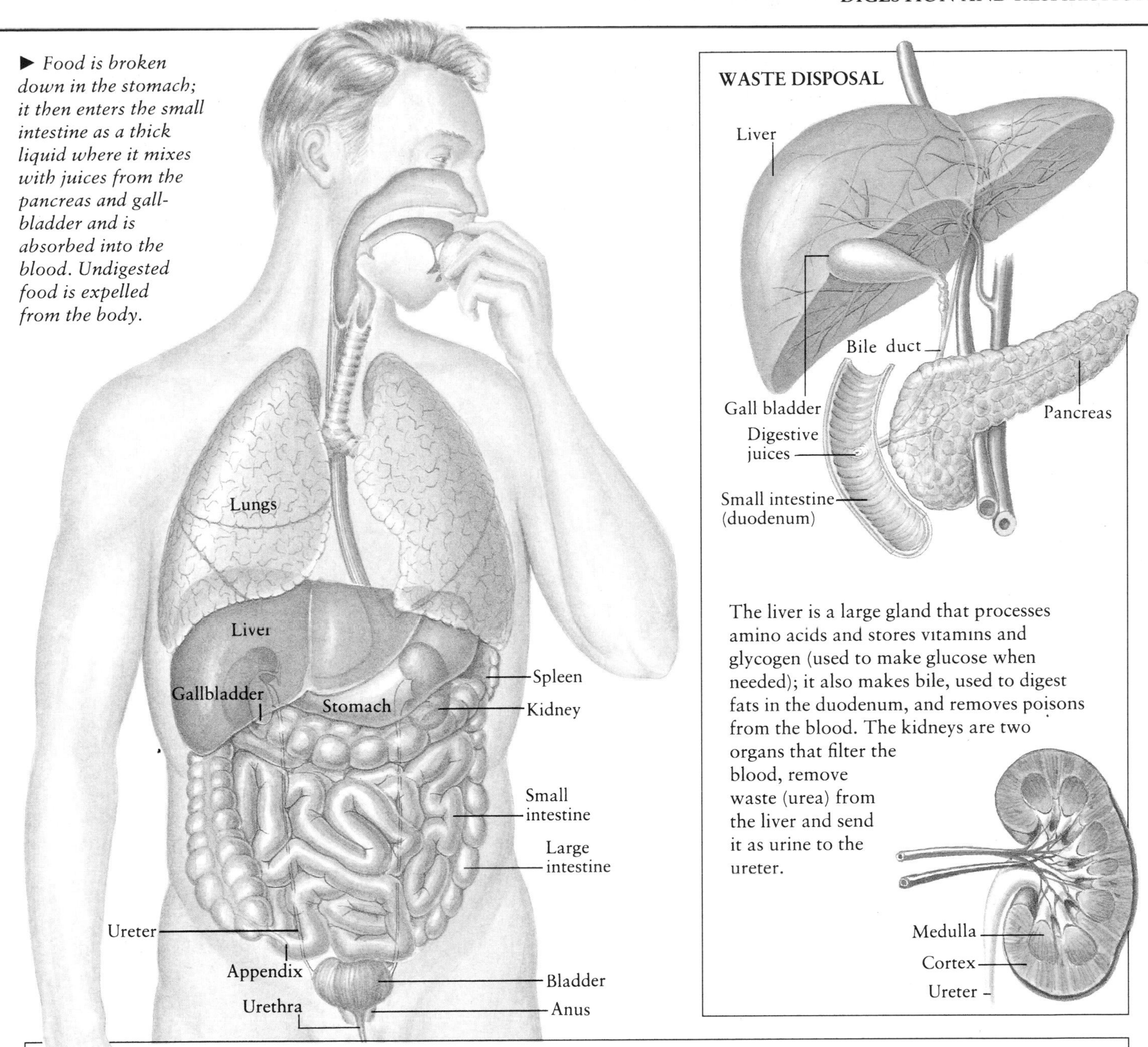

► *Food is broken down in the stomach; it then enters the small intestine as a thick liquid where it mixes with juices from the pancreas and gall-bladder and is absorbed into the blood. Undigested food is expelled from the body.*

WASTE DISPOSAL

The liver is a large gland that processes amino acids and stores vitamins and glycogen (used to make glucose when needed); it also makes bile, used to digest fats in the duodenum, and removes poisons from the blood. The kidneys are two organs that filter the blood, remove waste (urea) from the liver and send it as urine to the ureter.

CROSS SECTION OF A MOLAR TOOTH

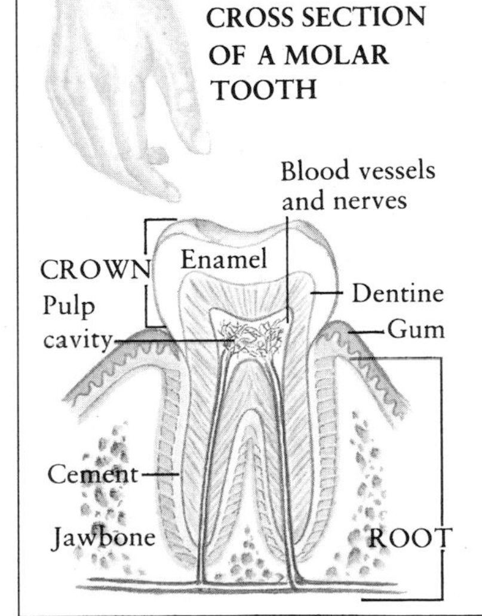

TEETH

Teeth prepare food for swallowing and digestion. Some teeth are cutters (incisors), others, grinders (molars). A tooth has three main layers; the outer is made of hard enamel to resist wear. Underneath lies a hard dentine, over an inner pulpy cavity, which contains nerves and blood vessels. Children have 20 baby (first) teeth; these gradually fall out to be replaced by 32 adult teeth.

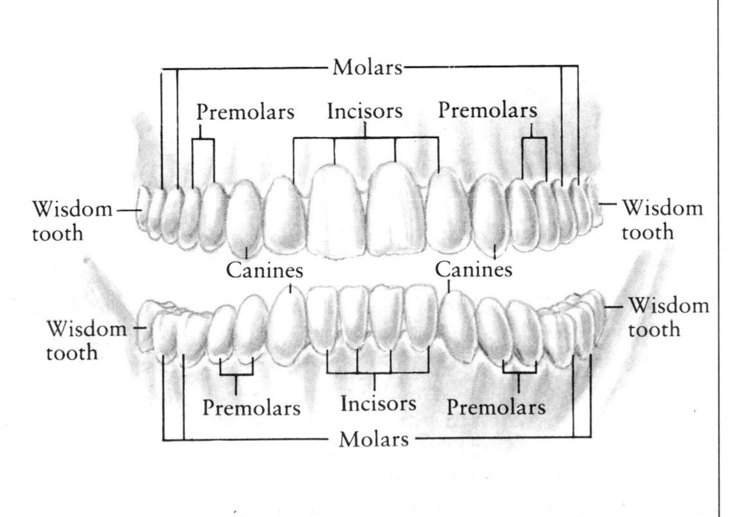

Reproduction

Humans reproduce sexually, like other mammals. The process of reproduction begins with conception—when sperm from a man fertilizes the egg of a woman. Both egg and sperm contain genetic information in chromosomes, and this information programs the development of the embryo. After about two months the embryo has most of its internal organs. It is now a fetus. At four months, it can move, and after about nine months, a new human being is ready to be born.

The DNA molecule

▲ *In the nucleus of each human cell are 23 pairs of chromosomes, made chiefly of proteins and the chemical deoxyribonucleic acid (commonly known as DNA). The DNA molecules contain coded instructions (genes) that control the workings of the cells. These genes also control how the cells develop into a body and carry the code for inherited characteristics.*

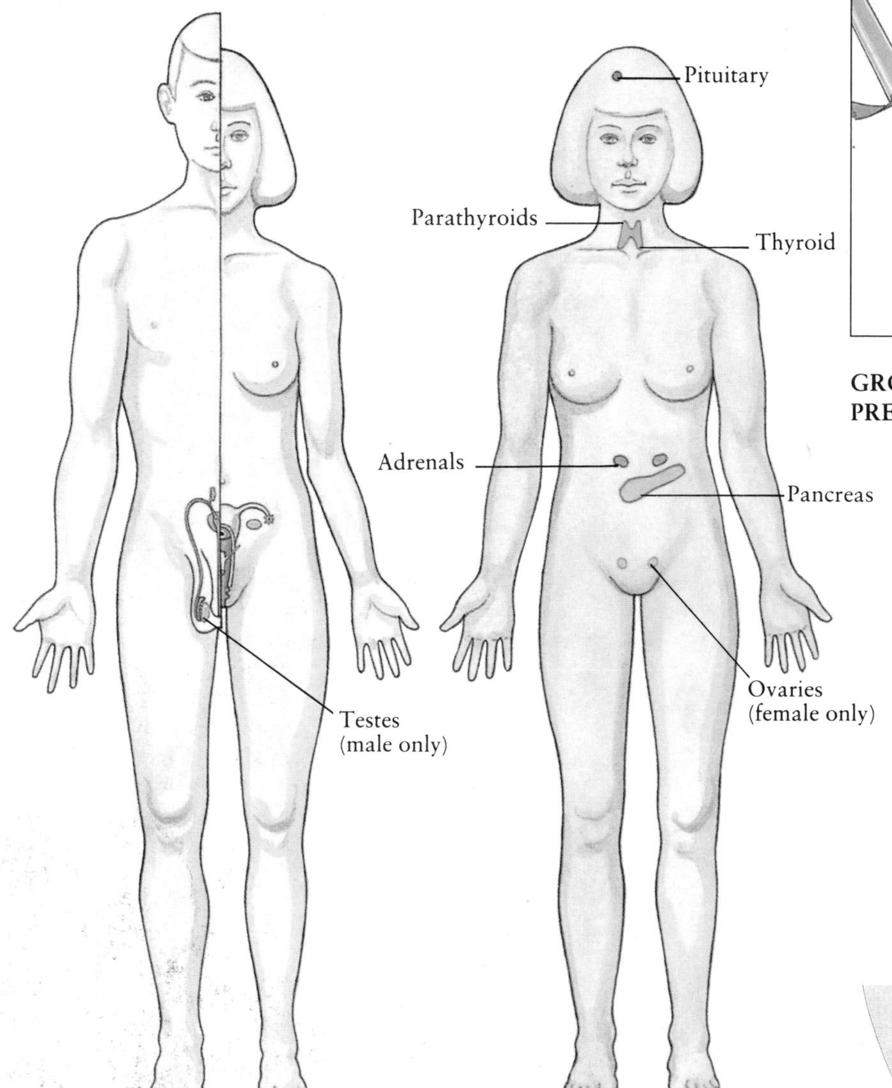

▲ *The endocrine glands produce hormones. The pituitary hormone regulates growth. Testes produce the male hormone testosterone; ovaries produce the female hormones estrogen and progesterone.*

GROWTH DURING PREGNANCY

▼ *A human pregnancy typically lasts 38–40 weeks. At 12 weeks the baby is about 4 inches (9 cm) long and weighs 0.5 oz. (14 g).*

5 weeks

8 weeks

12 weeks

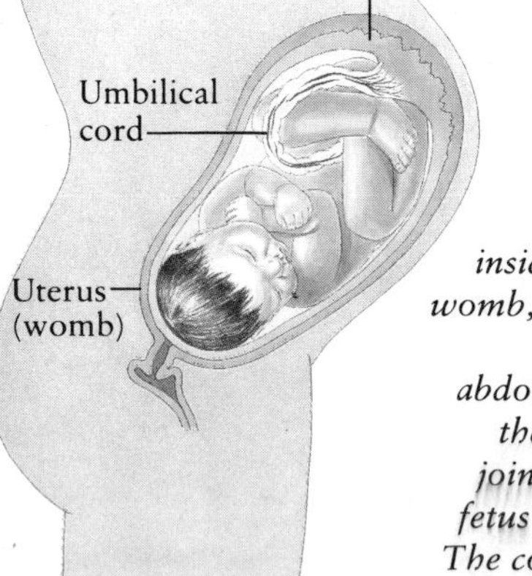

◀ *For 9 months the baby grows inside the uterus, or womb, a hollow organ in the mother's abdomen. Cells form the umbilical cord joining the growing fetus to the placenta. The cord provides the baby with air and food.*

FACTS ABOUT REPRODUCTION

- The greatest number of children born in a single birth was 8 girls and 2 boys to a Brazilian woman in 1946.
- The mother who has given birth the most times in recent decades was a woman in Chile who in 1981 produced a final total of 55 children; they included 5 sets of triplets.

REPRODUCTIVE SYSTEMS

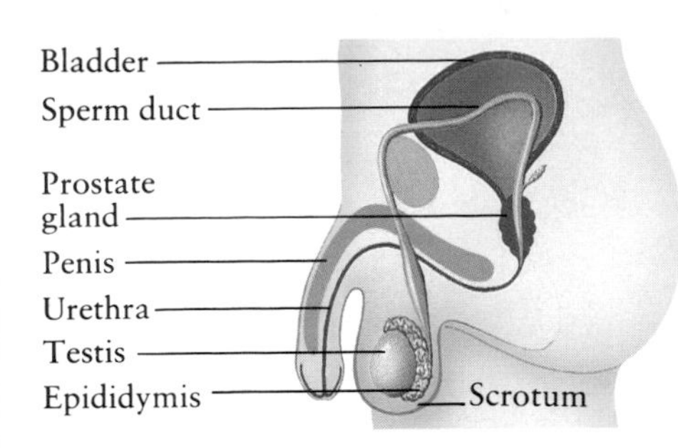

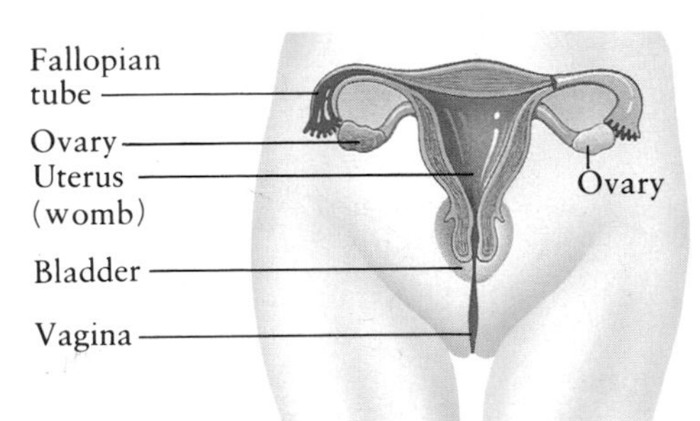

◀ *The male sex organs (genitals) produce the sexual cell, or sperm. Millions of sperm are made in the male's testes. During sexual intercourse, the sperm move through the urethra and out of the penis and then into the woman's body.*

◀ *An adult woman usually produces one egg a month from her ovaries. The egg passes into the fallopian tubes, and to the uterus. The lining of the uterus thickens, ready to nourish a fertilized egg.*

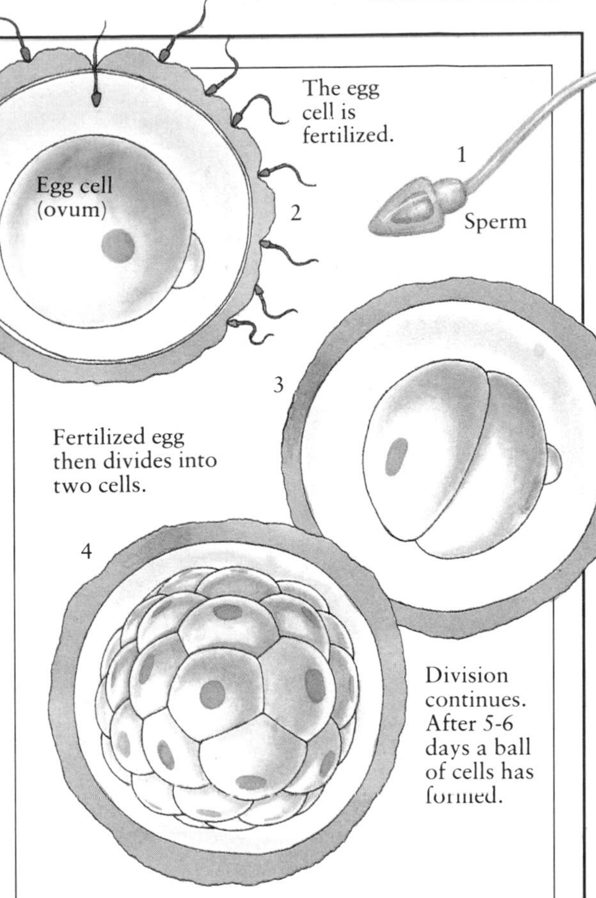

FERTILIZATION

During intercourse, millions of sperm pass from the man's body into the woman's, through the vagina. Only a few hundred reach the fallopian tube, and only one will fertilize the egg. The nuclei of the two cells (male and female) merge, and the cell begins its journey down the fallopian tube to the uterus. On the way, it grows by dividing: one cell becomes 2, 4, 8, 16, and so on.

▼ *By 4 months, the baby has doubled in size. It has well-developed features such as fingers and toes.*

▼ *At 7 months, the baby's lungs and most of its other body organs are working properly. This means that with modern care, the baby will usually survive if it is born prematurely.*

▼ *From 6 to 9 months of a mother's pregnancy, substances in her bloodstream are passed through the placenta that will help the baby to fight off diseases after its birth. At 9 months, the baby is ready to be born.*

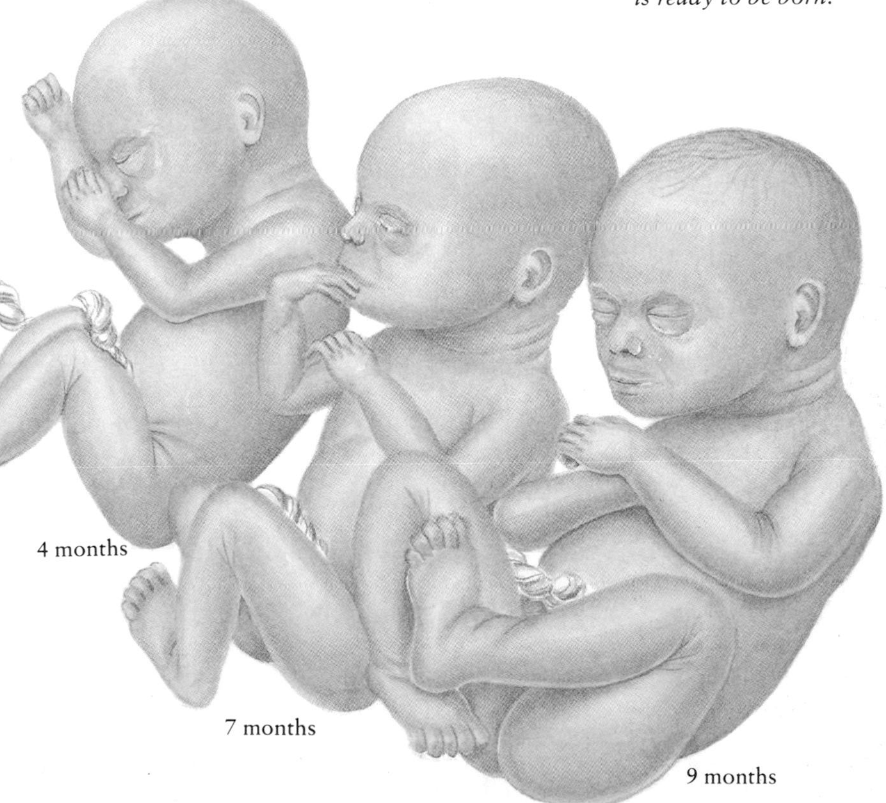

▲ *If the fertilized cell separates into two cells, two babies grow. Identical twins* (above) *have the same chromosomes, are the same sex and look alike. Two eggs fertilized at the same time by two sperm cells produce non-identical twins, with different chromosomes.*

Growth and Aging

The rate at which children grow never ceases to astonish parents. Humans grow from conception (before birth) until after puberty (about 18–20 years old). Although our bodies do not usually grow taller after that, they do go on changing—putting on or losing weight, for example. Between the ages of 20 and 30 people are at their strongest. As people get older, their body cells renew themselves more slowly, their senses become less acute and they may suffer loss of memory.

AVERAGE NORMAL GROWTH FOR BOYS AND GIRLS

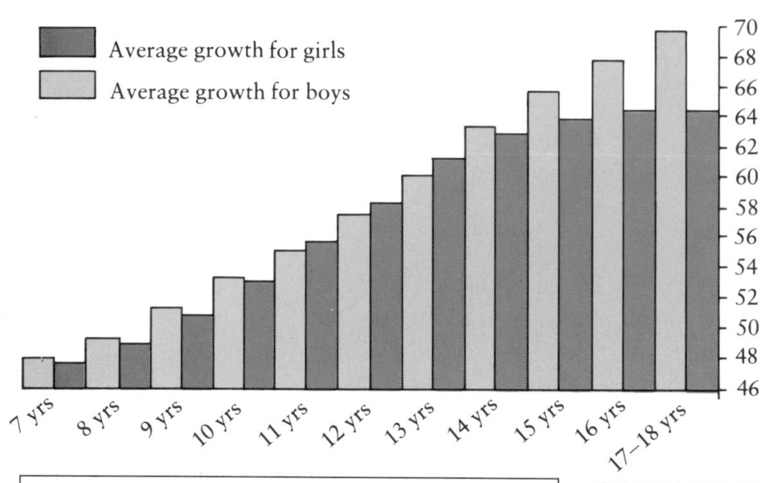

◀ *This chart shows the differences in growth between girls and boys. Girls are on average only heavier and taller than boys around the age of 12 when they start their adolescent growth spurt. By the age of 18 boys are both taller and heavier.*

▲ *Small babies learn to do many things by copying older people; By the time they are 12 months they have begun to stand and say a few words. By 18 months, infants have learned to walk, and play with simple toys such as balls and building blocks.*

Day 1: Uterus lining shed—period begins

Day 5: Uterus lining starts to build up, ready to receive a fertilized egg

Day 14–28: Most likely time for ovum to be fertilized

Day 14: Mature ovum released from ovary

MENSTRUATION

The changes to girls that occur at puberty (usually 9–14) are activated by sex hormones. A girl begins to have periods (a loss of blood). Every month the ovaries of most women of childbearing age release an egg cell. If the egg is not fertilized, it is discharged with some blood and other cells through the vagina; this period of time (3–7 days) is called the menstrual period.

GROWING UP

Humans develop slowly compared to other animals. At birth an average baby is about 20 inches (50 cm) long and weighs nearly 8 pounds (3.5 kg). The infant cannot move and depends on its mother for food, which at first is milk. By the age of two it has tripled in weight. Its hair has grown and it can walk and climb stairs. The baby has teeth and eats solid foods. It can talk and is learning rapidly. In growth, girls outstrip boys briefly around the age of 12, but after that boys grow taller and heavier. Puberty starts later in boys than in girls.

Age two

Age six

Age 10–12

SIGNS OF AGING

In boys, one visible sign of developing sexual maturity is the growth of hair on the face (a beard or mustache). This happens any time after the age of 12. Hair also grows around the sexual organs and elsewhere on the body. The voice becomes deeper. With advancing age, other body changes are common. Hair becomes gray, thins, or falls out. The skin wrinkles and muscles begin to sag.

Age 1

Age 30

Age 70

Single cell

Cell grows larger and prepares to divide.

Cell dividing into two

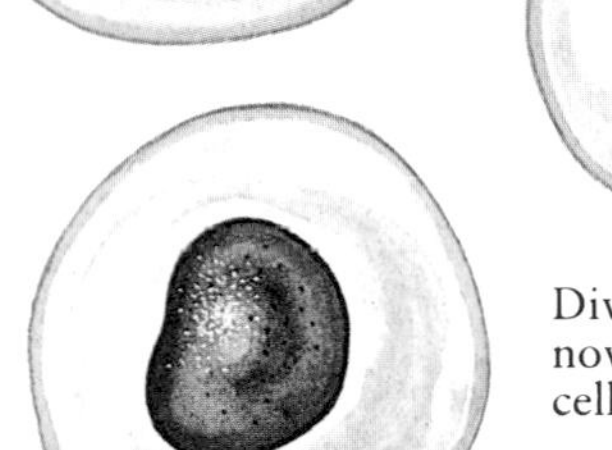
Division complete: now two identical cells

CELL RENEWAL

New cells are made by other cells dividing. The two grow to full size and divide again, and so on. This is how all living things grow and repair themselves. In our bodies, more than two million blood cells are made every second to replace old ones dying at the same rate. As we get older, cells renew themselves more slowly, and brain and nerve cells that die are not replaced.

Age 20–22

Age 30–34

FACTS ABOUT GROWTH

- The average life span in the West and Japan is over 70. Women tend to live longer than men.
- At the age of four a boy is 59 percent adult height, a girl about 64 percent. Boys may carry on growing until they are 23, most girls are fully grown by the age of 20.
- The ovum (female egg cell) is the biggest cell in the body: about the size of a period.
- The heaviest human ever weighed was an American, Jon Minnoch, (1941–1983) who weighed an estimated 1,400 pounds (635 kg) in 1978. He slimmed to 476 pounds (216 kg) by 1979.
- Only one in five people over 100 years old is a man.
- The first "test-tube" baby (conceived outside the mother's body) was Louise Brown, born in Oldham, England, in 1978.

Looking After Your Body

The body can look after itself—it has powerful defenses against disease and amazing powers of repair. But it needs sensible maintenance. Keeping healthy is mostly common sense. Eat a balanced diet of different foods, including fresh fruit and vegetables. Take exercise to keep your body fit and trim (and to enjoy yourself). Avoid harmful habits, such as smoking. Follow basic rules of hygiene (brushing your teeth, washing, taking baths and showers) to keep your whole body clean and healthy.

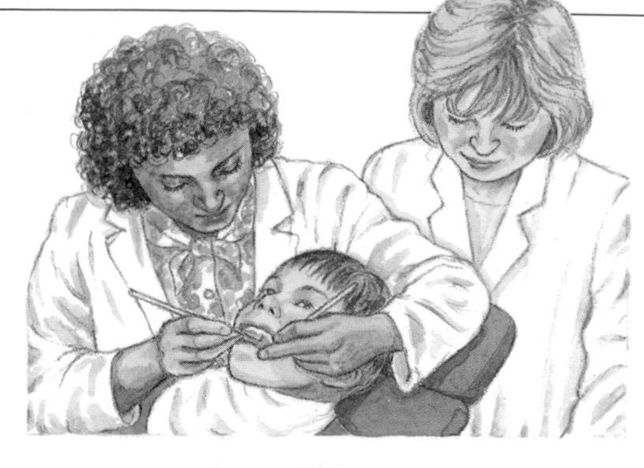

▲ *Regular checkups at the dentist's should ensure your teeth need only minor repairs—such as filling a small cavity.*

NUTRITION

Fiber: aids digestion; bread, cereals, vegetables.

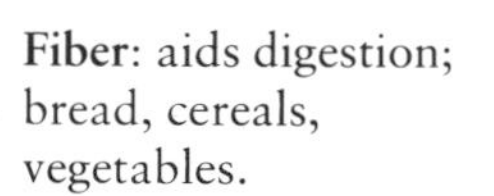

Carbohydrates: provide fuel; sugars, starches; bread, cereals, potatoes.

▲ *Nutrition is the process by which the body takes in and uses food. Our diet is the food and drink we eat. A balanced diet should contain some of each of the above foods.*

Fats: provide energy; butter, milk, cheese, eggs, meat, fish, vegetable oils, nuts.

Proteins: provide amino acids; meat, fish, eggs, milk, nuts, bread, potatoes, beans, peas.

FIGHTING TOOTH DECAY

Brushing teeth and gums regularly gets rid of tiny scraps of food sticking to them. This helps to stop tooth decay or cavities, which can cause toothache. Sugars in food and bacteria cause the tooth enamel to decay. Small holes can be filled by the dentist, but badly decayed teeth may have to be taken out.

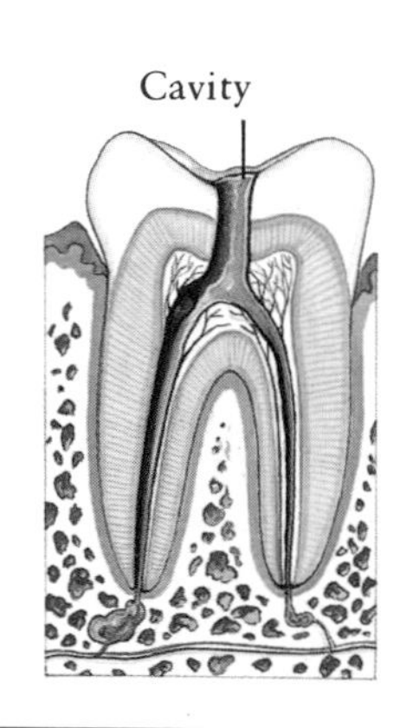

WHY WE NEED SLEEP

When you sleep, your heartbeat and breathing slow down, but the brain stays active. Most adults sleep between seven and eight hours a night, although children need more sleep. During sleep, you have periods of dreaming and often change body position.

During sleep the pattern of brain waves records dream periods.

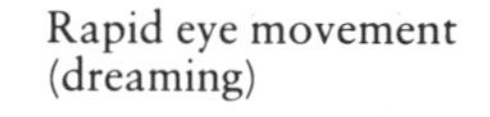

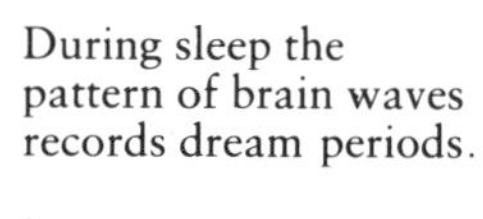

Rapid eye movement (dreaming)

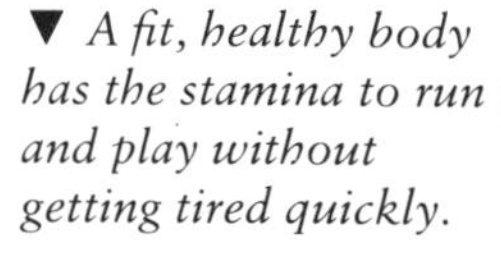

▼ *A fit, healthy body has the stamina to run and play without getting tired quickly.*

HYGIENE AND DISEASE

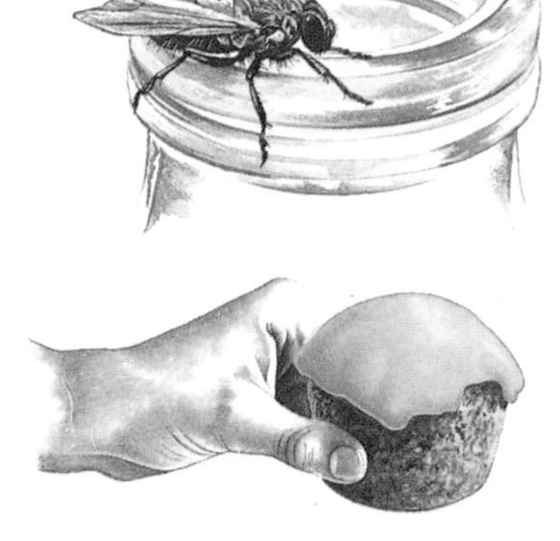

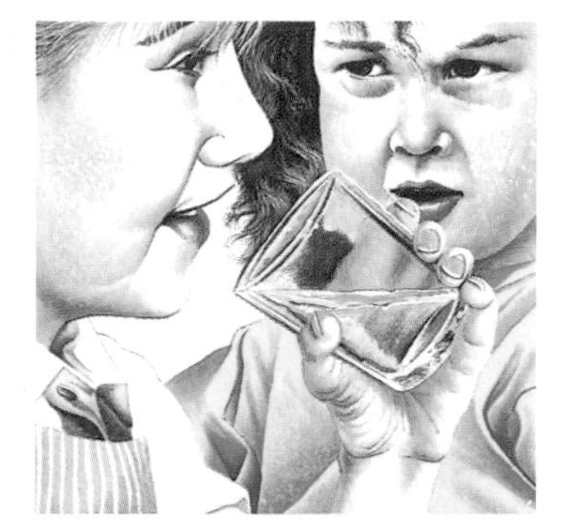

◀ *Following simple rules of hygiene can help to ward off illness—especially infectious disease. Remember to wash your hands before eating and don't share food and drink. Keep food protected from flies or mice, which leave harmful bacteria.*

DON'T POISON YOUR BODY

Putting poisonous or harmful substances into your body is not a good idea. Smoking cigarettes is known to be a cause of lung cancer. Drinking too much alcohol can seriously damage your body. Taking harmful, habit-forming drugs such as narcotics, barbiturates, tranquilizers, amphetamines, or hallucinogens can damage your health and spoil your enjoyment of life.

▶ *This sign on a bottle or container is a warning that a substance is poisonous and should never be eaten or drunk.*

VACCINATION

Vaccination or immunization works by injecting into the body either a harmless form of a poison produced by a disease-bacterium, or a weak form of the bacterium or virus itself. The body's immune system makes antibodies to fight the invading "disease," and these are stored until a real attack by the disease.

▼ *The energy from food builds the strong muscles you need to ride a bike.*

▼ *Keeping fit makes the body supple for energetic movements such as jumping.*

KEEPING FIT

You can keep fit in all sorts of enjoyable ways. Take regular exercise, rather than exhausting yourself in one outburst of energy a week. A brisk walk is good for people of all ages. Running, cycling, and swimming are other good ways to keep fit. So is dancing (but not in a smoke-filled room). Some people enjoy gymnastics, or a game of tennis, softball, or football, while others prefer activities such as canoeing, sailing, or climbing.

The greatest achievement in the history of space exploration has been landing astronauts on the Moon.

STARS AND PLANETS

A Timescale of the Universe **160** Beginnings **162** The Sun **164** The Planets **168** Mercury **170**
Venus **172** The Earth **174** The Moon **176** The Earth and the Moon **178** Mars **180**
Jupiter **182** Saturn **184** Uranus **186** Neptune **188** Pluto and Beyond **190**
Minor Planets and Meteoroids **192** Comets **194** The Milky Way **196**
Clusters and Superclusters **198** Cosmology **200** Life of a Star **202**
Extraordinary Stars **204** Black Holes and Neutron Stars **206**
Star Distance, Star Brightness **208** The Moving Sky **210**
The Constellations **212** The Birth of Astronomy **214** Optical Astronomy **216**
Radio Telescopes **218** Space Telescopes **220** A Rocket to the Moon **222**
Artificial Earth Satellites **224** Space Probes **226** Life in Space **228**
Space Shuttles and Space Stations **230** The Future in Space **232**

Our understanding of Earth and its place in the universe has progressed considerably from the days when people thought the world was flat. The first suggestion that the Earth orbited the Sun was made nearly 300 years before the birth of Christ. Nevertheless, it was not until the telescope was invented in the 17th century that the general belief that the Earth was the center of the universe was shattered forever. Our knowledge of the Solar System expanded further in the 18th and 19th centuries with the discovery of Uranus and then Neptune.

This century has seen not only the construction of even more powerful telescopes, but also the advent of space travel. Every day we receive messages from space probes that reveal the secrets of other planets and, as we receive more and more information, of the universe itself.

COSMIC TIME

A Timescale of the Universe

The universe is everything that exists. All the planets, stars, and the "star cities," or galaxies, are part of the universe, and so is all of space. The universe has no center, or edge—it seems to go on for ever. Most astronomers believe that the universe began about 15 billion years ago, in a huge explosion they call the Big Bang. They think that during the Big Bang the raw material of everything found in the universe was created in an instant of time, far shorter than anything we can measure or imagine. There is evidence to support this idea, because the galaxies in the universe seem to be flying apart, as if from an explosion. Scientists have also detected the faint heat-waves left over from a vast explosion. Since time began with the Big Bang, we cannot ask what caused the explosion, as nothing can exist without time. The Big Bang is the ultimate mystery.

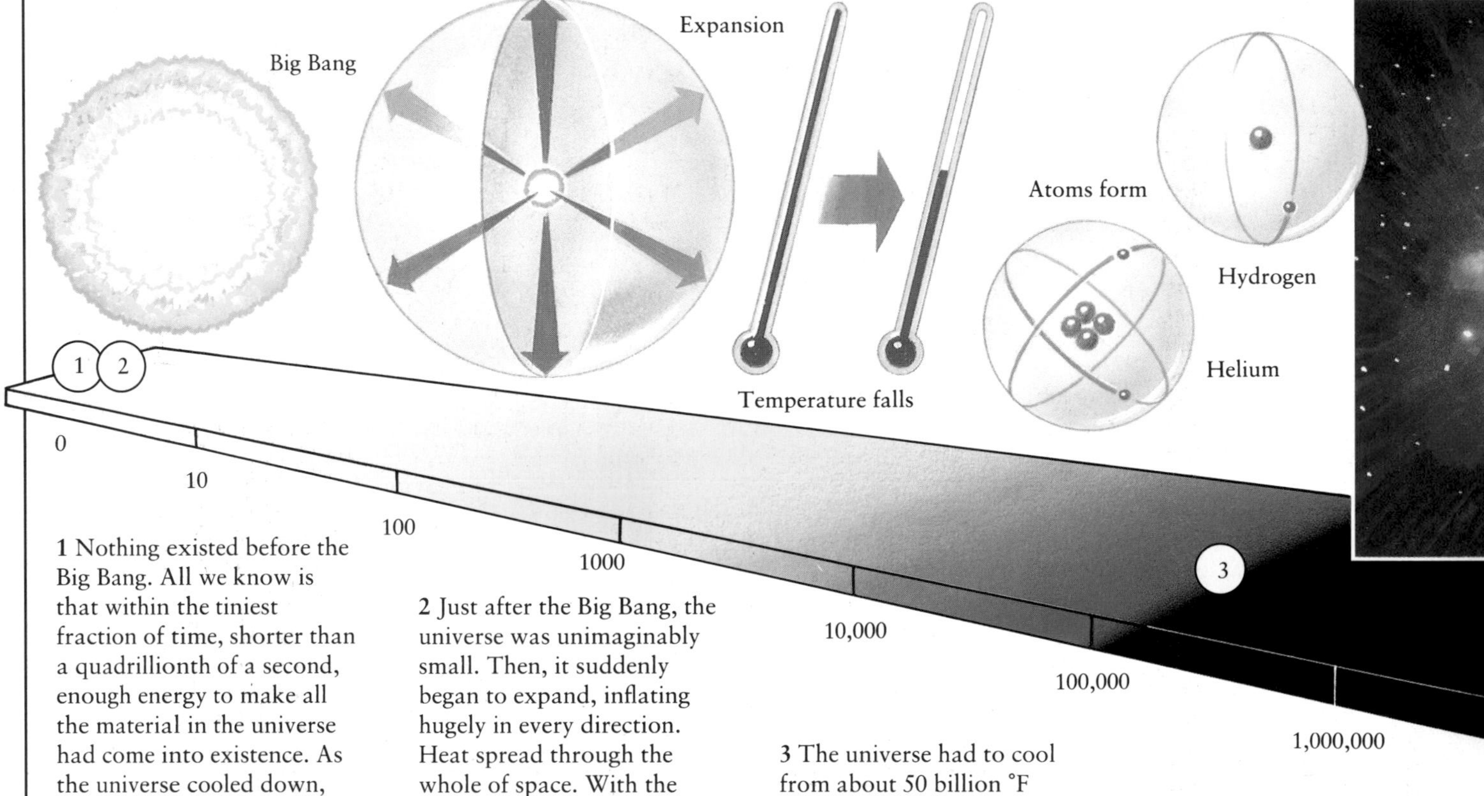

1 Nothing existed before the Big Bang. All we know is that within the tiniest fraction of time, shorter than a quadrillionth of a second, enough energy to make all the material in the universe had come into existence. As the universe cooled down, this energy was transformed into atomic particles.

2 Just after the Big Bang, the universe was unimaginably small. Then, it suddenly began to expand, inflating hugely in every direction. Heat spread through the whole of space. With the heat spread tiny ripples in the radiation given out by the explosion.

3 The universe had to cool from about 50 billion °F (10 billion °C) to 5,500°F (3,000°C) before atoms could form. Atoms are the minute units of matter. The atoms were mainly hydrogen, which is the simplest and most plentiful substance found in the universe. The rest were more complex atoms of helium.

4 Hydrogen and helium filled the universe with a thin, dark fog. The gas atoms in denser parts of the fog were pulled into separate, much smaller clouds by gravity. (Gravity is the force by which objects attract one another.) The centers of the clouds, where the gas atoms were packed together, heated up, giving birth to stars as the galaxies formed.

FACTS ABOUT MATTER AND ENERGY

- Matter, or mass, is all the material in the universe. The famous scientist Albert Einstein (1879–1955) suggested that energy can be turned into matter, and that matter can be turned into energy. His theory is the basis for all our ideas on the beginning of the universe.
- Radiation is a form of energy. Particles with high-energy are a form of radiation. Some radiation, like that from a fire, can be felt as heat.

◀ *Astronomers can examine the past and see what the universe was like long before the Earth formed. Light travels at great speed—186,000 mi./s (300,000 km/s). However, distances in space are so huge that light still takes years to reach us from the nearest stars—the ones that shine in the night sky. Light takes billions of years to reach us from remote galaxies. The light we receive from a distant galaxy now shows that galaxy, not as it is today, but as it used to be all that time ago. This means astronomers can observe objects formed when the universe was young.*

▼ *In 1992, it was announced that the* Cosmic Background Explorer *satellite* (COBE) *had traced the background radiation and "ripples" left over from the Big Bang, 15 billion years before.*

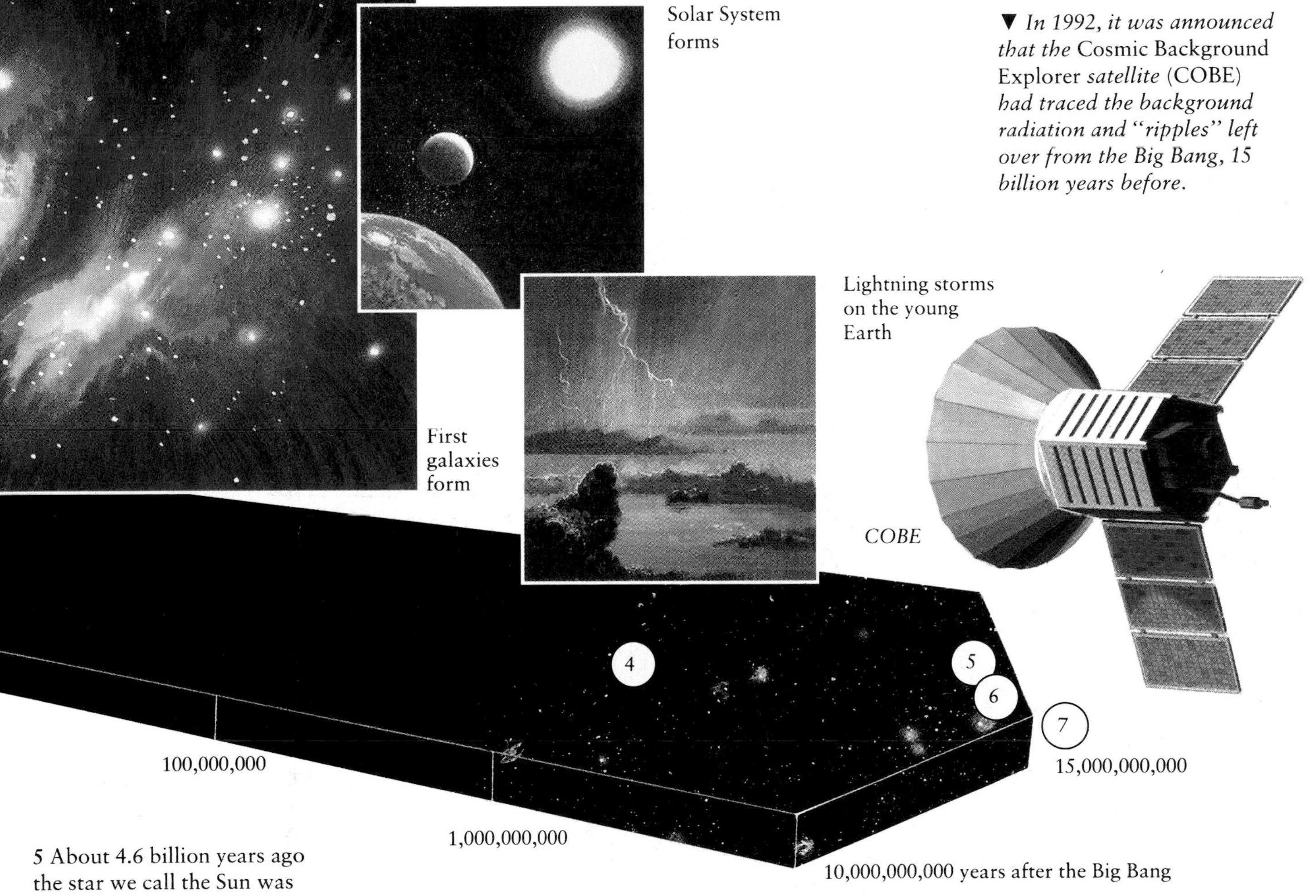

5 About 4.6 billion years ago the star we call the Sun was formed. Around it was a cloud of gas and dust containing substances such as carbon and oxygen. These had been formed in older stars and blasted out into space when the stars died. These substances came together to form the planets.

6 The first living cells appeared on Earth 3.5 billion years ago. How life began is uncertain. Maybe lightning storms provided the energy to start chemical reactions in the "soup" of elements on the young planet.

7 It has taken one ten-thousandth of the time since the Big Bang for recognizable humans to develop from apes. Today, scientists try to work out the story of the universe by sending into space satellites which look back across time.

THE SOLAR SYSTEM

Beginnings

We live on a small rocky planet we call the Earth, which travels through space on an orbit, or path, around a star we call the Sun. The Earth is part of the Solar System—the name we give to the Sun and the family of planets, asteroids, and comets that orbit it. This family reaches far into space—the Solar System is about a million times wider than Earth. "Solar" means "of the Sun," and the Sun is by far the most important member of the family. It is about 740 times more massive than all the planets put together. Because of its great size, it has a powerful gravitational pull, and this pull keeps the Solar System together and controls the movements of the planets. The Solar System began about 4.6 billion years ago, when a cloud of hydrogen, helium, and a tiny percentage of other elements started to condense into a cluster of stars. One of these stars was the Sun.

THE BIRTH OF THE SOLAR SYSTEM

2

3

4

The "empty" space between stars in fact contains hydrogen atoms and tiny grains of solid material. The grains lie very far apart from each other. However, in some regions of a galaxy the grains are found much closer together, forming dark clouds of gas and dust. These regions are called nebulae.

Stars are formed from a nebula when the material in the nebula is given a shake, making it break up into much smaller nebulae (**1**). For example, a star might explode nearby and send out powerful energy waves. In some colliding galaxies, stars are formed where nebulae meet and pass through each other.

The gravity of a small dark nebula, or globule, begins to pull itself inward (**2**). Pressure causes the center to heat up. The dark cloud may take less than 100,000 years to change into a shining star (**3**). The baby star spins on its axis, throwing off two spiral arms that form a ring around it (**4**).

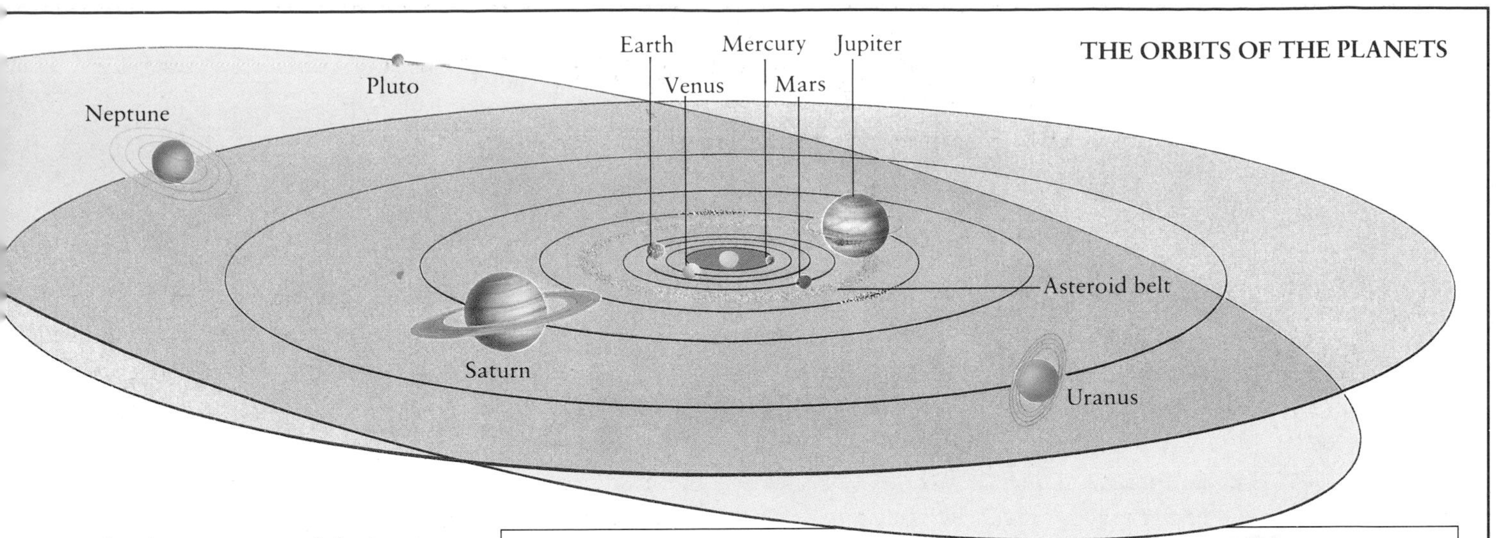

▲ *The planets go around the Sun in elliptical orbits, which means they follow a path shaped like a flattened circle. Pluto's orbit is also tilted.*

PLANET	SPEED IN ORBIT mi./s (km/s)
Mercury	29.7 (47.9)
Venus	21.7 (35.0)
Earth	18.5 (29.8)
Mars	15.0 (24.1)
Jupiter	8.1 (13.1)
Saturn	5.9 (9.6)
Uranus	4.2 (6.8)
Neptune	3.3 (5.4)
Pluto	2.9 (4.7)

The Sun's gravity pulls the planets inward. At the same time, the planets' own energy of motion is trying to fling them off into space. These two forces balance exactly. The closer a planet is to the Sun, the faster it has to move to maintain this balance.

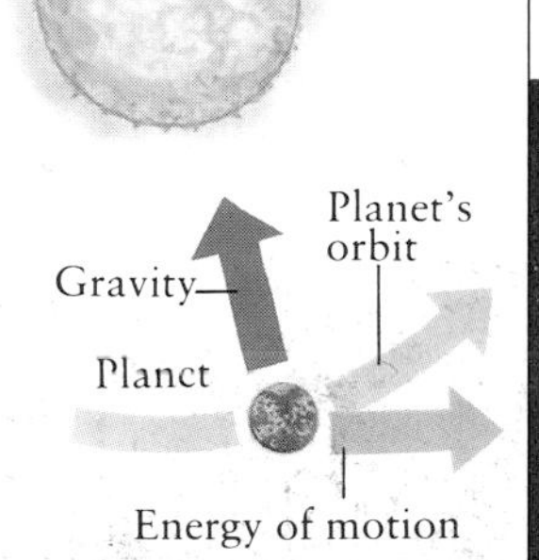

As a star like the Sun forms, the force of its inward collapse heats its center. When the temperature inside the star reaches millions of degrees, nuclear reactions are able to start. These send a new wave of energy outward (5), which keeps the star from shrinking further and blasts the remains of the nebula out into space.

The ring of gas and dust thrown out by our young Sun began to collect into fragments and solid grains of matter. Once these "planetesimals" reached a certain size, their gravity began to pull in other, smaller fragments, and they grew rapidly (6). Gradually they turned into the planets we know today (7).

The inner planets have hard surfaces. The giant outer planets grew into rocky globes too, but they then attracted the hydrogen and helium gases and icy particles that collected in the cold regions of the Solar System. Some smaller planetesimals became asteroids *(see pages 192–193)*; others became moons of the planets.

The Sun

The Sun lies at the center of our Solar System, a fiercely-hot ball of gas nearly 900,000 mi. (1.4 million km) wide. Its appearance changes all the time: prominences leap from the Sun into space, and dark spots appear on its surface. Since its birth, some 4.6 billion years ago, the Sun has been the power station for the Earth and the other planets, providing them with their light and heat. The source of the Sun's energy lies deep inside its center, where the nuclear reactions that keep it shining take place.

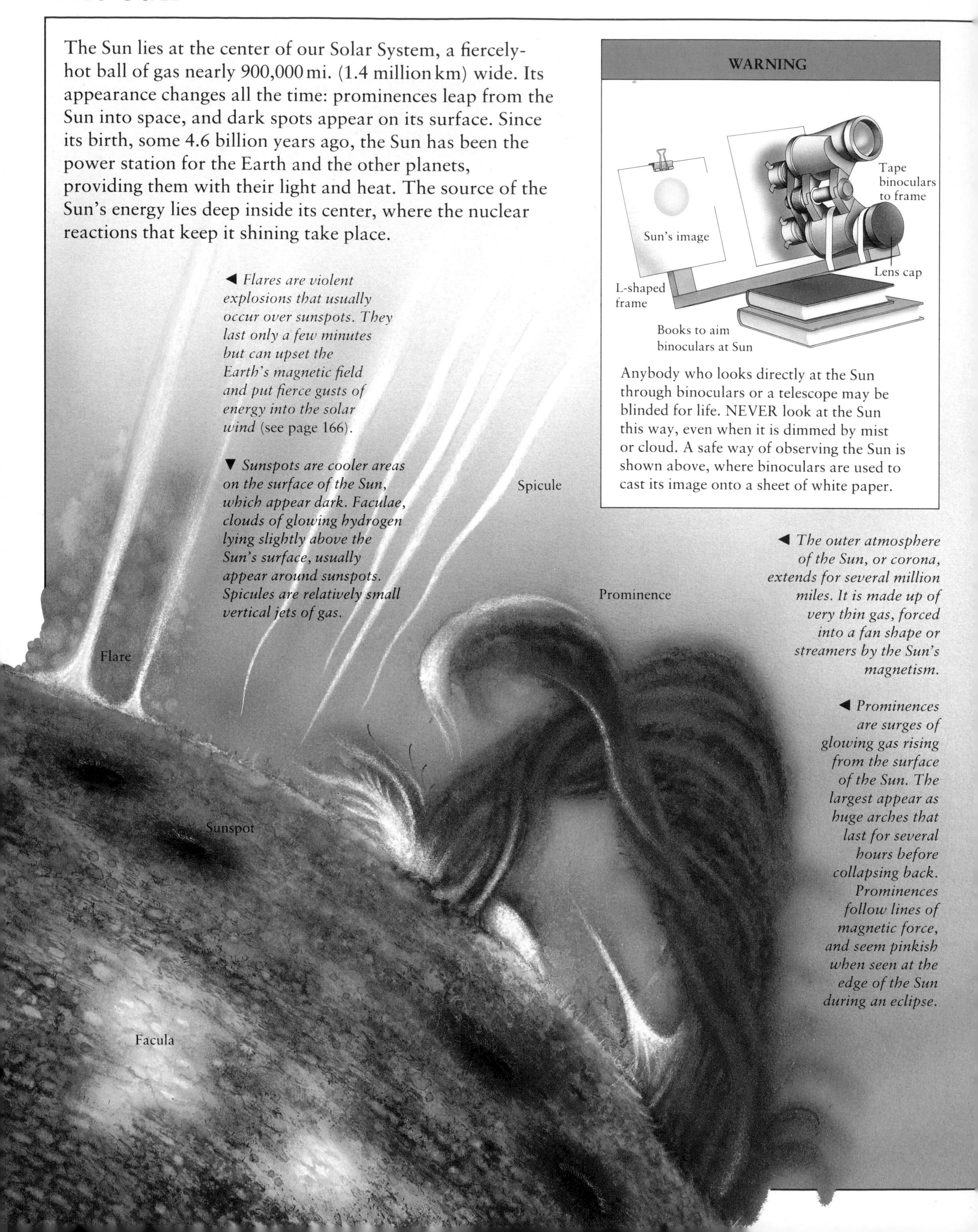

WARNING

Anybody who looks directly at the Sun through binoculars or a telescope may be blinded for life. NEVER look at the Sun this way, even when it is dimmed by mist or cloud. A safe way of observing the Sun is shown above, where binoculars are used to cast its image onto a sheet of white paper.

◀ *Flares are violent explosions that usually occur over sunspots. They last only a few minutes but can upset the Earth's magnetic field and put fierce gusts of energy into the solar wind* (see page 166).

▼ *Sunspots are cooler areas on the surface of the Sun, which appear dark. Faculae, clouds of glowing hydrogen lying slightly above the Sun's surface, usually appear around sunspots. Spicules are relatively small vertical jets of gas.*

◀ *The outer atmosphere of the Sun, or corona, extends for several million miles. It is made up of very thin gas, forced into a fan shape or streamers by the Sun's magnetism.*

◀ *Prominences are surges of glowing gas rising from the surface of the Sun. The largest appear as huge arches that last for several hours before collapsing back. Prominences follow lines of magnetic force, and seem pinkish when seen at the edge of the Sun during an eclipse.*

HOW THE SUN SHINES

The temperature at the center of the Sun is about 27 million °F (15 million °C). Here, the atoms that make up its main gas, hydrogen, have so much energy that they break apart, coming together again as helium gas. During this reaction, a burst of energy is given out. This energy drives the Sun. It is thought the Sun contains enough hydrogen to keep giving out energy for billions of years. As it uses up its fuel, our star will change *(below)*.

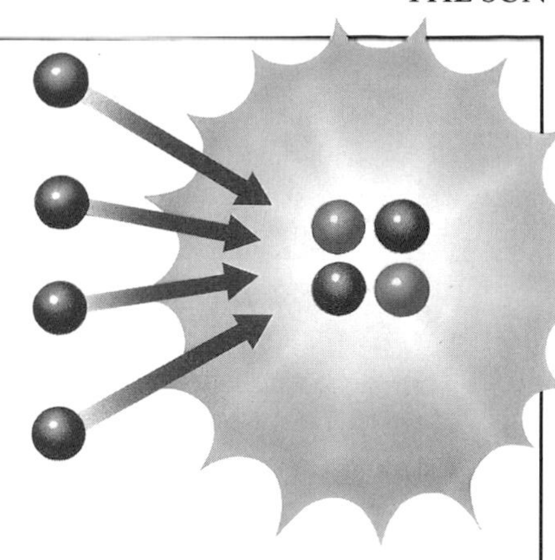

► *The Sun will continue to shine almost unchanged for several billion years. Meanwhile, the Earth will continue to pass through ice ages and warm periods as its orbit goes through a regular cycle of slight change.*

◄ *In about 5 billion years' time, energy from the Sun's huge core will make its outer layers expand. As our star swells and grows hotter, the water on Earth will start to boil away. Life forms will suffocate in the great heat.*

► *As the Sun turns into a red giant star, Earth will be scorched to a cinder, and its atmosphere will be stripped away. A few million years later the thin outer layers of the Sun will have consumed the Earth; Mars will probably escape.*

◄ *After the red giant stage, lasting about 100 million years, the Sun will run out of nuclear fuel. It will shrink and become a white dwarf star. From the surface of Mars* (left) *it will be a dim pinpoint. The Earth will no longer exist.*

THE STRUCTURE OF THE SUN

Light and heat are produced in the core of the Sun. This energy then flows in waves through the radiative zone, with sufficient force to stop the vast bulk of the Sun from collapsing inward under gravity. The energy waves are weakened by this journey so that when they reach the convective zone they can radiate no further. Instead, the energy waves reach the visible surface of the Sun (the photosphere) by a violent churning motion called convection.

Every second, the Sun loses 7 million tons of material, but all the material lost so far only amounts to less than 0.01 percent of its mass since it started shining. Of this amount, 4 million tons are turned into energy when protons and neutrons fuse together, giving out radiation and falling into the core. The Sun's lifetime is not limited by the amount of its fuel, but by the growth of its core. When the core reaches a certain size, the Sun will expand and it will start to destroy the Earth *(see page 165)*.

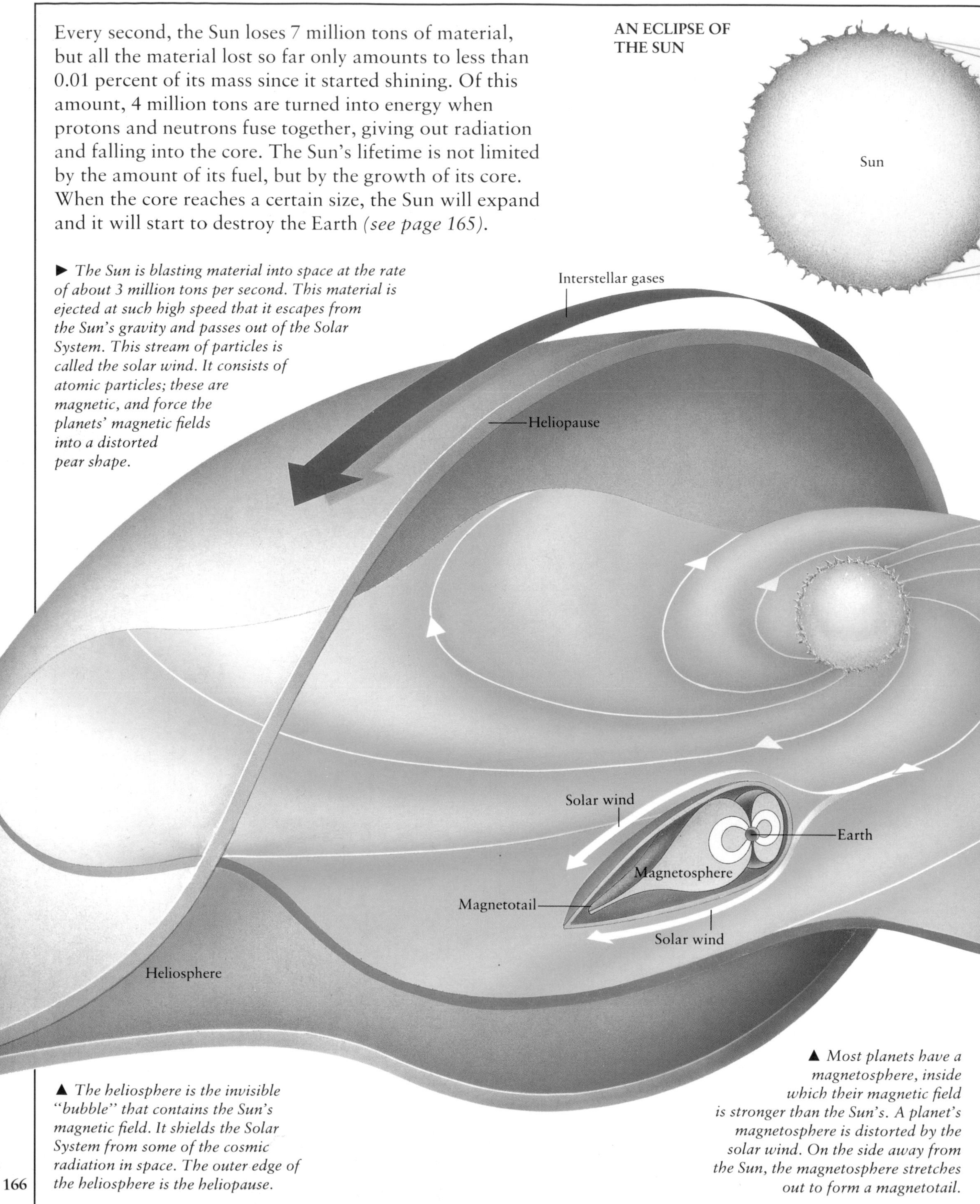

► *The Sun is blasting material into space at the rate of about 3 million tons per second. This material is ejected at such high speed that it escapes from the Sun's gravity and passes out of the Solar System. This stream of particles is called the solar wind. It consists of atomic particles; these are magnetic, and force the planets' magnetic fields into a distorted pear shape.*

▲ *The heliosphere is the invisible "bubble" that contains the Sun's magnetic field. It shields the Solar System from some of the cosmic radiation in space. The outer edge of the heliosphere is the heliopause.*

▲ *Most planets have a magnetosphere, inside which their magnetic field is stronger than the Sun's. A planet's magnetosphere is distorted by the solar wind. On the side away from the Sun, the magnetosphere stretches out to form a magnetotail.*

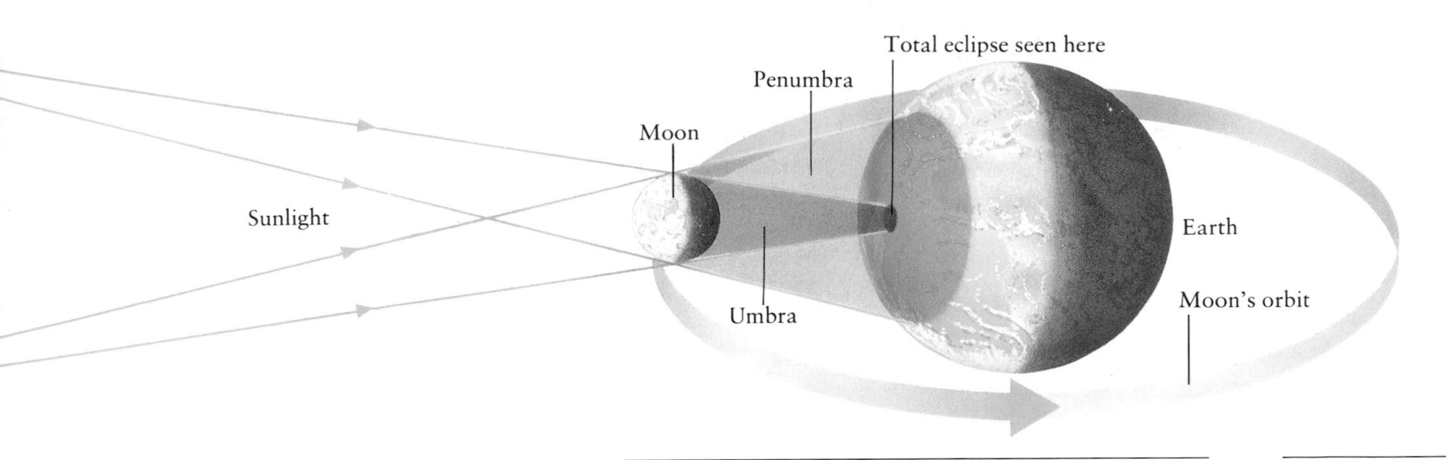

▲ *The Moon casts a tapering shadow in space. When it passes between the Earth and the Sun, the shadow's tip may cross the Earth, causing an eclipse. Inside the shadow, the Sun is blocked from view, and all that can be seen in the dark sky is the Sun's corona. An eclipse is total only if seen from within the Moon's umbra, or central shadow. Inside the penumbra, or outer shadow, the Sun's disk is not completely hidden.*

TYPES OF ECLIPSE

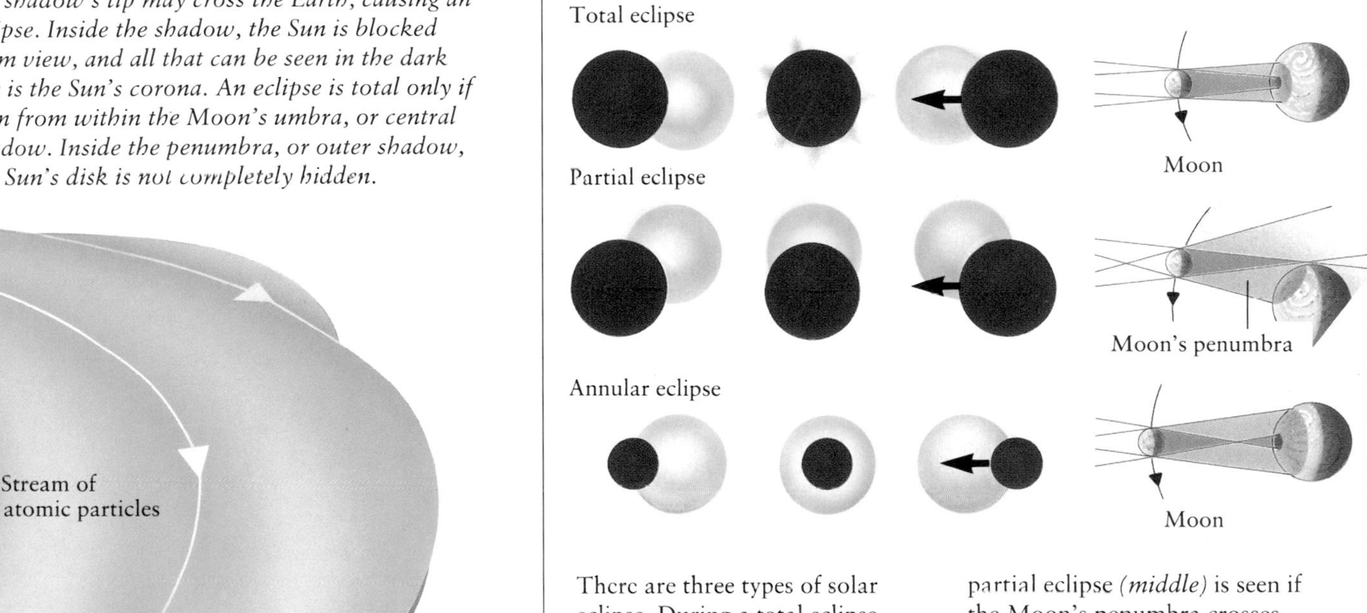

There are three types of solar eclipse. During a total eclipse *(top)* the Moon totally covers the Sun. This is visible from a narrow strip crossing the Earth's surface, usually about 90 mi. (150 km) wide. Beyond this strip a partial eclipse is seen. Only a partial eclipse *(middle)* is seen if the Moon's penumbra crosses Earth, instead of the umbra. An annular eclipse *(below)* happens when the Moon is at its greatest distance from Earth (called the apogee) and appears too small to cover the Sun completely.

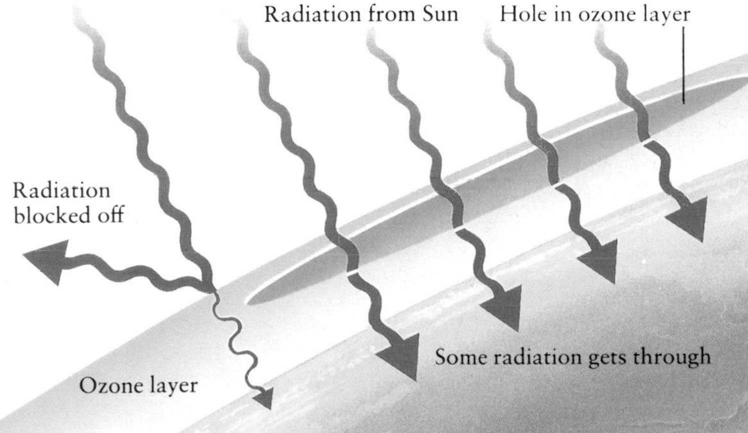

► *The Sun gives out invisible radiation which can destroy living tissue. This radiation is filtered by a thin layer of ozone in the stratosphere before it reaches Earth. However, a "hole" in the layer, near the South Pole, has been caused by gases known as chlorofluorocarbons (CFCs). CFCs are now being banned.*

▲ *This plant has been affected by the kind of radiation usually blocked out by the ozone layer.*

The Planets

A planet is a large body, made of gas, metal, or rock, that orbits a star. We know of nine planets that orbit our Sun. All of them were formed at the same time, from the same cloud of gas and dust around the Sun, but there are great differences between them. The four inner terrestrial planets (Mercury, Venus, Earth, and Mars) are rock and metal. Jupiter, Saturn, Uranus, and Neptune are mainly liquid and ice; these giant outer bodies are called the gaseous planets. Distant Pluto does not fit in either group.

THE DENSITY OF THE PLANETS
Compared with the gaseous planets, the inner rocky planets have no more than a thin skin of atmosphere. They contain much more material per unit of volume than the gaseous planets do, which means they are much denser. Water has a density of one gram per cubic centimeter. The Earth is five and a half times denser than water. Saturn, however, is *less* dense than water. It could float on a huge ocean.

▼ *Although Mercury is the innermost planet, it is not the hottest. It does have the shortest year, going around the Sun once every 88 days. It has a cratered surface and no satellites, or moons.*

▼ *Venus is the brightest object in the sky, but the planet is not shining by its own light. Venus is bright because the clouds that cover its surface reflect the Sun's light very well. All the planets shine by reflection.*

Mercury

Sun

Venus

Earth

Mars

Jupiter

▲ *The Earth is the only planet with liquid water on its surface. It is also the densest planet—almost eight times as dense as Saturn.*

▲ *The highest mountain and the deepest valleys in the Solar System are found on Mars, so it may once have been the most volcanically active planet.*

▲ *The largest planet, Jupiter, also spins fastest on its axis—its day lasts less than 10 Earth hours. This very rapid spin produces a force which makes the planet's liquid body bulge outward at the equator.*

MEASURING DISTANCES IN SPACE
Astronomers use special units to represent huge distances in space. These distances are measured in light-years: one light-year is equal to the distance traveled by a beam of light in one year, or 5.88 trillion mi. (9.46 trillion km). The basic unit of distance within our Solar System is the astronomical unit, or AU. One AU is the average distance from the Earth to the Sun —about 92,960,000 mi. (149,600,000 km).

Mercury 0.39 AU
Venus 0.72 AU
Earth 1 AU
Mars 1.52 AU
Jupiter 5.20 AU
Saturn 9.54 AU
Uranus 19.19 AU

0 1 2 3 4 5 6 7 8 9 10 11 12 13 14 15 16 17 18 19 20 21 22 23

FACTS ABOUT THE PLANETS

• We owe much of our knowledge about the Solar System to space probes. These unmanned spacecraft have investigated nearly every planet, sending back pictures to Earth. The *Voyager* probes were especially important. Between 1979 and 1989 the probes sent back close-up pictures of the four giant planets and their moons, and discovered rings around Jupiter, Uranus, and Neptune.

• Only Mercury and Venus have no natural satellites, or moons, at all. Saturn has the greatest number of moons: 18.

• The planets are named after ancient Greek and Roman gods; for example, Jupiter is named after the chief of the Roman gods.

• After the formation of the Solar System, leftover rocky debris in the cloud of particles around the Sun kept striking the planets as they formed. Mercury and our Moon still show the scars of impacts, but the Earth's craters have been smoothed out by weather and movements of the planet's surface. This main bombardment ended 3.9 billion years ago; some major impacts have occurred since.

Uranus

◀ *Almost nothing was known about Uranus until the space probe* Voyager 2 *flew past it in 1986. Uranus's axis is tipped over so that it spins almost sideways.*

▼ *Saturn used to be known as "the planet with the rings." Ring systems have now been discovered around all the giant planets, but Saturn's system is by far the largest and most complicated. Only Saturn's rings can be seen through a telescope from Earth.*

Saturn

Neptune

Pluto

▲ *Until 1999, Neptune will be the outermost planet in the Solar System, since Pluto's orbit has brought it briefly closer to the Sun. Neptune has a cold, windswept, cloudy surface.*

▲ *We know very little about distant Pluto, which was discovered only in 1930. It is the smallest planet, and is probably made of ice. No space probe has ever visited this mysterious world.*

Pluto 29.6 AU (at its closest) | Neptune 30.1 AU | Pluto 49.5 AU (at its farthest)

26 27 28 29 30 31 32 33 34 35 36 37 38 39 40 41 42 43 44 45 46 47 48 49 50

Mercury

Mercury is the innermost planet. It is a dead, airless world that whirls through space in the merciless glare of the Sun. The one spacecraft to have visited it was *Mariner 10* (in 1974), which photographed half the planet. During its daytime, which lasts for about three Earth months, it is so hot that lead would run like water over the rocks. But at night its surface is colder than icy Jupiter. Mercury holds two records for the major planets of the Solar System: the longest day and the shortest year.

▲ *Mercury is the second smallest of the planets, after Pluto. It is also smaller than two satellites, Jupiter's Ganymede and Saturn's Titan. Because it is so close to the Sun, it is bombarded by the solar wind.* (Distances not to scale.)

Earth

Mercury

◀ *Mercury's surface is heavily scarred with craters. They are the result of collisions during the first few hundred million years of the Solar System's history. There are also lava flows, which must have occurred early in Mercury's history, as they are also cratered.*

▼ *Mercury's distance from the Sun changes as it rounds its orbit. From the planet's surface, the Sun would appear 1.5 times bigger when Mercury is closest than when it is farthest away.*

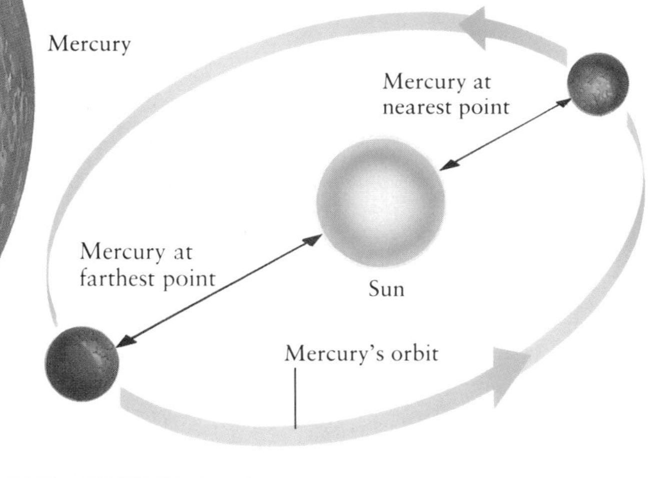

MERCURY DATAFILE

Diameter: 3,030 mi.
Mass: 0.06 × Earth
Density: 5.4 (water = 1)
Minimum distance from Sun: 28.5 million mi.
Maximum distance from Sun: 43.3 million mi.
Minimum distance from Earth: 28 million mi.
Day/night: 176 Earth days
Length of year: 88 Earth days
Tilt of axis: 0° 0′
Surface gravity: 0.38 × Earth
Temperature: −300°F to 800°F
Satellites: 0

STRUCTURE

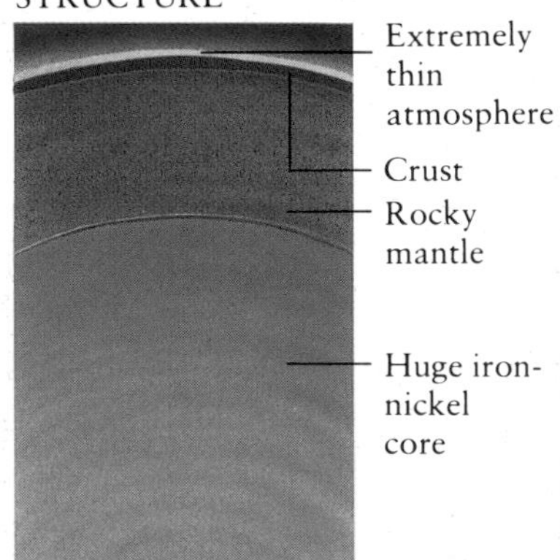

FACTS ABOUT MERCURY

- The length of Mercury's year is shorter than its day. The planet has the shortest year in the Solar System—only 88 Earth days—but sunrise to sunrise on Mercury takes 176 Earth days.
- It is possible that some of Mercury's rocky mantle was knocked off by an impact with another body. Most of the interior is taken up by a huge iron core.

▲ *The smooth plains between the craters on Mercury suggest the planet might once have been a volcanic world. Following the formation of the Solar System, lava flows would have filled in the heavily cratered areas remaining after the bombardment of the planets.*

Impact with huge object

Shock waves travel through the planet

Caloris Basin, 800 miles across

Ridges form on the opposite side of Mercury

THE FORMATION OF THE CALORIS BASIN

The dominant feature on Mercury is the Caloris Basin. It is an ancient lava-filled crater about 800 mi. (1,300 km) across —a quarter of the planet's diameter. It is thought that the basin was formed when a huge body, about 60 mi. (100 km) across, crashed into Mercury. Shock waves from the collision traveled around the planet, meeting on the opposite point and throwing up a confusion of ridges. Half the Caloris Basin was in darkness when visited by *Mariner 10*, so that it has never been seen clearly.

THE FORMATION OF RIDGES

Mercury is the second densest of all the planets, so it must have a large metallic core, presumably of iron. When heated or cooled, iron changes its size much more than rock. So as the hot core cooled over the hundreds of millions of years after the planet formed, it began to shrink, making the hard crust go loose and wrinkled like the skin on a dried apple. These wrinkle ridges are known as rupes, and are found all over Mercury. They are up to 2 mi. (3 km) high.

▼ *There is practically no atmosphere on Mercury to reflect the light of the Sun, and so the sky is always black. The Sun takes about three of our months to pass across the sky, but some of the deepest craters have been in darkness for billions of years. The surface is crossed with huge wrinkle ridges, such as Discovery Rupes, 2 mi. (3 km) high and about 300 mi. (500 km) long.*

Venus

Through a telescope Venus appears as a gleaming, silvery gem. But appearances deceive, because the planet is in fact a rocky waste, hotter than Mercury and spread out under a choking carbon dioxide atmosphere that is denser than water. Sulfuric acid droplets fall on the surface from clouds that permanently cast an orange gloom.
The surface features shown on these pages (mountains, craters, and volcanoes) were detected by spacecraft radar —the only way of probing the thick clouds.

▲ *Although the surface conditions are totally different, Venus is almost a twin of the Earth in size. Venus is very hot because its thick atmosphere is very efficient at holding in the Sun's heat.* (Distances not to scale.)

◀ *The orbit of Venus lies between the Earth and the Sun. This means it can be seen only in the twilight after sunset or before sunrise. Venus is sometimes called the morning star or the evening star.*

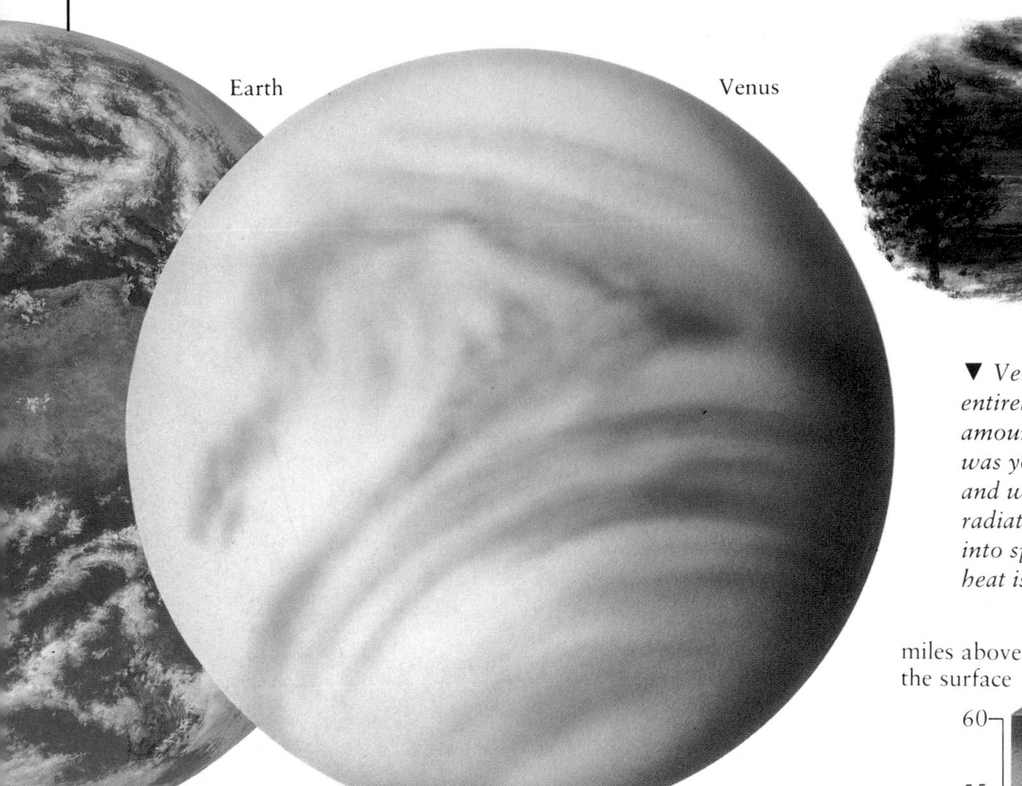

▲ *The surface of Venus is completely hidden by dense clouds.*

▼ *Venus's atmosphere is made up almost entirely of carbon dioxide, produced in vast amounts from erupting volcanoes when the planet was young. Sunlight penetrates the atmosphere and warms the surface of the planet. The ground radiates heat waves, but they cannot escape back into space because of the thick cloud layer. The heat is trapped, warming the planet still more.*

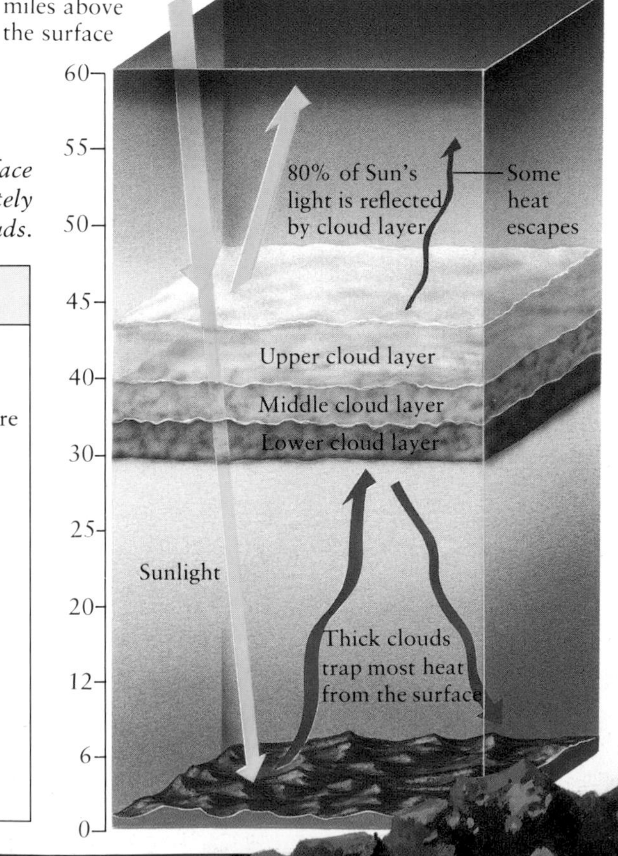

VENUS DATAFILE

Diameter: 7,545 mi.
Mass: 0.82 × Earth
Density: 5.2 (water = 1)
Minimum distance from Sun: 67 million mi.
Maximum distance from Sun: 68 million mi.
Minimum distance from Earth: 25 million mi.
Day/night: 117 Earth days
Length of year: 225 Earth days
Tilt of axis: 12° 42′
Surface gravity: 0.90 × Earth
Temperature: 900°F average
Satellites: 0

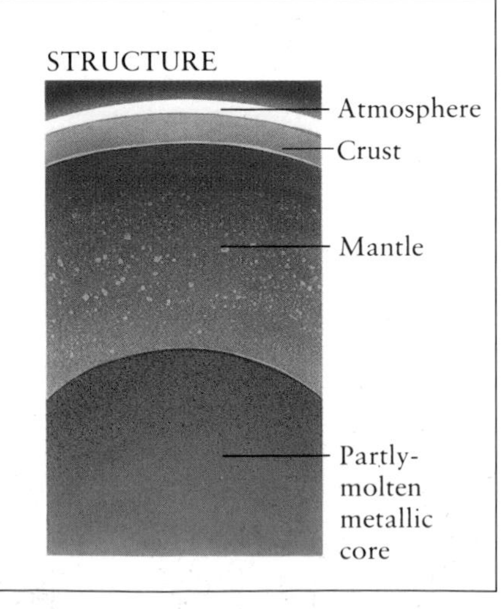

◄ *An artist's impression of Venus's surface, based on radar information. The highland areas, or continents, are colored yellow. The mountain Maxwell Montes is 7 mi. (12 km) high. Its peaks are the second highest in the Solar System. About 80 percent of the surface of Venus is covered with dusty, rocky lava plains, which have smothered most of the early craters. A well-known crater, Cleopatra, is 100 mi. (160 km) across.*

► *The* Magellan *spacecraft mapped 99 percent of the surface of Venus in 1990–1992. It recorded features as small in size as a football field.*

◄ *Maat Mons is a volcanic feature on Venus, 5 mi. (8 km) high. Radar information from* Magellan *was processed by computer technology to produce this image.*

► *Sulfur dioxide from early volcanic activity has helped to form dense sulfuric acid clouds. These spread in a corrosive mist over the surface of Venus.*

The Earth

As far as we know, the Earth is unique in the Solar System for two reasons: it has liquid water on its surface, and it supports life. Both are probably dependent on each other. Without water, the type of plant life we know could not have flourished; without plant life, oxygen would not have been released into the early carbon dioxide atmosphere to make air for animals to breathe. A permanent carbon dioxide covering might have created a similar atmosphere to that on Venus, turning Earth's surface into a desert.

▲ *The blue-white Earth is the largest of the four inner terrestrial planets. Its color contrasts strongly with that of its neighboring planets—jewel-bright Venus and reddish Mars.* (Distances not to scale.)

▼ *The Earth's atmosphere is about 78 percent nitrogen, 21 percent oxygen, and 1 percent other gases. Half of the Earth's atmospheric material is found in the troposphere, 6 mi. (10 km) high. The stratosphere contains the ozone layer which absorbs dangerous ultraviolet rays from the Sun. The mesosphere is where meteoroids, or small space bodies, burn up because of air resistance.*

EARTH DATAFILE

Diameter: 7,926 mi.
Mass: 6.5 sextillion tons
Density: 5.5 (water = 1)
Minimum distance from Sun: 91 million mi.
Maximum distance from Sun: 94.5 million mi.
Day/night: 24 h
Length of year: 365 days 5 h
Tilt of axis: 23° 27′
True rotation period: 23 h 56 min
Maximum temperature: 136°F
Minimum temperature: −128°F
Satellites: 1

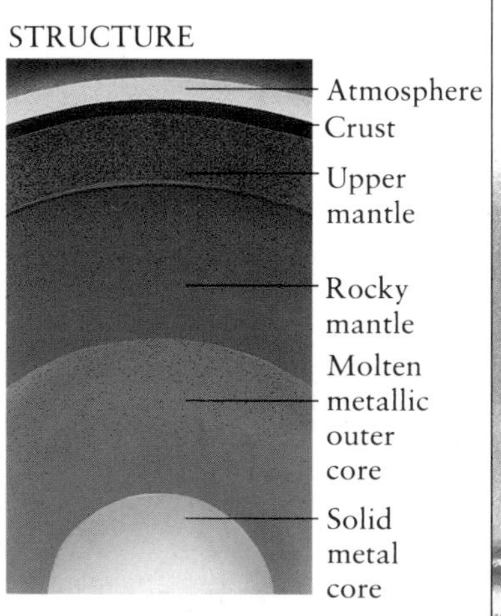

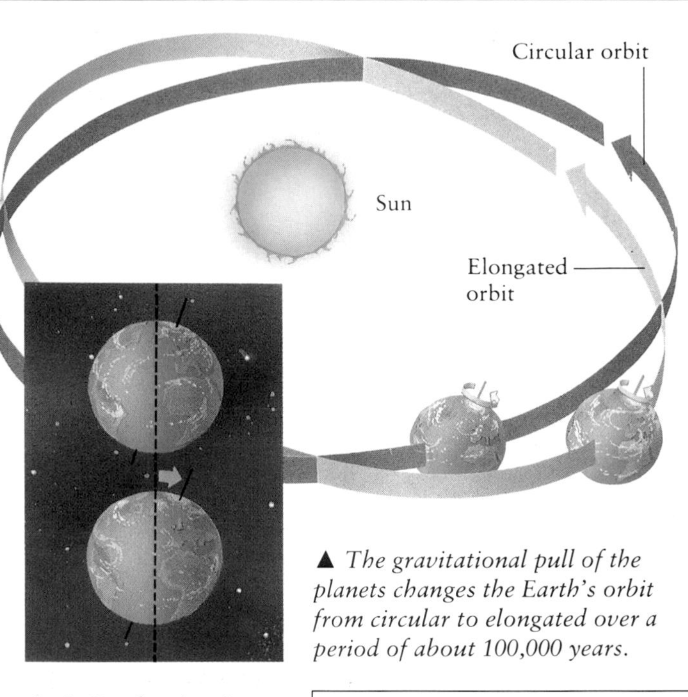

▲ *The gravitational pull of the planets changes the Earth's orbit from circular to elongated over a period of about 100,000 years.*

▲ *As Earth spins, its axis slowly rotates. One complete rotation takes 25,800 years. This is known as precession.*

► *The Earth has had a number of Ice Ages, the last one thousands of years ago. These may have been caused by changes in the shape of the Earth's orbit and the tilt of its axis. Today, summers are warm and winters cool. But changes to the Earth's orbit and tilt could mean that the temperature hardly changes and winter snow does not melt in the summer. Instead, ice reflects the Sun's radiation back into space, the temperature falls further, and an Ice Age begins.*

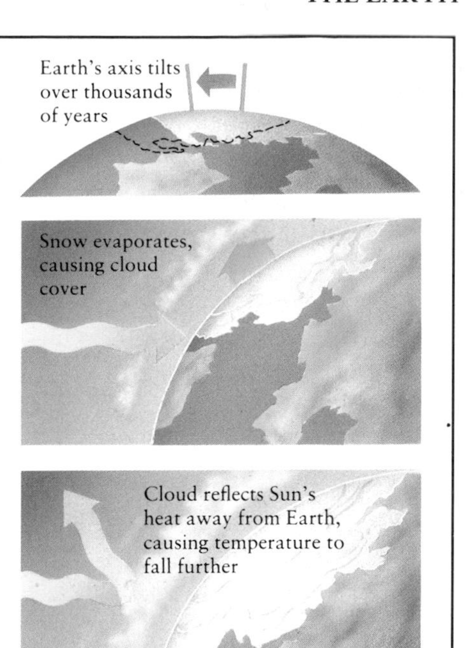

▼ *The Earth's surface water is in a delicate balance. Vapor rises from the sea, condenses as clouds, and falls back as rain or snow.*

THE CHANGING SURFACE

The crust of the Earth is not an unbroken shell; it is made up of several giant plates of solid rock. The plates are floating on the moving molten rock of the Earth's mantle beneath. The Earth's surface is constantly changing because of the great forces created by these plates as they drift. If two plates move against each other the crust may be forced up in mountain chains. An example is the Andes chain, which runs down South America. In the middle of the oceans, ridges form where the sea floor is spreading, as rock from the mantle wells up and forms new crust. The ridge marks the edge of a plate.

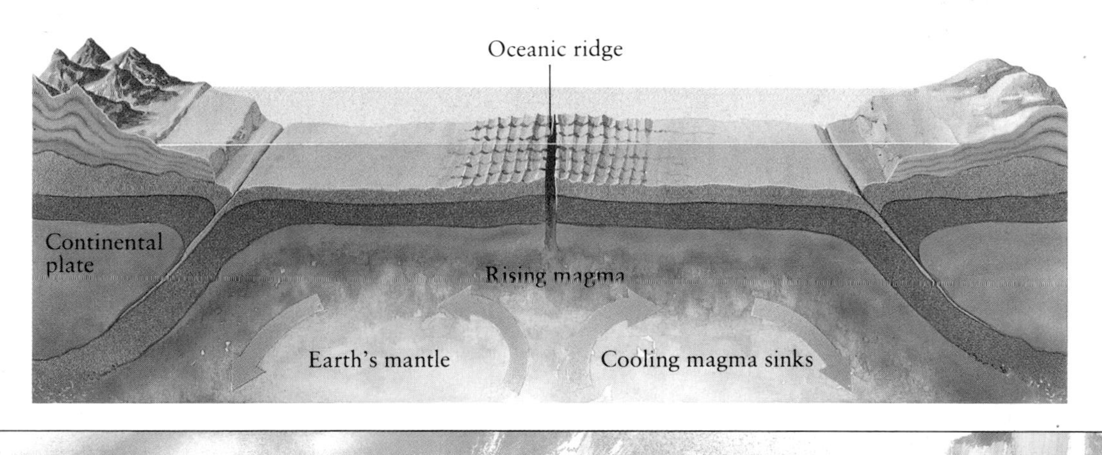

The Moon

The Moon is the Earth's satellite, a pitted rocky body orbiting our planet. It has no light of its own, but seems bright in the sky because it reflects the light of the Sun. The Moon is a fossilized world, where little has happened during the last 3 billion years—but it has had a violent past. The Moon's surface was cratered by bodies that crashed into it between 3 and 4 billion years ago, exploding on the surface. The craters have not worn away, but display the Moon's history for all to see.

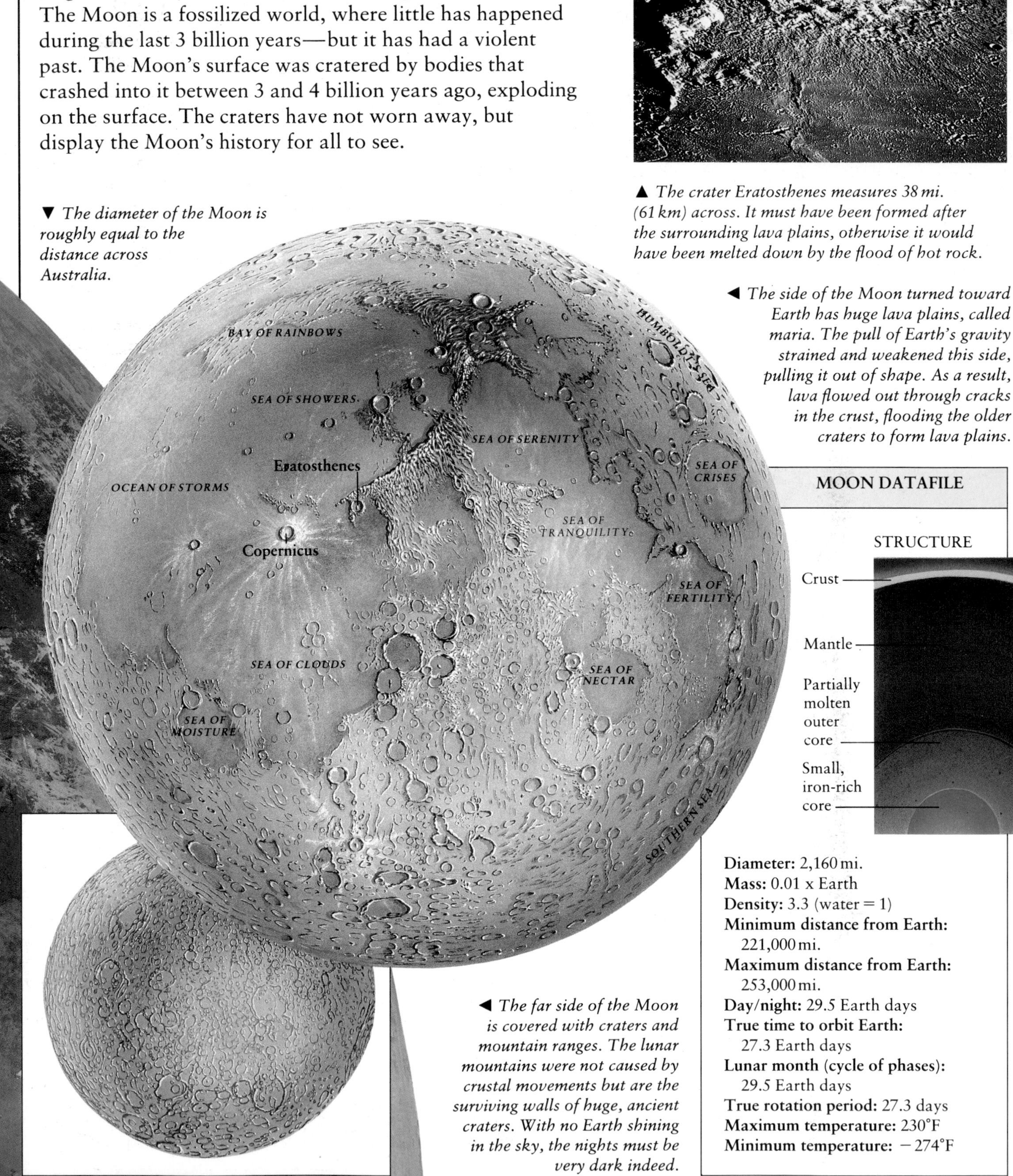

▲ *The crater Eratosthenes measures 38 mi. (61 km) across. It must have been formed after the surrounding lava plains, otherwise it would have been melted down by the flood of hot rock.*

▼ *The diameter of the Moon is roughly equal to the distance across Australia.*

◀ *The side of the Moon turned toward Earth has huge lava plains, called maria. The pull of Earth's gravity strained and weakened this side, pulling it out of shape. As a result, lava flowed out through cracks in the crust, flooding the older craters to form lava plains.*

◀ *The far side of the Moon is covered with craters and mountain ranges. The lunar mountains were not caused by crustal movements but are the surviving walls of huge, ancient craters. With no Earth shining in the sky, the nights must be very dark indeed.*

MOON DATAFILE

Diameter: 2,160 mi.
Mass: 0.01 x Earth
Density: 3.3 (water = 1)
Minimum distance from Earth: 221,000 mi.
Maximum distance from Earth: 253,000 mi.
Day/night: 29.5 Earth days
True time to orbit Earth: 27.3 Earth days
Lunar month (cycle of phases): 29.5 Earth days
True rotation period: 27.3 days
Maximum temperature: 230°F
Minimum temperature: −274°F

HOW THE CRATERS WERE FORMED

Most of the craters on the Moon were formed when smaller bodies, a few miles across, crashed into its surface. A body striking the surface of the Moon at a speed greater than 6 mi./s (10 km/s) would explode, blasting out a crater about 10 times its own width. Silica, a glassy substance found in rock, might be sprayed out for hundreds of miles. The silica would form bright rays, extending out of the crater. The violent impact on the floor of the crater would create a shock wave, making the floor spring back up as a central mountain.

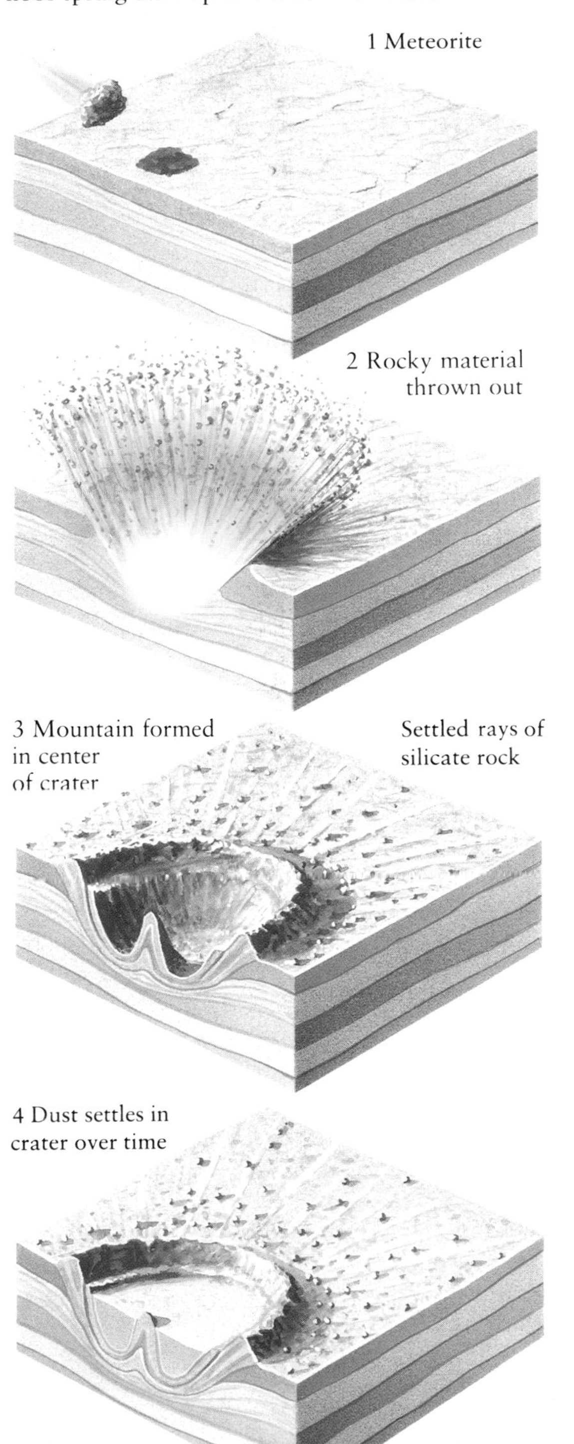

▲ *The U.S.* Apollo *project answered a number of questions about the Moon, confirming that its surface is firm and the core is still hot.*

MOON ROCKS

Anorthosite

Vasicular basalt

Typical basalt

◀ *Compared with Earth rocks, the samples brought back from the Moon by the* Apollo *project contain more of some elements such as titanium, and less of others such as gold. The ages of the Moon rock samples range from about 4.5 billion years (not long after the birth of the Moon) to 3.1 billion years—the time when the lava plains formed.*

FACTS ABOUT THE MOON

- The Moon has no atmosphere, and so there is no wind or weather either. As a result, the astronauts' footprints will last for centuries.
- Today, laser beams can measure the distance to the Moon with an accuracy of about 4 in. (10 cm).

The Earth and the Moon

It would be hard to imagine two rocky worlds more different than the Earth and the Moon—yet they orbit each other almost as a double planet. Most satellites are much smaller than the planet controlling them, but the diameter of the Moon is a quarter of the Earth's diameter. The Moon is a very useful body for us, illuminating half the nights every month as it reflects the Sun's light. Yet it may also block the light sent from the Sun during a solar eclipse, casting a black shadow on part of the Earth.

FACTS ABOUT THE MOON

- Does the Moon influence living things? The word *lunatic*, which means moonstruck, used to mean a person affected by the phase of the Moon.
- Some gardeners think Full Moon is the best time to plant seeds—in northern countries it is traditional to plant potatoes at Easter, when the Moon is always near full.

THE FORMATION OF THE MOON
Despite the closeness of the Moon to the Earth, we are still unable to say for certain how our satellite formed. A century ago it was believed that the molten Earth, spinning very fast, became unstable and threw a large blob of material into orbit. This theory has now been abandoned because, for this to happen, the Earth would have to spin once in only two and a half hours, which seems impossible. Three other theories have been put forward; they are illustrated here.

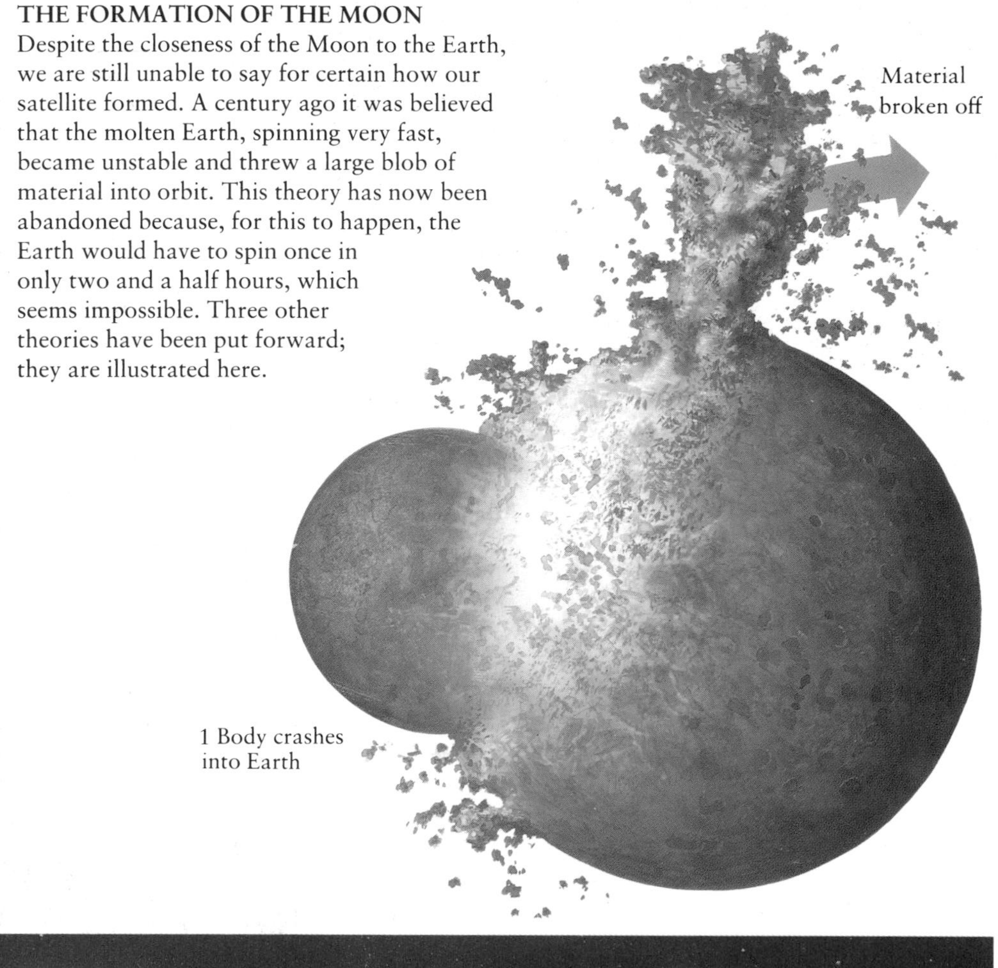

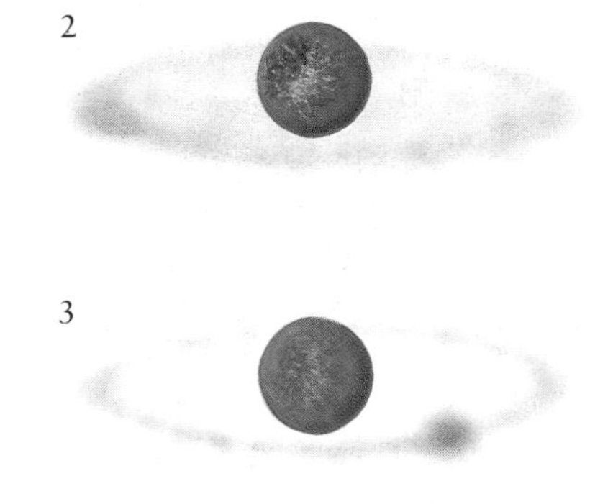

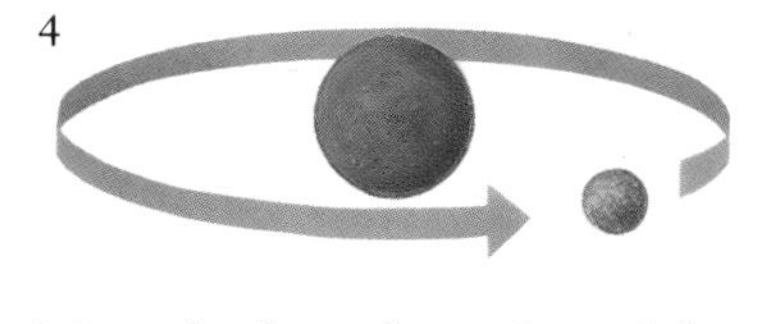

◄ *Samples from the surface of the Moon show that its rocks have a very different composition from those on the Earth. If the Moon ever did form part of our planet, it could have been broken off by a collision with a planet-sized body (**1**). This body added different material to the debris thrown off into space (**2**). The debris formed an orbiting cloud (**3**) which finally condensed into the Moon (**4**).*

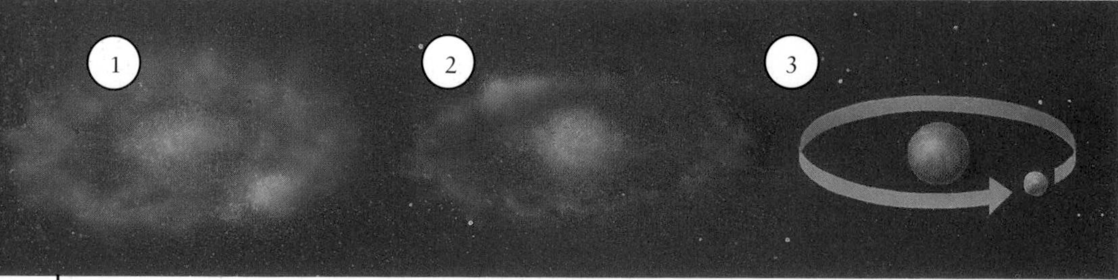

◄ *The Earth and the Moon may have formed together as a double planet from the cloud of debris left over after the formation of the Sun. This idea is attractive because it does not assume an unlikely event, such as a collision. But if the two bodies formed so closely together, why are their surface rocks so different? And why does the Moon have such a small iron core compared with the Earth's?*

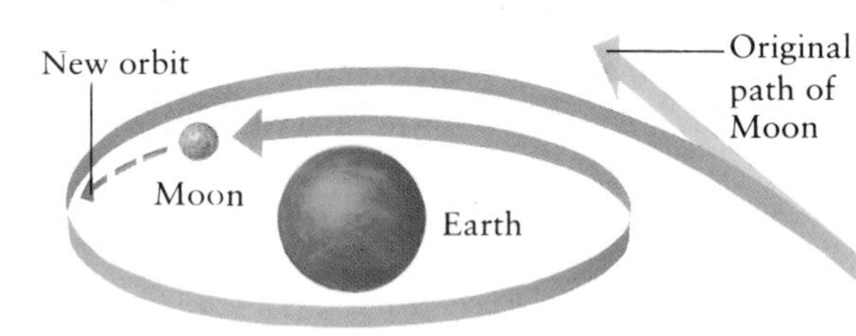

► *The capture theory assumes that the Moon was a passing body caught by the Earth's gravity. This explains its different composition. However, calculations show that a capture is far less likely than a collision with another body.*

◀ *In this photograph, taken by* Apollo 11, *the blue-and-white Earth contrasts with the bare lunar surface. The Moon's weak gravity could not hold on to any atmosphere, and so it became just a huge ball of stone, exposed to fierce temperature extremes of day and night. The photo is often called "Earthrise." However, this is not accurate. The Earth never rises or sets in the Moon's sky, since the same hemisphere of the Moon is always turned toward the Earth.*

THE PHASES OF THE MOON

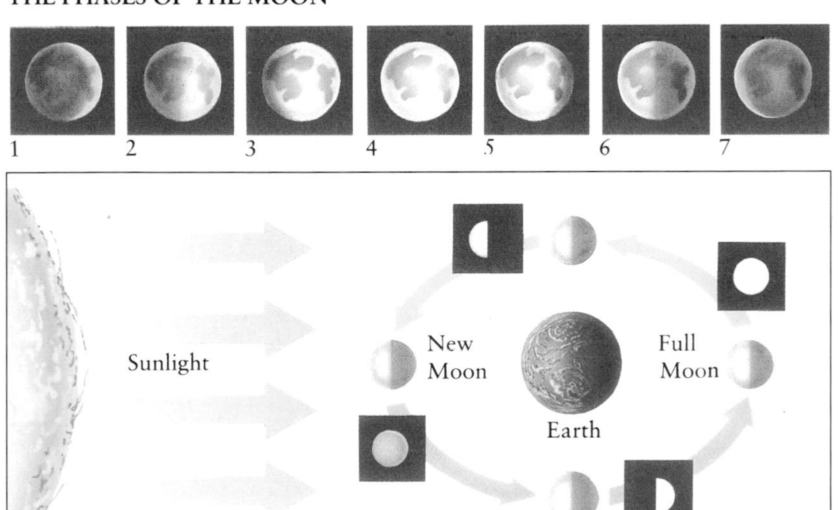

◀ *At New Moon, the unlit side of the Moon is toward the Earth, so it is invisible. As it moves through the first quarter of its orbit, a crescent moon* (**1**)*, half* (**2**)*, and then nearly all the unlit side* (**3**) *are seen. At Full Moon the sunlit side faces the Earth, and the Moon appears round* (**4**)*. The phases then continue in the reverse order until the Moon is new again* (**5, 6, 7**)*. The phase cycle takes 29.5 days—the lunar month.*

ECLIPSE OF THE MOON

The Full Moon sometimes passes through the Earth's shadow. The Moon does not become invisible, because a little sunlight is diffused into the shadow by the Earth's atmosphere.

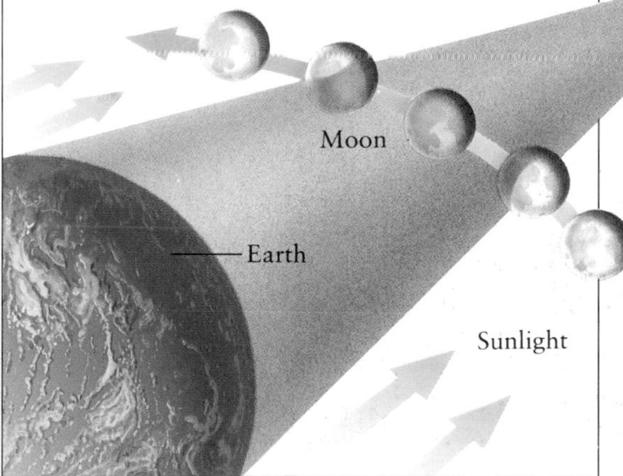

► *Because the Moon has an elliptical orbit it is not always the same distance from the Earth. The point when the Moon is at its closest to the Earth is called its perigee. The point when the Moon is farthest from the Earth is its apogee.*

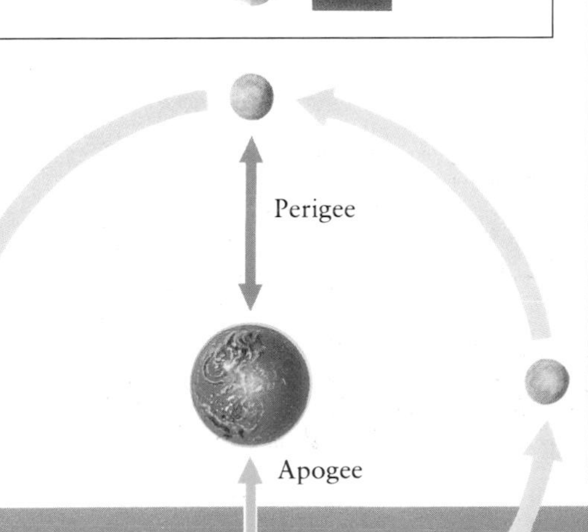

Mars

This mysterious planet has intrigued skywatchers for centuries. It shines very brightly when closest to the Earth, moving quickly in front of the stars, and it has a bright, reddish color. Less than a century ago, many people believed in Martians, and the possibility of finding some kind of life form inspired the *Viking* missions in 1976. Although apparently dead and inhospitable, Mars is the only planet selected for possible human exploration, and further visits by unmanned spacecraft are planned.

▲ *Mars is the outermost of the rocky planets. A vast gap, twice the diameter of its own orbit, separates Mars from Jupiter. Most of the asteroids, or minor planets, are found in this gap.* (Distances not to scale.)

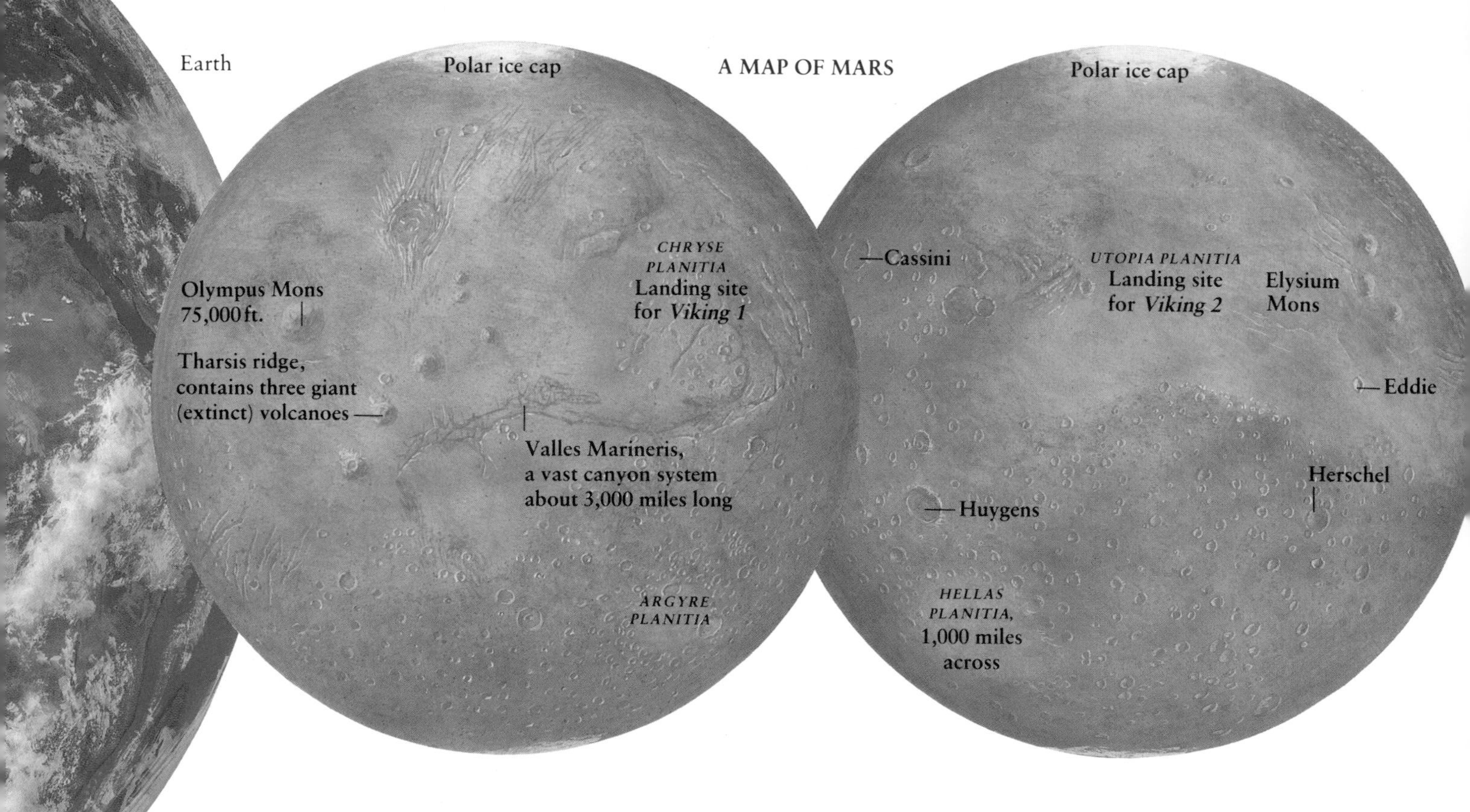

MARS DATAFILE

Diameter: 4,217 mi.
Mass: 0.11 × Earth
Density: 3.9 (water = 1)
Minimum distance from Sun: 128 million mi.
Maximum distance from Sun: 155 million mi.
Minimum distance from Earth: 35 million mi.
Day/night: 24 h 37 min
Length of year: 687 Earth days
Tilt of axis: 25° 12′
Surface gravity: 0.38 × Earth
Temperature: − 116°F to 32°F
Satellites: 2

STRUCTURE

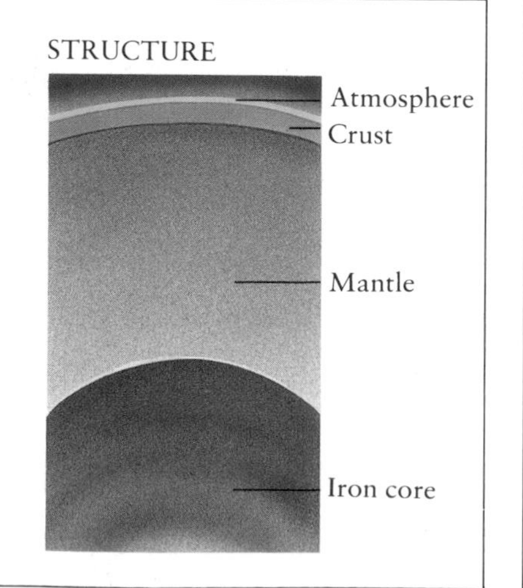

MARTIAN ATMOSPHERE

Mars's atmosphere is primarily carbon dioxide, with nitrogen, argon, and small amounts of other gases. At 20 mi. (30 km) up, winds can raise huge dust-clouds. At night, carbon dioxide freezes on parts of the surface as hoar frost. Traces of water can form thin hazes at dusk and dawn.

▲ *This view of the surface of Mars was taken by the* Viking 2 *lander* (left). *It shows a desert of rocks, each up to about 12 in. (30 cm) across, under an orange sky.*

▼ *The color of Mars's sky comes from wind-borne dust blown from the reddish surface. Early in the planet's history, when it was warmer and damp, the iron-rich surface rusted, turning Mars into a red planet.*

SATELLITES OF MARS

Deimos is about 7 mi. (11 km) long, and covered with small craters. From Mars, Deimos would appear as an almost star-like object, remaining above the horizon for two nights at a time.

Deimos

Phobos is larger and closer to Mars than Deimos. It is about 12 mi. (19 km) long. It orbits Mars in only seven and a half hours, so that it rises in the west, and then sets in the east about four hours later.

FACTS ABOUT MARS

- About 4 billion years ago, Mars may have been warm enough for water to run in rivers over its surface.
- Phobos, the inner moon, goes around the planet three times while Mars spins on its axis once. This means that Phobos's "month" is shorter than Mars's day.
- The Valles Marineris is an enormous canyon system running for some 2,500 mi. (4,000 km) along the surface of Mars. In some places it is 4 mi. (6 km) deep.
- Did a *Viking* orbiter record a landslide on Mars in 1978? Two close-up pictures of part of the Valles Marineris, taken two and a half minutes apart, appear to show a sudden cloud of dust 2,000 ft. (600 m) high.

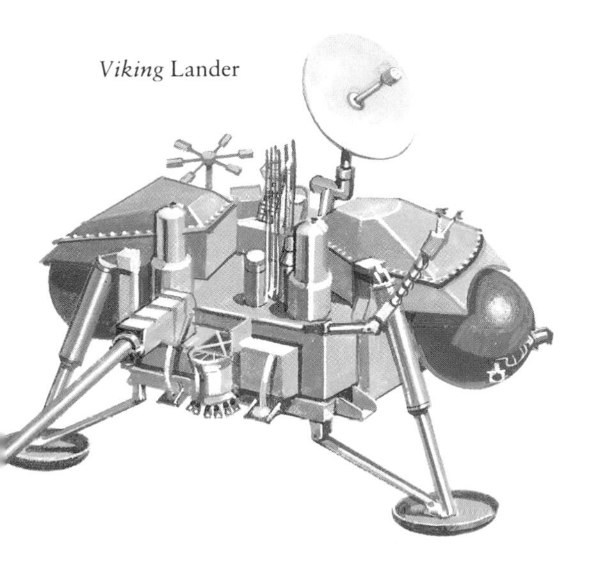

► *Roughly half of the Martian surface shows signs of past volcanic activity. The volcanic mountain Olympus Mons rises 14 mi. (23 km) above the plain. It is the highest peak in the Solar System and shows how violent the volcanic activity must have been. Elsewhere on Mars, there are winding valleys that look just like dried-up river beds.*

Dust cloud storm

Giant volcano

Crater

◄ *Some craters on Mars seem almost as unchanged as those found on the Moon.*

Jupiter

Jupiter is by far the largest planet in the Solar System—it is so huge that all the other planets could be squeezed inside it. It spins faster than any of the other planets, too, so that its day lasts less than 10 hours. Jupiter is made up of about 90 percent hydrogen and 10 percent helium, with traces of other elements. Its core must be hotter than the surface of the Sun, but the clouds exposed to space are bitterly cold. Its vivid stripes are cloud markings, drawn out into dark belts and light zones by Jupiter's rapid spin.

▲ *Although Jupiter is the largest planet, it has only a thousandth of the Sun's mass. Even the smallest and dimmest known stars contain about a hundred times as much material as Jupiter.* (Distances not to scale.)

► *The details of Jupiter's appearance are always changing. The cloud belts are caused by clouds of frozen ammonia, ammonium hydrosulfide, water ice, and other compounds being swept through the outer layers of the planet at up to 250 mph (400 km/h). Some markings have lasted for decades.*

Earth

▲ *Jupiter's ring is about 4,000 mi. wide and 20 mi. thick. It is surrounded by a fainter 12,500-mi.-wide halo. The ring is made up of particles which measure about 0.01 mm across. Another ring of sulfur particles lies in Io's orbit.*

► *The Great Red Spot is a vast whirlpool on the surface of Jupiter. About twice the Earth's diameter, it draws material up from below as it rotates every six days. This cloud feature has been observed for a century, and possibly longer. The color, which sometimes fades away for several years, may be caused by sunlight reacting with chemicals in the clouds to release red phosphorus.*

JUPITER DATAFILE

Diameter: 88,700 mi.
Mass: 318 × Earth
Density: 1.3 (water = 1)
Minimum distance from Sun: 460 million mi.
Maximum distance from Sun: 507 million mi.
Minimum distance from Earth: 367 million mi.
Day/night: 9 h 50 min (equator)
Length of year: 11.9 Earth years
Tilt of axis: 3°
Surface gravity: 2.7 × Earth
Temperature: −238°F
Satellites: 16 known

STRUCTURE

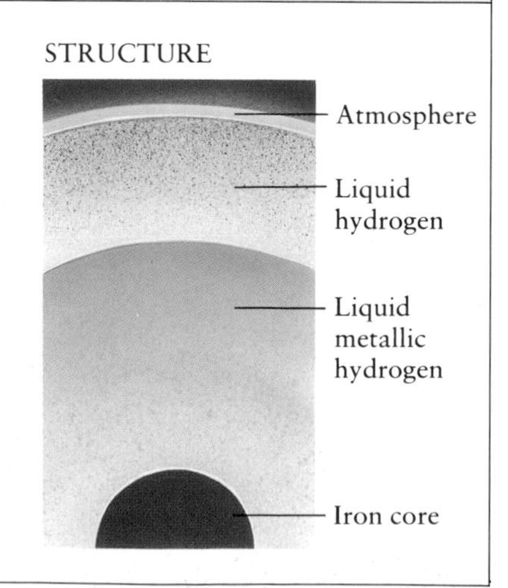

JUPITER'S ATMOSPHERE

Jupiter's clouds lie in the upper 150 mi. (200 km) of the atmosphere—less than one percent of the distance to the center. The planet is made up of almost pure hydrogen and helium, compressed to a fiercely hot liquid. This churning liquid generates a powerful magnetic field and electrical currents, which produce radio waves.

Tops of clouds
Hydrogen gas
Crystals of ammonia ice
Clouds of ammonium sulfide
Droplets of water ice
Liquid hydrogen

EXPLORING JUPITER

The *Galileo* probe should reach Jupiter in 1995. The main craft will study the planet and its satellites. A probe will be released into Jupiter's atmosphere to record the conditions and chemical make-up of the outer layers, before the probe is destroyed by the fierce pressure. *Galileo* was launched in 1989. It used the gravity of Venus and the Earth to speed it on its way to Jupiter

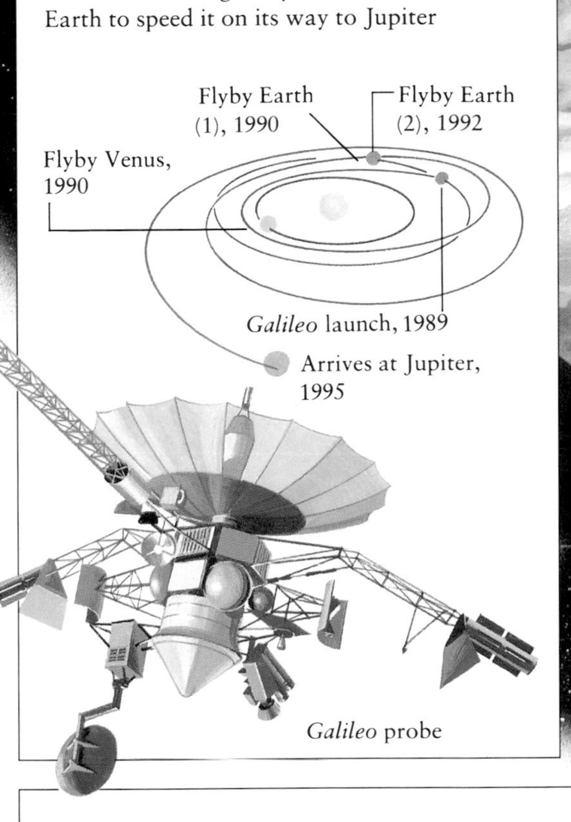

Galileo probe

THE GALILEAN SATELLITES

Jupiter's four large satellites were discovered by Galileo Galilei (1564–1642) with his primitive telescope in 1610. They include Ganymede, the largest satellite in the Solar System. Although the satellites can be seen using binoculars, the *Voyager* spacecraft produced our first detailed view of their surfaces in 1979. *(Their sizes are shown to scale here.)*

▲ *Ganymede is the largest satellite in the Solar System, 3,270 mi. (5,262 km) across. Parts of its crust may have drifted over the hot rock beneath, as happened on Earth.*

▶ *Europa is covered in ice many miles thick, making it the smoothest known body anywhere. The dark streaks may be fractures in the surface, filled with ice.*

▶ *Io is a sulfur-covered volcanic world. Constant eruptions shoot material 125 mi. (200 km) into space before it falls back onto the orange-yellow surface.*

◀ *Callisto, the outermost Galilean satellite, is covered with ancient craters. However, unlike our Moon, it shows no sign of any later lava flows. All the satellites keep the same face turned inward toward Jupiter.*

A volcano explodes on Io

Saturn

Saturn has been known for centuries as the ringed planet. In fact, due to the *Voyager* space probes we now know that all four giant planets have rings, though Saturn's rings are by far the most impressive. There is less cloud detail visible on Saturn than on Jupiter. This may be because a layer of haze makes it hard to see what cloud belts lie lower down. However, about three times every century violent storms disturb the calm surface and cause brilliant white "spots" to break out on the planet.

▲ *Like the other giant planets, Saturn spins so quickly that its equator now bulges outward noticeably. This gives the planet a slightly squashed look. The bulge is due to centrifugal force.* (Distances not to scale.)

► *The cloud features on Saturn are difficult to make out, but* Voyager 2 *recorded details including a "Great Brown Spot" near the north pole.*

Saturn

► *Under this calm-looking surface, material is being swept around at speeds of up to 1,120 mph (1,800 km/h). In 1990 a huge white eruption was easily visible from Earth through telescopes.*

Earth

▲ *Saturn's rings extend for over 46,000 mi. (74,000 km), but are only a few miles deep. Although they seem to be several wide zones, they are really thousands of separate narrow ringlets. The rings could be the remains of a small satellite, 60 mi. (100 km) across.*

Cassini division

▲ *The Cassini Division is a wide zone in the rings where the particles are scattered. They have been pulled into different orbits by Saturn's moons.*

SATURN DATAFILE

Diameter: 74,500 mi.
Mass: 95 × Earth
Density: 0.7 (water = 1)
Minimum distance from Sun: 840 million mi.
Maximum distance from Sun: 938 million mi.
Minimum distance from Earth: 746 million mi.
Day/night: 10 h 14 min (equator)
Length of year: 29.5 Earth years
Tilt of axis: 26° 42′
Surface gravity: 1.2 × Earth
Temperature: −274°F
Satellites: 18 known

STRUCTURE

Atmosphere
Liquid hydrogen
Liquid metallic hydrogen
Iron core

SATURN'S ATMOSPHERE

The cloud layers form a skin as thin as apple peel over the hydrogen and helium body of Saturn. *Voyager 2* had a much clearer view of the cloud formations than *Voyager 1* did, when it passed nine months earlier. This suggests that the upper ammonia haze varies in transparency.

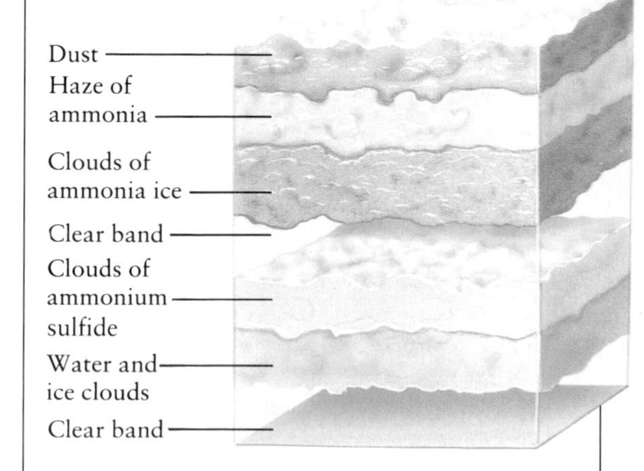

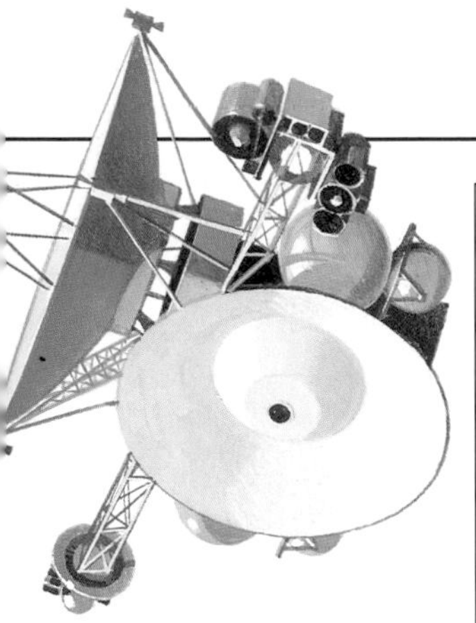

▲ *The* Cassini *spacecraft will be launched in 1997 and is planned to reach Saturn in 2002. It will orbit Saturn and study Titan, the largest moon. A probe will be dropped into Titan's atmosphere.*

SATURN'S SATELLITES
Saturn has more moons than any other planet. There are 18 satellites confirmed; 12 were discovered by the *Voyager* probes. The existence of other satellites is suspected. The moons range from the second largest in the Solar System (Titan, 3,200 mi. [5,150 km] across) to the second smallest (Pan, at 12 mi. [20 km] across, is slightly larger than Mars's Deimos). Titan is the only moon in the Solar System to have its own thick atmosphere, which makes it a prime target for the *Cassini* mission. Phoebe is the farthest moon from the planet, orbiting at 8 million mi. (13 million km) away.

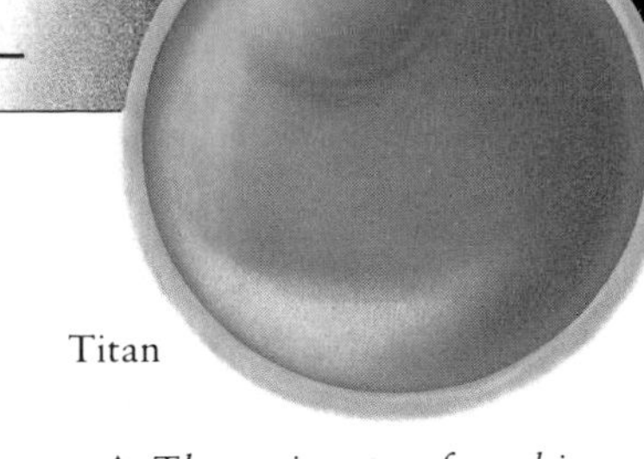

▲ *The main gases found in Titan's thick orange atmosphere are nitrogen and methane, with traces of organic compounds including ethane and acetylene. These compounds form the crude beginnings of molecules found in living cells. Titan's hidden surface could be a deep-frozen record of how more advanced organic molecules formed naturally out of simple gases.*

▲ Voyager 1 *visited Saturn in 1980. It sent back this image of the unique oval red cloud feature in Saturn's southern hemisphere.*

► *Saturn's rings are made up of billions of particles in orbit around the planet, ranging in size from grains to large rocks.*

RINGS AND SHEPHERD MOONS
Only rings A, B, and C are visible from the Earth. The particles in the brightest ring, B, seem to be icy. Calculations suggest that ice would eventually be dulled by fine space dust. Since this has not yet happened, the rings are probably less than a billion years old. The gravity of tiny satellites, known as shepherd moons, appears to control the position of some of the orbiting particles. For example, Prometheus and Pandora, each less than 93 mi. (150 km) across, orbit on either side of the narrow F ring.

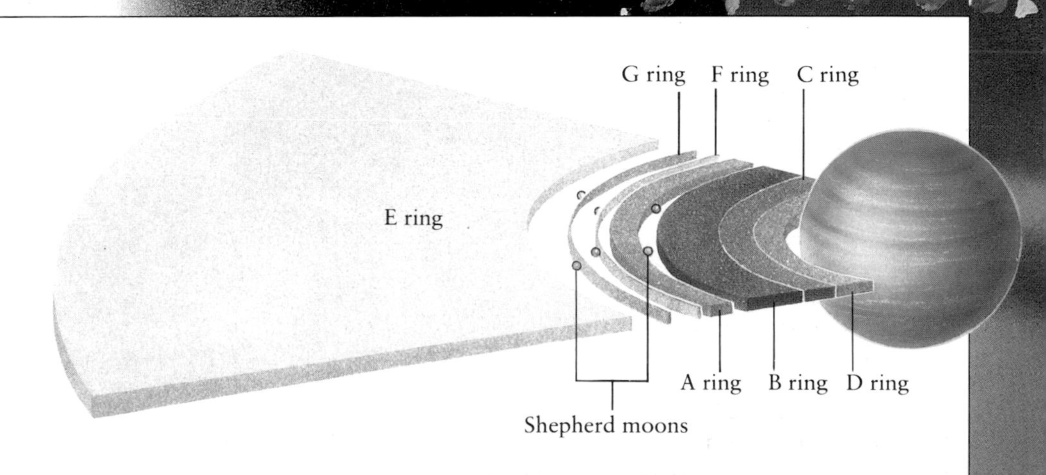

Uranus

Before the last decade, chance played a large part in our understanding of Uranus. Barely visible with the naked eye, the planet was discovered by accident in 1781. Two centuries later its faint ring system was also discovered by accident. Apart from the fact that it orbits the Sun on its side and has a family of satellites, little else was known. In the course of a few days in January 1986, pictures taken by the space probe *Voyager 2* transformed our knowledge of this remote giant.

▲ *Uranus has the volume of more than 60 Earths. The tilt of its axis means that its "north" pole is pointing slightly south. Uranus has a backward, or retrograde, spin—as do Venus and Pluto.* (Distances not to scale.)

Earth

Voyager 2

Uranus

▲ *Uranus has many rings—nine major ones and many more faint ones. They are made up of roughly yard-long pieces of rock, with little ice or dust. Some of Uranus's rings have been created or controlled by shepherd moons orbiting beside them* (see page 185).

DISCOVERING URANUS

Uranus was discovered by William Herschel (1738–1822), a musician from Germany who settled in Bath, England, and became fascinated by astronomy. In 1781 he was looking at the sky with his homemade telescope when he noticed a "star" that showed as a small disk. The discovery earned Herschel royal recognition.

URANUS DATAFILE

Diameter: 32,000 mi.
Mass: 14.5 × Earth's mass
Density: 1.3 (water = 1)
Minimum distance from Sun: 1.7 billion mi.
Maximum distance from Sun: 1.87 billion mi.
Minimum distance from Earth: 1.6 billion mi.
Day/night: 17 h 14 min
Length of year: 84 Earth years
Tilt of axis: 82°
Surface gravity: 0.93 × Earth
Temperature: −328°F
Satellites: 15

STRUCTURE

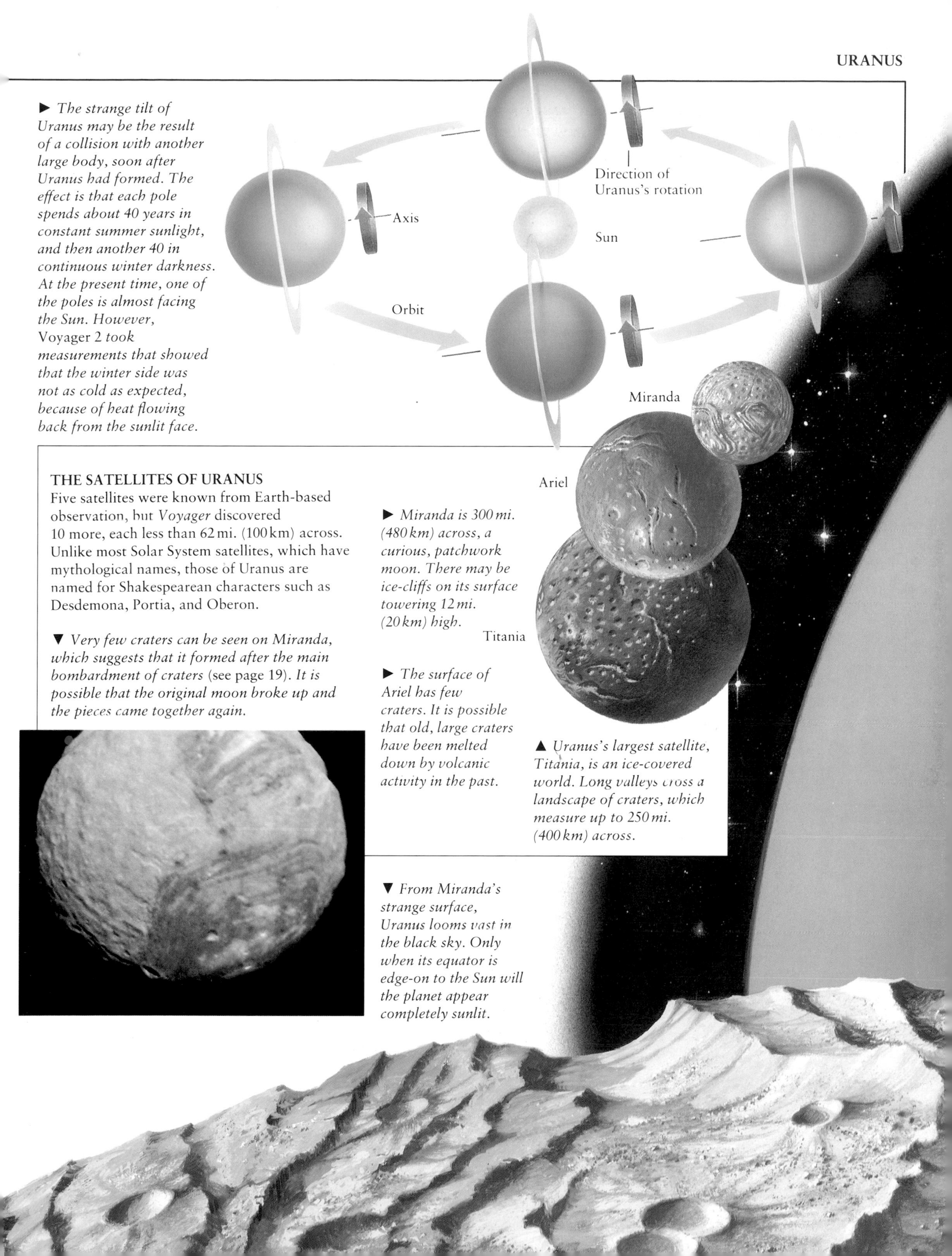

► *The strange tilt of Uranus may be the result of a collision with another large body, soon after Uranus had formed. The effect is that each pole spends about 40 years in constant summer sunlight, and then another 40 in continuous winter darkness. At the present time, one of the poles is almost facing the Sun. However,* Voyager 2 *took measurements that showed that the winter side was not as cold as expected, because of heat flowing back from the sunlit face.*

THE SATELLITES OF URANUS

Five satellites were known from Earth-based observation, but *Voyager* discovered 10 more, each less than 62 mi. (100 km) across. Unlike most Solar System satellites, which have mythological names, those of Uranus are named for Shakespearean characters such as Desdemona, Portia, and Oberon.

▼ *Very few craters can be seen on Miranda, which suggests that it formed after the main bombardment of craters* (see page 19). *It is possible that the original moon broke up and the pieces came together again.*

► *Miranda is 300 mi. (480 km) across, a curious, patchwork moon. There may be ice-cliffs on its surface towering 12 mi. (20 km) high.*

► *The surface of Ariel has few craters. It is possible that old, large craters have been melted down by volcanic activity in the past.*

▲ *Uranus's largest satellite, Titania, is an ice-covered world. Long valleys cross a landscape of craters, which measure up to 250 mi. (400 km) across.*

▼ *From Miranda's strange surface, Uranus looms vast in the black sky. Only when its equator is edge-on to the Sun will the planet appear completely sunlit.*

Neptune

Neptune was discovered in 1846 as a result of the effect of its gravitational pull on Uranus. Its brilliant blue methane atmosphere looks calm and cold (it has a temperature of −346°F [−210°C]). But *Voyager 2* discovered winds ripping through Neptune's atmosphere at 1,400 mph (2,200 km/h), the fiercest winds in the Solar System. The winds travel in a different direction to the planet's spin. Neptune's large satellite, Triton, the coldest land surface in the Solar System, has plumes of gas erupting into space.

▲ *Neptune is a little smaller than Uranus. It is the outermost giant planet, and for 20 years in every 250 years, when Pluto comes closest to the Sun (as in 1979–1999), it is also the outermost planet.* (Distances not to scale.)

Earth

► *Methane gas, which reflects blue light well, gives Neptune its intense color.*

Neptune

► Voyager 2 *found four faint rings around Neptune. The three brighter rings have been called Galle, Adams, and Leverrier, after the three people involved in the discovery of Neptune.*

◄ *Some cloud features appear and vanish in just a few minutes, rising up from the warmer layers and sinking again as they cool down. Long-lasting features, such as the Great Dark Spot and some of the white clouds, are being forced along by fierce currents.*

▲ *Neptune's rings are very narrow. Galle is 9 mi. (15 km) wide, and Adams, the widest, is less than 30 mi. (50 km) across.*

NEPTUNE DATAFILE

Diameter: 30,760 mi.
Mass: 17.2 × Earth
Density: 1.8 (water = 1)
Minimum distance from Sun: 2.77 billion mi.
Maximum distance from Sun: 2.82 billion mi.
Minimum distance from Earth: 2.68 billion mi.
Day/night: 17 h 6 min
Length of year: 165 Earth years
Tilt of axis: 29° 36′
Surface gravity: 1.2 × Earth
Temperature: −346°F
Satellites: 8

STRUCTURE

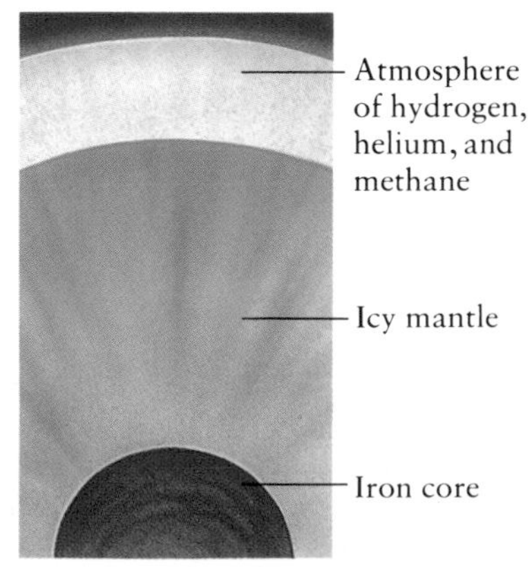

Leverrier

▲ *The Sun's gravity rules the planets, but their own much weaker gravity also pulls at each other. The independent calculations of John Couch Adams in England and Urbain Leverrier in France in 1845 showed that an unknown planet was pulling Uranus from its true path. These calculations led to the discovery of Neptune by Johann Galle in Germany.*

► *The Great Dark Spot was discovered by* Voyager 2. *Large enough to contain the Earth, the spot changes in size and shape as the material inside it is slowly churned around once every 16 days. A much smaller dark spot, known as D2, was also seen in Neptune's southern hemisphere; both spots were observed for several weeks.*

SATELLITES OF NEPTUNE

Neptune has eight satellites, six of them discovered by *Voyager 2*. The five tiny inner satellites orbit Neptune faster than it spins. The next, Proteus, is 258 mi. (415 km) across. Nereid, the small outermost satellite discovered from Earth, takes a year to make one orbit, sweeping close to Neptune and then far out into space, like a comet.

TRITON

The largest of Neptune's satellites, Triton, is 1,681 mi. (2,705 km) across, with a crust of rock hard ice at a temperature of −391°F (−235°C). Although its gravity is too weak to hold a proper atmosphere, a thin layer of nitrogen is fed by plumes of gas from the surface. Triton orbits Neptune backward; this suggests that the satellite was captured by the planet's gravity, or suffered a space accident that altered its path.

▲ *This close-up of Triton's south pole was taken by* Voyager 2. *The probe was only 120,000 mi. (190,000 km) from Triton's surface. The white material may be frost; the long dark streaks may be dust, blown by winds.*

Pluto and Beyond

Uranus and Neptune have slightly erratic orbits. The cause of this was suspected to be the pull of gravity from a ninth planet. This led to the discovery of Pluto in 1930 during a thorough search of the sky for "Planet X" by the American astronomer Clyde Tombaugh. We still know very little about Pluto. It is the only planet we have not observed in close-up from a spacecraft and is too far away to see in detail with a telescope.

▲ *Tiny Pluto, on the frontier of the Solar System, is smaller than our Moon and only twice the diameter of the largest asteroid. Its eccentric orbit carries it closer than Neptune to the Sun at perihelion.* (Distances not to scale.)

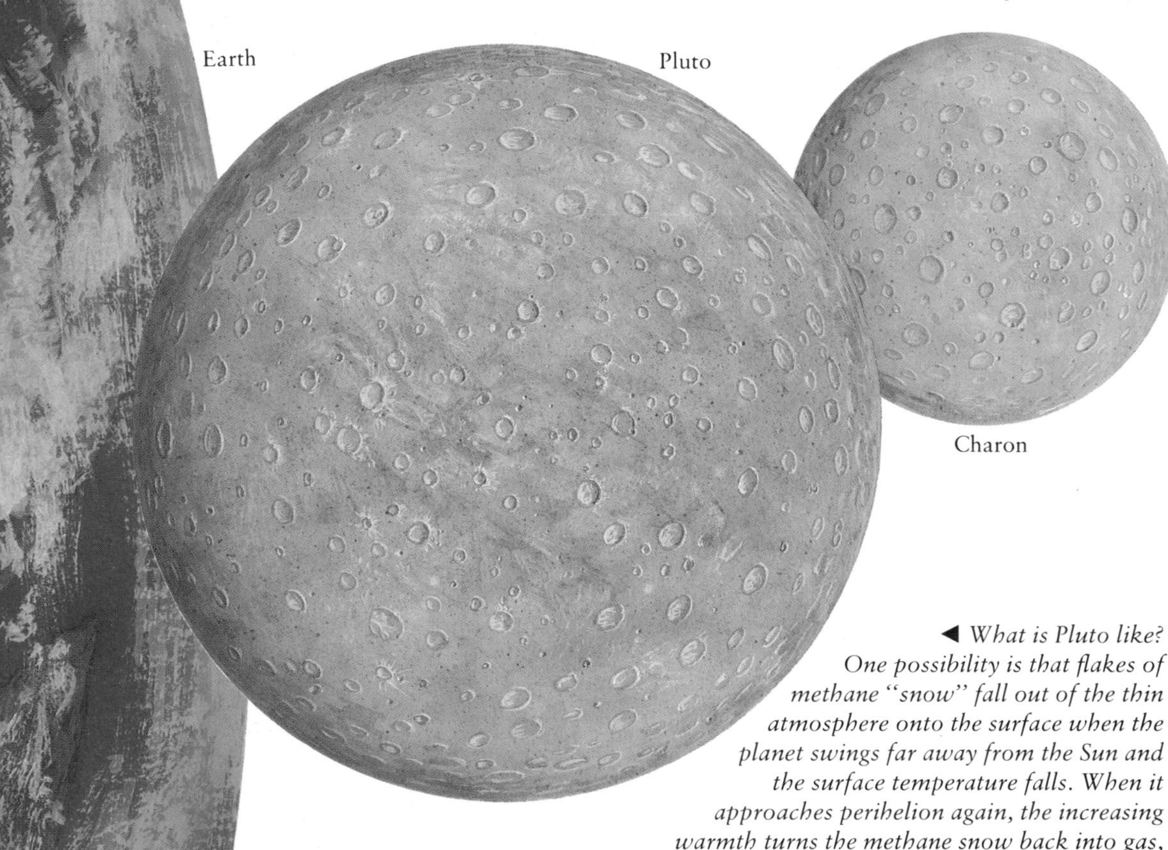

◀ *No telescope has seen Pluto, or its moon Charon, in detail. Charon was detected from Earth in 1978. It is about half the size of Pluto, with a diameter of 750 mi. (1,200 km). Pluto and Charon orbit each other like a double planet. Each one keeps the same hemisphere facing toward the other, and at certain periods during Pluto's year Charon is eclipsed at each revolution.*

◀ *What is Pluto like? One possibility is that flakes of methane "snow" fall out of the thin atmosphere onto the surface when the planet swings far away from the Sun and the surface temperature falls. When it approaches perihelion again, the increasing warmth turns the methane snow back into gas, and the atmosphere is restored.*

▼ *The clearest image we have of Pluto and Charon comes from the HST* (see pages 70–71). *The photo shows their relative sizes, but they are farther apart than it appears here.*

PLUTO DATAFILE

Diameter: 1,430 mi.
Mass: 0.002 × Earth
Density: Approx 2 (water = 1)
Minimum distance from Sun: 2.7 billion mi.
Maximum distance from Sun: 4.6 billion mi.
Minimum distance from Earth: 2.7 billion mi.
Day/night: 6 Earth d 9 h
Length of year: 248 Earth years
Tilt of axis: 62° 24′
Surface gravity: 0.03 × Earth
Temperature: −380°F
Satellites: 1

STRUCTURE

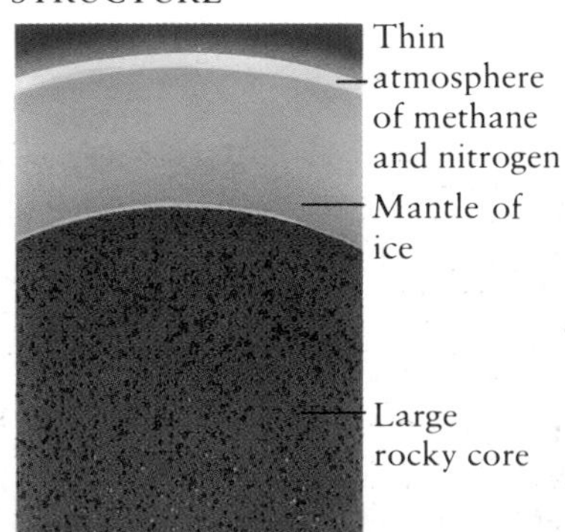

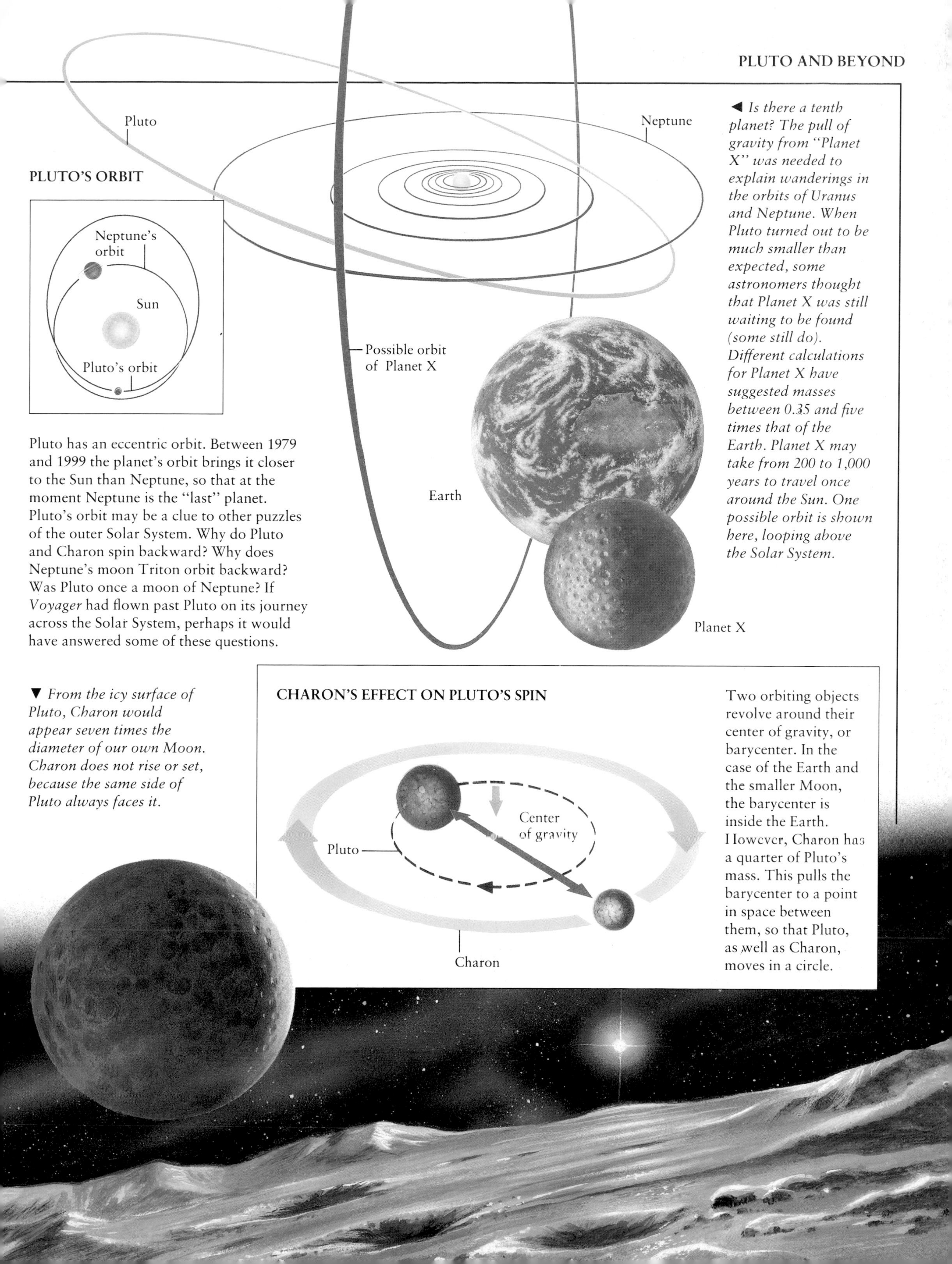

◀ *Is there a tenth planet? The pull of gravity from "Planet X" was needed to explain wanderings in the orbits of Uranus and Neptune. When Pluto turned out to be much smaller than expected, some astronomers thought that Planet X was still waiting to be found (some still do). Different calculations for Planet X have suggested masses between 0.35 and five times that of the Earth. Planet X may take from 200 to 1,000 years to travel once around the Sun. One possible orbit is shown here, looping above the Solar System.*

Pluto has an eccentric orbit. Between 1979 and 1999 the planet's orbit brings it closer to the Sun than Neptune, so that at the moment Neptune is the "last" planet. Pluto's orbit may be a clue to other puzzles of the outer Solar System. Why do Pluto and Charon spin backward? Why does Neptune's moon Triton orbit backward? Was Pluto once a moon of Neptune? If *Voyager* had flown past Pluto on its journey across the Solar System, perhaps it would have answered some of these questions.

Two orbiting objects revolve around their center of gravity, or barycenter. In the case of the Earth and the smaller Moon, the barycenter is inside the Earth. However, Charon has a quarter of Pluto's mass. This pulls the barycenter to a point in space between them, so that Pluto, as well as Charon, moves in a circle.

▼ *From the icy surface of Pluto, Charon would appear seven times the diameter of our own Moon. Charon does not rise or set, because the same side of Pluto always faces it.*

Minor Planets and Meteoroids

In the early days of the Solar System there was debris everywhere, as grains of solid matter grew together into larger objects. The planets and the larger satellites were the most successful, but countless other smaller bodies formed as well. Some passed too near a planet, especially Jupiter, whose gravity kept them from growing into major planets. Other bodies, moving in the wide space between the orbits of Mars and Jupiter, collided and broke up. The remains of these are known as asteroids, or minor planets.

FACTS ABOUT THE MINOR PLANETS

- Ceres was the first asteroid, or minor planet, to be discovered, in 1801.
- Some asteroids have orbits that take them very close to Earth. One, called Hermes, has come within 483,000 mi. (777,000 km) of us. However, the threat of an asteroid colliding with Earth is very small.

► *The asteroid belt is found between the orbits of Mars and Jupiter. It probably contains about 100,000 bodies larger than 0.6 mi. (1 km) across. Over 3,000 of the largest have been given names. The largest, Ceres, has a diameter of 620 mi. (1,000 km). Asteroid 1991 DA travels well past Jupiter. The asteroids known as the Trojans share Jupiter's orbit. At its closest point to the Sun the asteroid Phaethon glows red-hot, as it passes twice as close to the Sun as the planet Mercury. The orbits of some other unusual asteroids are shown here.*

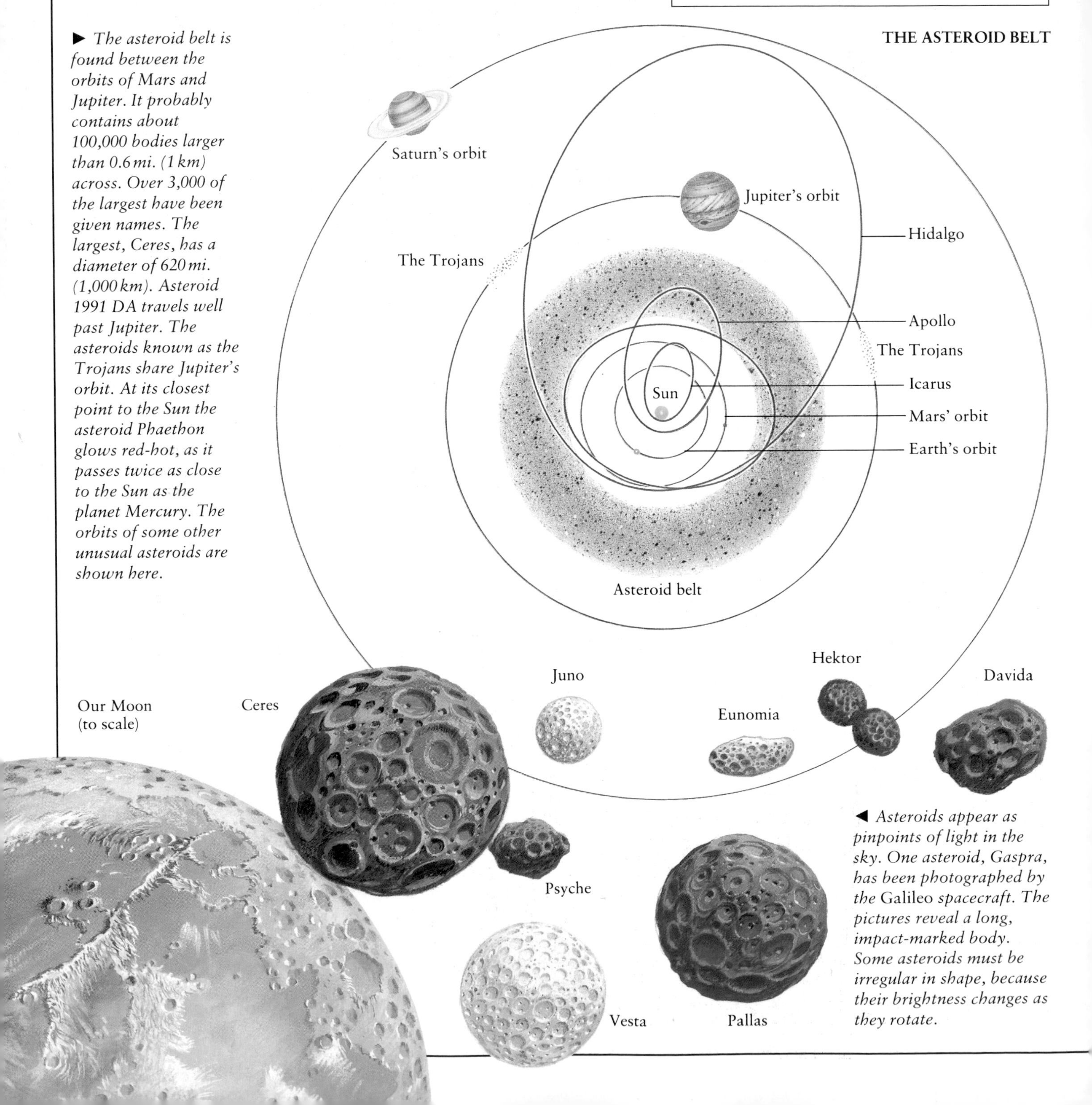

◄ *Asteroids appear as pinpoints of light in the sky. One asteroid, Gaspra, has been photographed by the Galileo spacecraft. The pictures reveal a long, impact-marked body. Some asteroids must be irregular in shape, because their brightness changes as they rotate.*

◀ *An especially bright meteor is called a fireball. Any meteor large enough to travel through the Earth's atmosphere and hit the ground is called a meteorite. These are thought to be pieces of minor planets or comets. Some meteorites are stony, others are metallic—presumably from the crust or core of a broken mini-planet. A few meteorites are made of very crumbly rock.*

METEOROIDS

Many meteoroids are particles thrown out from the crumbly nucleus of a comet *(see pages 194–195)*. They exist by the million, traveling along the comet's orbit in huge swarms, though the grains may be miles apart. Meteoroids themselves are invisible, but if they collide with the Earth's atmosphere (at speeds from 6 to 25 mi/s. [10 to 40 km/s]) they evaporate in a streak of light—forming meteors, or "shooting stars."

▼ *If the Earth passes through a large swarm of meteoroids, many meteors may be seen every hour. This is a meteor shower. The meteor shower below is connected with the comet Tempel Tuttle.*

▶ *Ordinary meteors usually burn up at about 30 mi. (50 km) above the Earth. Anything larger than a small stone will light up the sky as a fireball. It may explode, or hit the ground as a meteorite.*

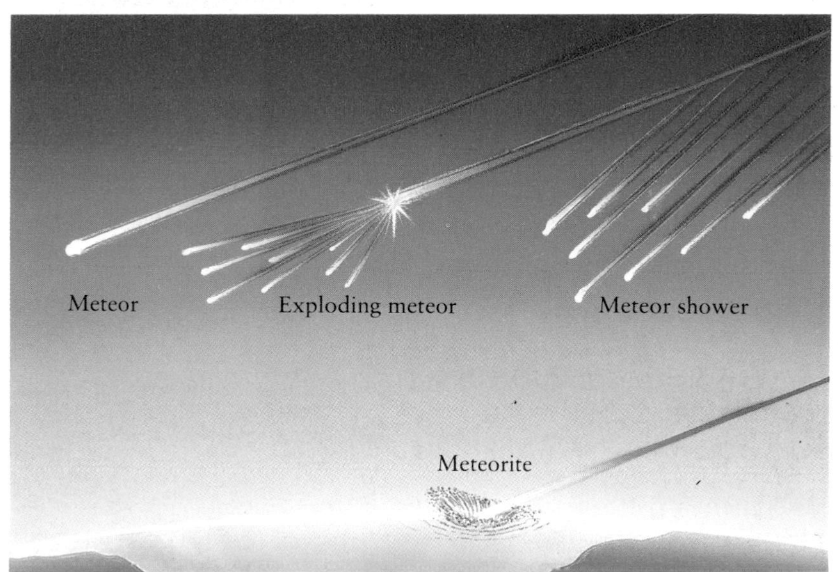

▼ *The Old Woman meteorite was discovered in California in 1976. It weighs 6,080 lb. (2,758 kg)—the second largest meteorite ever found in the United States. It is made of iron and nickel and may once have formed part of the molten core of a small planet that broke up about 4 billion years ago.*

Comets

It is easy to understand how the unexpected sight of a comet's long, bright tail in the sky must have terrified ancient civilizations. Today we know that comets are just lumps of ice and rock, traveling from the far outer Solar System to orbit our Sun. As a comet nears the Sun the ice melts, giving off jets of gas and releasing clouds of dusty rock particles. From the Earth this gas and dust is seen as a dramatic tail, shining by reflected sunlight, and stretching for millions of miles.

▲ *Comets are named after the person who discovers them. Comet West, discovered by Richard West in 1976, was one of the brightest of recent times.*

Crumbling particles of rock and ice

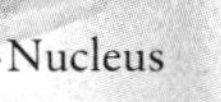

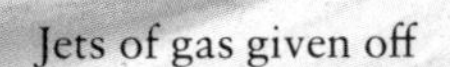

◀ *The solid part of a comet, called its nucleus, may be only a couple of miles across, but it may produce a coma, or cloud, of dust and gas, 10 times the Earth's diameter.* (Illustration not to scale.)

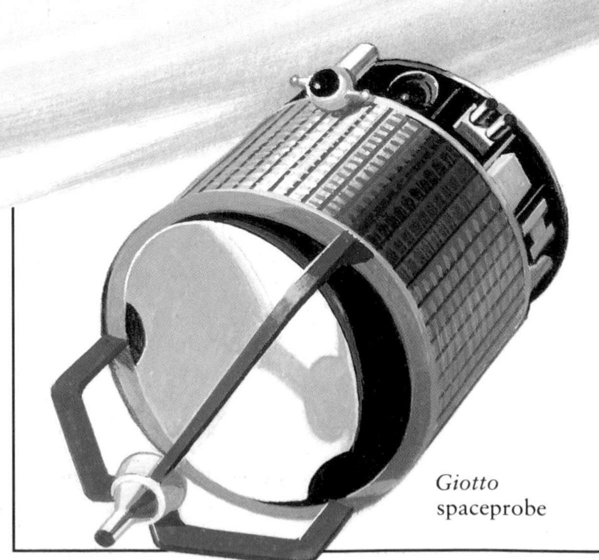

Giotto spaceprobe

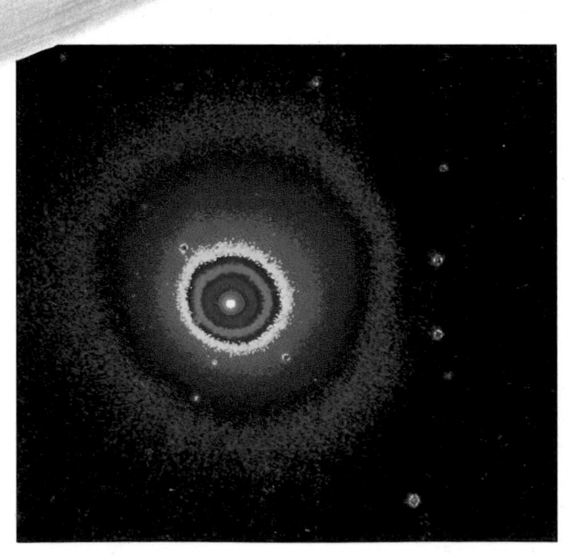

◀ *Probably the most famous comet of all is Halley, which returns every 76 years. On its appearance in 1985–1986, it was visited by the* Giotto *probe* (far left). *As Halley neared our planet this false color image was taken from Earth. The different colors show different levels of brightness (white is brightest). The nucleus gives out gas and dust in circular rings.*

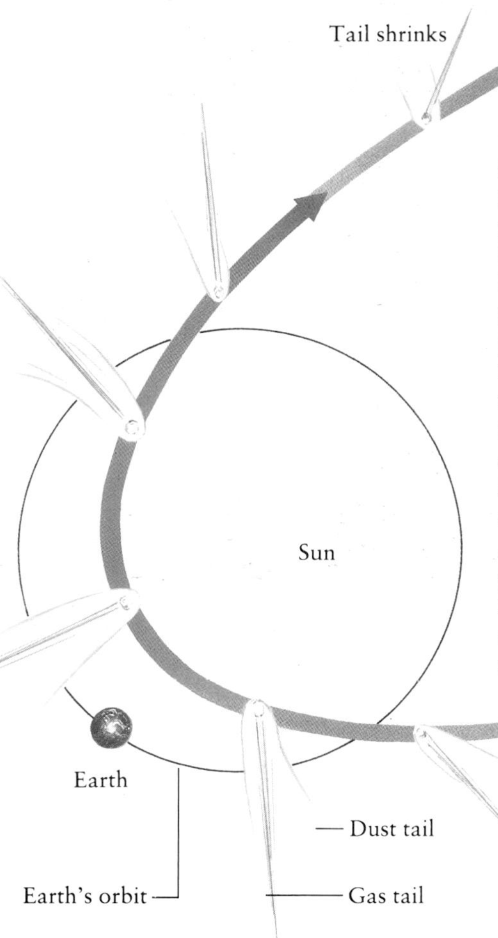

◄ As a comet nears the Sun, gas sent out from its nucleus is forced away from the Sun by the solar wind. Dust particles are pushed back by the pressure of the Sun's radiation on them, and they form a separate tail. The closer the comet comes to the Sun, the faster the gas and dust are released. At Halley's nearest point to the Sun, about 14 tons of its ice was turning into gas every second.

▲ Halley's Comet appeared over Jerusalem in A.D. 66. The first recorded sighting of Halley's Comet dates from 240 B.C.

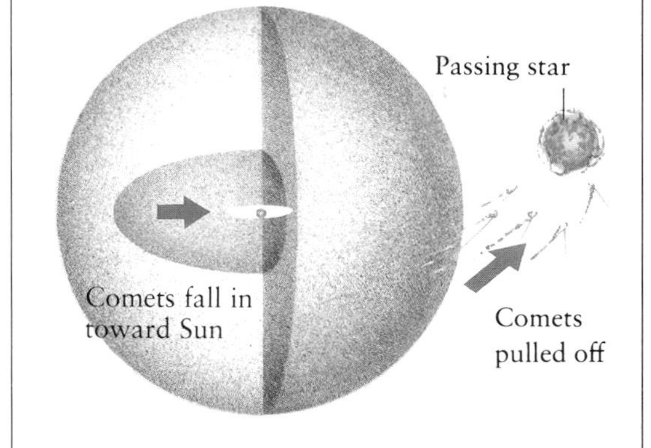

WHERE ARE COMETS FROM?

No one knows for sure where comets come from. One suggestion is that they may form a vast cloud surrounding the Solar System. The gravitational effect of passing stars pulls comets from the cloud to fall finally toward the Sun and shine briefly in the sky.

FACTS ABOUT COMETS

- Some comets take thousands of years to go around the Sun once, but Comet Encke takes just over three years.
- Comet Biela was first seen in 1806. When it returned in 1845, it divided into two comets.
- The comet Schwassmann-Wachmann orbits the Sun almost in a circle. It revolves beyond Mars and can always be seen.
- In 1983 the comet IRAS-Araki-Alcock passed within 3 million mi. (5 million km) of the Earth—the closest a comet has come since 1770.
- Comet Ikeya-Seki, which was first seen in 1965, is said to be a Sun-grazer, passing the Sun about 40 times closer than Mercury.
- Halley's Comet was named after Edmund Halley (1656–1742), an English scientist. Halley believed that comets seen in 1531, 1607 and 1682 were in fact the same comet. He suggested that this comet would return in 1758. His prediction was proved right on Christmas Day 1758. Halley's discovery was important proof that the law of gravity worked for comets.

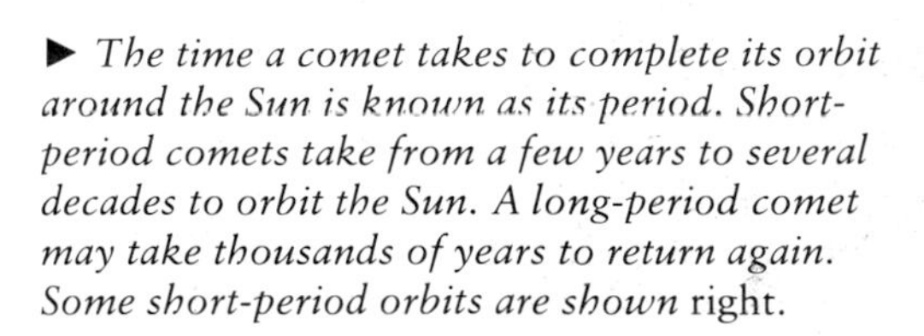

► The time a comet takes to complete its orbit around the Sun is known as its period. Short-period comets take from a few years to several decades to orbit the Sun. A long-period comet may take thousands of years to return again. Some short-period orbits are shown right.

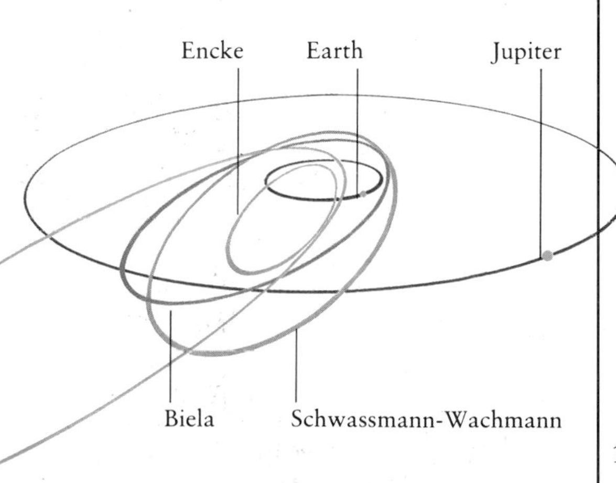

BEYOND THE SOLAR SYSTEM

The Milky Way

The Sun is just one of a hundred billion stars, existing in space in a vast "star-city." This city is our galaxy, which we call the Milky Way. The softly-shining band of light in our night sky is also called the Milky Way, and is the edge-on view of our own galaxy. The Milky Way is spiral in shape, and the Sun is found out toward its "edge." The galaxy is so huge that it would take a beam of light about 130,000 years to cross from one edge to the other, even though light travels at almost 186,000 mi. (300,000 km) a second. Huge areas of the Milky Way are unexplored because our view of them is so poor—it is like trying to see people at the other side of a crowd. Dark clouds block the light from the center of the galaxy, increasing the astronomers' problems. However, observing other galaxies has helped to build up a picture of what our own is like.

THE MILKY WAY
The Milky Way is a barred spiral galaxy, with two arms that rotate slowly. The area in which our Sun is found takes about 225 million years to go around once. At the center is a bright halo of old stars that formed with the galaxy, 14 billion years ago. The arms contain vast nebulae of gas and dust where new stars are being born.

HOW A GALAXY BEGINS
Galaxies begin as huge masses of dark gas. As they shrink under the pressure of gravity (**1**), the gas at the center becomes dense enough to start forming stars. Some galaxies start spinning (**2**), and if the spin is fast enough it forces the outer areas into a flat disk, forming a spiral (**3**) or barred spiral galaxy. Galaxies that spin slowly or not at all become spherical or elliptical in shape.

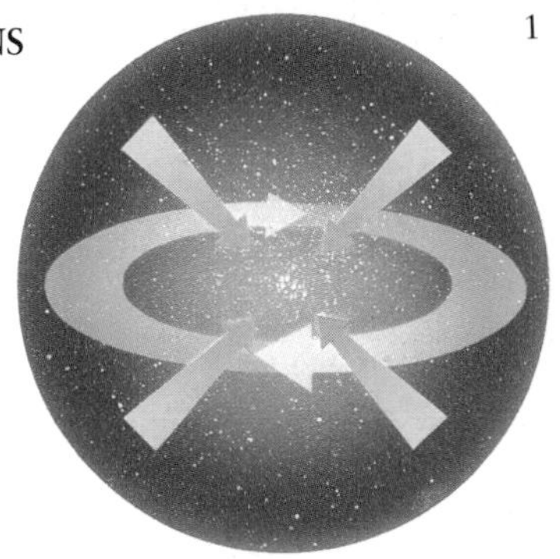

▲ *A slowly-spinning mass of gas starts to collapse, and the first stars are formed at the center. As the cloud shrinks, its turning speed increases.*

▲ *Gas clouds meet in the swirling disk, and attract more clouds because of their extra gravity. Stars start to form here too.*

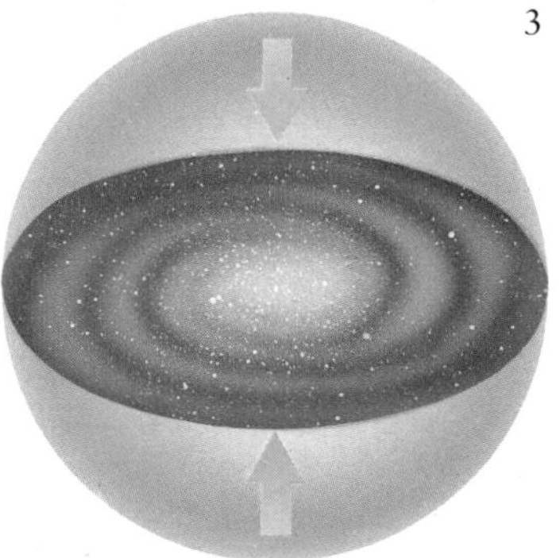

▲ *There is no gas left at the center to make new stars, but the arms are rich in raw star material. The galaxy is now in its prime of life.*

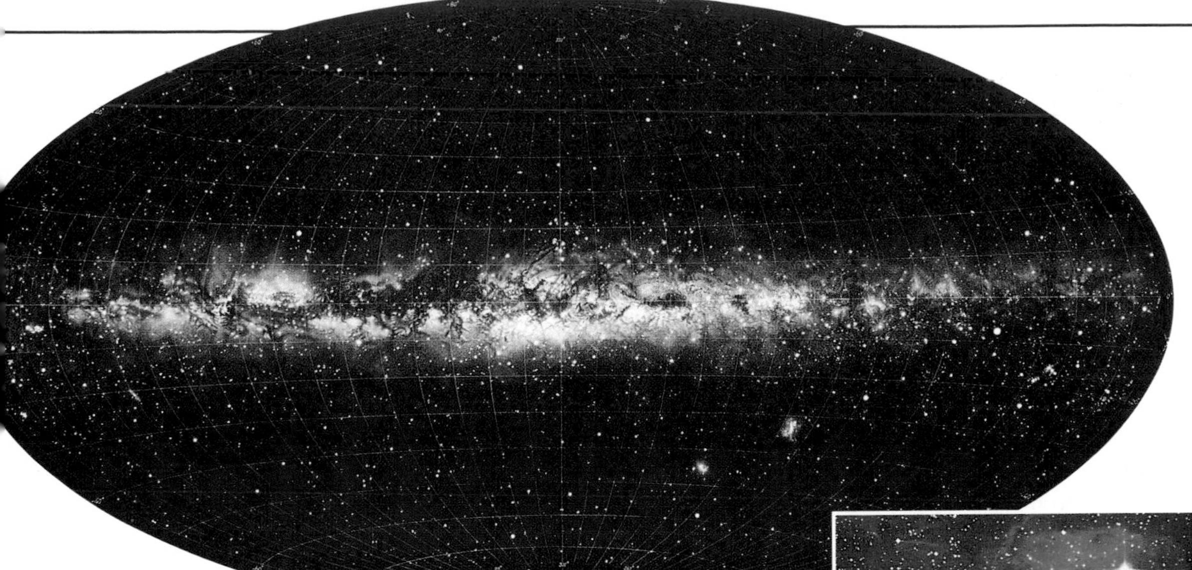

◀ *This panorama of our galaxy was obtained by combining several photographs taken of the Milky Way from different parts of the world. If you imagine the right and left ends joined together, with your head inside the ring, this is how the Milky Way would appear to someone floating in space.*

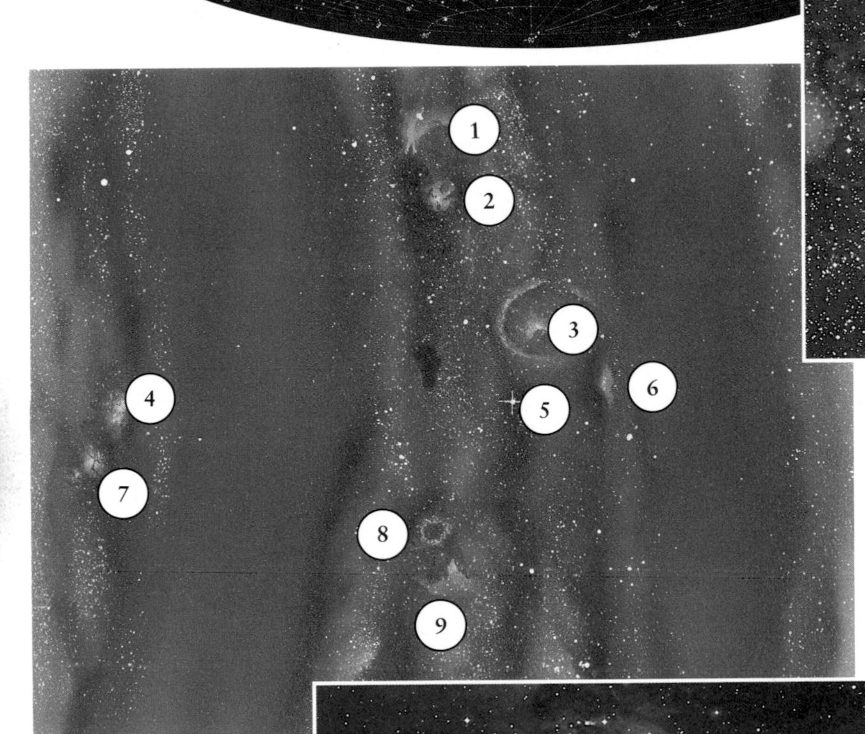

▲ *Within the main arms of the Milky Way are smaller arms where stars and nebulae are more closely connected. This illustration shows some of the Sun's neighbors in the galaxy.*

1 Cone Nebula
2 Rosette Nebula
3 Orion Nebula
4 Lagoon Nebula
5 Solar System
6 California Nebula
7 Trifid Nebula
8 Vela Supernova Remnant
9 N. American Nebula

▼ *The Orion Nebula is about 30 light-years across and 1,600 light-years away. It was once dark, but millions of years ago stars began forming inside it, and their radiation makes the nebula glow.*

▲ *The Milky Way is rich in dark nebulae. However, the galaxy is foggy with tiny particles that act as a color filter. The effect causes objects like this nebula to appear red to us on Earth.*

MILKY WAY DATAFILE

Diameter:
130,000 light-years
Thickness of spiral arms:
3,000 light-years (approx.)
Thickness of central bulge:
10,000 light-years
Diameter of central bulge:
20,000 light-years
Total mass: 110 billion × Sun
Average density (estimate):
0.00000000000000000007
(water = 1)
Age: 14 billion years
Time to rotate once:
(Position of Sun)
225 million years
Distance of Sun from center:
30,000 light-years
Satellite galaxies: 2

Clusters and Superclusters

Just as stars are born in groups or clusters, so do galaxies often exist in groups. Our galaxy belongs to a cluster of some 32 galaxies, called the Local Group. Most of the galaxies in the Local Group are small and faint and would be invisible to us if they were much farther away. Beyond the edge of the Local Group, astronomers have discovered thousands of other groups of galaxies in the universe. These groups in turn seem to form looser groups in space, known as superclusters.

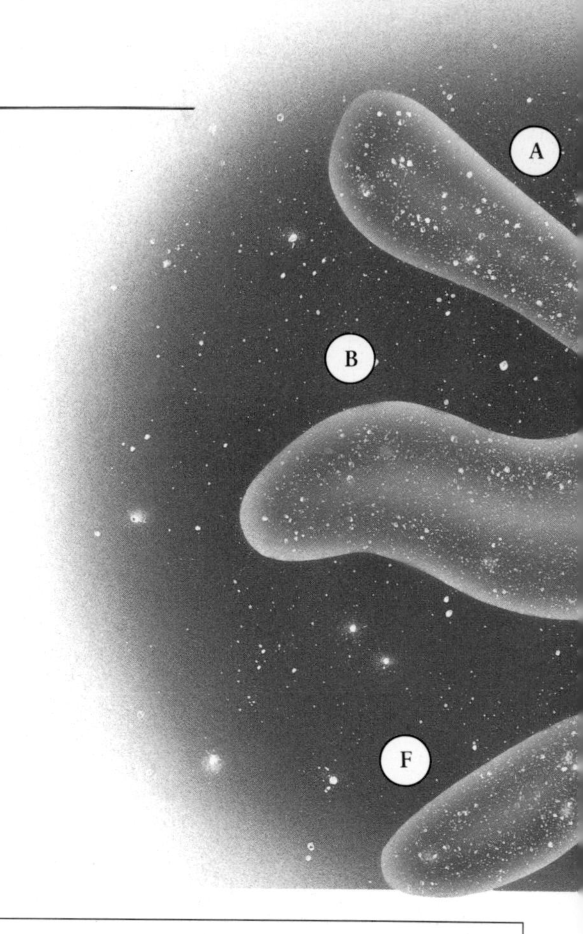

TYPES OF GALAXY

◀ *Galaxies are put into classes according to their shape. Irregular galaxies (class Irr) are usually smaller than the Milky Way. The gravity of much larger galaxies nearby may have pulled them out of shape—for example, the Magellanic Clouds, in the grip of the Milky Way.*

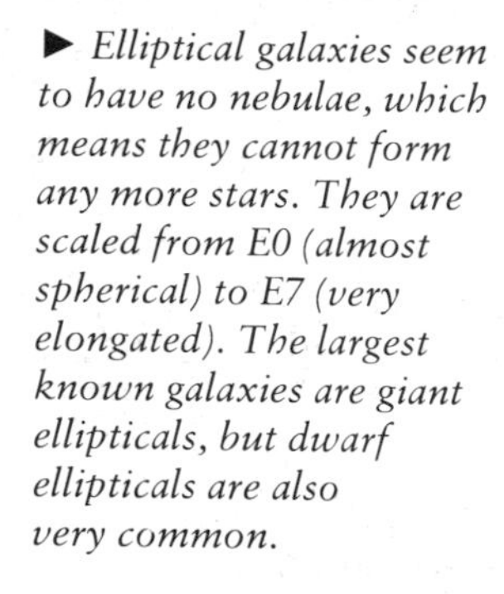

► *Elliptical galaxies seem to have no nebulae, which means they cannot form any more stars. They are scaled from E0 (almost spherical) to E7 (very elongated). The largest known galaxies are giant ellipticals, but dwarf ellipticals are also very common.*

◀ *Ordinary spiral galaxies are classed from Sa (very tight arms) to Sc (very loose arms). Another type, S0, has a very large nucleus, or center, which is more like an elliptical galaxy's. Until very recently the Milky Way galaxy was thought to be an Sb or Sc type.*

► *Barred spiral galaxies, classed from SBa (tight arms) to SBc (loose arms), have centers with a short bar of stars across them. The spiral arms begin at the ends of this bar. Astronomers have recently found evidence that there is such a bar in our own Milky Way galaxy.*

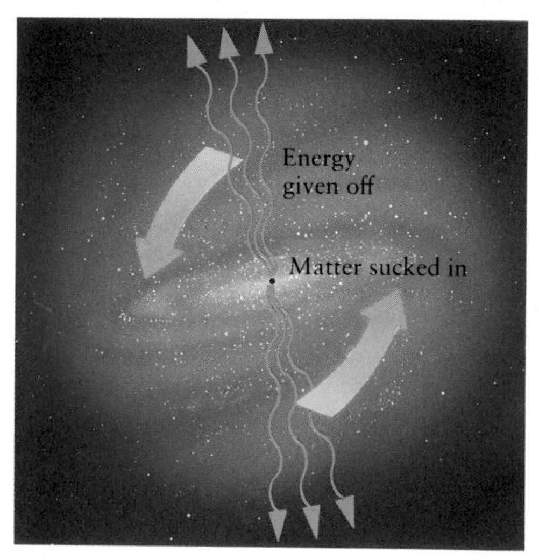

QUASARS

The strangest galaxies known are the far-off objects called quasars. Although they cannot be seen in any detail, they seem to be galaxies that are sending out huge amounts of energy from a small area of space near their center. Some quasars are as bright as thousands of galaxies like the Milky Way. The source of this energy could be the presence of a black hole at the galaxy's center, which is sucking gas and whole stars into it *(see pages 206–207)*. As it is sucked in, this material spins around the black hole almost at the speed of light, sending huge amounts of energy into space.

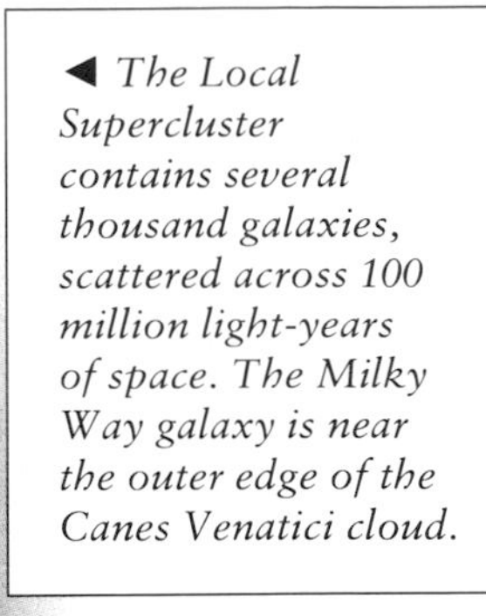

◄ *The Local Supercluster contains several thousand galaxies, scattered across 100 million light-years of space. The Milky Way galaxy is near the outer edge of the Canes Venatici cloud.*

KEY
A Virgo III cloud
B Virgo II cloud
C Virgo I cloud
D Canes Venatici cloud
E Canes Venatici spur
F Crater cloud
G Leo II cloud

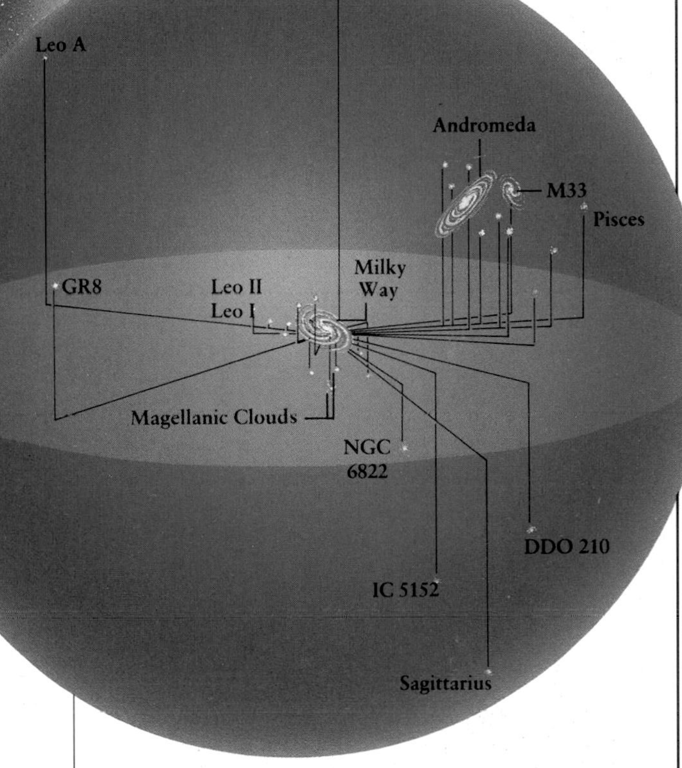

▲ *The 32 galaxies detected in our Local Group range from the Andromeda galaxy, the largest, to tiny irregular galaxies. The Milky Way is second in order of size, one of only three spiral galaxies in the Group. The galaxies can be pictured scattered in space, inside an imaginary sphere about 5 million light-years across.*

SUPERCLUSTERS

The Local Group is a fairly small cluster of galaxies, belonging to a collection of other small clusters called the Canes Venatici cloud. A much larger group of galaxies, the Virgo cluster, is about 60 million light-years away. It lies at the center of our Local Supercluster—a collection of major clouds of galaxies. The entire supercluster measures over 100 million light-years across. In the diagram *(above)*, the galaxy clusters are shown with sharp boundaries to help make their shapes more clear. In reality, the clusters are much more scattered throughout space. Notice how the central Virgo I cloud is round, while the others are elongated, pointing inward toward Virgo I as if stretched by the pull of its gravity. This has occurred because Virgo I, although not particularly large, contains about 20 percent of all the galaxies in the supercluster.

FACTS ABOUT THE LOCAL GROUP

- The Andromeda galaxy is the largest in the Local Group. It may contain 400 billion stars.
- The Milky Way's nearest neighbors in space are the two Magellanic Clouds, 200,000 light-years away. They are satellites of the Milky Way.
- A belt of fast-moving gas, called the Magellanic Stream, connects our galaxy to the two Magellanic Clouds.
- M33 in Triangulum is the third-largest member of the Local Group.
- The most distant galaxy in our Local Group is Leo A, 5 million light-years away.
- Dwarf elliptical galaxies are probably the commonest type of galaxy in the Universe, but they are so dim that we can only make out the 15 or so near to us in the Local Group.

► *A supernova, or exploding star, plays an important part in a galaxy's development. Such stars scatter extra elements into space, besides the original hydrogen and helium, to build up new stars. The last nearby supernova was seen in the Large Magellanic Cloud in 1987.*

Cosmology

Cosmology is the study of how the universe began and how it will be in the future. The universe may go on getting bigger forever. Alternatively, the galaxies may come together, until finally they collide and explode in the violence of a "Big Crunch." Cosmologists try to work out the likely fate of the universe. They believe they have traced the development of the universe back to a fraction of a second after the Big Bang, but they do not know what caused the Big Bang itself.

THE SCALE OF THE UNIVERSE

Suppose that the edge of the visible universe is a hollow sphere as large as the Earth (diameter 7,926 mi. [12,756 km]). On this scale the Milky Way would be 130 ft. (40 m) across; the distance to the nearest star, 4.2 light-years away, would be 1.3 mm; the Earth would be a speck 0.0000000004 mm across.

THE FATE OF THE UNIVERSE

What is the future of the universe? It may keep on expanding; it may collapse and end; or it may even be one of a series of universes. It all depends upon how much material the universe contains. As the galaxies fly apart after the Big Bang, their gravitational attraction is slowing them down. But the gravity between bodies in space becomes weaker as they move farther apart, so the slowing-down effect becomes less as the universe expands. There is a thin dividing line between there not being enough matter in the universe—so that gravity is too weak ever to stop the expansion—and there being too much matter—so that everything rushes together in a "Big Crunch."

AN EVER-EXPANDING UNIVERSE

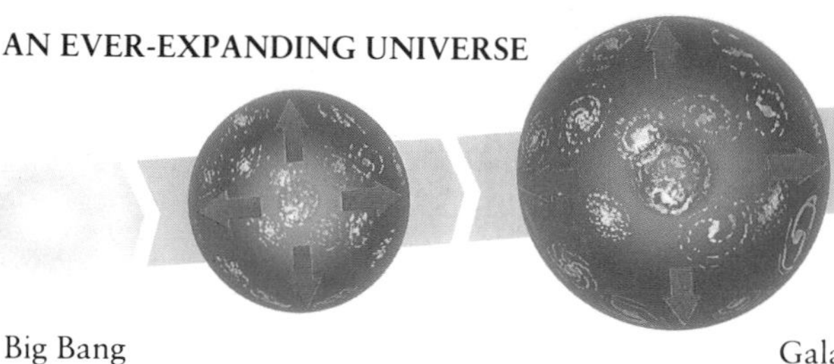

Big Bang

Galaxies fly apart after the Big Bang

A FINITE UNIVERSE

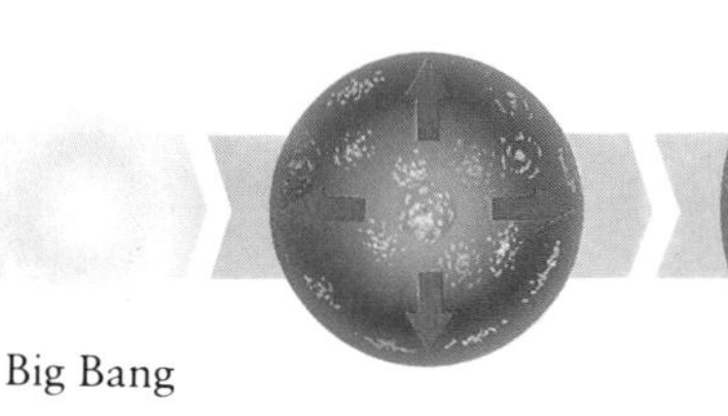

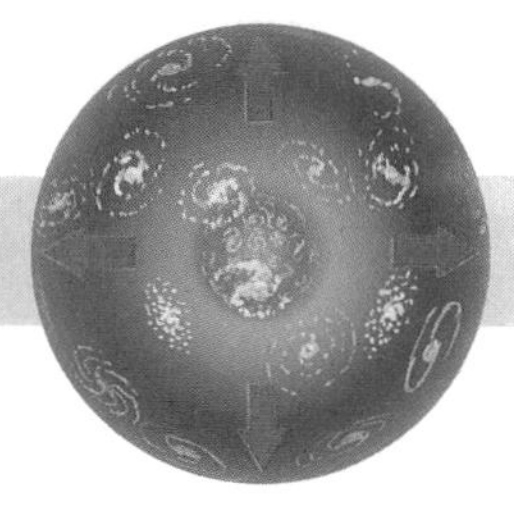

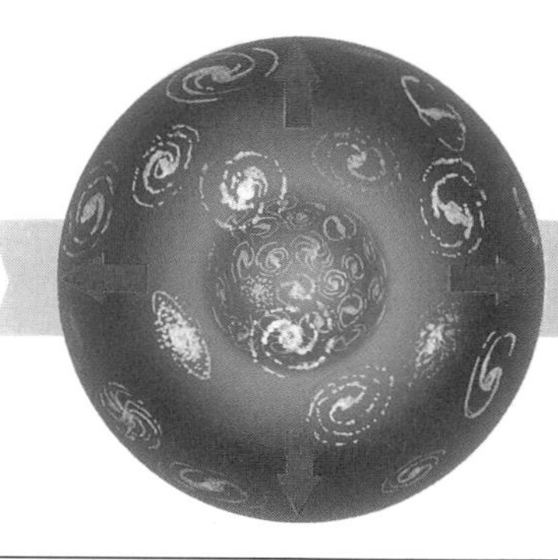

Big Bang

FACTS ABOUT THE UNIVERSE

- There are clues to show that the Big Bang occurred. For one thing, the galaxies are still flying apart from the explosion. In 1965 astronomers found a very feeble warmth in space, which is a trace of the fantastic heat created by the Big Bang.
- Some astronomers believe the universe must contain "missing mass." This is invisible material that astronomers believe exists because of the effect of its gravity. If the universe does not contain this material, then after the Big Bang it should have expanded so fast that galaxies could not have formed at all. The missing mass has not been detected yet, but it may add up to roughly ten times the mass of the stars and nebulae that are visible.

RED SHIFT

Light travels in waves, and the color of light depends upon wavelength. Waves of blue light are much closer together than waves of red light. If a very fast-moving object is sending out light, the light waves will be squashed ahead of it and stretched out behind. This means that the light from the approaching object seems bluish. Light from an object traveling fast away from someone appears reddish. This is called a red shift.

Reddish light detected on Earth

Galaxy moving away from Earth

◀ *Distant galaxies show a color shift toward red. These galaxies are moving away at speeds of thousands of miles a second. Their light waves are stretched by this speed, making them look redder in color than they really are. This change of color is not seen by the eye, but is detected with an instrument called a spectrograph. Astronomers can use the red shift to measure the speed at which a galaxy travels.*

▼ *Galaxies and clusters of galaxies twist and loop themselves through space like strings of frog spawn (eggs). Is there almost as much invisible matter in the supposedly empty space between the strings?*

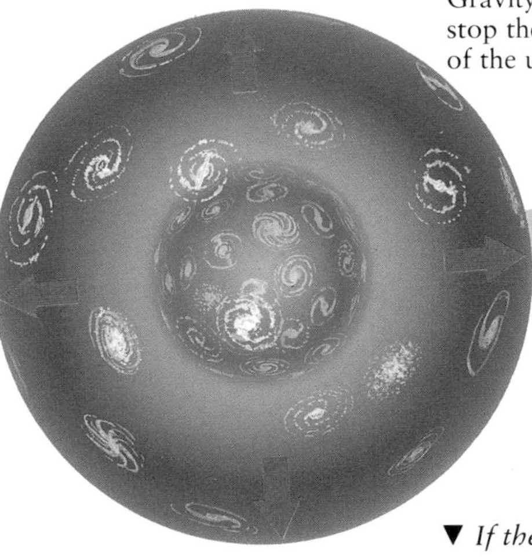

Gravity cannot stop the expansion of the universe

▼ *If the Big Crunch occurred, the sky would grow as hot as the Sun. Finally, everything would vanish into a black hole.*

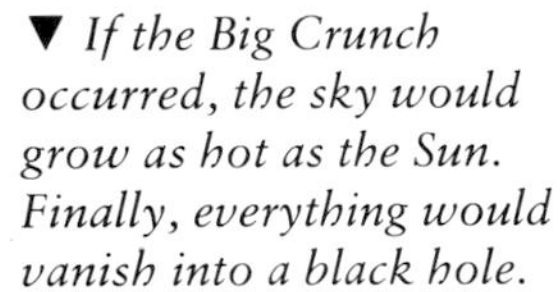

Gravity stops expansion of the universe

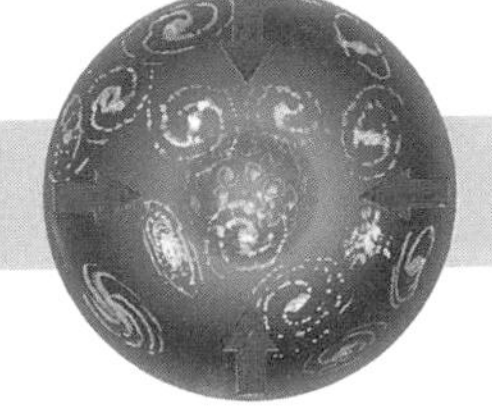

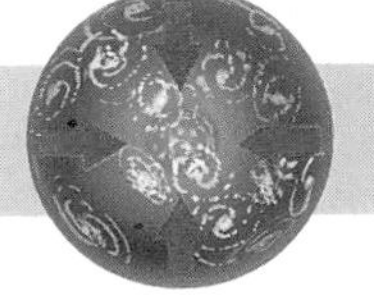

Big Crunch

Big Bang

A SERIES OF UNIVERSES
A completely new universe begins

▼ *Time and space would end if the universe ended. A new universe would have to start everything again. It would not be the old one recycled.*

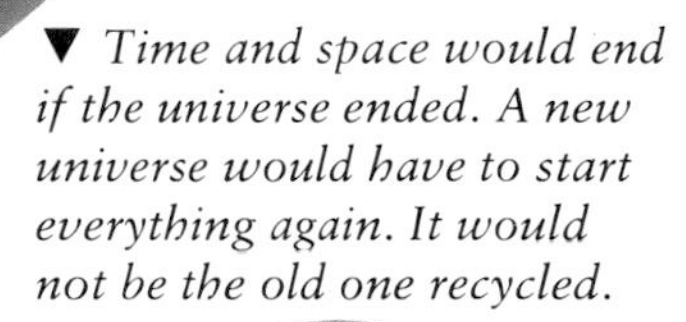

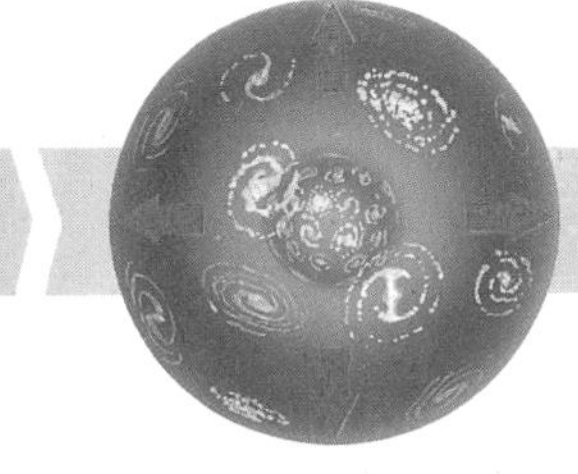

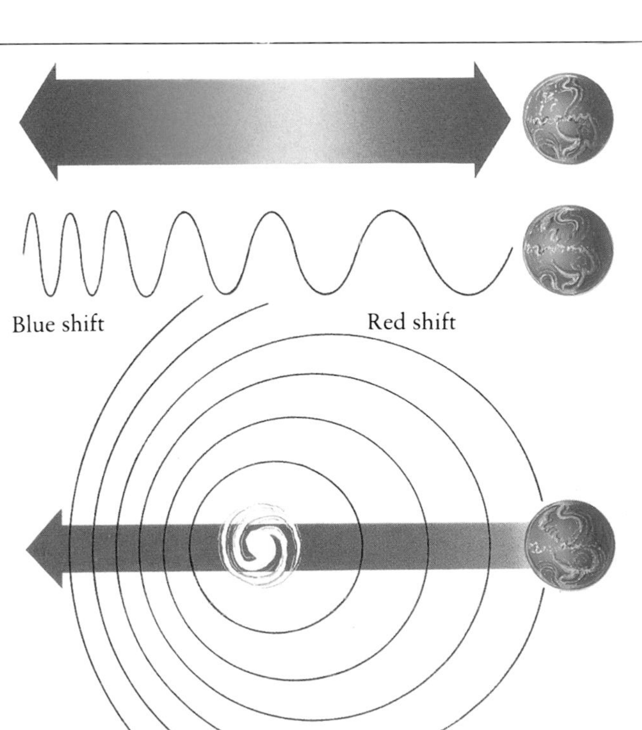

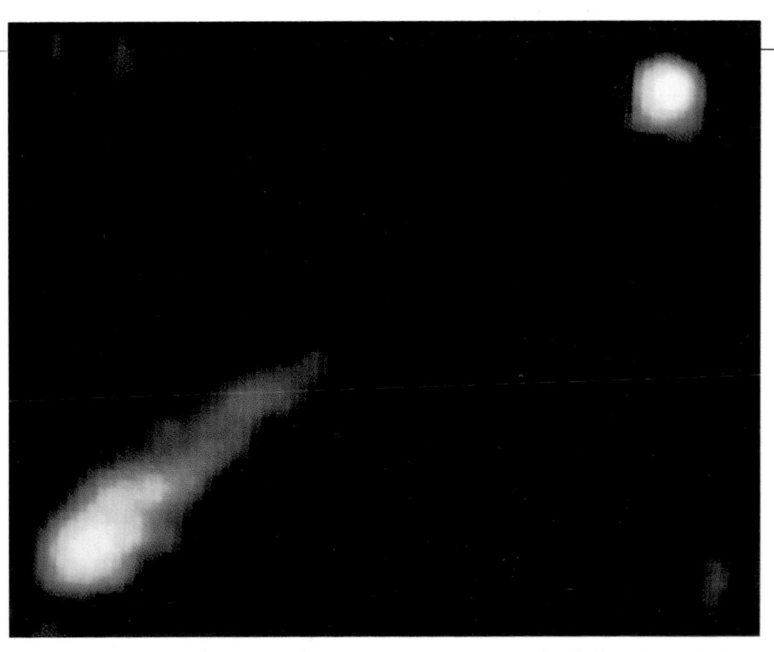

▲ *Red shift measurements of the quasar 3C 273 show that it is 2.1 billion light-years away, traveling 30,000 mi./s (50,000 km/s). As the universe expands, distant galaxies move away faster than galaxies nearby. Their distance can be calculated from their red shift.*

THE STARS

The Life of a Star

A star is born inside a huge nebula of gas and dust (called "gust"). A nebula starts to shrink into much smaller nebulae when it receives a "shake," perhaps after colliding with another nebula, or being hit by the shock waves of an exploding star. Eventually the nebula will break up into a cluster of baby stars, containing a mixture of bright, medium, and dim stars. Although all these stars are born at the same time, their lifetimes and endings will be very different, according to the amount of material they contain. Generally, a massive star has a shorter life than a less massive star. Stars are sources of light and heat. They also process and recycle material, turning some of the hydrogen and helium that were created at the beginning of the universe into other elements. These include carbon and oxygen, which are the building blocks of life.

► *A medium-hot star, like the Sun, gives out much less energy than the brightest stars. This means it will shine steadily for billions of years before it begins to expand. The dimmer the star, the longer its life.*

▲ *The main sequence period for both a medium-hot star and a very hot, bright star* (below) *is the time when the star is shining steadily by turning hydrogen into helium. The star grows slightly hotter and brighter during this time. When the star's core of used helium reaches a certain size it is a crisis point in its life. The core of the star collapses under the force of its gravity and becomes still hotter. Now even helium starts nuclear reactions.*

▲ *This new energy source is so powerful that part of the star is blown outward. The outer layers cool to a reddish color. At this stage in its life the star is known as a "red giant." At its greatest size, the diameter of a red giant may be a hundred times that of the original star, which forms the core of the red giant.*

► *A very bright, very hot star may shine at full power for only a few million years before it uses up all its fuel. Although it has more fuel than a medium star, it uses it up much faster, and has only a brief life.*

▲ *Although our Sun shines with a yellowish light, not all stars are yellow. Stars vary in size, brightness, and temperature. The hottest stars shine with a white-blue light; they have a diameter up to 20 times greater than our Sun, and so are known as blue giants. The dim, cool stars known as red dwarfs are about a quarter of the Sun's diameter. Huge red giant stars have outer layers that have blown away from the core of energy, becoming cooler and so redder. They can be 500 times the Sun's width. A dim white dwarf will give out little light and may be only a few percent of the Sun's width.*

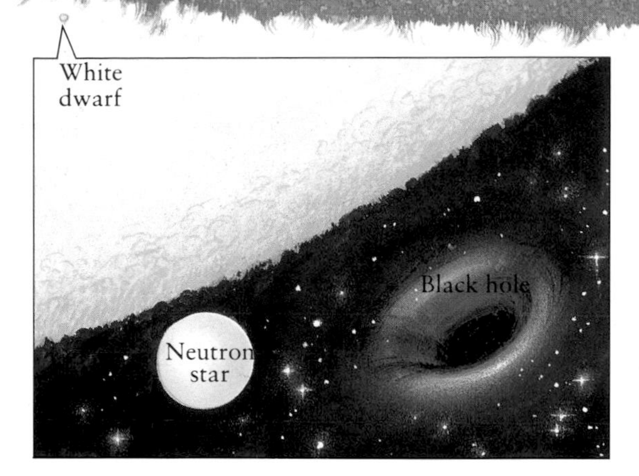

◀ *In a white dwarf, the star's atoms are crushed together hundreds of times more tightly than normal. A neutron star is a dead star, made from the solid nuclei of atoms, the densest material in the universe. If it is massive enough, a neutron star will become a black hole.*

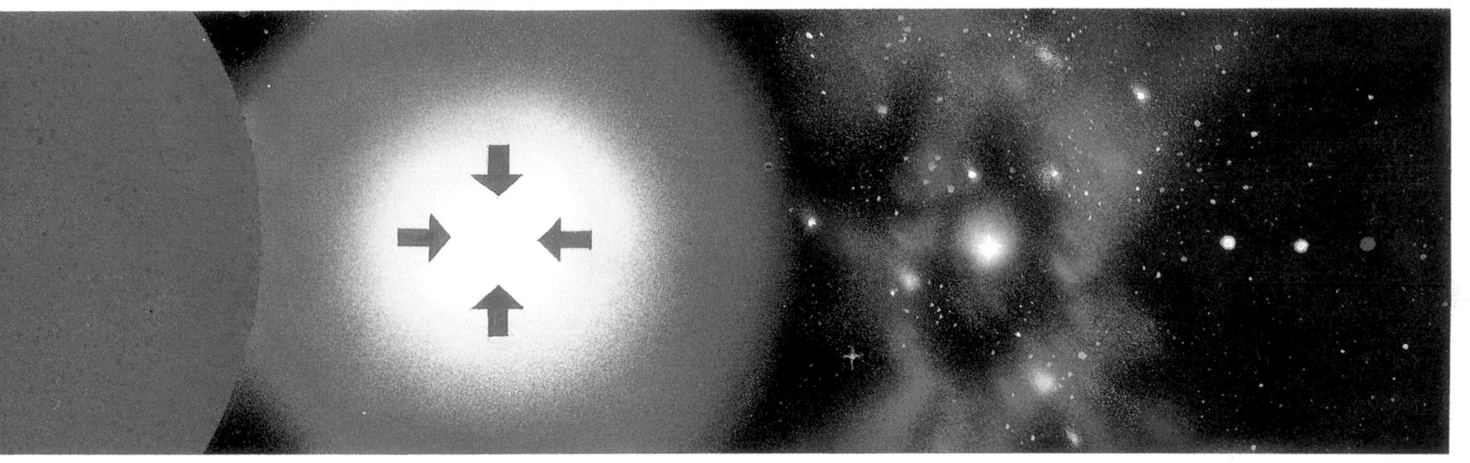

▲ *Eventually the core runs out of fuel and collapses completely, because it is no longer producing enough energy to balance the shrinking force of gravity. This is the end of nuclear reactions in the star's core. If the star is much more massive than our Sun, its collapse gives out so much energy that the star is blown to pieces in a supernova explosion* (below).

▲ *The collapse of a star like the Sun will not result in a dramatic supernova explosion. Instead, the star will shrink to a hot body the size of a small planet. The star is now called a white dwarf. The gravity of a massive star, however, will be so strong it will crush the nuclei of the star's atoms together, making a tiny, very dense, neutron star* (below).

▲ *An ordinary star, after becoming a white dwarf, cools into a dead black dwarf.*

▼ *If a neutron star has a mass greater than five Suns it will form a black hole.*

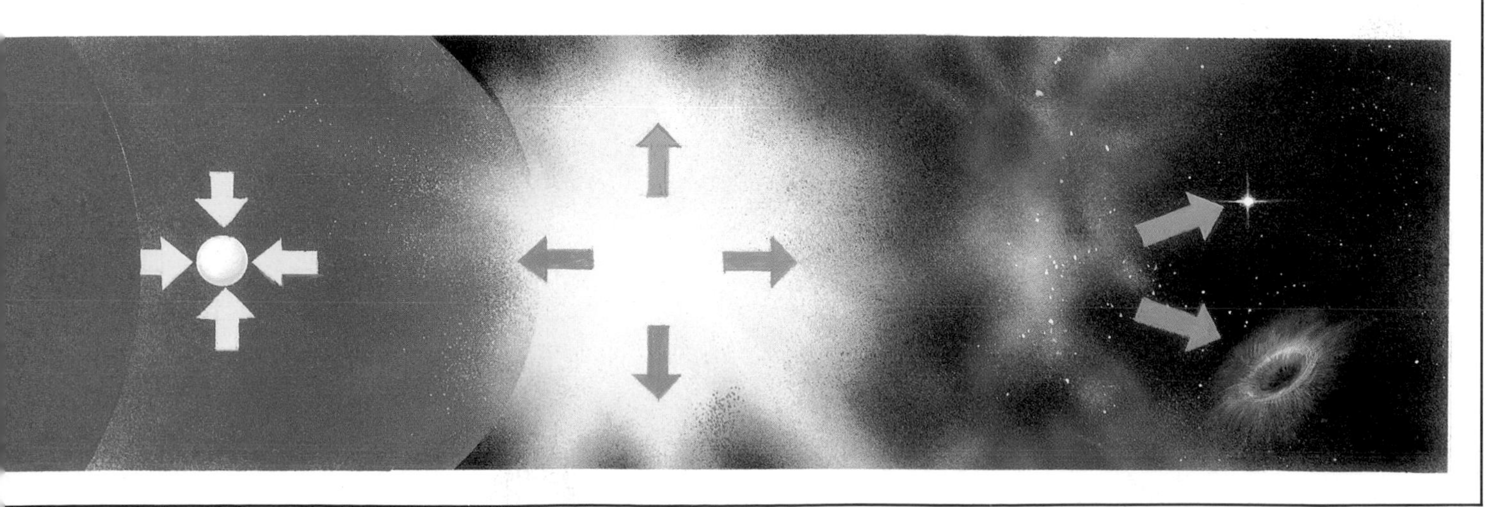

Extraordinary Stars

If you think of a star as a body similar to our Sun, the different stars shown here may surprise you. Not all stars shine at constant levels. Some, known as variable stars, change greatly in brightness during years or even days. Other stars, called pulsars, spin faster than a washing machine. Twenty percent of all known stars have a partner, around which they orbit in a binary (two-part) system. Occasionally the transfer of gas in a binary system can lead to the drama of a supernova, or exploding star.

VARIABLE STARS
A variable star may be an unstable star that goes through a stage when it starts to swell and shrink, changing its brightness. Other variable stars may occur in binary systems where gas passes from one star to another, causing a sudden flare. A few variable stars do not actually give out less light at all, but seem to do so because they orbit each other in a binary system, and the light from one star is blocked for a time by its "partner."

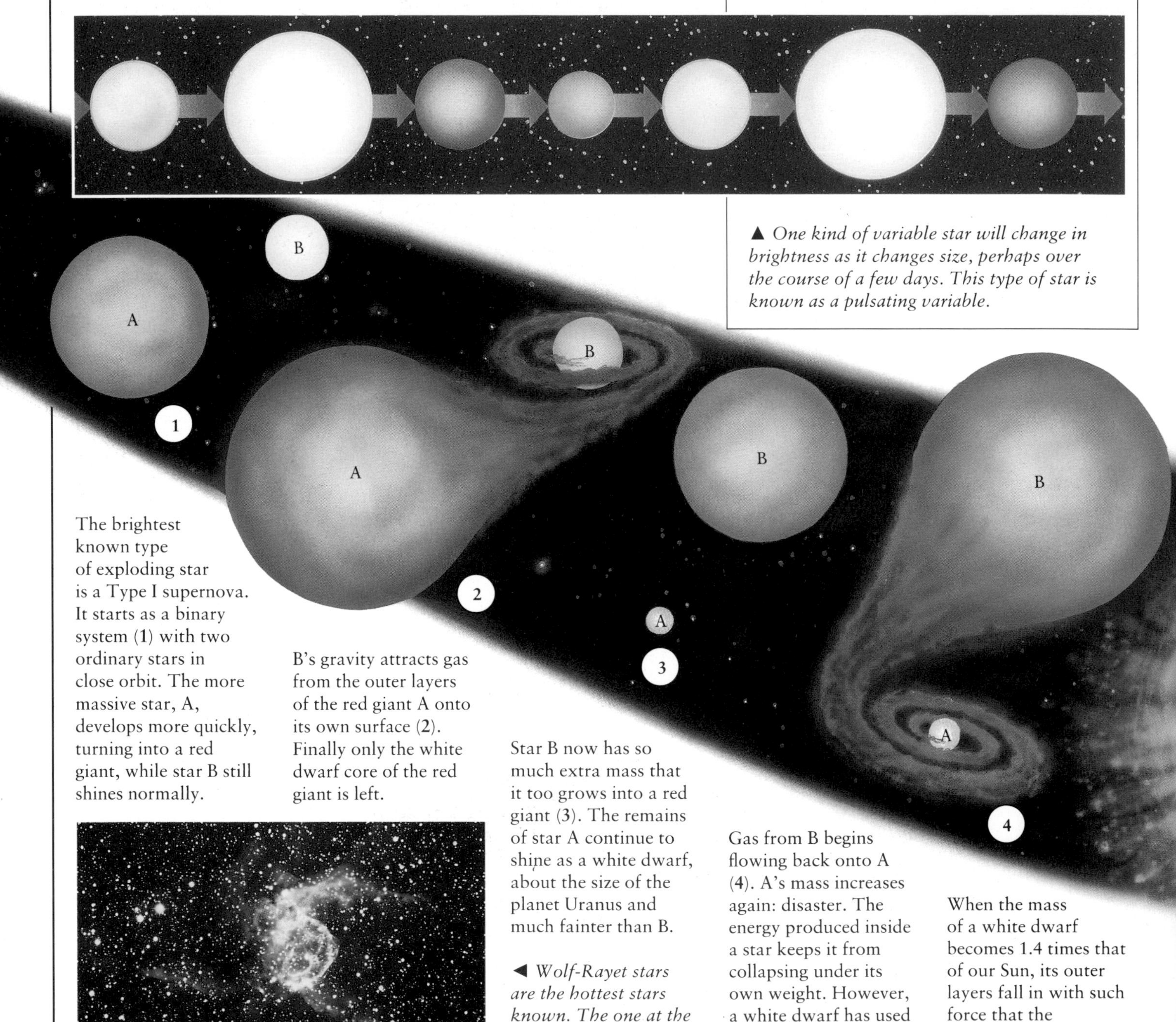

▲ *One kind of variable star will change in brightness as it changes size, perhaps over the course of a few days. This type of star is known as a pulsating variable.*

The brightest known type of exploding star is a Type I supernova. It starts as a binary system (1) with two ordinary stars in close orbit. The more massive star, A, develops more quickly, turning into a red giant, while star B still shines normally.

B's gravity attracts gas from the outer layers of the red giant A onto its own surface (2). Finally only the white dwarf core of the red giant is left.

Star B now has so much extra mass that it too grows into a red giant (3). The remains of star A continue to shine as a white dwarf, about the size of the planet Uranus and much fainter than B.

Gas from B begins flowing back onto A (4). A's mass increases again: disaster. The energy produced inside a star keeps it from collapsing under its own weight. However, a white dwarf has used up most of its nuclear fuel. If too much mass is added, A won't be able to hold its shape.

When the mass of a white dwarf becomes 1.4 times that of our Sun, its outer layers fall in with such force that the temperature rises to several billion degrees. The blast of energy blows the star apart (5).

◀ *Wolf-Rayet stars are the hottest stars known. The one at the center of this nebula has a surface temperature of about 90,000°F (50,000°C).*

PULSAR

In 1967, radio astronomers at Cambridge, England, were trying out a new telescope. It began to record bursts of radiation that repeated every couple of seconds. At first the Cambridge astronomers thought that this might be an intelligent message from space, but soon they realized that they had discovered a new kind of star—a pulsar. From that chance discovery, more than 400 radio pulsars were discovered in the next 20 years.

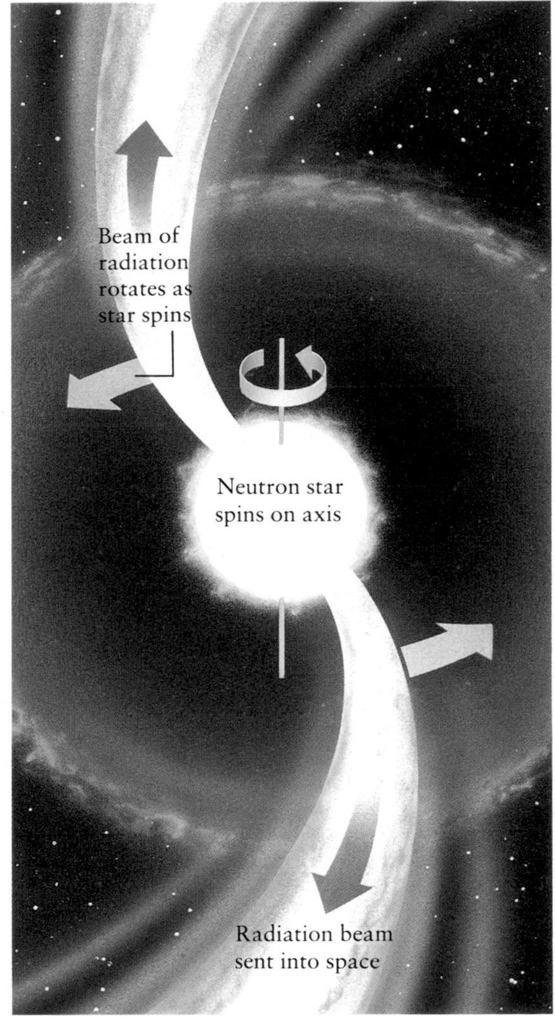

► *A pulsar is a rapidly spinning neutron star. Its fierce magnetic field squirts light and radio waves into beams of energy that sweep around as the star turns. If the beam crosses Earth the star is detected by the pulse of its radiation.*

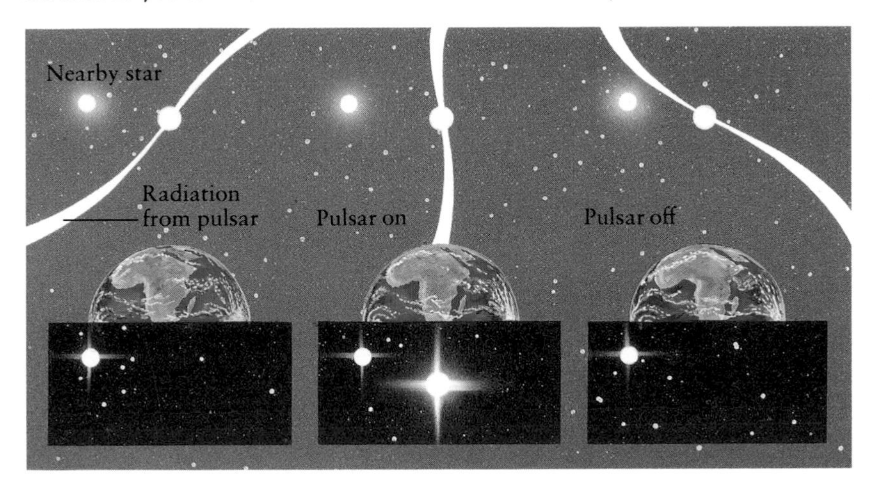

▲ *Pulsars flash on and off when their radiation beam passes across the Earth. Even the slowest pulsar sends out one pulse every four seconds. The most rapid one spins 622 times a second. Why do neutron stars spin so much faster than the original star? Find the answer by watching an ice skater spin on the ice. As the skater shrinks by folding her arms up close to her body, she speeds up. A star collapsing into a tiny neutron star has the same effect.*

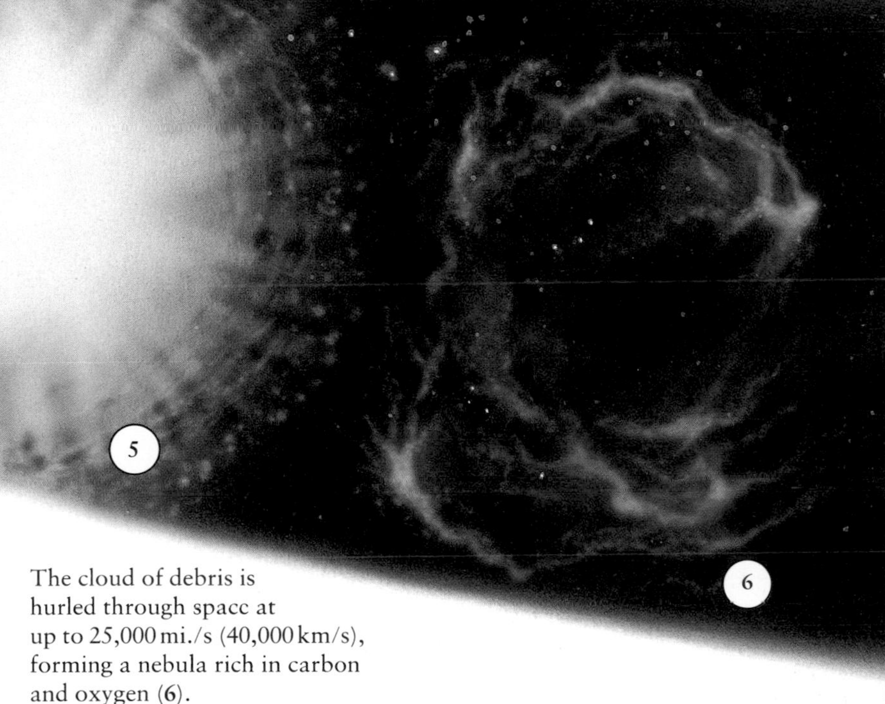

The cloud of debris is hurled through spacc at up to 25,000 mi./s (40,000 km/s), forming a nebula rich in carbon and oxygen (6).

BINARY STARS

In some binary systems the two stars (1) may be very close, orbiting each other in a few hours; in others, the stars are millions of miles apart. Often, one star is more massive than the other and has evolved, perhaps into a red giant (2). Although the stars in a binary system appear to revolve around each other, they are really moving around their common center of gravity (3).

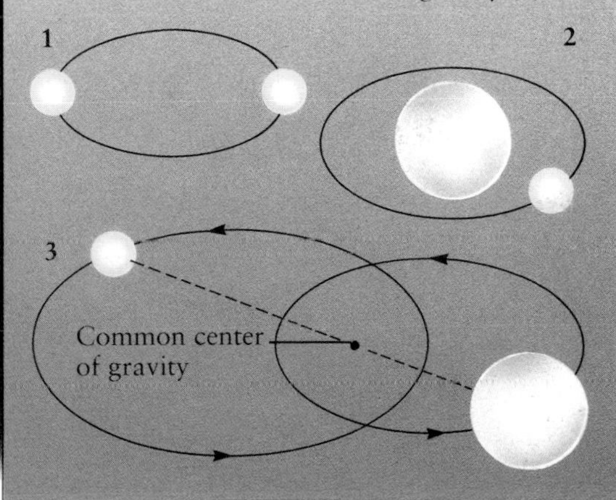

Black Holes and Neutron Stars

When a massive star dies, it may leave behind a heavy core which becomes a neutron star—the smallest, densest kind of star. If the force of gravity around the neutron star is strong enough, a black hole could be the result. These fascinating objects occur when gravity is so strong that space acts like a one-way funnel. Anything—light rays or moving solid matter—that passes into this funnel is compressed to nothing and disappears from our universe. A black hole is the end of space and time.

► *A binary system containing a black hole will be ablaze with radiation, because material from the other star is energized as it spirals into the black hole.*

WHAT IS A BLACK HOLE LIKE?

A black hole is caused by a very dense object, such as a massive neutron star, which creates a powerful gravitational field—a kind of space funnel—in a small space. An object passing close enough will be pulled into the space funnel; once something has been pulled into a black hole it can never escape.

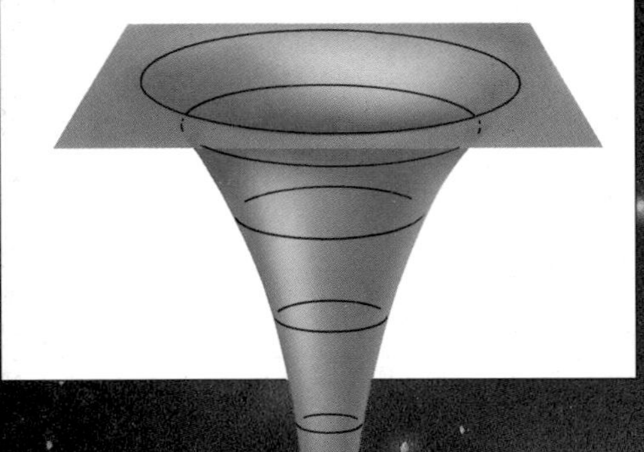

▲ *A black hole on its own in space would be difficult to detect. But if it is near another star, the pull of the black hole's gravity may draw material from the star into a whirling ring that gives off bursts of X-rays. If astronomers detect a starlike object that is giving out high-energy radiation they have probably found a black hole.*

A BLACK HOLE IN THE MILKY WAY?

A faint object in the Milky Way galaxy, normally about a million times too faint to be seen with the naked eye, sometimes gives out bursts of X-rays. Known as V404 Cygni, it is the most likely black hole in our galaxy. V404 Cygni is a binary system. The two stars orbit each other in 6.5 days. One star may be similar to the Sun, the other is a very dense object about six times as massive as the Sun—as massive as a black hole should be.

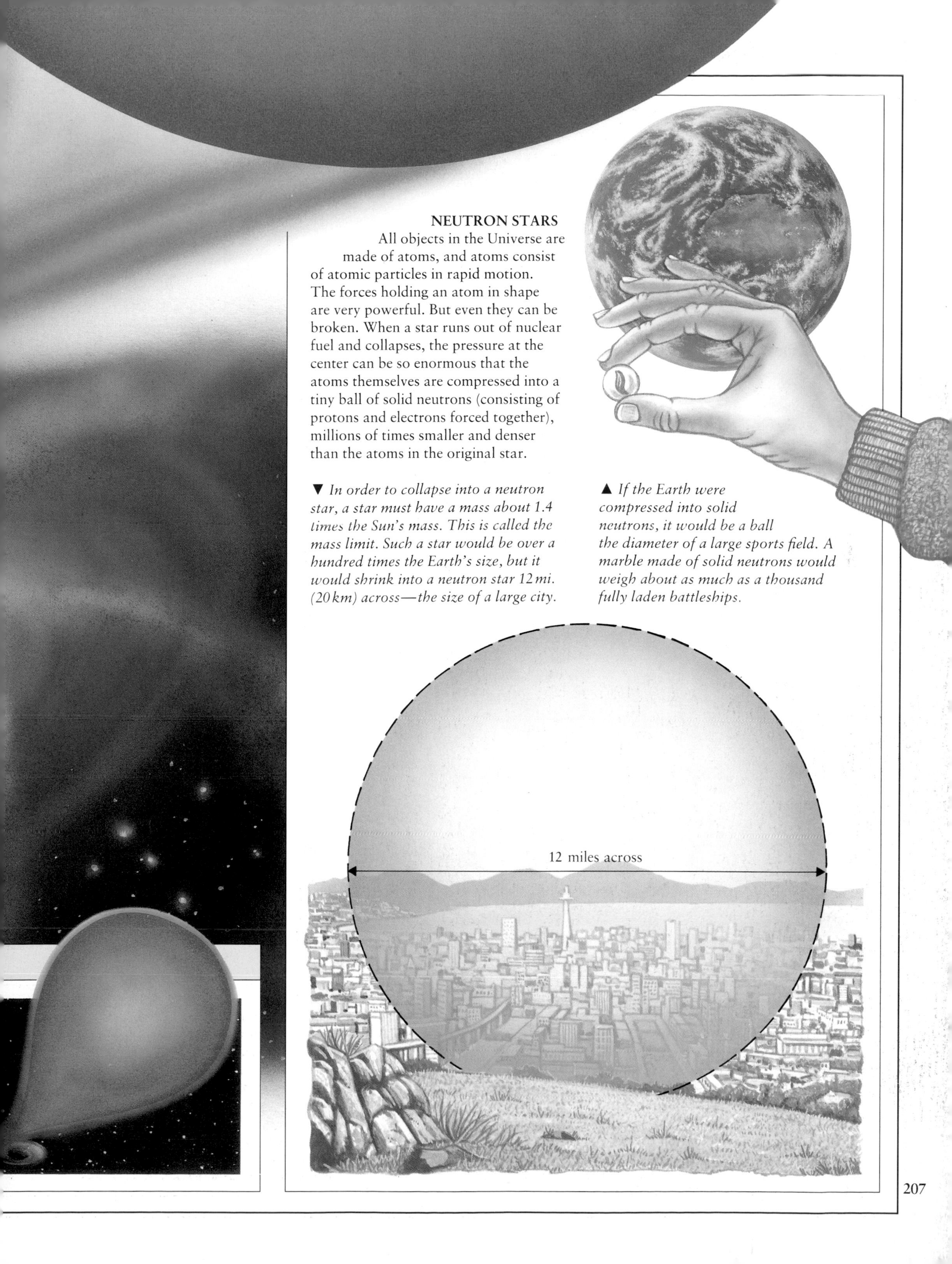

NEUTRON STARS

All objects in the Universe are made of atoms, and atoms consist of atomic particles in rapid motion. The forces holding an atom in shape are very powerful. But even they can be broken. When a star runs out of nuclear fuel and collapses, the pressure at the center can be so enormous that the atoms themselves are compressed into a tiny ball of solid neutrons (consisting of protons and electrons forced together), millions of times smaller and denser than the atoms in the original star.

▼ *In order to collapse into a neutron star, a star must have a mass about 1.4 times the Sun's mass. This is called the mass limit. Such a star would be over a hundred times the Earth's size, but it would shrink into a neutron star 12 mi. (20 km) across—the size of a large city.*

▲ *If the Earth were compressed into solid neutrons, it would be a ball the diameter of a large sports field. A marble made of solid neutrons would weigh about as much as a thousand fully laden battleships.*

Star Distance, Star Brightness

To measure the distance of a star from Earth, astronomers use parallax (*see opposite*). The first star distance was measured in 1838, marking a milestone in astronomy. As more star distances were cataloged, astronomers realized that some nearby stars appeared fainter than more distant ones, instead of the other way around as expected. It was proof that stars had different brightnesses, or luminosities. This formed the basis for the classification of stars into the families that are recognized today.

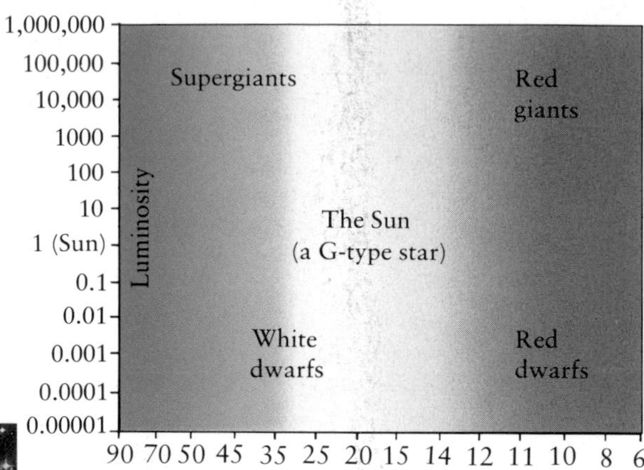

◄ *A group of stars close together in the sky is known as a star cluster. Clusters such as the Pleiades* (left) *are very useful for astronomers, because the stars are all the same distance away from Earth. So if one star looks brighter than another, it must be more luminous by that amount. The Pleiades are very young, maybe a hundred million years old, and the brightest members are very hot main sequence stars.*

▲ *The Hertzsprung-Russell Diagram plots the temperature and luminosity of stars. Generally, the more material a star has, the hotter it is. A blue giant star is found in the upper left of the diagram. The Sun, a yellow star of average size, lies in the middle, or main sequence. Few stars do not belong to the main sequence. Red giant stars lie in the upper right of the diagram. Lower left stars are white dwarfs —very hot, but dim.*

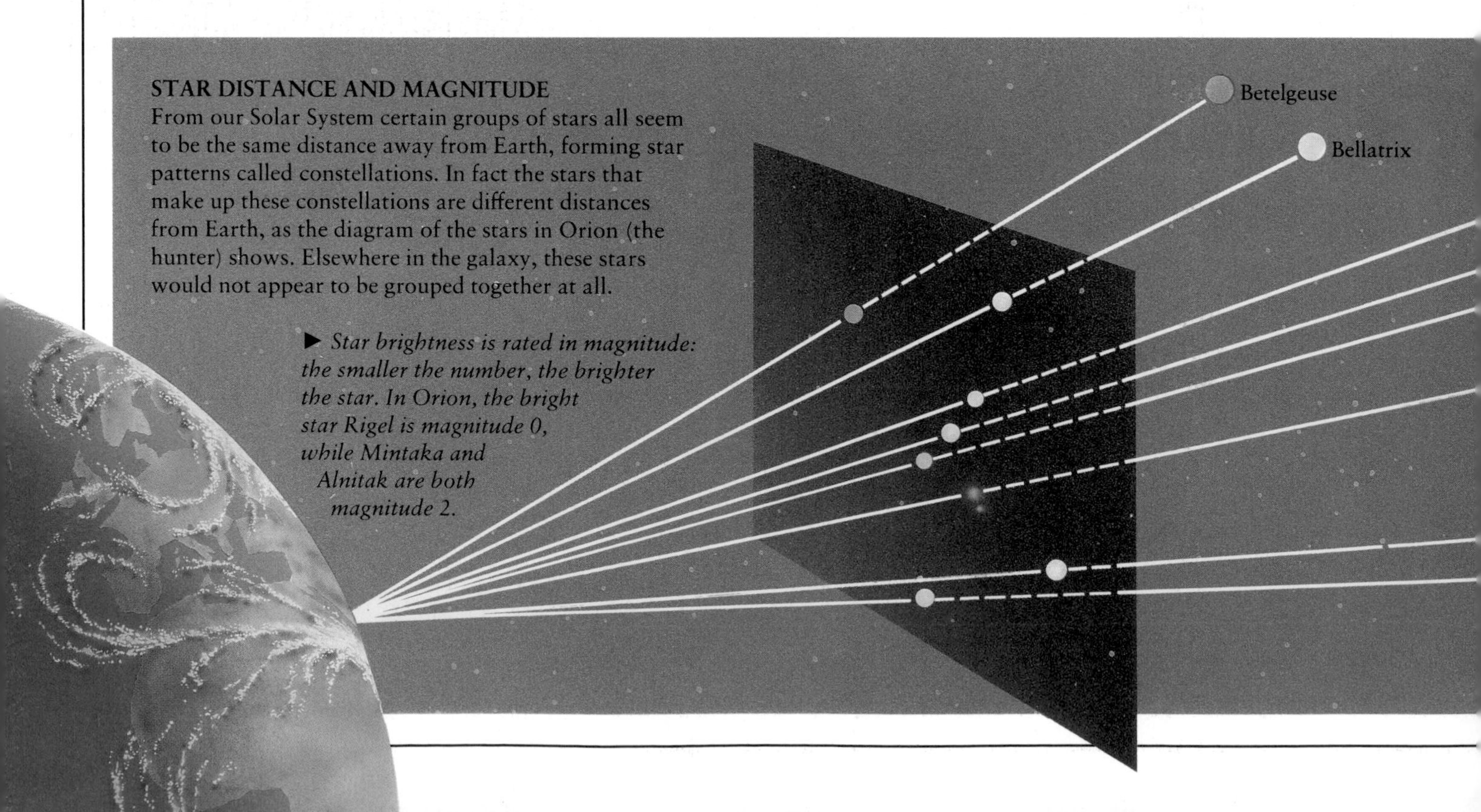

STAR DISTANCE AND MAGNITUDE

From our Solar System certain groups of stars all seem to be the same distance away from Earth, forming star patterns called constellations. In fact the stars that make up these constellations are different distances from Earth, as the diagram of the stars in Orion (the hunter) shows. Elsewhere in the galaxy, these stars would not appear to be grouped together at all.

► *Star brightness is rated in magnitude: the smaller the number, the brighter the star. In Orion, the bright star Rigel is magnitude 0, while Mintaka and Alnitak are both magnitude 2.*

PARALLAX

When you move your head from side to side, nearby objects seem to move more than distant ones. This is called parallax. Astronomers use parallax to find out how far away stars are from Earth. They take two measurements of the direction of a star about six months apart, when the Earth is on opposite sides of its orbit and has moved 186 million mi. (300 million km) through space. From the slight change in the star's position, its distance can be calculated. The farther the star, the smaller the change.

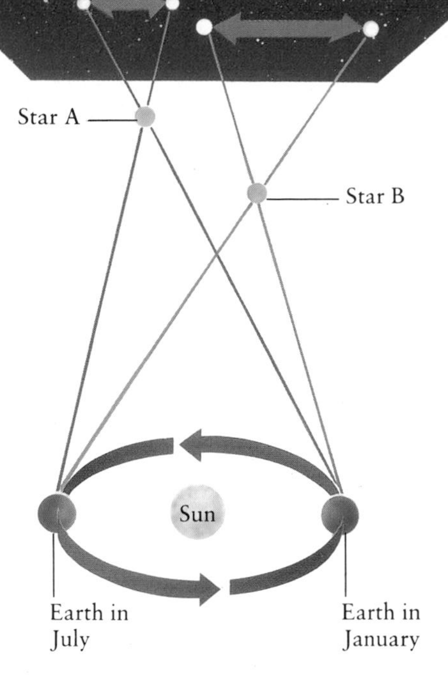

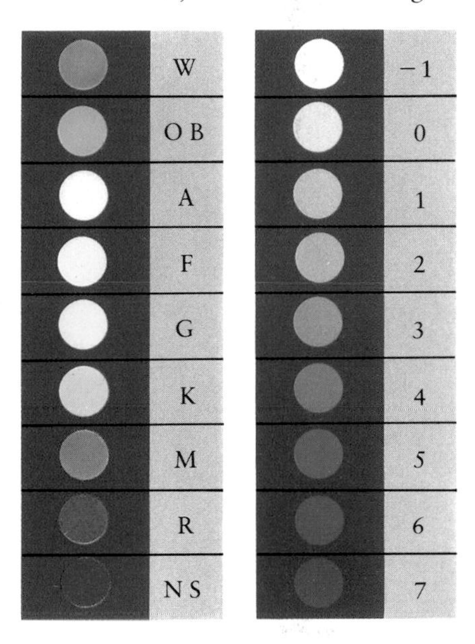

◀ *A very hot star will look bluish-white, and a cooler one, yellow. Stars may be classified according to temperature* (far left). *Letters stand for different temperature bands. Our yellow Sun is a G star. The table* (left) *shows the luminosity of average stars in these classes.*

FACTS ABOUT STARS

- The nearest star to our Sun is Proxima Centauri, 4.2 light-years away. It is a dim red dwarf star.
- Sirius is the brightest star in our sky.
- The dimmest stars we are able to detect from Earth are about 1×10^{27} times fainter than the dimmest star visible with the naked eye.
- The oldest known star is thought to be CS 22876 −32. It may be 15 billion years old.
- There are many newborn stars visible. An example is L1551, which is being born now in a nebula 500 light-years away.
- Pulsar PSR 1957 +20 is the fastest spinning star. It revolves 622 times a second.
- The supernova seen in A.D. 1006 appeared about 40 times brighter than Venus.

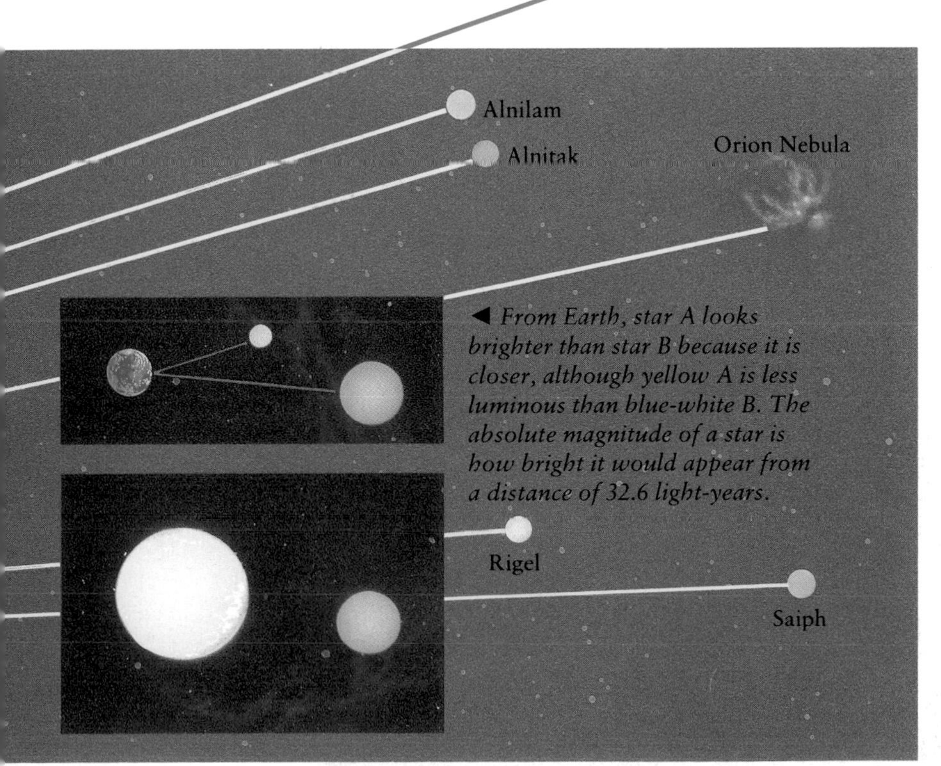

◀ *From Earth, star A looks brighter than star B because it is closer, although yellow A is less luminous than blue-white B. The absolute magnitude of a star is how bright it would appear from a distance of 32.6 light-years.*

▲ *From Earth, all the stars of Orion seem to be the same distance away, forming a particular pattern in our night sky. But Betelgeuse is 330 light-years from Earth, and Mintaka is 2,300 light-years away.*

The Moving Sky

To an observer on Earth, the stars appear to be attached to the inside of a vast hollow globe which spins around the Earth from east to west once a day. This view is not true. It is the Earth, and not the sky, that is spinning. All the same, it is often useful for astronomers to pretend that this celestial sphere, or globe, in the sky really does exist. The Earth's poles, and lines of latitude and longitude, can then all be marked on the celestial sphere. This helps astronomers to map the position of stars in the sky.

FURTHER FACTS

Homing pigeons use the Sun and stars to find their way. Experiments in planetariums have shown that the pigeons recognize certain stars or star patterns, but the Sun and stars appear in different parts of the sky at different times, so they must also have a built-in "clock" to allow for this effect.

► The celestial sphere appears to spin around the north and south celestial poles, which line up with the Earth's axis. It is divided up into 24 segments, running from the north to the south pole. In an hour it turns through one of these segments, carrying the stars steadily around the sky. The stars form easily recognizable patterns in the night sky. These patterns are called constellations. They are always in the same place on the celestial sphere, so their position can be noted. Astronomers can use the constellations to know where to find a particular star.

► The celestial equator is a projected circle, in line with the Earth's equator. To an observer standing on the Equator, stars lying near the celestial equator pass overhead.

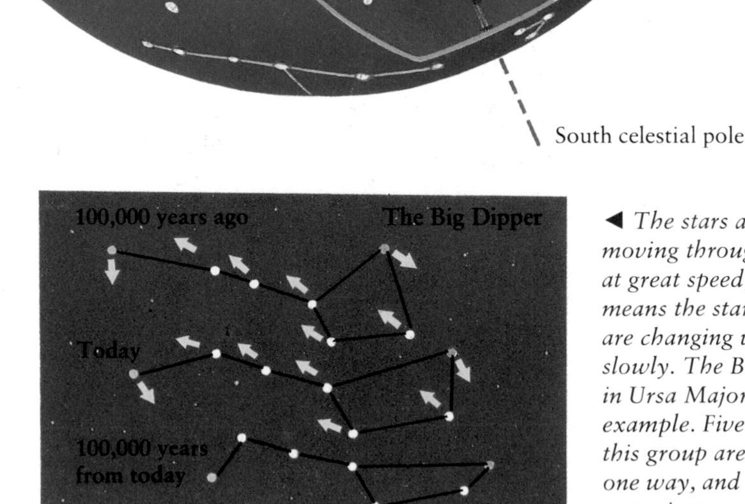

FACTS ABOUT THE MOVING SKY

- An observer always has part of the celestial sphere blocked out by the Earth. For example, the famous constellation Crux (the Cross), which lies near the south celestial pole, cannot be seen from Europe or from much of North America.
- Different constellations are seen from Earth in summer and winter. For example, in Europe, Orion is high in the midnight sky in January but cannot be seen in July at all. This is because the Earth's movement has positioned the Sun between it and Orion. The best time to look for a particular star is during the time of year when it is on the opposite side of the Earth to the Sun.

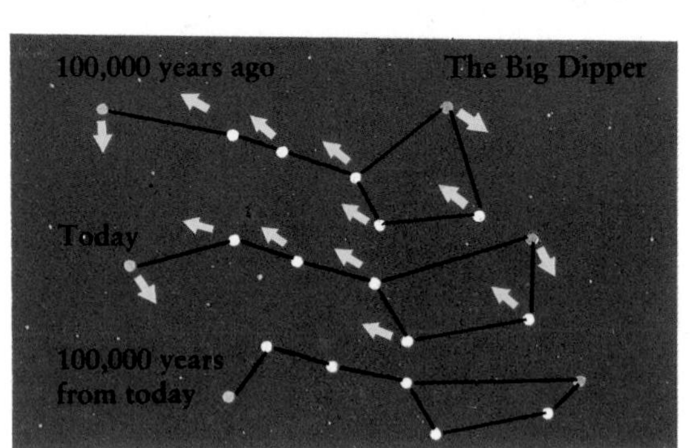

◄ The stars are moving through space at great speed, which means the star patterns are changing very slowly. The Big Dipper in Ursa Major is an example. Five stars in this group are moving one way, and two the opposite way.

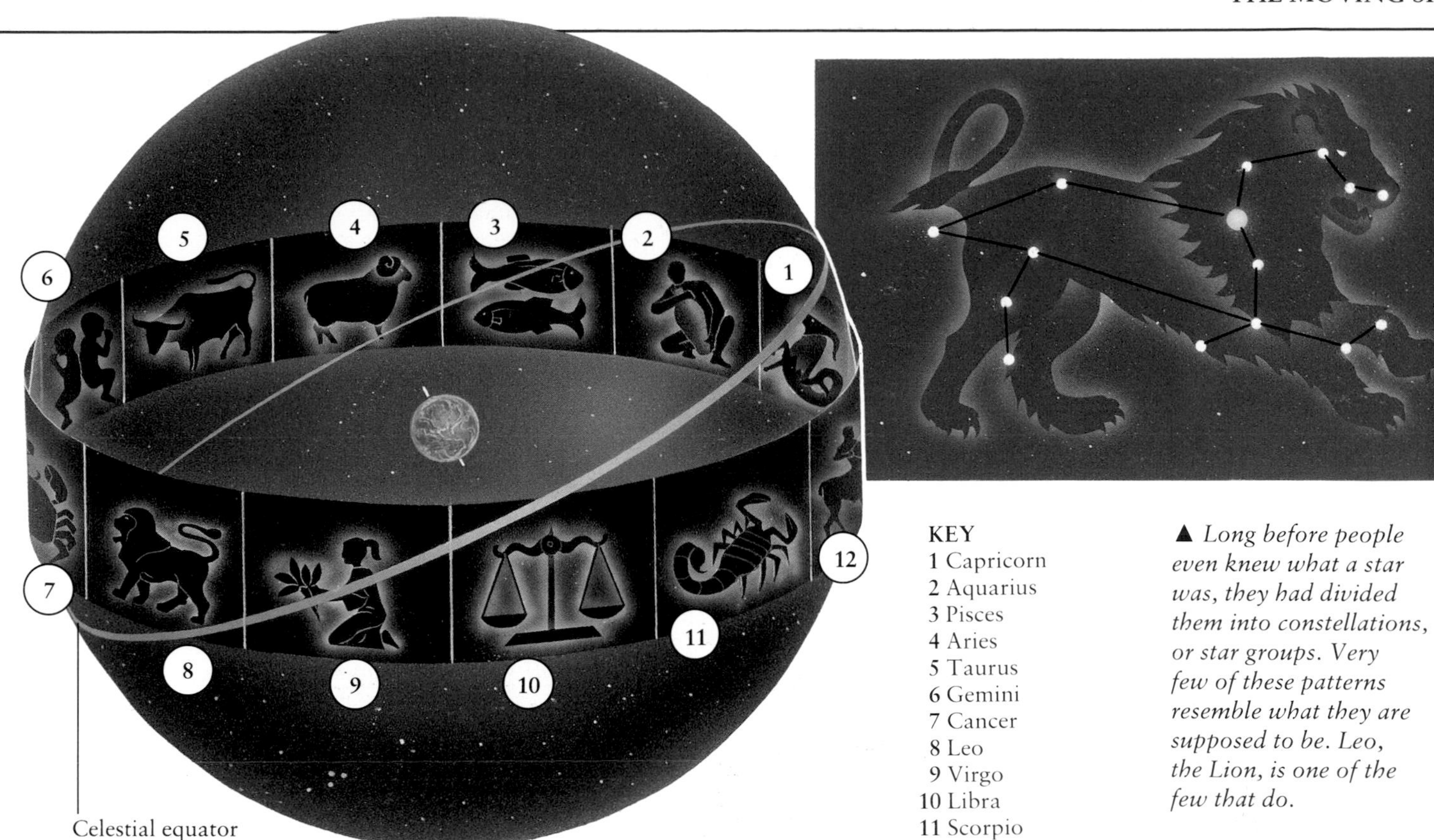

KEY
1 Capricorn
2 Aquarius
3 Pisces
4 Aries
5 Taurus
6 Gemini
7 Cancer
8 Leo
9 Virgo
10 Libra
11 Scorpio
12 Sagittarius

▲ *Long before people even knew what a star was, they had divided them into constellations, or star groups. Very few of these patterns resemble what they are supposed to be. Leo, the Lion, is one of the few that do.*

▲ *During a year, the Sun appears to take a particular path through the celestial sphere. This path marks the center of the band of sky known as the Zodiac. There are 12 divisions of the Zodiac, each one represented by a constellation. The Sun seems to spend about one month in each constellation.*

► *This old star map shows the southern part of the celestial sphere. The dashed circle marks the path of the Sun, and the South Pole is shown a little way above the center. Unlike the north celestial pole, the South Pole does not have a bright star near it. At the foot of the map the celestial equator is marked, in line with Earth's equator.*

The Constellations

Most people think of a constellation as a group of stars. In fact a constellation is a definite area of the celestial sphere, with internationally-agreed boundaries. The areas fit together to make up the sky. The maps show major constellations of the sky. The faintest stars are those with the smallest dots; these should just be visible with the naked eye from well-lit cities. The outline of the Milky Way is shown, but this can be seen properly only under dark country skies.

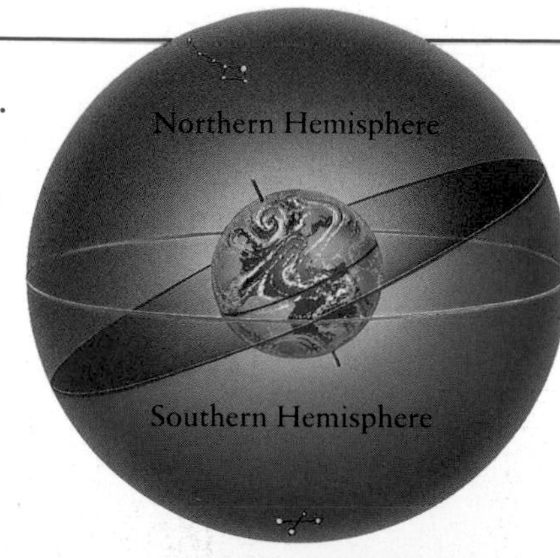

◀ *Each map shows one complete celestial hemisphere, as seen by someone standing at the North or South Pole. The celestial equator is the dividing line between the two maps, and the Earth's axis points to the celestial poles, at the center of each.*

CONSTELLATION FACTS

- In total, the sky contains 88 constellations. Most of these were named in ancient times. In A.D. 150 the Greek astronomer Ptolemy described 48 star patterns, including well-known ones such as Ursa Major (the Great Bear). Many of these had been recorded by Babylonian astronomers before 2000 B.C. Between the 16th and 18th centuries A.D., when explorers began venturing into the Southern Hemisphere, new parts of the celestial sphere came into view, and more constellations were added to the ancient ones.
- The largest constellation is Hydra (the Water Snake); the smallest constellation is Crux (the Cross).
- The faintest space object visible with the naked eye appears dimly in the Andromeda constellation. It is the Andromeda galaxy, 2.2 million light-years away.

▲ *Ursa Minor (the Little Bear) is a constellation of the Northern Hemisphere. The bright star at the very tip of the bear's tail is the North Star, Polaris. Ursa Minor is visible from northern Europe and North America all year round.*

NORTHERN HEMISPHERE

NORTHERN HEMISPHERE

1 *Equuleus* Little Horse
2 *Delphinus* Dolphin
3 *Pegasus* Pegasus
4 *Pisces* Fishes
5 *Cetus* Whale
6 *Aries* Ram
7 *Triangulum* Triangle
8 *Andromeda* Andromeda
9 *Lacerta* Lizard
10 *Cygnus* Swan
11 *Sagitta* Arrow
12 *Aquila* Eagle
13 *Lyra* Lyre
14 *Cepheus* Cepheus
15 *Cassiopeia* Cassiopeia
16 *Perseus* Perseus
17 *Camelopardus* Giraffe
18 *Auriga* Charioteer
19 *Taurus* Bull
20 *Orion* Orion (hunter)
21 *Lynx* Lynx
22 *Polaris* North Star
23 *Ursa Minor* Little Bear
24 *Draco* Dragon
25 *Hercules* Hercules
26 *Ophiuchus* Serpent Holder
27 *Serpens* Serpent
28 *Corona Borealis* Northern Crown
29 *Boötes* Herdsman
30 *Ursa Major* Great Bear

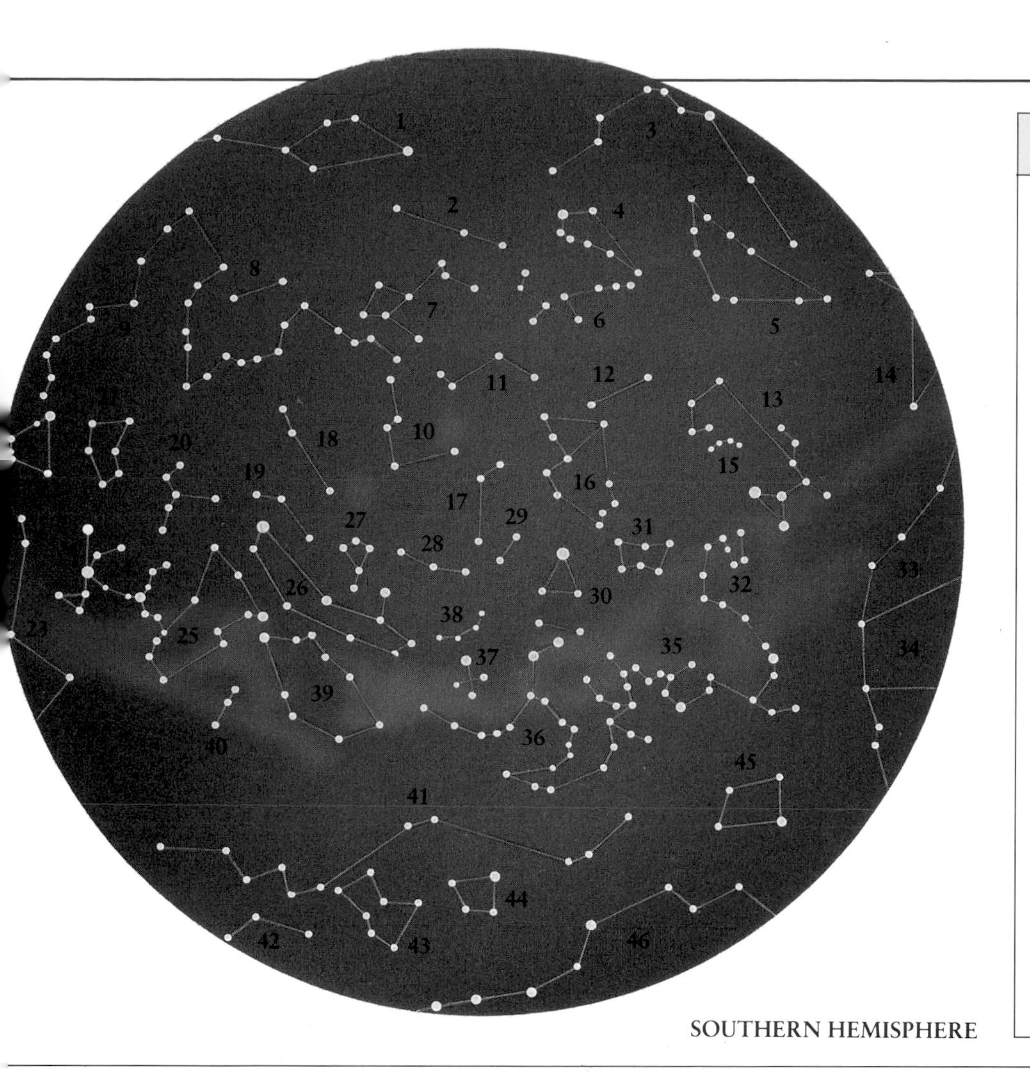

SOUTHERN HEMISPHERE

THE SOUTHERN CROSS

The constellation known as Crux, or the (Southern) Cross, appears on the flags of both Australia and New Zealand. The constellation contains five stars of different brightness. On Australia's flag the smallest star is shown with only five points, two less than the other stars. Only four stars appear on New Zealand's flag, all with the same number of points.

31 *Gemini* Twins
32 *Cancer* Crab
33 *Canis Minor* Little Dog
34 *Hydra* Water Monster
35 *Leo* Lion
36 *Leo Minor* Little Lion
37 *Canes Venatici* Hunting Dogs
38 *Coma Berenices* Berenice's Hair
39 *Virgo* Virgin

SOUTHERN HEMISPHERE
1 *Cetus* Whale
2 *Sculptor* Sculptor
3 *Aquarius* Water Bearer
4 *Piscis Austrinus* Southern Fish
5 *Capricornus* (Sea) Goat
6 *Grus* Crane
7 *Phoenix* Phoenix
8 *Fornax* Furnace
9 *Eridanus* Eridanus (a river)
10 *Hydrus* Water Snake
11 *Tucana* Toucan
12 *Indus* Indian
13 *Sagittarius* Archer
14 *Aquila* Eagle
15 *Corona Australis* Southern Crown
16 *Pavo* Peacock
17 *Octans* Octant
18 *Dorado* Swordfish
19 *Pictor* Painter
20 *Columba* Dove
21 *Lepus* Hare
22 *Orion* Orion (hunter)
23 *Monoceros* Unicorn
24 *Canis Major* Great Dog
25 *Puppis* Stern (of Argo)
26 *Carina* Keel (of Argo)
27 *Volans* Flying Fish
28 *Chamaeleon* Chameleon
29 *Apus* Bird of Paradise
30 *Triangulum Australe* Southern Triangle
31 *Ara* Altar
32 *Scorpius* Scorpion
33 *Serpens* Serpent
34 *Ophiuchus* Serpent Holder
35 *Lupus* Wolf
36 *Centaurus* Centaur
37 *Crux* (Southern) Cross
38 *Musca* Fly
39 *Vela* Sails (of Argo)
40 *Pyxis* Mariner's Compass
41 *Hydra* Water Monster
42 *Sextans* Sextant
43 *Crater* Cup
44 *Corvus* Crow
45 *Libra* Scales
46 *Virgo* Virgin

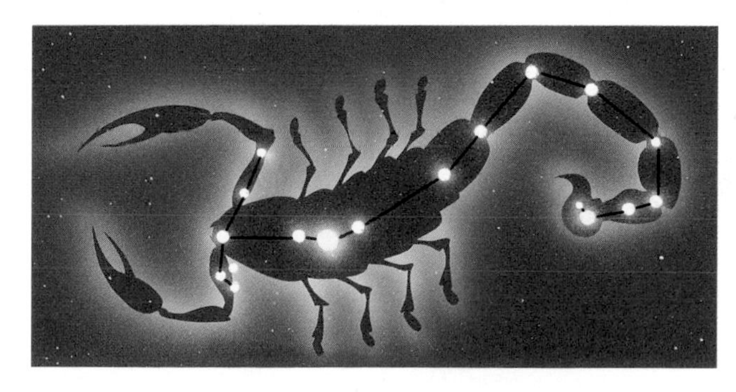

◄ *The constellation Scorpius (the Scorpion) can be seen in the Southern Hemisphere. The most prominent star in its body is Antares, a giant star about 500 light-years away. Antares is 10,000 times more luminous than our Sun.*

OBSERVING THE SKIES

The Birth of Astronomy

The people of early civilizations must have been aware of the fixed patterns of the stars in the sky, their repeated appearance and disappearance with the seasons, and the strange "wanderings" of the objects that we now call the planets. However, with no way of understanding their true nature, their astrological significance seemed the most important thing about them. The night sky was divided up into constellations, and they and the planets were given names. But right up until the beginning of the 17th century most people thought that the Earth was at the center of the universe.

Astronomical knowledge took a major leap forward in the early 17th century. The newly invented telescope was first turned to the sky, and Johannes Kepler proved that the planets move around the Sun, not the Earth. Modern astronomy was born.

ANCIENT ASTRONOMY
In ancient civilizations astronomy was closely linked to astrology, the belief that events in the sky can affect the lives of people on Earth. The positions of the planets were observed for astrological predictions. The duties of priests and astronomers overlapped. For example, the Babylonian ziggurat below was half-temple and half-observatory.

A ziggurat

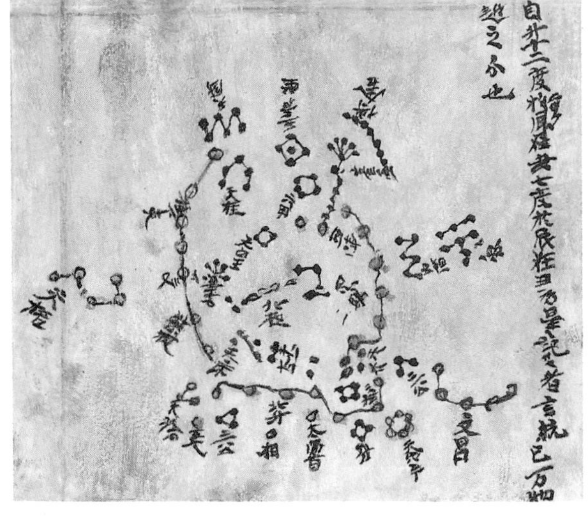

▲ *Astronomers in China were charting the positions of the stars as early as the 1300s B.C. The Big Dipper appears on this old star map.*

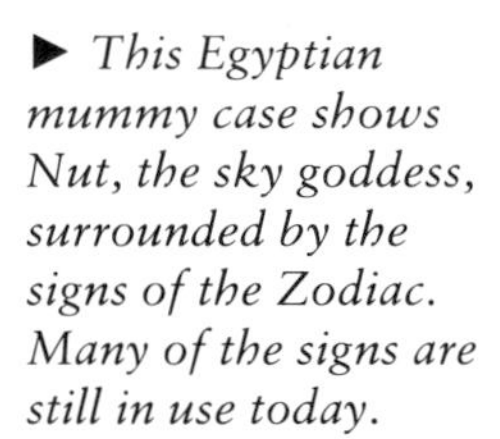

► *This Egyptian mummy case shows Nut, the sky goddess, surrounded by the signs of the Zodiac. Many of the signs are still in use today.*

◄ *Stonehenge, in England, dates from about 5,000 years ago. The Hele Stone (in the distance) seems to have been accurately placed to mark the position of midsummer sunrise. This event was important. In an age without calendars, the Sun's position as it rose and set was the simplest guide to the year's seasons.*

VIEWS OF THE UNIVERSE

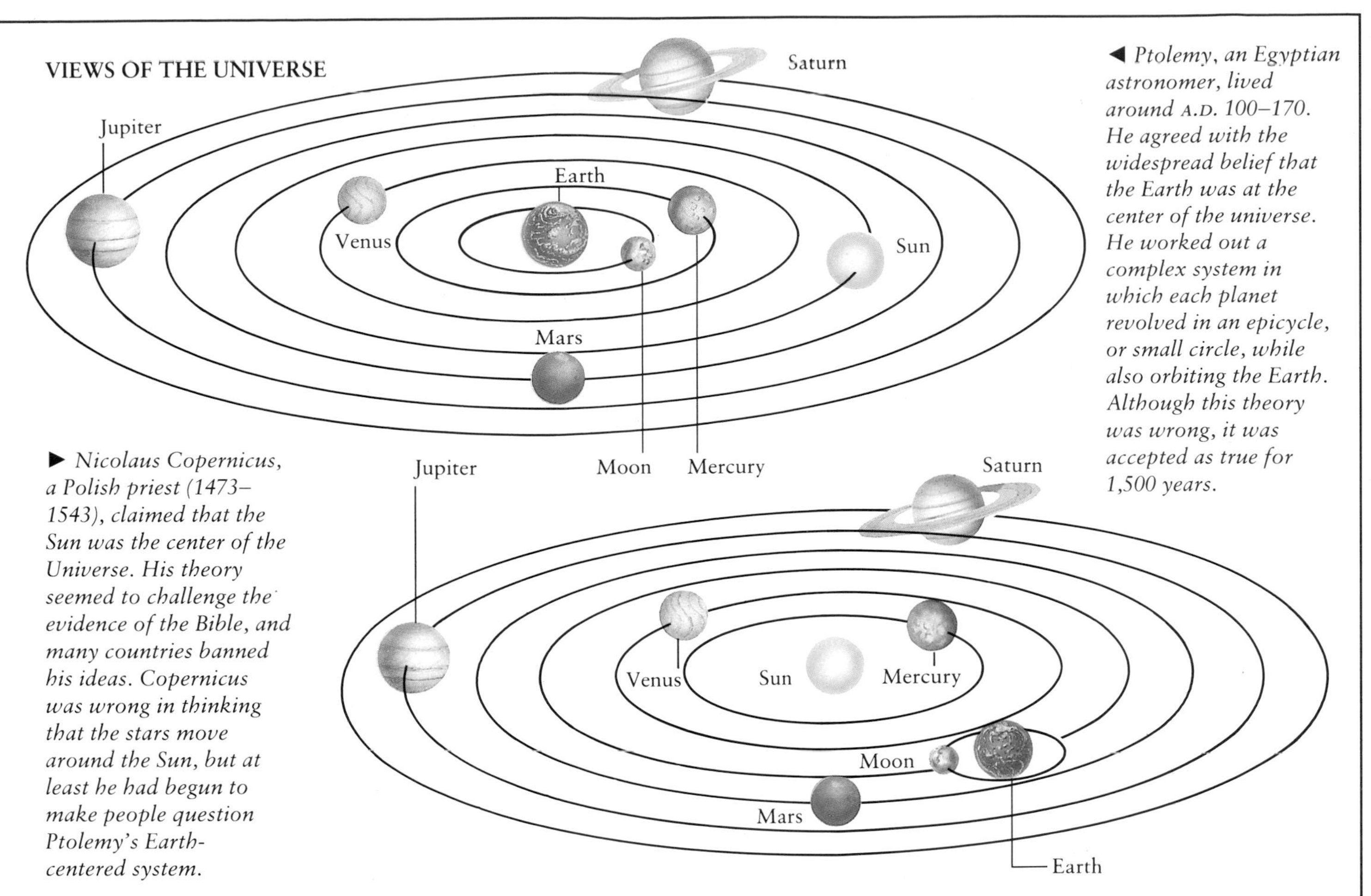

◀ *Ptolemy, an Egyptian astronomer, lived around A.D. 100–170. He agreed with the widespread belief that the Earth was at the center of the universe. He worked out a complex system in which each planet revolved in an epicycle, or small circle, while also orbiting the Earth. Although this theory was wrong, it was accepted as true for 1,500 years.*

► *Nicolaus Copernicus, a Polish priest (1473–1543), claimed that the Sun was the center of the Universe. His theory seemed to challenge the evidence of the Bible, and many countries banned his ideas. Copernicus was wrong in thinking that the stars move around the Sun, but at least he had begun to make people question Ptolemy's Earth-centered system.*

JOHANNES KEPLER

The German mathematician Johannes Kepler (1571–1630) analyzed naked-eye observations of the planet Mars to prove that the planets move around the Sun in elliptical orbits.

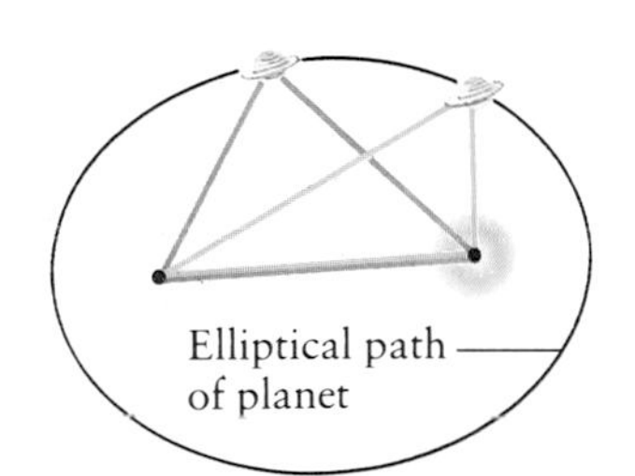

▲ *In an elliptical orbit there are two points, equally distant from the center. The Sun is at one point; there is nothing at the other. The farther apart the two points, the more elliptical the orbit.*

▼ *Kepler's Second Law states that an imaginary line between the Sun and a planet covers equal areas in equal times. This means that the planet moves faster when near the Sun.*

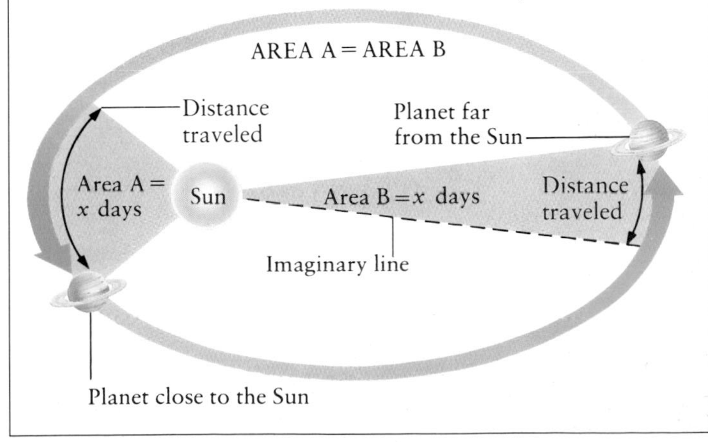

DATE	EARLY EVENTS IN ASTRONOMY (Astronomer in parentheses)
3000 B.C.	Earliest Babylonian astronomical records.
2900	Construction of the stone circle begins at Stonehenge.
2137	Two Chinese astronomers beheaded for failing to predict a solar eclipse.
280	First suggestion that the Earth orbits the Sun (Aristarchus).
270	First accurate measurement of the size of the Earth (Eratosthenes).
130	First comprehensive naked-eye star catalog drawn up (Hipparchus).
A.D. 140	Theory put forward showing the Earth at the center of the universe (Ptolemy).
903	Accurate star positions measured with naked eye (Al-Sufi).
1054	Supernova seen in the constellation Taurus (Chinese observers).
1433	The most complete star catalog yet published (Ulugh Beigh).
1543	Theory published suggesting the Sun is the center of the Universe (Nicolaus Copernicus).
1572	Supernova visible in daylight in Cassiopeia (studied by Tycho Brahe).

Optical Astronomy

In dim light, the pupil of the human eye opens up to about 0.3 in. (7 mm) across. Even this tiny "light collector" is able to see the Andromeda galaxy, 2 million light-years away. A telescope is an artificial eye with a larger opening, or aperture, which collects more light than the human eye. Therefore it makes stars look brighter and can reveal fainter and more distant objects. In 1608 a new age of astronomy began when the first telescope was made by Dutchman Hans Lippershey.

▼ *William Herschel (1738–1822) built large reflecting telescopes in order to study very faint objects. His largest telescope, with an aperture of 47 in. (120 cm), was set up in his garden in England. The observer stood on the platform below the mouth of the tube.*

GALILEO'S DISCOVERIES

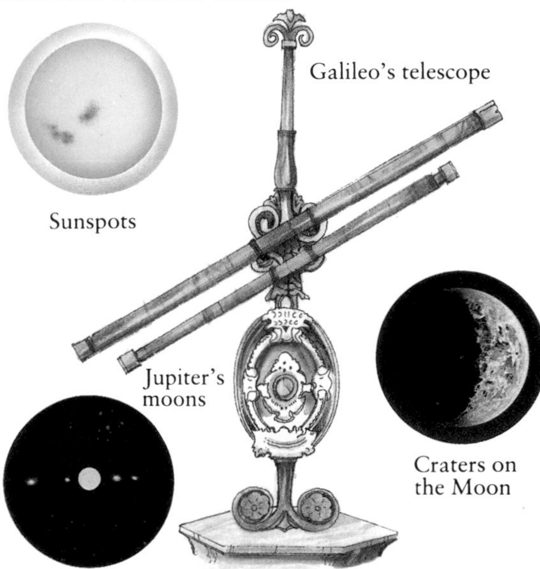

Galileo Galilei (1564–1642) was one of the first people to turn a telescope to the night sky. His first discovery was that the Moon was heavily cratered, instead of being perfectly smooth, as Ptolemy had thought. He also discovered four large moons in orbit around Jupiter, and he observed spots on the surface of the Sun.

▼ *Early refracting telescopes had to be very long in order to give a clear image. The telescope and observer had no protection from the weather. These fragile structures were called aerial telescopes.*

KEY DISCOVERIES IN OPTICAL ASTRONOMY

▼ *Edwin Hubble used this 100-inch (2.4-meter) reflecting telescope to discover the expansion of the universe.*

- In 1609 Galileo first turned a telescope to the sky; his observations of the phases of Venus confirmed Copernicus's theory that the Earth is not the center of the universe.
- The discovery of Uranus in 1781 extended our knowledge of the Solar System. Uranus was the first planet to be discovered by telescope.
- In 1838 telescopic observation meant that the distance to a star could be measured for the first time—by F. W. Bessel in Germany.
- In the 1920s Edwin Hubble used his observations at Mt. Wilson to show that the universe is expanding.

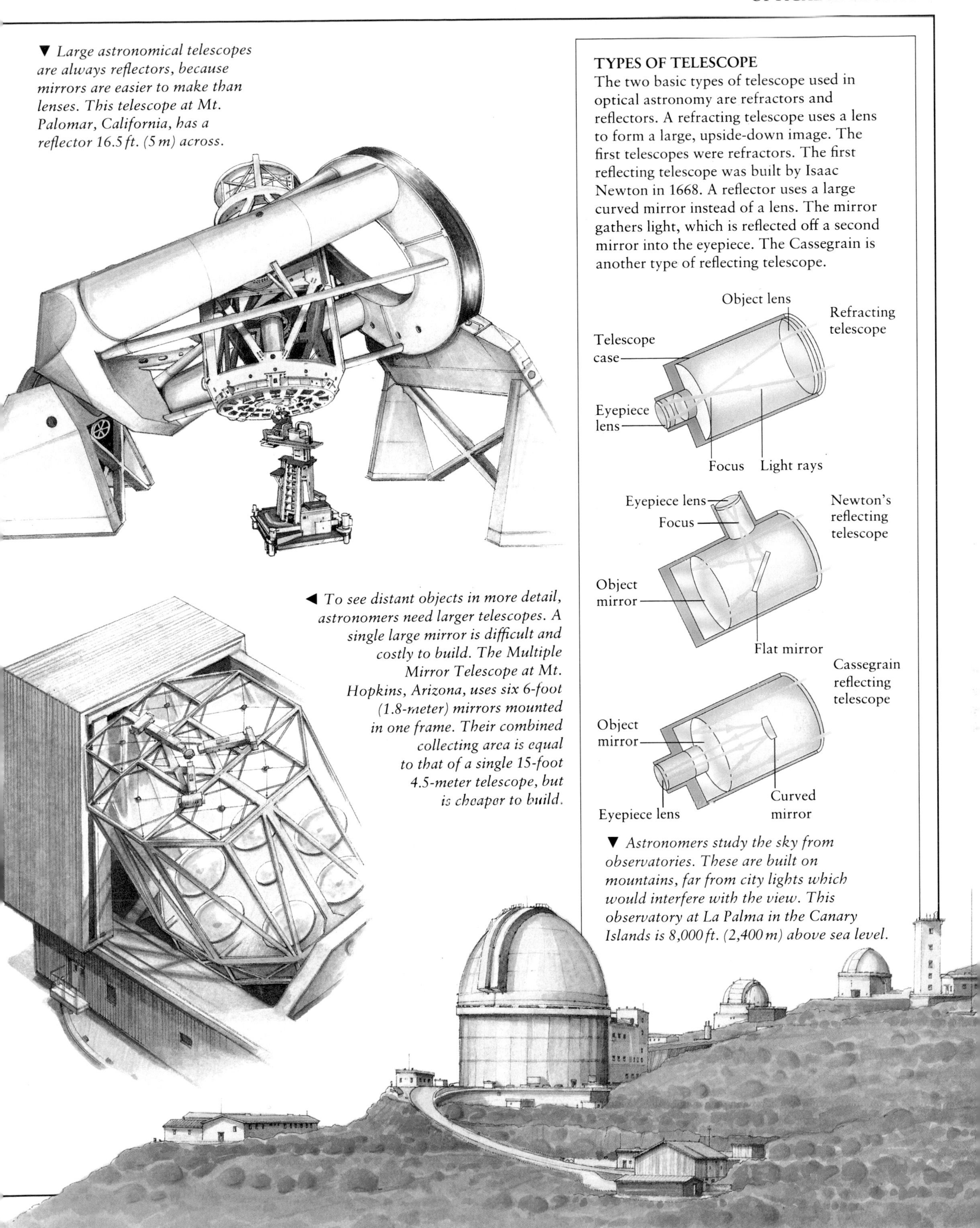

▼ *Large astronomical telescopes are always reflectors, because mirrors are easier to make than lenses. This telescope at Mt. Palomar, California, has a reflector 16.5 ft. (5 m) across.*

◀ *To see distant objects in more detail, astronomers need larger telescopes. A single large mirror is difficult and costly to build. The Multiple Mirror Telescope at Mt. Hopkins, Arizona, uses six 6-foot (1.8-meter) mirrors mounted in one frame. Their combined collecting area is equal to that of a single 15-foot 4.5-meter telescope, but is cheaper to build.*

TYPES OF TELESCOPE

The two basic types of telescope used in optical astronomy are refractors and reflectors. A refracting telescope uses a lens to form a large, upside-down image. The first telescopes were refractors. The first reflecting telescope was built by Isaac Newton in 1668. A reflector uses a large curved mirror instead of a lens. The mirror gathers light, which is reflected off a second mirror into the eyepiece. The Cassegrain is another type of reflecting telescope.

▼ *Astronomers study the sky from observatories. These are built on mountains, far from city lights which would interfere with the view. This observatory at La Palma in the Canary Islands is 8,000 ft. (2,400 m) above sea level.*

Radio Telescopes

Besides sending out visible light, many astronomical objects emit radio waves, which are invisible. The Earth's atmosphere is completely penetrable to radio waves, which can pass through even the thickest clouds. For this reason, radio telescopes are very important in astronomy. Radio telescopes are a special kind of telescope that can collect radio waves and so "see" objects that are too dim or distant to be seen with ordinary telescopes. Quasars and pulsars were discovered by radio telescopes.

HOW A RADIO TELESCOPE WORKS

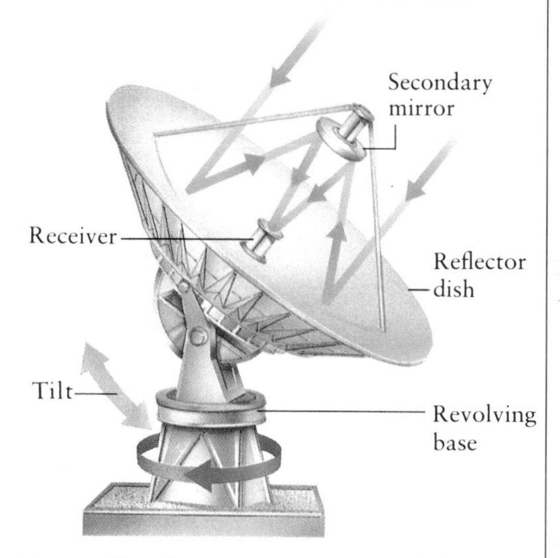

Most radio telescopes use a concave dish to collect radio waves from space. The radio waves are reflected onto an antenna, which sends the signal to an amplifier to be strengthened. The signal will be processsed by a computer which can turn it into images. Radio waves have less energy than light waves, so radio telescopes must have a large dish to detect faint objects.

▲ *This unusual radio telescope at Nançay, France, has a fixed curved reflector. A second, movable reflector directs the radio waves from space onto the fixed reflector's surface.*

► *The main radio telescope at the Parkes Observatory, Australia, has a 210-foot (64-meter) dish. The largest "steerable" radio telescope in the world is at Effelsberg, Germany, and has a 330-foot (100-meter) dish.*

► *Only a little of the energy sent out by space objects passes completely through the Earth's atmosphere. This energy includes the visible light waves detectable by optical telescopes, as well as the shortwave radio waves picked up by radio telescopes. Some infrared radiation reaches the Earth's surface, but our atmosphere blocks most other forms of radiation from space.*

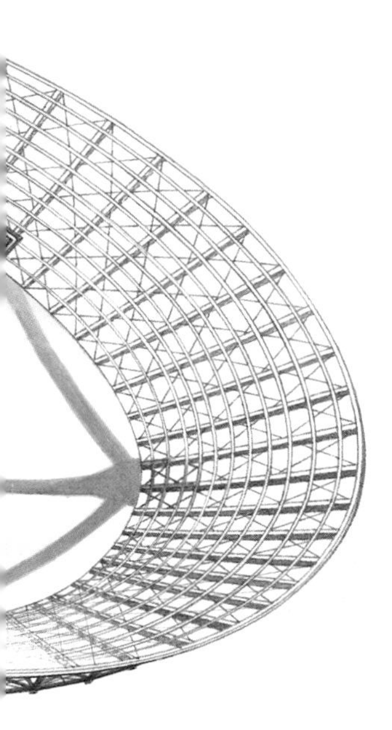

MAJOR RADIO OBSERVATORIES

- National Radio Astronomy Observatory, Socorro, New Mexico, includes the Very Large Array (VLA) *(below)*.
- Arecibo, Puerto Rico (1,000-foot dish, pointing directly upward).
- Green Bank, West Virginia (new 330-foot dish being installed by 1995).
- Australia Telescope, Culgoora, New South Wales (six 72-foot dishes, linked by computer to act like a much larger telescope).
- Effelsberg, Germany (330-foot dish).

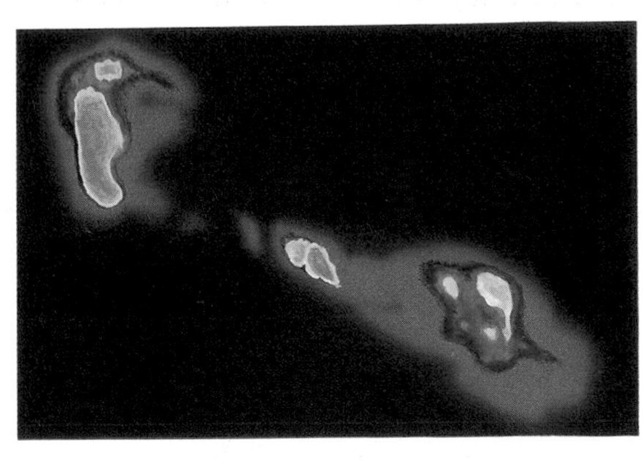

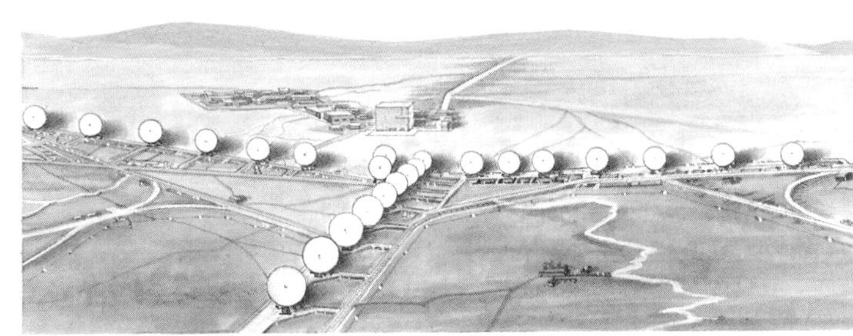

▲ *The clouds emitted by galaxy Centaurus A were recorded by the Very Large Array and turned into images by computer. They do not send out visible light.*

◄ *The VLA consists of 27 antennas, each 82 ft. across.*

► *Before 1900 no one had any idea how the Sun or other stars shone, and many people believed that the Milky Way was the most important galaxy in the universe, since no other galaxies had been identified. Some of the important astronomical advances made this century appear here.*

FURTHER FACTS

A radio telescope need not be a metal dish. The radio telescope that led to the discovery of the first pulsar was a collection of wires and poles, built in a field in Cambridge, England, in 1967. A simple metal bar may act as an antenna for radio waves.

DATE	KEY EVENT
1908	Ejnar Hertzsprung (Denmark) and Henry Russell (U.S.) discover that most stars belong to an orderly family, the "main sequence."
1912	Cepheid variable stars discovered by Henrietta Leavitt (U.S.). They are important, as their distance can be measured very accurately.
1915	Albert Einstein (Switzerland) publishes the *General Theory of Relativity*.
1923	dwin Hubble (U.S.) observes Cepheids in the Andromeda Galaxy and ıeasures the first distance between galaxies.
1929	Hubble proves that the galaxies are moving away from each other, ınd that the universe is expanding.
1930	Clyde Tombaugh (U.S.) discovers Pluto while searching for "Planet X."
1932	Karl Jansky first detects radio waves from space at Holmdel, New Jersey.
1937	First radio telescope dish built, measuring 31 ft. (9.4 m) across.
1958	The Earth's radiation belts discovered by James Van Allen's equipment on the American satellite *Explorer 1*.
1961	Quasar 3C 273 is the first to be discovered, by radio astronomers at Cambridge, England.
1965	Big Bang background radiation discovered by Arno Penzias and Robert Wilson at Holmdel, New Jersey.
1967	Pulsars discovered by Jocelyn Bell and Anthony Hewish at Cambridge, England.
1990	Light recorded from farthest-ever point to date.
1990	Hubble Space Telescope is launched, then found to be defective. In 1993, imperfections are largely corrected while HST is in orbit.

Space Telescopes

Putting a telescope into space, above the blocking, blurring effect of the Earth's atmosphere, sounds like an astronomer's dream. The Hubble Space Telescope (HST) is the latest and largest satellite designed to observe space objects. However, there are problems too. Carrying a telescope into space is an expensive and difficult business. If anything goes wrong while the telescope is in orbit there is no one to fix it, and its power supplies will eventually run down.

TELESCOPES IN SPACE

Some space observations are impossible from the Earth's surface because our atmosphere blocks out some types of radiation. X-rays and gamma rays from the Sun, red dwarfs, and exploding galaxies such as Centaurus A are blocked, as well as infrared rays from cool objects such as comets and nebulae. The atmospheric currents around Earth even affect visible light, making the stars twinkle. Images obtained in space are perfectly steady, and so more detail can be seen. An orbiting telescope can also observe the whole sky, which cannot be done from anywhere on the Earth's surface.

Centaurus A galaxy

Red dwarf

Comet

▲ *The Earth's atmosphere blocks much radiation sent out by space objects. Radio waves are reflected back into space at a height of about 200 mi. (300 km). Gamma rays and X-rays are absorbed at about 30 mi. (50 km); few infrared rays descend below 12 mi. (20 km). Only visible light and shortwave radio waves can reach the telescopes on Earth's surface.*

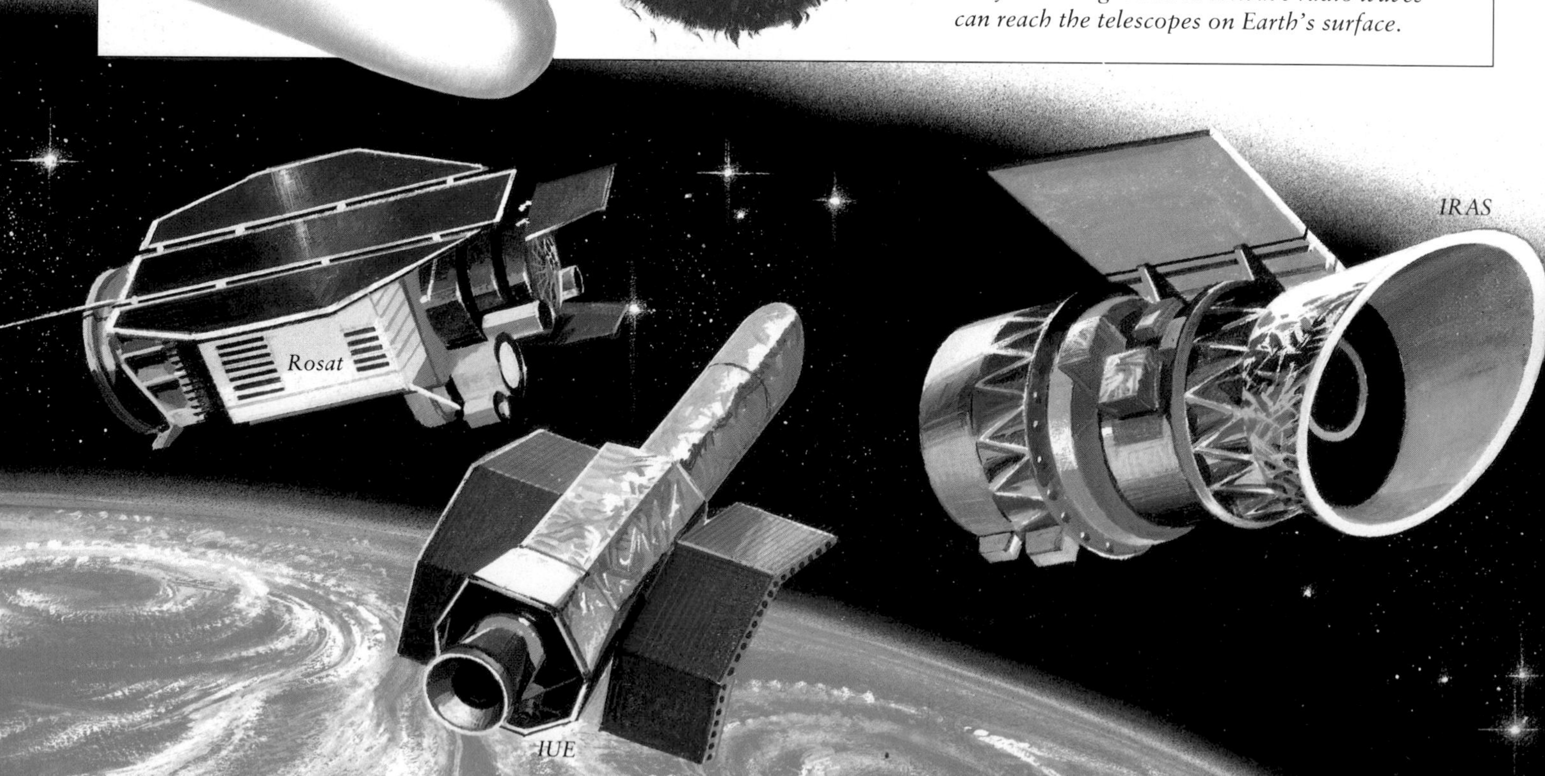

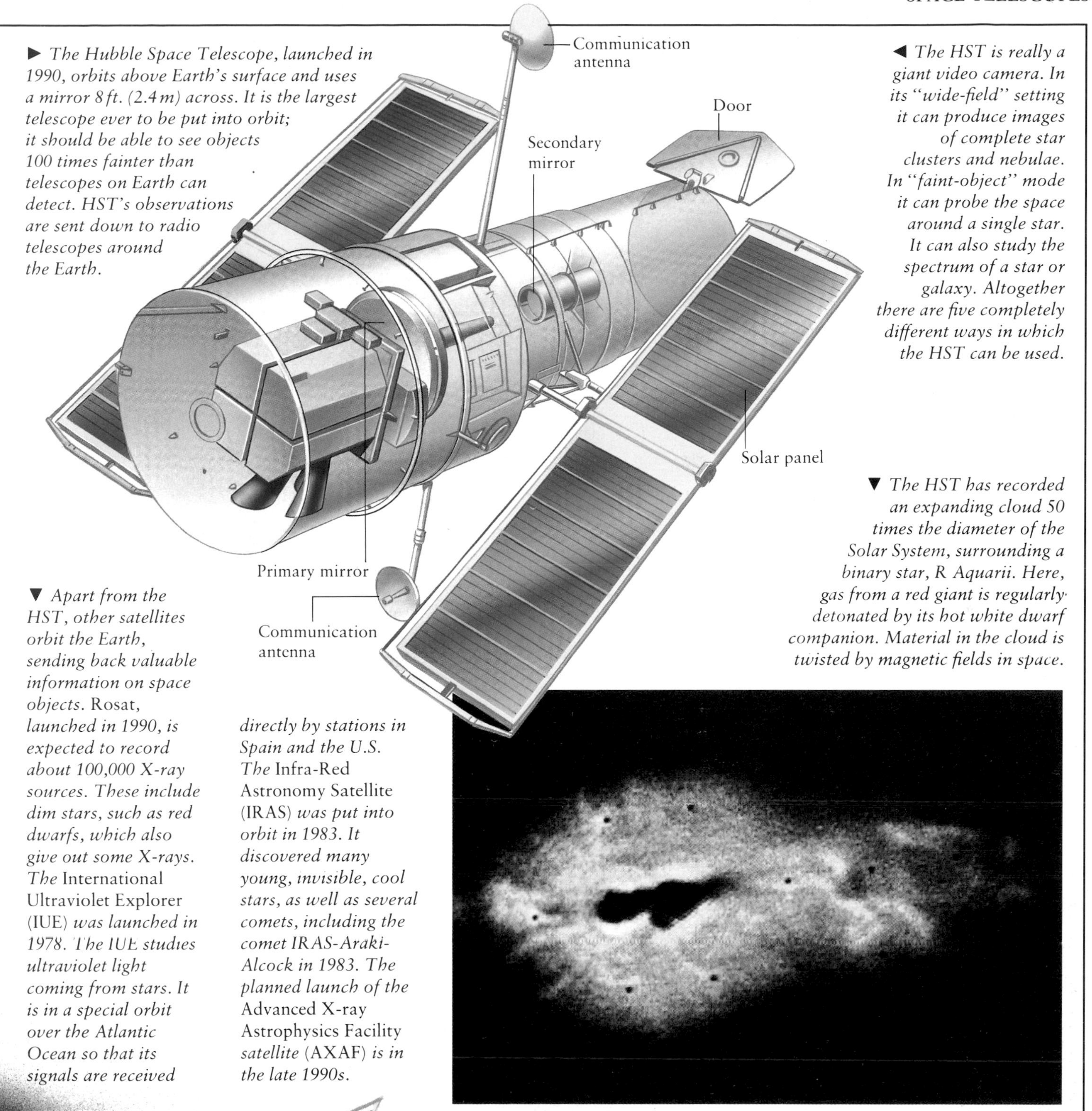

► *The Hubble Space Telescope, launched in 1990, orbits above Earth's surface and uses a mirror 8 ft. (2.4 m) across. It is the largest telescope ever to be put into orbit; it should be able to see objects 100 times fainter than telescopes on Earth can detect. HST's observations are sent down to radio telescopes around the Earth.*

◄ *The HST is really a giant video camera. In its "wide-field" setting it can produce images of complete star clusters and nebulae. In "faint-object" mode it can probe the space around a single star. It can also study the spectrum of a star or galaxy. Altogether there are five completely different ways in which the HST can be used.*

▼ *The HST has recorded an expanding cloud 50 times the diameter of the Solar System, surrounding a binary star, R Aquarii. Here, gas from a red giant is regularly detonated by its hot white dwarf companion. Material in the cloud is twisted by magnetic fields in space.*

▼ *Apart from the HST, other satellites orbit the Earth, sending back valuable information on space objects.* Rosat, *launched in 1990, is expected to record about 100,000 X-ray sources. These include dim stars, such as red dwarfs, which also give out some X-rays. The* International Ultraviolet Explorer (IUE) *was launched in 1978. The IUE studies ultraviolet light coming from stars. It is in a special orbit over the Atlantic Ocean so that its signals are received directly by stations in Spain and the U.S. The* Infra-Red Astronomy Satellite (IRAS) *was put into orbit in 1983. It discovered many young, invisible, cool stars, as well as several comets, including the comet IRAS-Araki-Alcock in 1983. The planned launch of the* Advanced X-ray Astrophysics Facility *satellite* (AXAF) *is in the late 1990s.*

AXAF

FACTS ABOUT THE SPACE TELESCOPE

• After launch, the HST would not focus properly. Studies showed that the mirror had been polished to the wrong curve. In 1993, space shuttle *Endeavour* was sent to make corrections. Specially trained astronauts installed corrective optics between HST's primary mirror and three astronomical instruments to refocus slightly blurred starlight. A 610-pound wide-field planetary camera was installed. Several of HST's six gyroscopes and some of its solar panels were also replaced.

SPACE EXPLORATION

A Rocket to the Moon

Space travel was a fantastic dream for centuries. But to leave the Earth behind, it was necessary to build an engine powerful enough to travel at 6.8 mi./s (11 km/s), the speed needed to beat the pull of Earth's gravity. During this century, the invention of liquid-fuel rockets has made space exploration possible. Rockets are the only vehicles powerful enough to carry spacecraft away from the surface of the Earth. Since 1957, rockets have carried hundreds of satellites into orbit, where they can gather information about our universe. Rockets have launched space probes to other planets. Early Soviet rockets were powerful enough to launch cosmonauts (Soviet astronauts) into orbit, where they started building experimental space stations. And to fly astronauts to the Moon, U.S. space scientists constructed the three-stage *Saturn V*, the largest rocket ever built.

TO THE MOON AND BACK

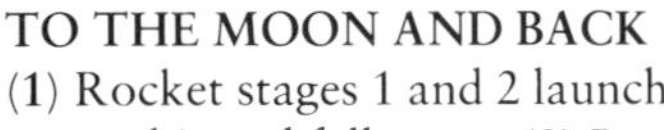

(1) Rocket stages 1 and 2 launch craft into orbit and fall away. (2) Rocket orbits Earth then carries on to Moon. (3) Command and Service Module (CSM) turns around; locks with Lunar Module (LM). (4) Astronauts enter LM; LM separates from CSM. (5) LM lands; CSM orbits Moon. (6) Astronauts carry out tasks; LM blasts off and re-connects with CSM. (7) LM ditched. (8) Re-entry.

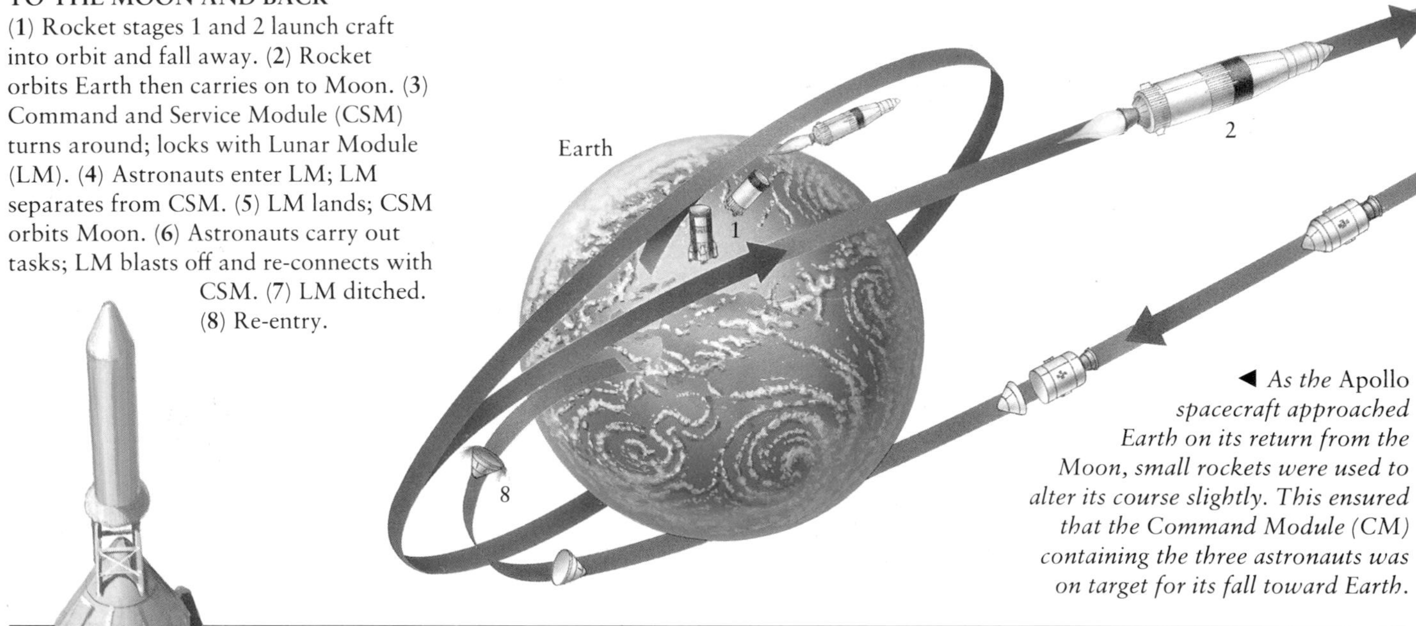

◀ *As the* Apollo *spacecraft approached Earth on its return from the Moon, small rockets were used to alter its course slightly. This ensured that the Command Module (CM) containing the three astronauts was on target for its fall toward Earth.*

LAUNCH TO SPLASHDOWN

The 360-foot (110-meter) rocket *Saturn V* launched *Apollo 11* to the Moon on July 16, 1969. On July 20, Buzz Aldrin and Neil Armstrong stood on the Moon's surface while the third crew member, Michael Collins, orbited in the Command Module (CM).

The CM was the only part of the spacecraft to return to Earth. Once it re-entered Earth's atmosphere its parachutes opened, slowing the CM down. Early U.S. missions would splashdown into the ocean, where the crew would be rescued by helicopters.

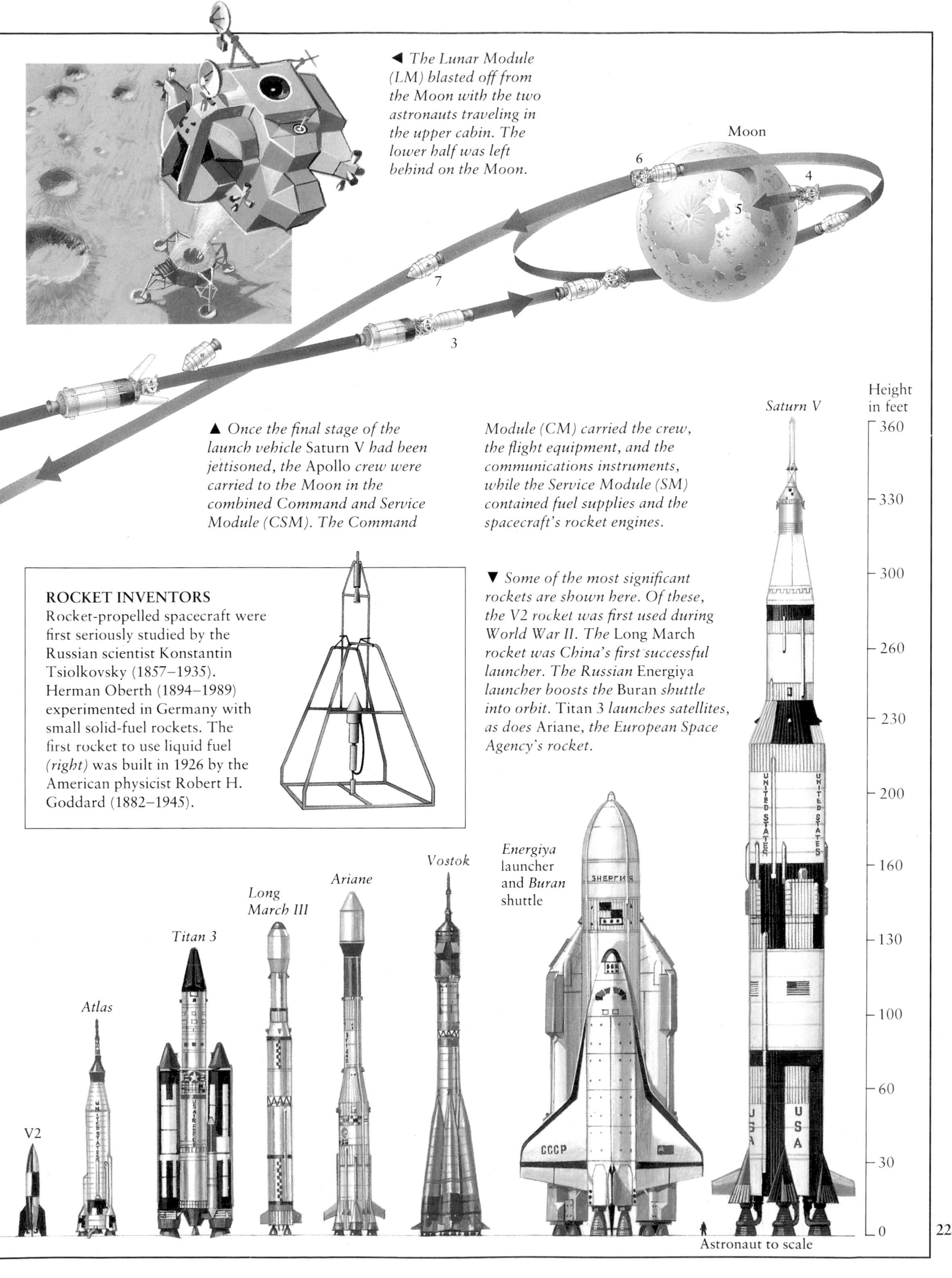

◀ *The Lunar Module (LM) blasted off from the Moon with the two astronauts traveling in the upper cabin. The lower half was left behind on the Moon.*

▲ *Once the final stage of the launch vehicle* Saturn V *had been jettisoned, the* Apollo *crew were carried to the Moon in the combined Command and Service Module (CSM). The Command Module (CM) carried the crew, the flight equipment, and the communications instruments, while the Service Module (SM) contained fuel supplies and the spacecraft's rocket engines.*

ROCKET INVENTORS

Rocket-propelled spacecraft were first seriously studied by the Russian scientist Konstantin Tsiolkovsky (1857–1935). Herman Oberth (1894–1989) experimented in Germany with small solid-fuel rockets. The first rocket to use liquid fuel *(right)* was built in 1926 by the American physicist Robert H. Goddard (1882–1945).

▼ *Some of the most significant rockets are shown here. Of these, the V2 rocket was first used during World War II. The* Long March *rocket was China's first successful launcher. The Russian* Energiya *launcher boosts the* Buran *shuttle into orbit.* Titan 3 *launches satellites, as does* Ariane, *the European Space Agency's rocket.*

Artificial Earth Satellites

An artificial satellite is a spacecraft placed in orbit around a planet. About 2,000 satellites have now been launched, for a number of purposes. Military satellites can spy, guide missiles, and even be weapons themselves. Communications satellites relay television, radio, and telephone transmissions. Weather satellites help with weather forecasting and also give immediate warning of cyclones. Scientific satellites may carry out studies of the Earth and its environment, or observe space objects.

FACTS ABOUT SATELLITES

- *Telstar 1* was the first communications satellite. Launched on July 10, 1962, it carried one television channel. Live television images could be sent to Europe from the U.S. for the first time.
- The hole in the Earth's ozone layer was discovered by the environmental satellite *Nimbus 7*.

► *Satellites are put into orbit aboard rockets or the space shuttle. The speed at which a satellite is launched has to be exactly right, or the satellite will fly off into space* (**1**) *or return to the ground* (**2**). *An orbiting satellite launched at the right speed will keep "falling" at the same rate as the Earth's surface curves beneath it* (**3**), *and so will never land.*

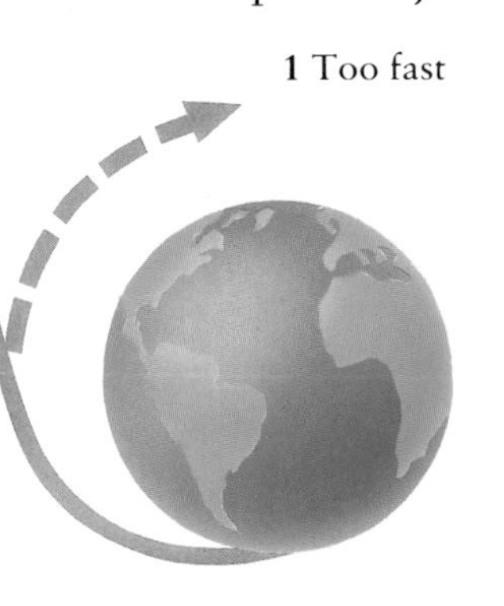

1 Too fast

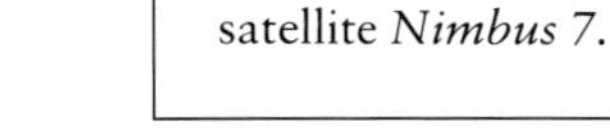

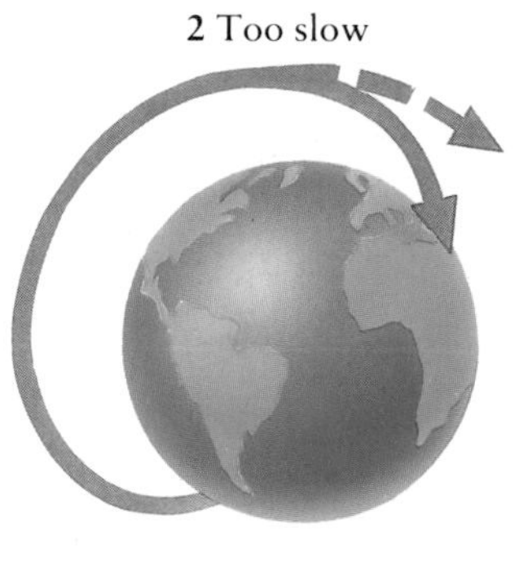

2 Too slow

3 Right speed

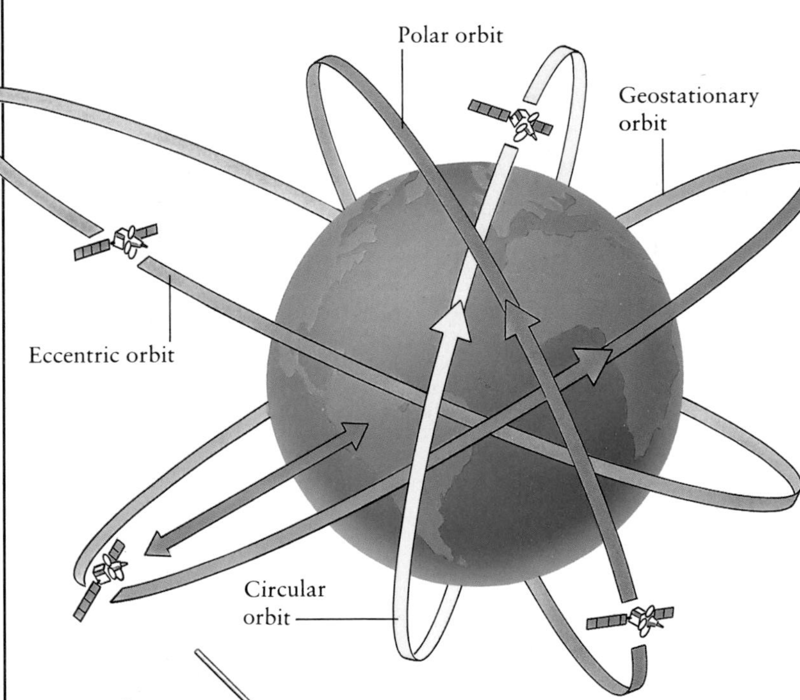

◄ *Satellite orbits can pass right over both the poles so that all the Earth's surface can be surveyed. In a geostationary orbit, the satellite always faces the same part of the Earth. A circular orbit means that the satellite is always the same height above the surface of Earth, but if the satellite's orbit is eccentric this distance will keep changing.*

SPUTNIK
On October 4, 1957, the Soviet Union began the age of space exploration with the launch of *Sputnik 1*, the first artificial satellite. *Sputnik* orbited the Earth in 90 minutes and stayed aloft for 6 months.

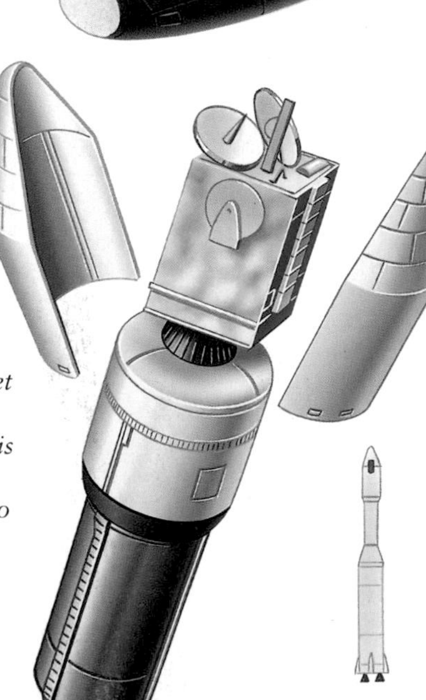

► *Unless launched in the space shuttle, all satellites are carried in the upper stage of a launch rocket. When the rocket reaches orbit, the satellite is released. This means that discarded rocket parts continue to collect as extra, unwanted satellites above the Earth, and could one day cause a space accident.*

▼ *The advantage of launching a satellite from the space shuttle, such as the* SBS-4 *satellite, is that astronauts can make last-minute adjustments before the satellite is released. Rocket motors on the satellite itself can be used to send it into an eccentric or geostationary orbit if required.*

SATELLITE FIRSTS

- *Sputnik 2* (U.S.S.R., 1957) launched the first living creature into space—a dog called Laika, who spent a week in orbit.
- *Explorer 1* (U.S., 1958) was the first successful U.S. satellite. It detected belts of radiation around the Earth.
- *Solar Max* (U.S., 1980), a satellite designed to study the Sun by taking X-ray pictures of solar flares, failed nine months after launch. It became the first satellite to be repaired in space, by a historic space shuttle mission in 1984.

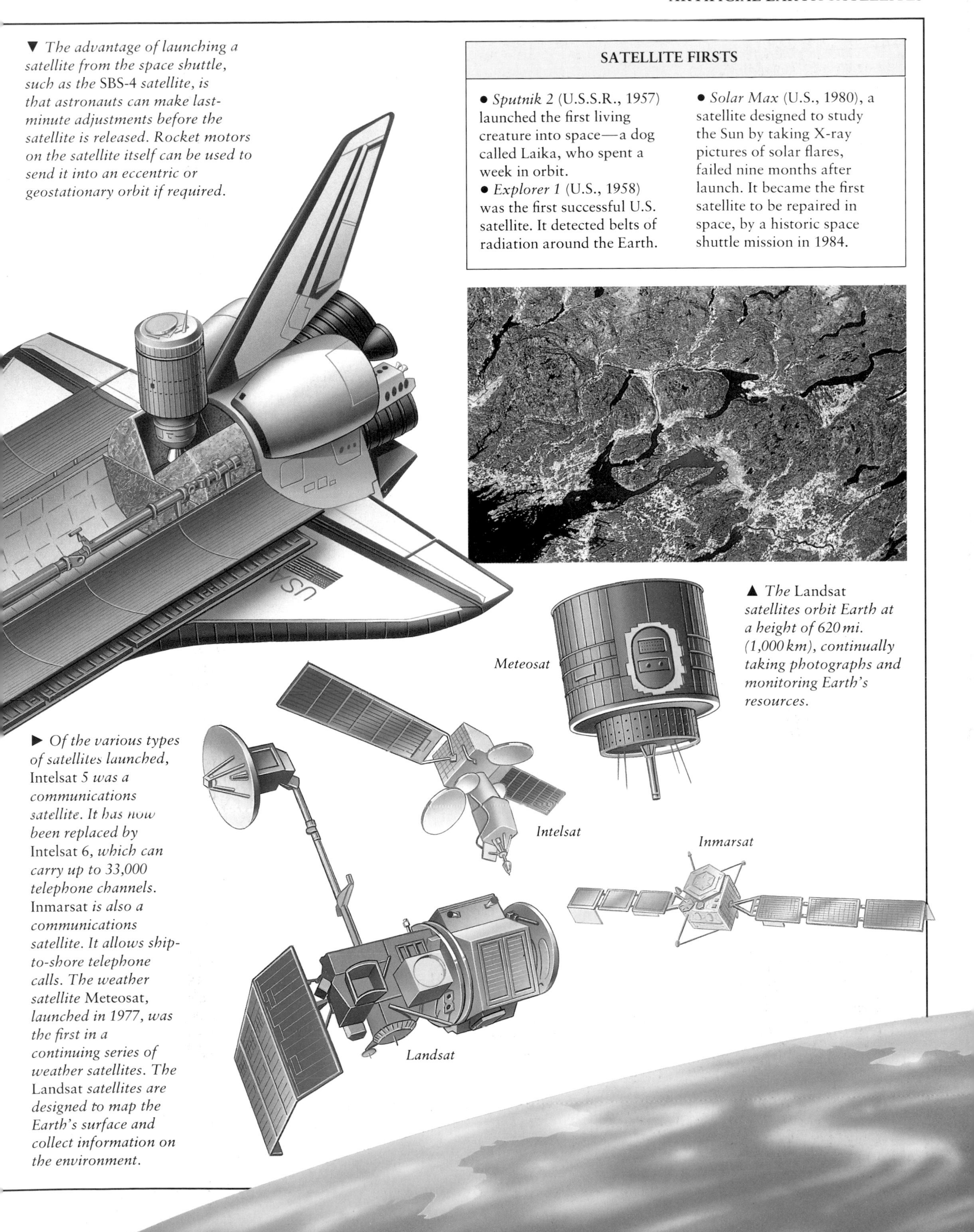

▲ *The* Landsat *satellites orbit Earth at a height of 620 mi. (1,000 km), continually taking photographs and monitoring Earth's resources.*

► *Of the various types of satellites launched,* Intelsat 5 *was a communications satellite. It has now been replaced by* Intelsat 6, *which can carry up to 33,000 telephone channels.* Inmarsat *is also a communications satellite. It allows ship-to-shore telephone calls. The weather satellite* Meteosat, *launched in 1977, was the first in a continuing series of weather satellites. The* Landsat *satellites are designed to map the Earth's surface and collect information on the environment.*

Space Probes

The first successful space probe was *Luna 2* (U.S.S.R.), which hit the Moon in 1959. Since then, probes have visited every planet in the Solar System apart from Pluto. These robot explorers have provided close-up pictures of worlds billions of miles away. The images are so clear that it is hard to appreciate the technical problems that were overcome. For example, the *Voyager 2* pictures from the edge of the Solar System were transmitted using the same voltage as two car batteries!

SPACE PROBE DATAFILE

The first successful visits by spacecraft (date in parentheses marks encounter with planet):

Moon: *Luna 2* (U.S.S.R. 1959); Lander *Luna 9* (U.S.S.R. 1966); Orbiter *Luna 10* (U.S.S.R. 1966)
Mercury: *Mariner 10* (U.S. 1974)
Venus: *Mariner 2* (U.S. 1962); Lander *Venera 7* (U.S.S.R. 1970); Orbiter *Venera 9* (U.S.S.R. 1975)
Mars: *Mariner 4* (U.S. 1965); Orbiter *Mariner 9* (U.S. 1971); Lander *Viking 1* (U.S. 1976)
Jupiter: *Pioneer 10* (U.S. 1973)
Saturn: *Pioneer 11* (U.S. 1979)
Uranus: *Voyager 2* (U.S. 1986)
Neptune: *Voyager 2* (U.S. 1989)
Asteroid: *Galileo* (U.S. 1991) examined Gaspra during its flight to Jupiter
Comet: *Giotto* (Europe 1986) examined Halley

▼ *The first artificial object to reach another world was the Soviet probe* Luna 2, *which crashed into the Moon on September 13, 1959.* Luna 1 *had missed.*

▼ Luna 9 *was the first spacecraft to make a soft landing on the Moon, on February 3, 1966. It returned the first pictures of the Moon's surface.*

Luna 2

Luna 9

Mariner 10

◀ Mariner 10 *was the first probe to visit two planets. It flew past Venus and then Mercury in 1974.*

THE FIRST VEHICLE ON THE MOON
The Lunokhods were unmanned Moon cars launched by the U.S.S.R. in 1970 and 1973. They were controlled from Earth by radio, and traveled the surface collecting data.

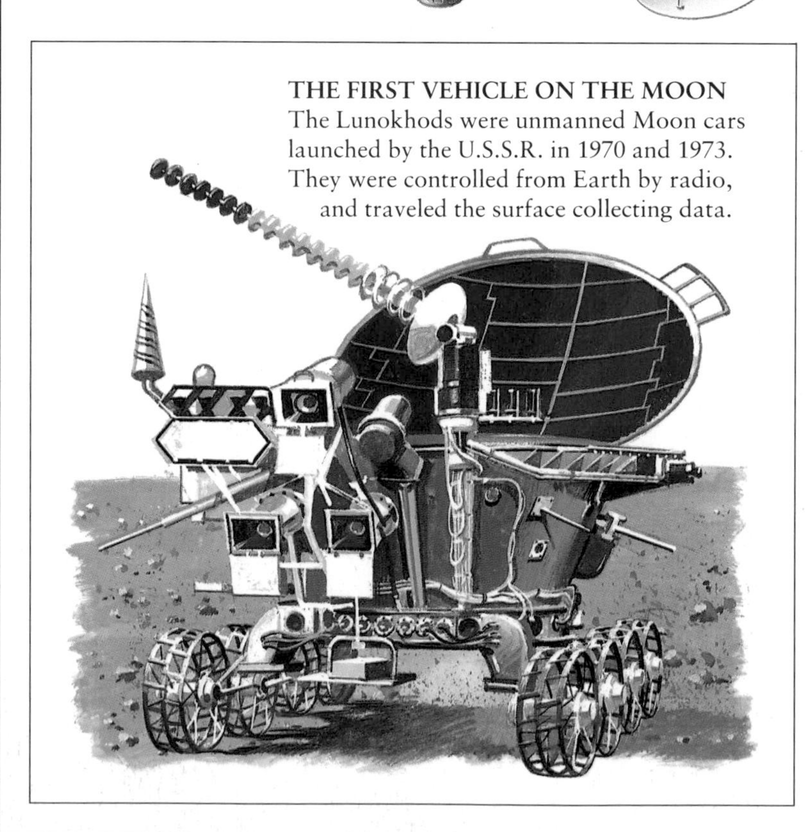

◀ Pioneer-Venus 2 *sent four probes into the atmosphere of Venus on December 9, 1978, while* Pioneer-Venus 1 *went into orbit.*

Pioneer-Venus 2

Mariner 2

▶ Mariner 2 *was the first successful probe to visit a planet. It flew past Venus on December 14, 1962, and made temperature measurements.*

SLINGSHOT LAUNCH

To the Sun via Jupiter! Space probes can be speeded on their way, or have their course changed, by using the pull of another planet's gravity as a free energy source. The *Ulysses* probe, launched in 1990, used Jupiter's gravity to swing it into a vertical path that will pass above the Sun's poles.

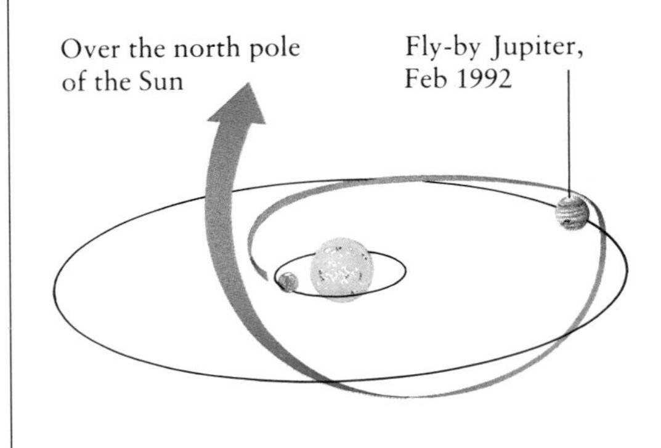

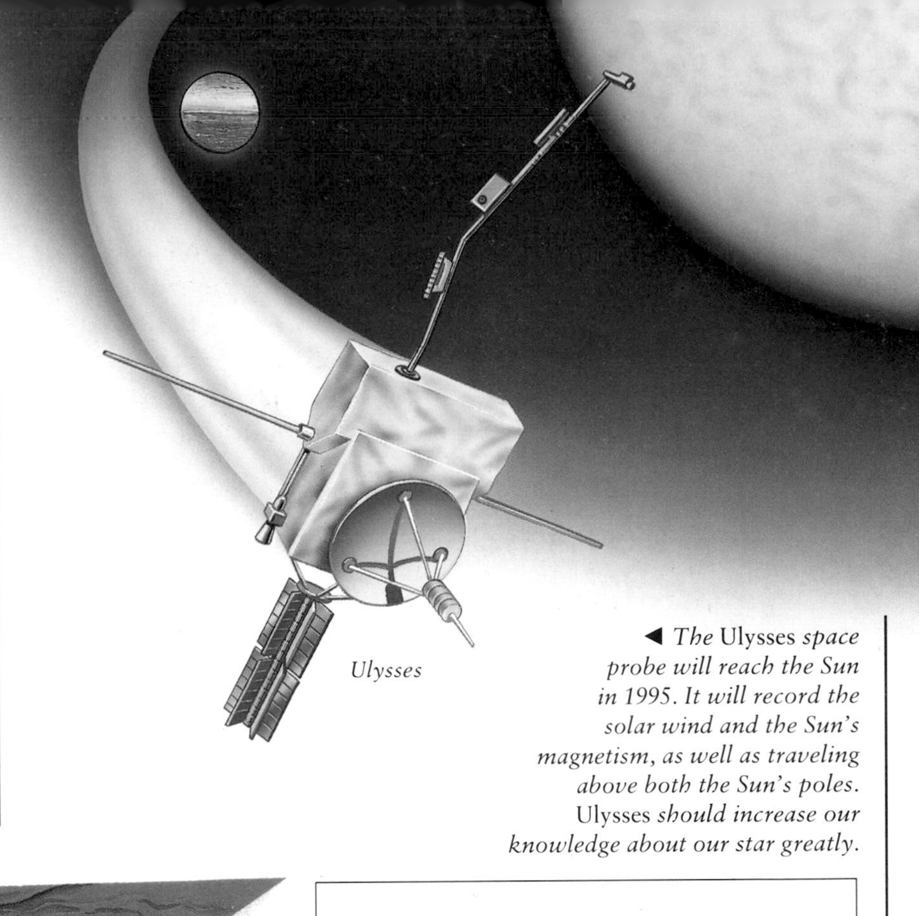

◀ *The* Ulysses *space probe will reach the Sun in 1995. It will record the solar wind and the Sun's magnetism, as well as traveling above both the Sun's poles.* Ulysses *should increase our knowledge about our star greatly.*

DEEP SPACE NETWORK

There is no second chance to receive data transmitted by a space probe, so there must always be a receiving station on Earth able to make contact. The Deep Space Network (DSN) has three main stations around the world so that all directions in the Solar System are covered.

Red

Blue

Green

▲ *Color pictures were made from separate red, blue, and green images taken by* Voyager.

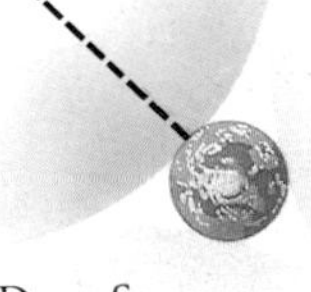

Deep Space Network, Australia

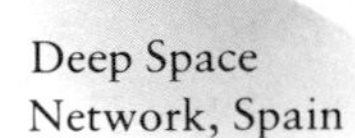

Deep Space Network, Spain

Deep Space Network, West Coast, U.S.

▲ *A telescope in Australia uses a 210-foot (64-meter) diameter antenna to receive signals.*

▲ *Eight hours later, the Spanish station near Madrid is facing the same direction in space.*

▲ *Commands to spacecraft are usually sent from the Goldstone station in California.*

PIONEER-VENUS 2

Pioneer-Venus 2 acted as a carrier to four smaller probes on the way to Venus in 1978. As it approached Venus, *Pioneer* launched its four probes toward different parts of the planet. The main probe was launched first, the smaller probes four days later.

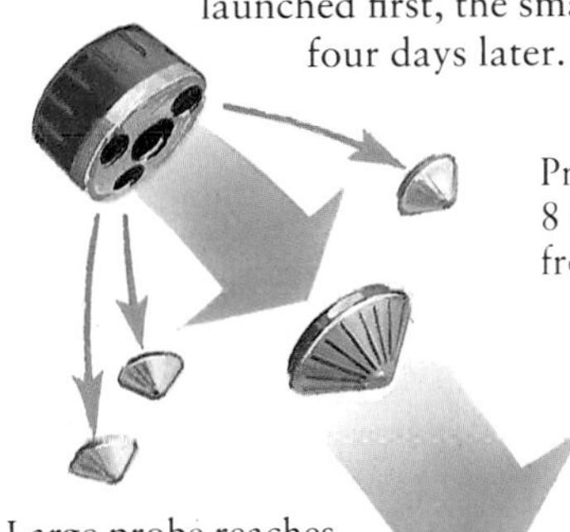

Probe reaches surface 56 minutes after entering atmosphere

Life in Space

When an astronaut travels into space, the familiar background to life—day and night, gravity, natural air, sunshine, and exercise—is suddenly cut off. Astronauts undergo vigorous training to cope with the artificial space environment. Just to endure lift-off, which makes the body feel like it is being squashed, astronauts have to be extremely fit. As everything in the cabin is weightless, astronauts have to learn how to eat, sleep, move, and keep healthy in a world without gravity.

FACTS ABOUT PEOPLE IN SPACE

April 12, 1961: First man in space (Yuri Gagarin, U.S.S.R.).
June 16, 1963: First woman in space (Valentina Tereshkova, U.S.S.R.).
March 18, 1965: First spacewalk (Alexei Leonov, U.S.S.R.).

Valentina Tereshkova

U.S./U.S.S.R. link-up

Yuri Gagarin

Challenger explosion

December 1968: First manned flight around the Moon (*Apollo 8*, U.S.A.).
July 20, 1969: First Moon landing (*Apollo 11*, U.S.A.).
May 1973: First fully successful space station (*Skylab*, U.S.A.).
July 1975: First docking of U.S. and Soviet spacecraft (*Apollo 18* and *Soyuz 19*).
April 12, 1981: First space shuttle launched (*Columbia*, U.S.A.).
January 28, 1986: *Challenger* shuttle explodes, killing all seven people on board. Crew includes teacher Christa McAuliffe.

▼ *A spacesuit must provide a supply of oxygen, remove carbon dioxide and other waste products, maintain atmospheric pressure and keep the astronaut comfortably warm. The latest suits worn on Shuttle flights allow the astronauts to spend many hours outside the spacecraft. These are known as Extra Vehicular Activity (EVA) suits.*

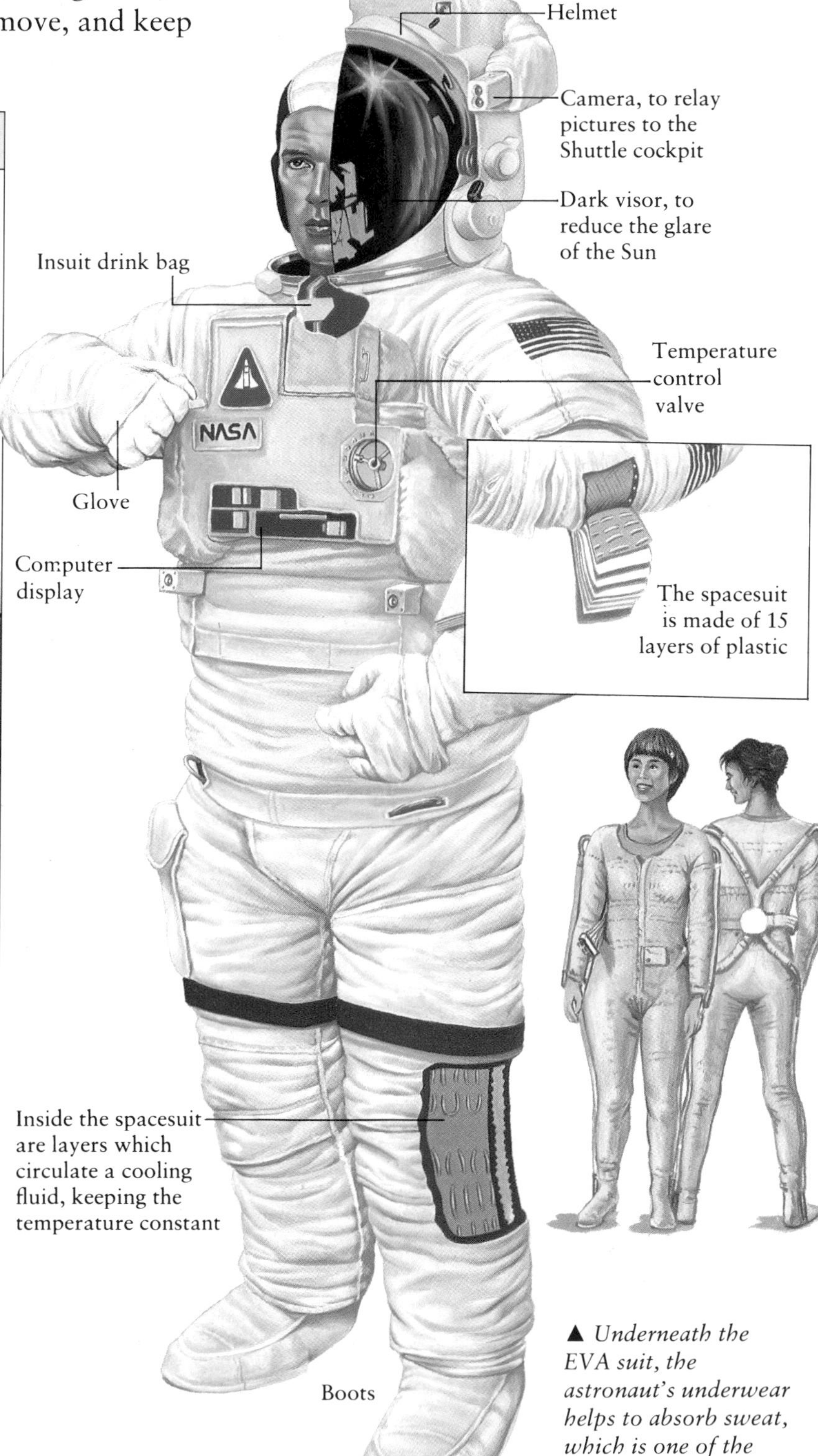

▲ *Underneath the EVA suit, the astronaut's underwear helps to absorb sweat, which is one of the main problems with an airtight garment.*

▼ *There is no gravity in space, so liquid and food particles from an astronaut's dinner would float around the cabin. To prevent this, astronauts' food and drink are specially packaged.*

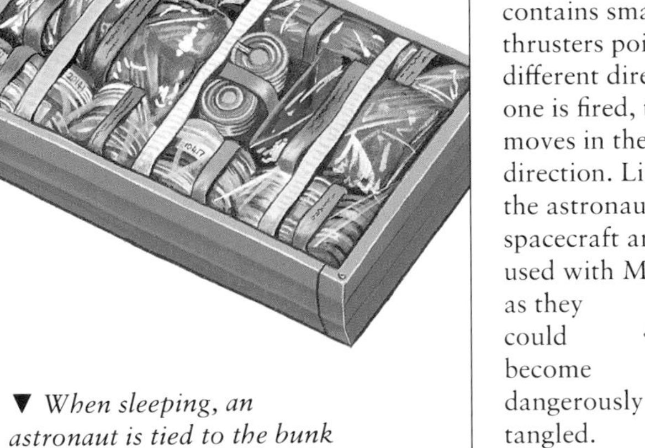

SPACE WALKING

When floating free in space, astronauts have nothing to push against to start themselves moving and no way of stopping either. The Manned Maneuvering Unit (MMU) used by U.S. astronauts contains small rocket thrusters pointing in different directions. When one is fired, the astronaut moves in the opposite direction. Lines tethering the astronaut to the spacecraft are no longer used with MMUs, as they could become dangerously tangled.

▼ *When sleeping, an astronaut is tied to the bunk or uses a secured sleeping suit. Sleep is timetabled, because there is no day or night in space.*

Shuttle lavatory

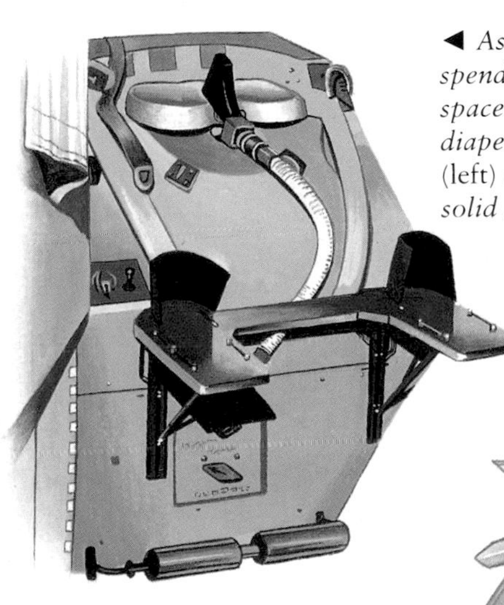

◀ *Astronauts who are spending a long time in a spacesuit wear a kind of diaper. The shuttle lavatory (left) is designed to contain solid and liquid waste.*

Growing crystals

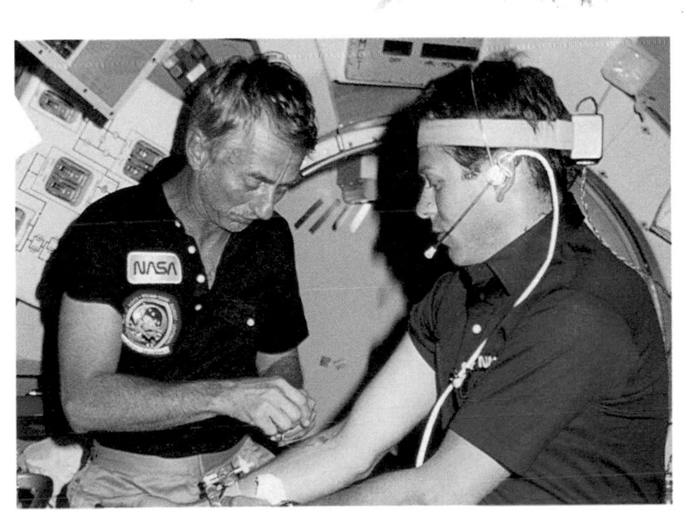

▲ *The health of astronauts is very important. The blood's circulation is affected by weightlessness, and this can cause sickness. Muscles may also start to deteriorate.*

▶ *Some space shuttle flights have carried a laboratory called Spacelab into orbit. Spacelab is used for experiments on the effects of weightlessness. Some of these experiments have been carried out on living things to discover how zero gravity affects them. Spacelab scientists have studied the growth of pine seedlings and the way a spider spins a web. Other experiments are related to industrial processes, such as the growth of crystals.*

Gravity and plants

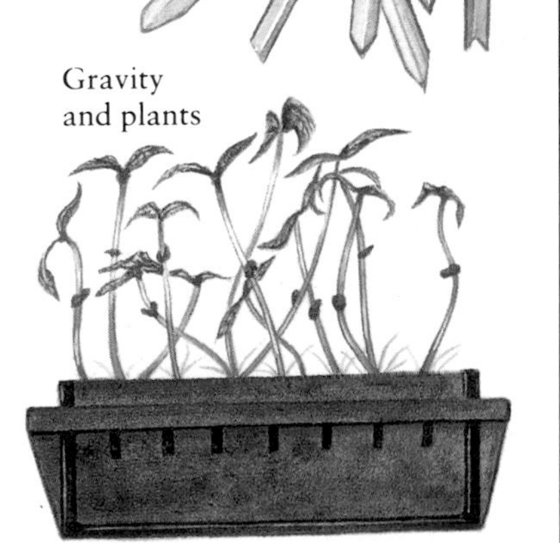

Space Shuttles and Space Stations

The flight of the first space shuttle, *Columbia*, in 1981 heralded a new era in space exploration—the launch of reusable spacecraft. Before then, all spacecraft were used only once, making a space mission an expensive business. The shuttle can take off like a rocket but land like a glider, and may be used many times. Soon shuttle spacecraft will be involved with the ambitious *Freedom* project. This is planned to be the biggest-ever permanently occupied space station; it may be completed by A.D. 2000.

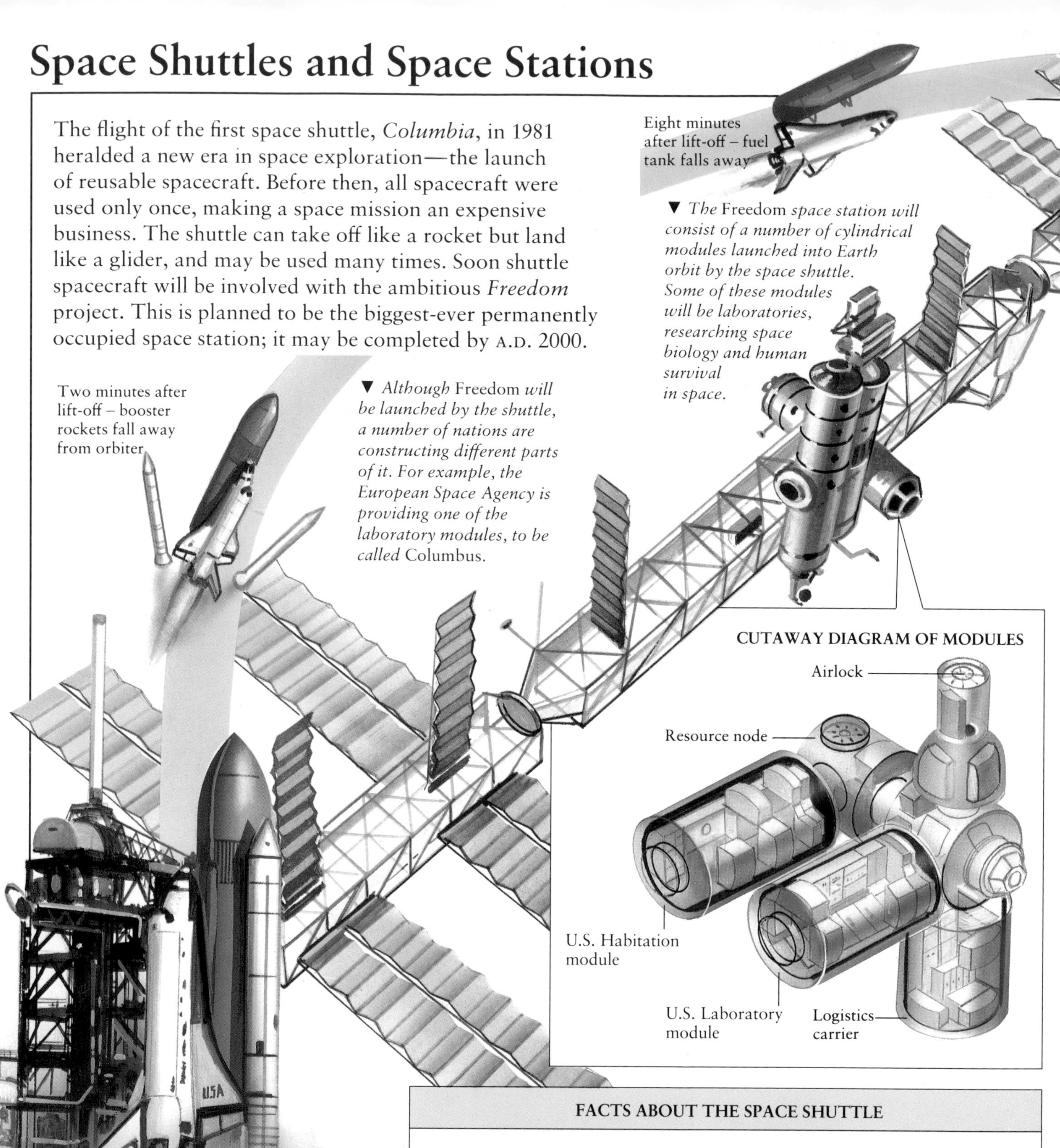

▼ *The* Freedom *space station will consist of a number of cylindrical modules launched into Earth orbit by the space shuttle. Some of these modules will be laboratories, researching space biology and human survival in space.*

▼ *Although* Freedom *will be launched by the shuttle, a number of nations are constructing different parts of it. For example, the European Space Agency is providing one of the laboratory modules, to be called* Columbus.

LAUNCHING THE SHUTTLE
The shuttle is launched on the back of a huge fuel tank. The two side rocket boosters use solid fuel and are recovered. The main tank contains liquid fuel and burns up as it falls back to Earth over the Indian Ocean. The shuttle's own rockets take it into final orbit.

FACTS ABOUT THE SPACE SHUTTLE

- The weight of the shuttle at launch is 2,200 tons—as much as 50 fully-laden trucks.
- Each of the four shuttles is designed to make up to 100 launches in a 20-year lifetime.
- On re-entry into the atmosphere, parts of the shuttle's surface reach a temperature of 2,900°F (1,600°C).
- Seventy percent of the shuttle is covered with heat-resistant tiles.
- The shuttle normally orbits 140 mi. (220 km) above the Earth, but to launch the Hubble Space Telescope in 1990 the *Discovery* shuttle reached the record height of 381 mi. (614 km).
- An extra piece of cargo on this flight was an eyepiece that Edwin Hubble used when taking the photographs that showed the expansion of the Universe!

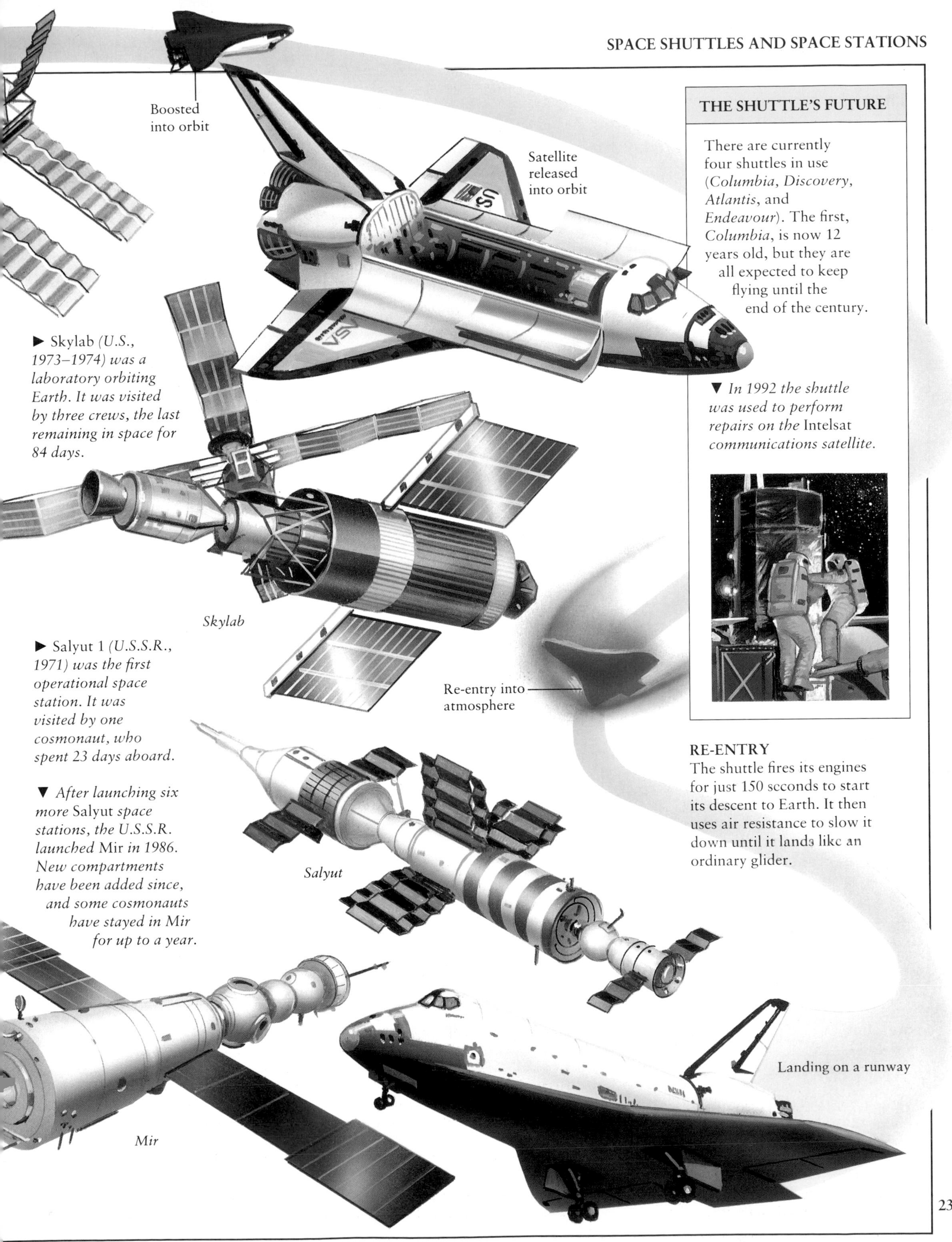

THE SHUTTLE'S FUTURE

There are currently four shuttles in use (*Columbia*, *Discovery*, *Atlantis*, and *Endeavour*). The first, *Columbia*, is now 12 years old, but they are all expected to keep flying until the end of the century.

▼ *In 1992 the shuttle was used to perform repairs on the* Intelsat *communications satellite.*

► Skylab *(U.S., 1973–1974) was a laboratory orbiting Earth. It was visited by three crews, the last remaining in space for 84 days.*

► Salyut 1 *(U.S.S.R., 1971) was the first operational space station. It was visited by one cosmonaut, who spent 23 days aboard.*

▼ *After launching six more* Salyut *space stations, the U.S.S.R. launched* Mir *in 1986. New compartments have been added since, and some cosmonauts have stayed in* Mir *for up to a year.*

RE-ENTRY

The shuttle fires its engines for just 150 seconds to start its descent to Earth. It then uses air resistance to slow it down until it lands like an ordinary glider.

The Future in Space

The biggest problem facing the future of space exploration is not technology but cost. Because the shuttle is still a very expensive craft to operate, several nations are working on a space plane which could take off and land using an ordinary airfield. Highly advanced jet engines would lift the space plane through the atmosphere, and a rocket would then carry it into orbit. Another cost-cutting project is the solar sail, which would use energy sent out by the Sun to "blow" spacecraft between the planets.

CODED MESSAGES
This message to possible other life forms was transmitted in radio code in 1974 toward a star cluster in the constellation Hercules. It will arrive in about the year 26,000.

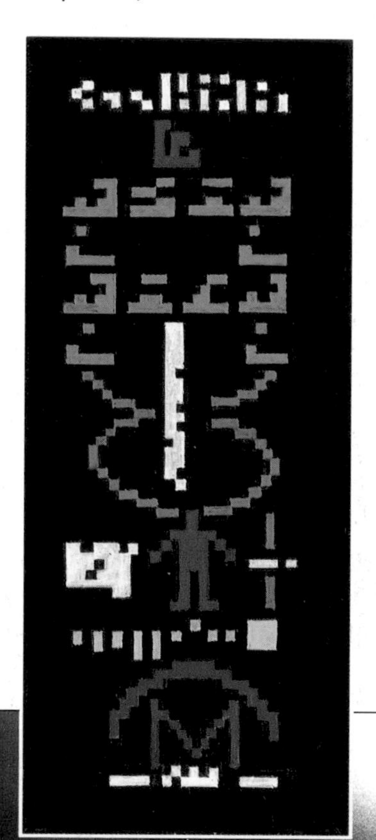

▼ *The solar wind of protons and electrons travels very fast, up to 560 mi./s (900 km/s), but has very little pressure. However, if large enough, a very light sail would gradually build up speed and could tow a load at spaceship velocity. This illustration shows a possible design.*

◀ *Next century, there may be a permanent human settlement on the Moon. The building materials would be mined from the Moon itself. Radio telescopes would search the sky for signals from space, free from Earth's radio noise. The settlement would be an important research base for scientists.*

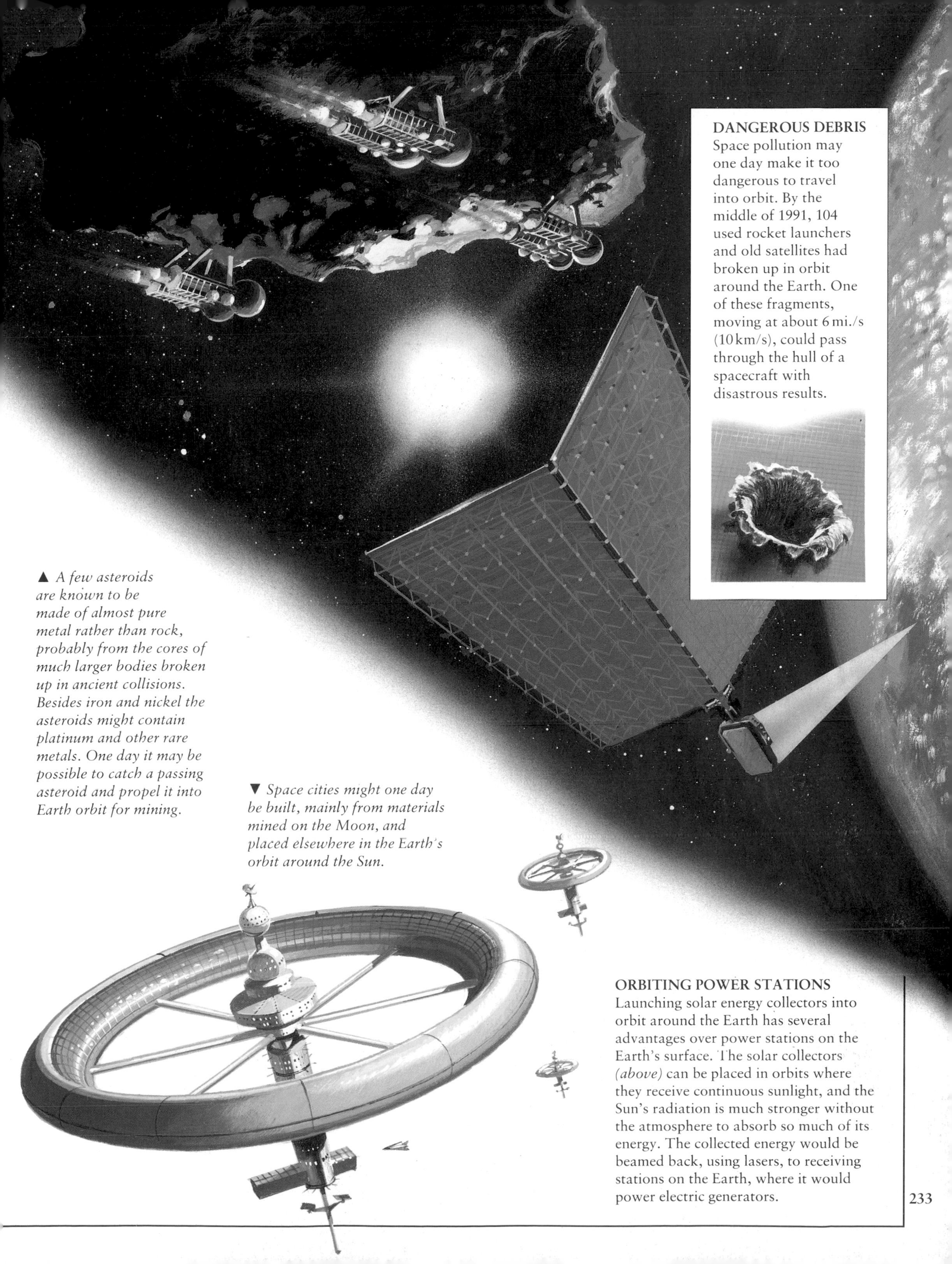

DANGEROUS DEBRIS

Space pollution may one day make it too dangerous to travel into orbit. By the middle of 1991, 104 used rocket launchers and old satellites had broken up in orbit around the Earth. One of these fragments, moving at about 6 mi./s (10 km/s), could pass through the hull of a spacecraft with disastrous results.

▲ *A few asteroids are known to be made of almost pure metal rather than rock, probably from the cores of much larger bodies broken up in ancient collisions. Besides iron and nickel the asteroids might contain platinum and other rare metals. One day it may be possible to catch a passing asteroid and propel it into Earth orbit for mining.*

▼ *Space cities might one day be built, mainly from materials mined on the Moon, and placed elsewhere in the Earth's orbit around the Sun.*

ORBITING POWER STATIONS

Launching solar energy collectors into orbit around the Earth has several advantages over power stations on the Earth's surface. The solar collectors *(above)* can be placed in orbits where they receive continuous sunlight, and the Sun's radiation is much stronger without the atmosphere to absorb so much of its energy. The collected energy would be beamed back, using lasers, to receiving stations on the Earth, where it would power electric generators.

A building site in the Canadian city of Toronto shows the materials and technology needed to build a skyscraper.

SCIENCE AND TECHNOLOGY

The Branches of Science 236 Atoms and Molecules 238
What Are Things Made Of? 240 The Periodic Table 242
Solids, Liquids, and Gases 244 Energy 246 Heat 248 Fuels 250
Gravity and Mass 252 Motion 254 Machines 256 Flying and Floating 258
Counting and Measuring 260 Mathematics 262 Measuring Time 264
The Spectrum 266 Light 268 Sound 270 Sound Applications 272
Electromagnetism 274 Electricity in Action 276 Electronics 278
Computers 280 Engineering 282 Buildings 284
Bridges, Tunnels, Dams, and Roads 288 Transportation 290
Land Transportation 292 Sea Transportation 294
Air Transportation 296 Materials 298 Farming 300 Medicine 302
Communications 304 Inventions, Discoveries, and Inventors 306

People have always observed the natural world. Disciplines such as mathematics, astronomy, and biology are all based on these observations. These and countless other subjects overlap to form the broad area we call science. Although the many branches of science have distinct characteristics, no single subject can act without the others. We cannot, for instance, have biology without chemistry, nor can we have chemistry without physics. A simple understanding of each of these areas helps make the others more clear.

Technology is born from an understanding of how things work. In addition to helping us find out more about the sciences themselves, the application of technology has enriched our lives in many ways. We have developed efficient farming methods, effective transportation networks, and new ways of obtaining energy. Technology also helps us to make drinking water safe, construct sturdy places to live, conquer disease, and live longer.

DISCOVERING SCIENCE

The Branches of Science

Science means knowledge (from the Latin word *scientia*), and therefore it covers every field of human inquiry. The desire to find out how and why things work is one of humankind's most distinctive qualities. Science has given people power, for example, to make their lives more comfortable and to change their environment. It has also given people technology – the ability to make tools, to build with materials, and to harness sources of energy. Science has various branches concerned with different areas of knowledge. They include the physical sciences, such as astronomy, chemistry, physics, and geology; mathematics; life sciences, such as biology, botany, and zoology; and social sciences, such as anthropology, economics, and psychology. The pages that follow concentrate on the ways in which science and technology have shaped the modern world.

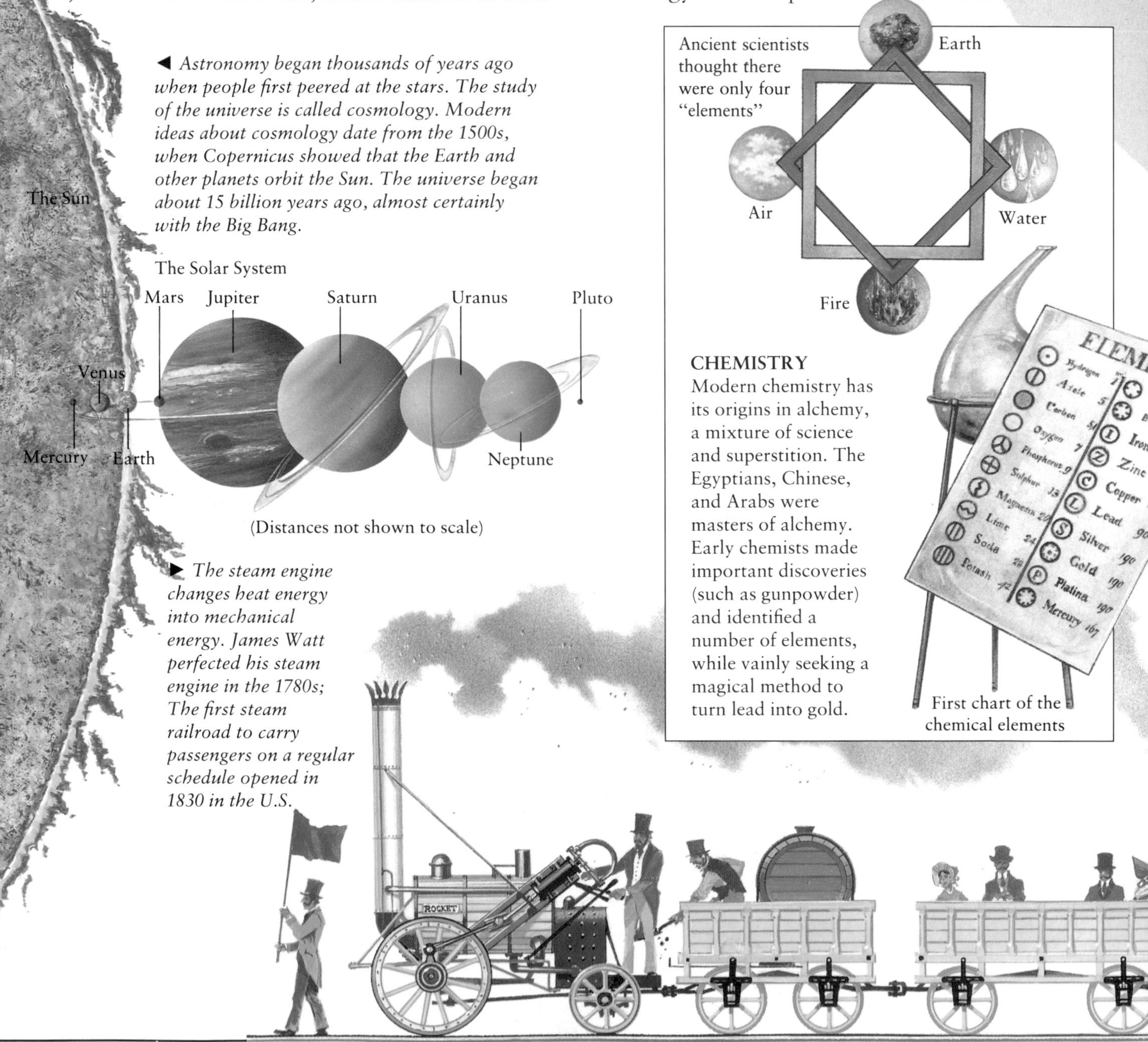

◀ *Astronomy began thousands of years ago when people first peered at the stars. The study of the universe is called cosmology. Modern ideas about cosmology date from the 1500s, when Copernicus showed that the Earth and other planets orbit the Sun. The universe began about 15 billion years ago, almost certainly with the Big Bang.*

CHEMISTRY
Modern chemistry has its origins in alchemy, a mixture of science and superstition. The Egyptians, Chinese, and Arabs were masters of alchemy. Early chemists made important discoveries (such as gunpowder) and identified a number of elements, while vainly seeking a magical method to turn lead into gold.

First chart of the chemical elements

▶ *The steam engine changes heat energy into mechanical energy. James Watt perfected his steam engine in the 1780s; The first steam railroad to carry passengers on a regular schedule opened in 1830 in the U.S.*

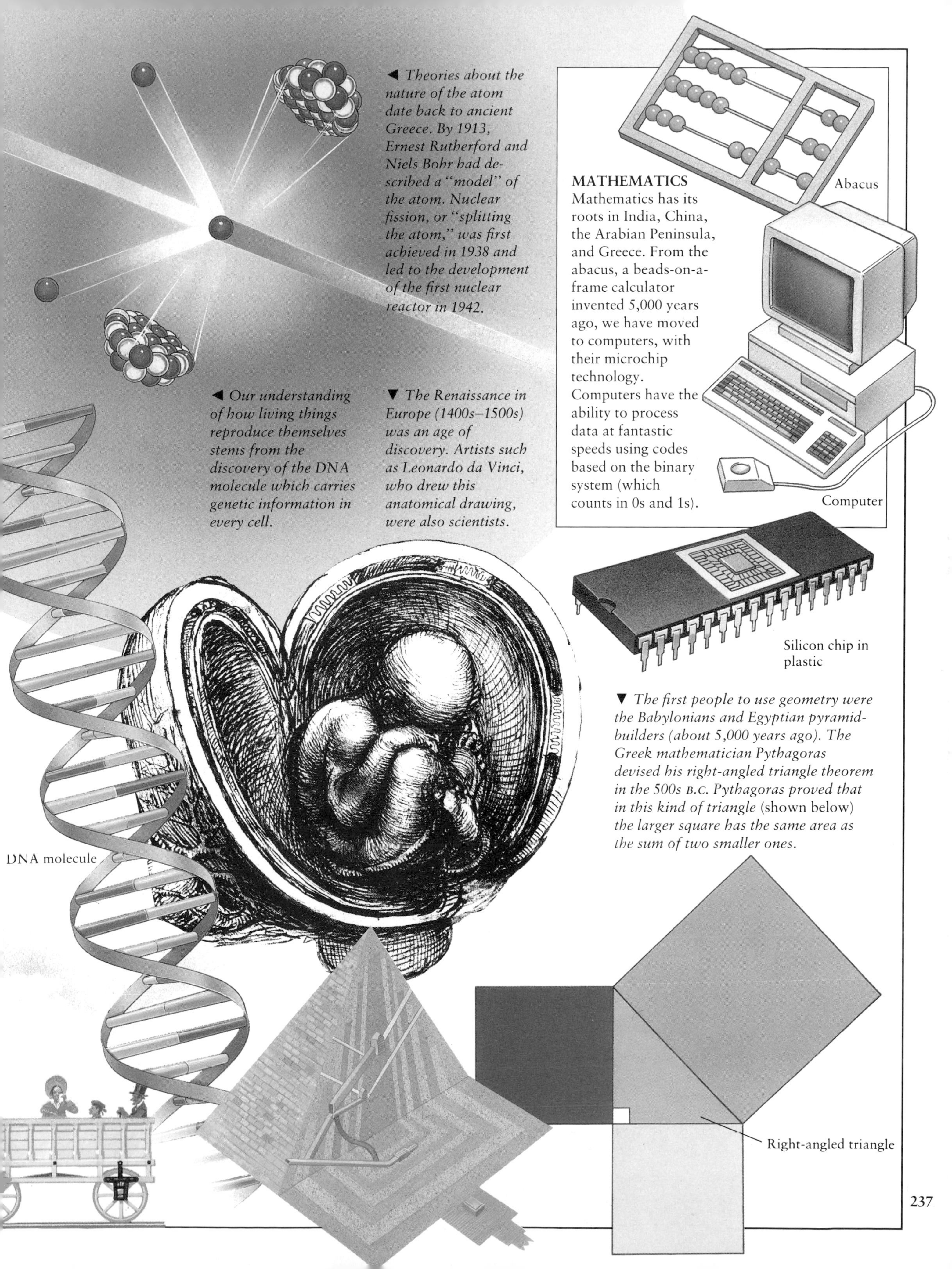

◄ Theories about the nature of the atom date back to ancient Greece. By 1913, Ernest Rutherford and Niels Bohr had described a "model" of the atom. Nuclear fission, or "splitting the atom," was first achieved in 1938 and led to the development of the first nuclear reactor in 1942.

◄ Our understanding of how living things reproduce themselves stems from the discovery of the DNA molecule which carries genetic information in every cell.

▼ The Renaissance in Europe (1400s–1500s) was an age of discovery. Artists such as Leonardo da Vinci, who drew this anatomical drawing, were also scientists.

MATHEMATICS
Mathematics has its roots in India, China, the Arabian Peninsula, and Greece. From the abacus, a beads-on-a-frame calculator invented 5,000 years ago, we have moved to computers, with their microchip technology. Computers have the ability to process data at fantastic speeds using codes based on the binary system (which counts in 0s and 1s).

▼ The first people to use geometry were the Babylonians and Egyptian pyramid-builders (about 5,000 years ago). The Greek mathematician Pythagoras devised his right-angled triangle theorem in the 500s B.C. Pythagoras proved that in this kind of triangle (shown below) *the larger square has the same area as the sum of two smaller ones.*

MATTER AND ENERGY

Atoms and Molecules

Matter is all the material in the universe—animals, plants, rocks, air, and water. Matter can exist in three familiar states—solid, liquid, and gas. A fourth state, plasma, is formed at very high temperatures—for example, within stars. In all its states, matter is made of the same basic units—atoms. The smallest piece of a substance that can exist on its own is called an atom. An element is a substance made up of only one kind of atom. Atoms are made of even smaller particles, called electrons, moving around a center called a nucleus made up of neutrons and protons, in much the same way as the planets orbit the Sun. Within the atom is immense energy, which can be released by splitting the atom to produce a chain reaction, called nuclear fission. Atoms can combine with other atoms to form molecules.

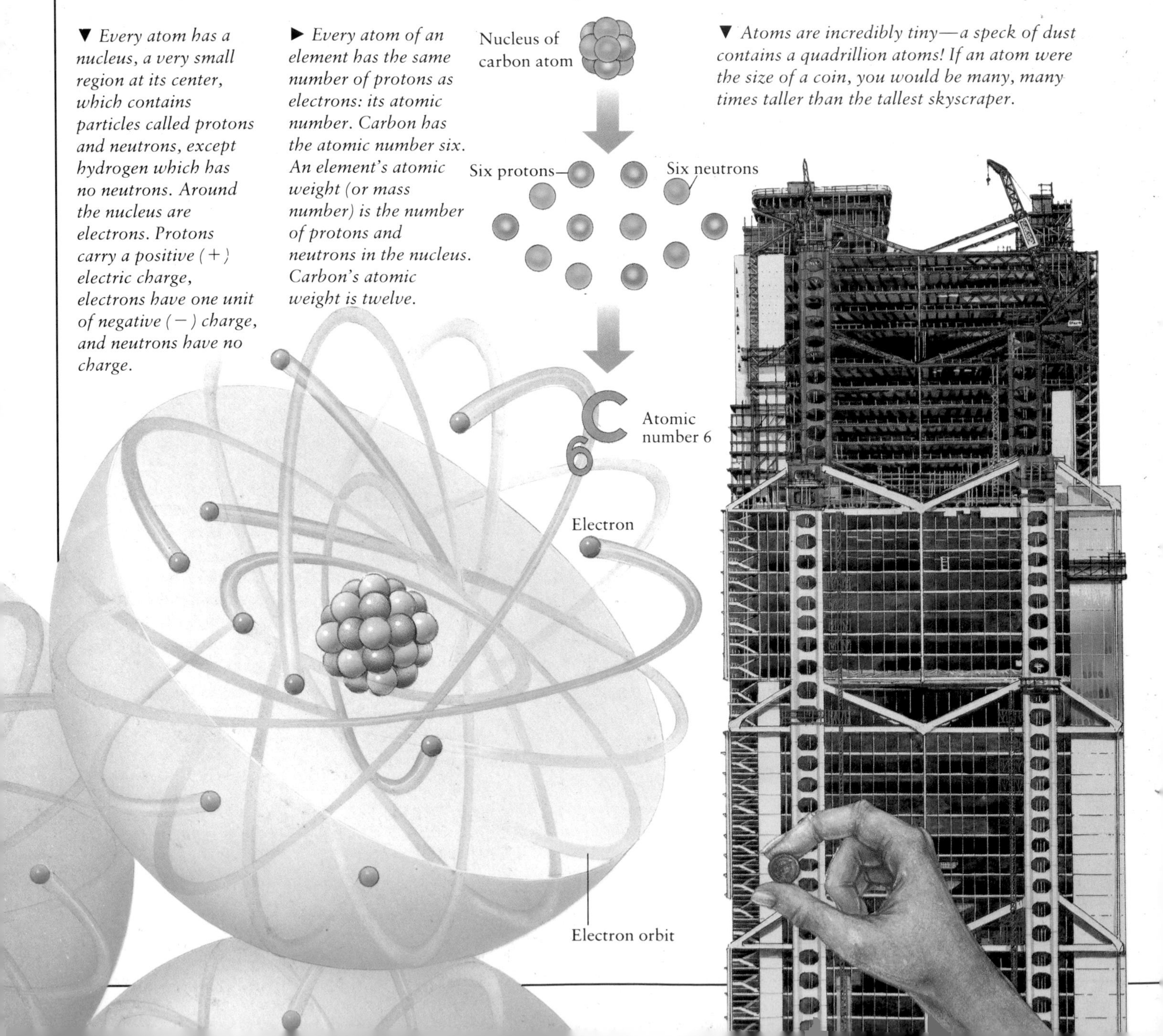

▼ *Every atom has a nucleus, a very small region at its center, which contains particles called protons and neutrons, except hydrogen which has no neutrons. Around the nucleus are electrons. Protons carry a positive (+) electric charge, electrons have one unit of negative (−) charge, and neutrons have no charge.*

► *Every atom of an element has the same number of protons as electrons: its atomic number. Carbon has the atomic number six. An element's atomic weight (or mass number) is the number of protons and neutrons in the nucleus. Carbon's atomic weight is twelve.*

▼ *Atoms are incredibly tiny—a speck of dust contains a quadrillion atoms! If an atom were the size of a coin, you would be many, many times taller than the tallest skyscraper.*

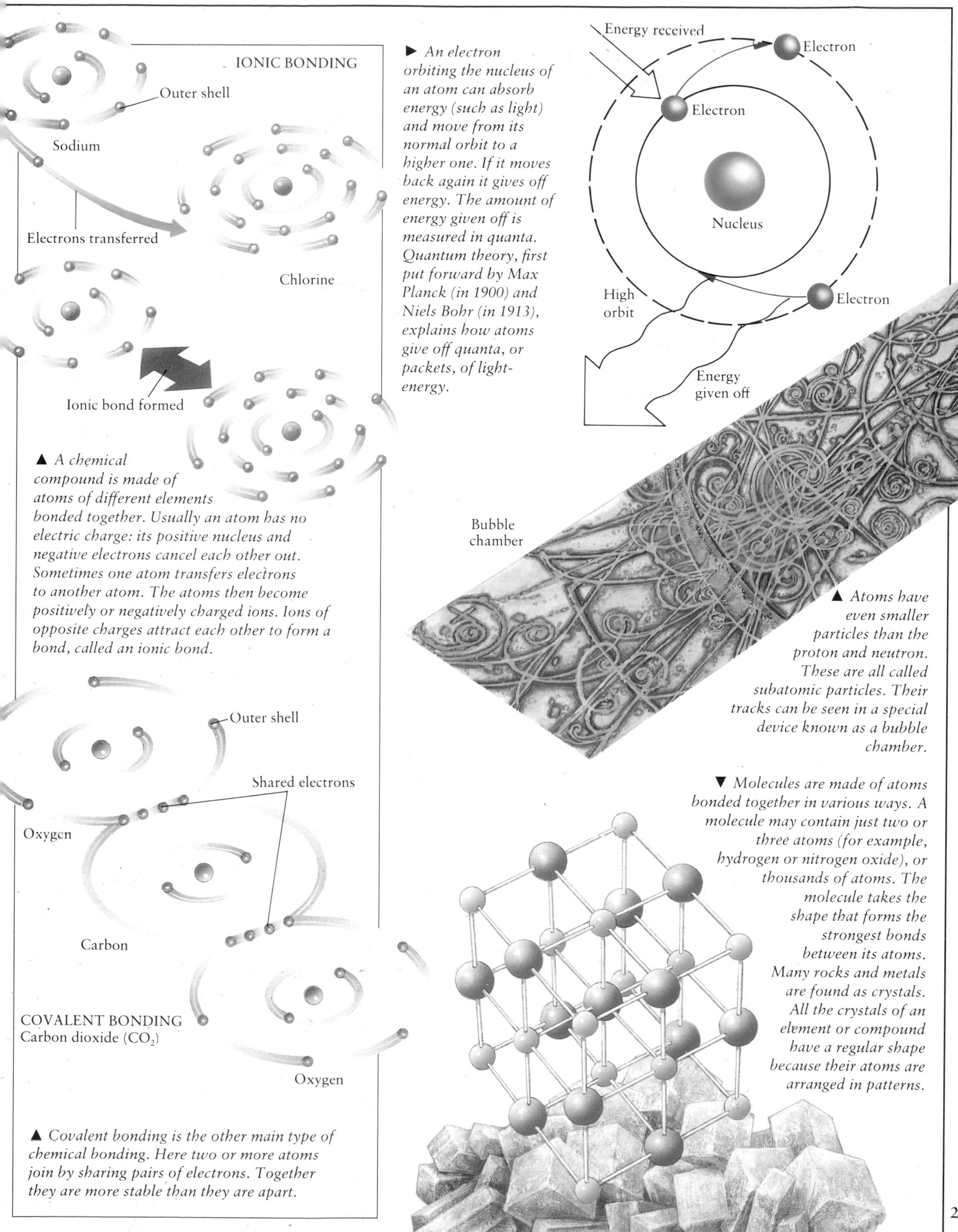

▶ An electron orbiting the nucleus of an atom can absorb energy (such as light) and move from its normal orbit to a higher one. If it moves back again it gives off energy. The amount of energy given off is measured in quanta. Quantum theory, first put forward by Max Planck (in 1900) and Niels Bohr (in 1913), explains how atoms give off quanta, or packets, of light-energy.

▲ A chemical compound is made of atoms of different elements bonded together. Usually an atom has no electric charge: its positive nucleus and negative electrons cancel each other out. Sometimes one atom transfers electrons to another atom. The atoms then become positively or negatively charged ions. Ions of opposite charges attract each other to form a bond, called an ionic bond.

▲ Atoms have even smaller particles than the proton and neutron. These are all called subatomic particles. Their tracks can be seen in a special device known as a bubble chamber.

▼ Molecules are made of atoms bonded together in various ways. A molecule may contain just two or three atoms (for example, hydrogen or nitrogen oxide), or thousands of atoms. The molecule takes the shape that forms the strongest bonds between its atoms. Many rocks and metals are found as crystals. All the crystals of an element or compound have a regular shape because their atoms are arranged in patterns.

▲ Covalent bonding is the other main type of chemical bonding. Here two or more atoms join by sharing pairs of electrons. Together they are more stable than they are apart.

What Are Things Made Of?

Every substance is either a chemical element or a combination of elements. The atoms in a substance are held together by chemical bonds that form molecules. Different elements can bond together to make compounds. As bonds form or break, a chemical reaction takes place. Carbon is found in all living things. It combines freely with hydrogen, nitrogen, and oxygen. Organic chemistry concentrates on substances that have carbon-to-carbon bonds: there are over one million organic compounds.

HYDROGEN BONDS
In water, hydrogen and oxygen atoms join by sharing pairs of electrons—an example of covalent bonding. Water molecules are also held together by hydrogen bonds. These weak bonds are important in building large molecules such as DNA (deoxyribonucleic acid) and protein structures.

Non-bonding pairs of electrons

Oxygen atom

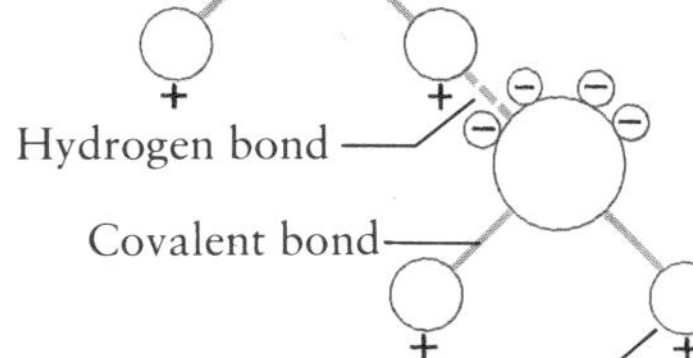

Positive hydrogen atom

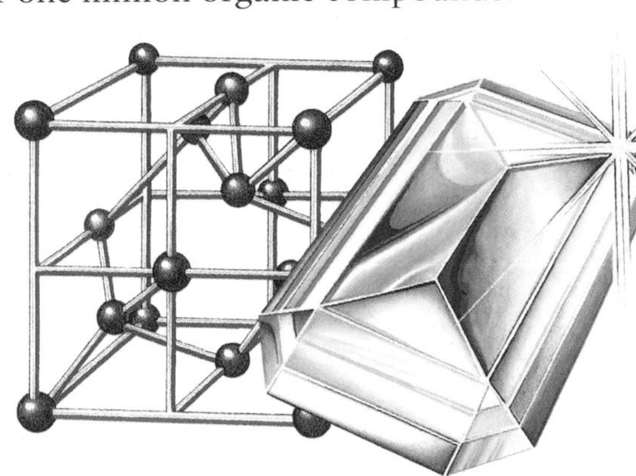

► *Diamond is a form of pure carbon. The atoms in diamond are arranged in a dense lattice framework. This is why diamond is so hard. Carbon makes up less than 0.03 percent of the Earth's crust. Most of this carbon is combined with other elements.*

◄ *Graphite, once called plumbago, is another form of pure carbon. Its atoms are arranged in layers which slide easily over one another. Graphite is one of the softest solids. It is also greasy and makes a useful lubricant. The "lead" in a pencil is a mixture of graphite and clay.*

Synthetic rubber gloves

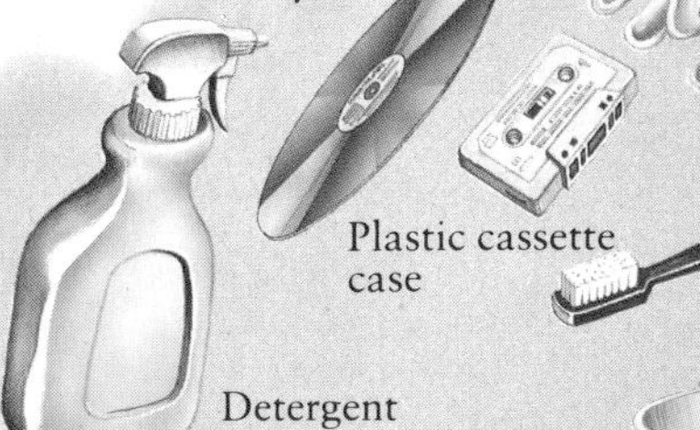

Toothbrus

Plastic bucket

THE USES OF HYDROCARBONS

Ethane

Benzene

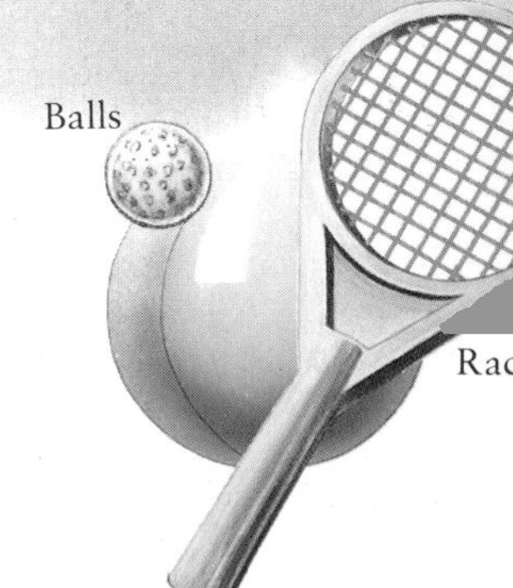

▲ *Hydrocarbons are compounds made only of hydrogen and carbon atoms. They are found in petroleum and natural gas. Hydrocarbons found in nature also provide the raw material for making plastics, solvents, and other synthetic materials. In hydrocarbons, carbon atoms are arranged either in chains (as in ethane) or rings (as in benzene).*

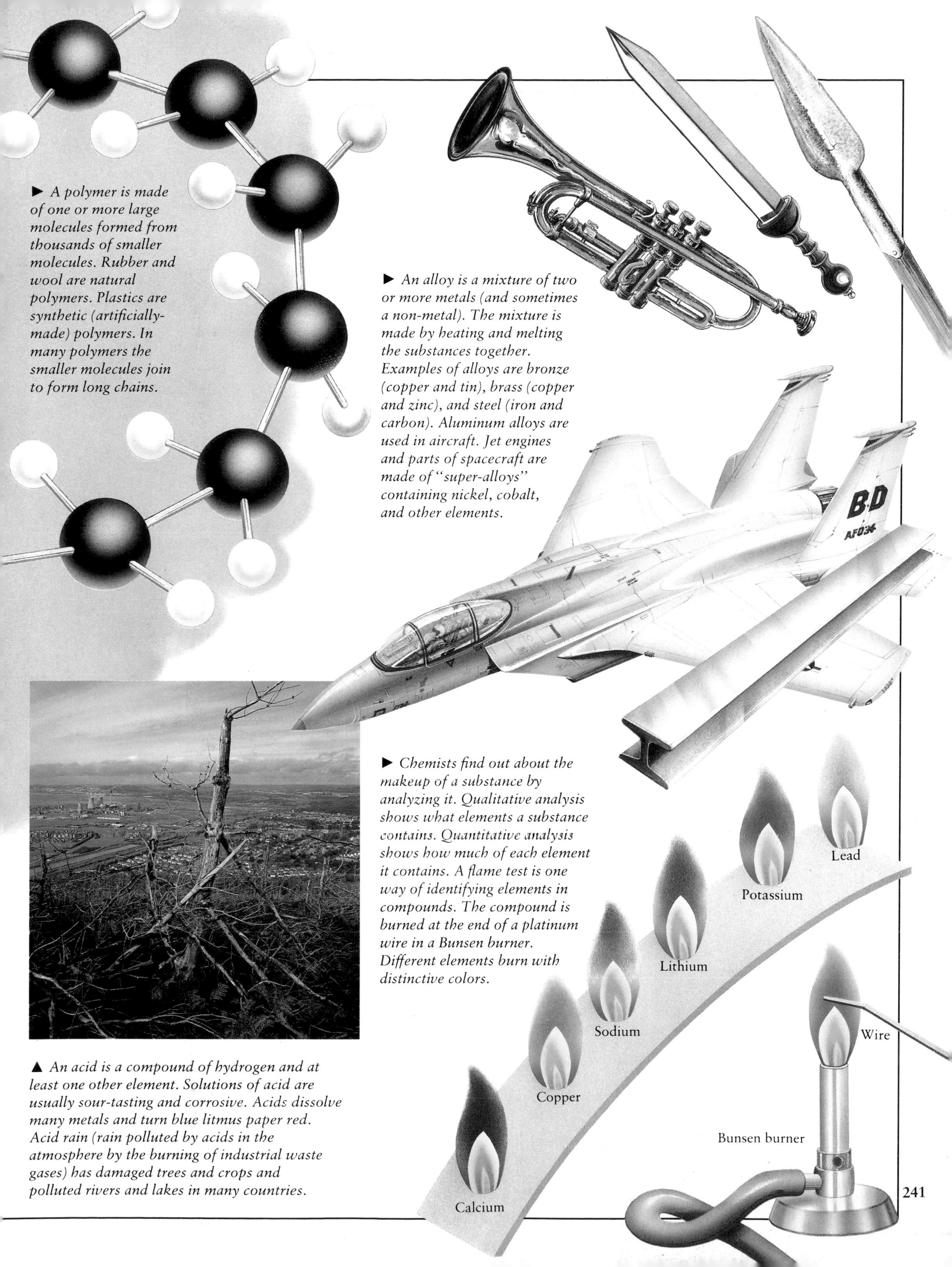

► *A polymer is made of one or more large molecules formed from thousands of smaller molecules. Rubber and wool are natural polymers. Plastics are synthetic (artificially-made) polymers. In many polymers the smaller molecules join to form long chains.*

► *An alloy is a mixture of two or more metals (and sometimes a non-metal). The mixture is made by heating and melting the substances together. Examples of alloys are bronze (copper and tin), brass (copper and zinc), and steel (iron and carbon). Aluminum alloys are used in aircraft. Jet engines and parts of spacecraft are made of "super-alloys" containing nickel, cobalt, and other elements.*

► *Chemists find out about the makeup of a substance by analyzing it. Qualitative analysis shows what elements a substance contains. Quantitative analysis shows how much of each element it contains. A flame test is one way of identifying elements in compounds. The compound is burned at the end of a platinum wire in a Bunsen burner. Different elements burn with distinctive colors.*

▲ *An acid is a compound of hydrogen and at least one other element. Solutions of acid are usually sour-tasting and corrosive. Acids dissolve many metals and turn blue litmus paper red. Acid rain (rain polluted by acids in the atmosphere by the burning of industrial waste gases) has damaged trees and crops and polluted rivers and lakes in many countries.*

The Periodic Table

In all, 103 elements have been officially named. Ninety-two elements occur naturally on the Earth and others have been made in laboratories. Scientists claim to have discovered a further six elements, known as 104 to 109. Twelve elements were known to the ancient world (before A.D. 1000). Seventy-six more were identified between the 1500s and 1920s. Each element from 1 (Hydrogen: H) to 103 (Lawrencium: Lr) has a symbol. Chemists use these symbols when writing formulas for compounds.

EVERYDAY USE OF ELEMENTS

◀ *Neon (used for artificial lighting) is a noble gas.*

Neon-lit sign

▶ *Phosphorus (used in matches) is another non-metal.*

Phosphorus match

▶ *The periodic table presents information about the elements: their name, atomic number, and similarities with other elements. The elements are arranged in periods, or rows, in order of increasing atomic number. Groups of elements share certain characteristics because of the way the electrons in their atoms are arranged in shells. Elements with the same number of electrons in their outermost shells behave similarly. There are two main groups: the non-metals and the metals; certain similar elements fall into families, e.g. alkali metals, transition metals, and inner transition metals.*

Alkali metals

Transition metals

Non-metals

Inner transition series

3 Lithium Li	4 Beryllium Be							
11 Sodium Na	12 Magnesium Mg							
19 Potassium K	20 Calcium Ca	21 Scandium Sc	22 Titanium Ti	23 Vanadium V	24 Chromium Cr	25 Manganese Mn	26 Iron Fe	27 Cobalt Co
37 Rubidium Rb	38 Strontium Sr	39 Yttrium Y	40 Zirconium Zr	41 Niobium Nb	42 Molybdenum Mo	43 Technetium Tc	44 Ruthenium Ru	45 Rhodium Rh
55 Cesium Cs	56 Barium Ba	57–71 Lanthanide series	72 Hafnium Hf	73 Tantalum Ta	74 Tungsten W	75 Rhenium Re	76 Osmium Os	77 Iridium Ir
87 Francium Fr	88 Radium Ra	89–103 Actinide series	Element 104	105 Element 105	106 Element 106	107 Element 107	108 Element 108	109 Element 109

57 Lanthanum La	58 Cerium Ce	59 Praseodymium Pr	60 Neodymium Nd	61 Prometheum Pm	62 Samarium Sm	63 Europium Eu	64 Gadolinium Gd	65 Terbium Tb
89 Actinium Ac	90 Thorium Th	91 Protactinium Pa	92 Uranium U	93 Neptunium Np	94 Plutonium Pu	95 Americium Am	96 Curium Cm	97 Berkelium Bk

DOWN
Going down, the size of atoms increases; elements in the same group behave similarly.

▼ *Water is a molecule made of two atoms of hydrogen (H) and one atom of oxygen (O). Hydrogen is the simplest and most abundant element in the universe. Methane gas is a hydrocarbon made of one atom of carbon and four atoms of hydrogen. Carbon dioxide is made of two atoms of oxygen to one of carbon.*

Carbon dioxide (CO_2)

▶ *Gold is a metallic element. It is soft, but heavy, and forms few compounds.*

Gold bar

Water (H_2O)

◀ *Uranium exists in several isotopes, or varieties. It is used as a nuclear fuel and in weapons.*

Nuclear explosion

Methane (CH_4)

▼ *By using the periodic table we can predict the properties of elements (or compounds of two or more elements), from what is known about neighboring elements in the table. The noble gases, for example, (helium, neon, argon, and others in the group) are gases that do not combine easily with other elements to form compounds. They are all "inert." The first periodic table was drawn up by the Russian chemist Dmitri Mendeleyev (1834–1907). He charted the known elements and predicted the existence of undiscovered ones.*

Water

Oxygen

◀ *Rusting metal, iron or steel, is an example of a chemical reaction between elements. In a reaction, atoms either lose or gain electrons. Rust is the visible evidence of a reaction known as oxidation. The oxygen in air or water takes electrons from the atoms in the metal, and this produces the familiar signs of rust. Salt water speeds rusting.*

Rusted metal

1 Hydrogen H								2 Helium He
			5 Boron B	6 Carbon C	7 Nitrogen N	8 Oxygen O	9 Fluorine F	10 Neon Ne
			13 Aluminum Al	14 Silicon Si	15 Phosphorus P	16 Sulfur S	17 Chlorine Cl	18 Argon Ar
28 Nickel Ni	29 Copper Cu	30 Zinc Zn	31 Gallium Ga	32 Germanium Ge	33 Arsenic As	34 Selenium Se	35 Bromine Br	36 Krypton Kr
46 Palladium Pd	47 Silver Ag	48 Cadmium Cd	49 Indium In	50 Tin Sn	51 Antimony Sb	52 Tellurium Te	53 Iodine I	54 Xenon Xe
78 Platinum Pt	79 Gold Au	80 Mercury Hg	81 Thallium Tl	82 Lead Pb	83 Bismuth Bi	84 Polonium Po	85 Astatine At	86 Radon Rn

ACROSS

Going across, the size of atoms increases; elements change from metals through metal-like elements to non-metals.

66 Dysprosium Dy	67 Holmium Ho	68 Erbium Er	69 Thulium Tm	70 Ytterbium Yb	71 Lutetium Lu
98 Californium Cf	99 Einsteinium Es	100 Fermium Fm	101 Mendelev-ium Md	102 Nobelium No	103 Lawrencium Lr

Detergent molecule

Dirt

◀ *Detergents are compounds of elements. They weaken the surface tension of the water, so that fabrics get thoroughly wet. The detergent molecules dislodge dirt particles by sticking to them and squeezing them out of the fibers.*

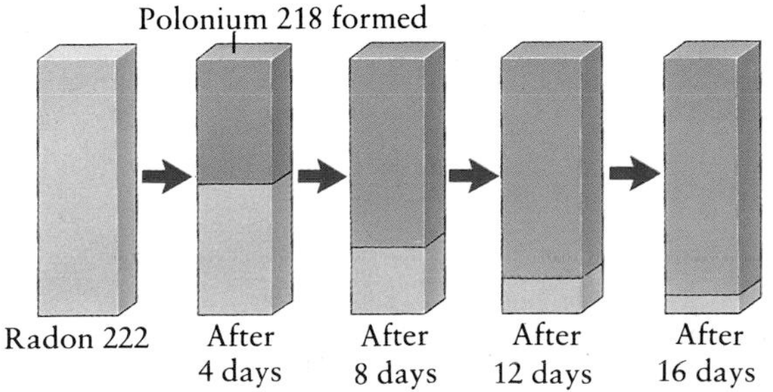

▲ *Certain kinds of atoms give out radioactivity (alpha, beta, and gamma radiation). A radioactive element decays into another element. The time taken for half of it to decay (from seconds to millions of years) is called its "half-life."*

▼ *Certain elements, such as uranium, are radioactive. The three forms of radiation in radioactivity are alpha, beta, and gamma radiation. Their penetrating abilities are illustrated in the diagram below.*

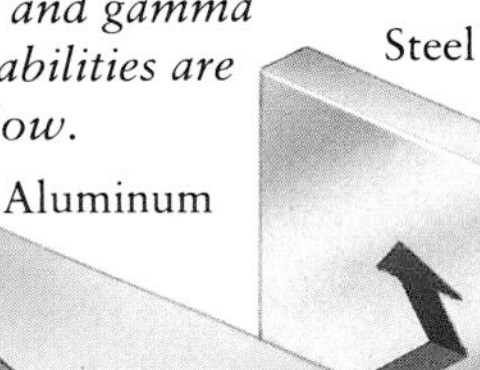

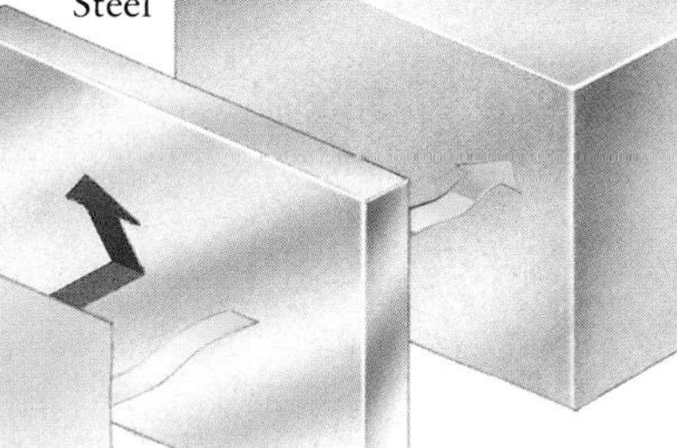

Solids, Liquids, and Gases

The three states in which matter can exist under normal conditions are solid, liquid, and gas. Solids have both shape and volume; their molecules are squashed together tightly. Liquids also have volume, but no shape. Their molecules are held together less tightly, so that a liquid will flow into a container. A gas has neither volume nor shape. Its molecules are spread very thinly, and a gas will fill any container that encloses it. Many substances change their state when cooled or heated.

GAS
Because the molecules are only loosely held together, they move around faster than in liquids or solids.

CHANGES OF STATE
Most matter can exist as solid, liquid, or gas. When matter is heated, it can change its state or "phase." The heat and pressure beneath a volcano *(below)* turn solid rock within the Earth into liquid lava or gases.

▲ *When a liquid is heated, it changes to a vapor or gas. This is evaporation. Steam from a volcano or boiling kettle is an example of water vapor.*

LIQUID
Molecules are more mobile than in a solid, less mobile than in a gas.

Warm air rises and cools

Condensation

SOLID
A solid is hard and rigid; atoms and molecules cluster in a regular pattern.

▲ *As a solid such as a rock is heated, its atoms vibrate more and move apart. The rigid structure breaks down and it melts. Nearly all solids melt eventually.*

► *Sublimation is a direct change from solid to gas, without a liquid stage between. Solid iodine does this when heated. Sublimation is part of the process of making freeze-dried coffee.*

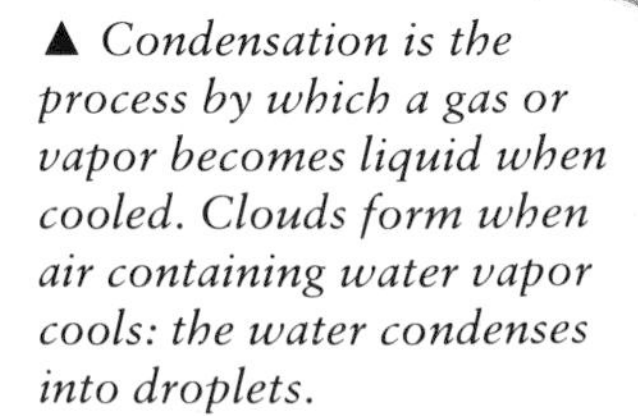

▲ *Condensation is the process by which a gas or vapor becomes liquid when cooled. Clouds form when air containing water vapor cools: the water condenses into droplets.*

MIXTURES

Mixtures are different from compounds because the substances in them are not chemically bonded. For example, iron filings mixed with sulfur can be separated using a magnet. But heating them produces a compound, iron sulfide, from which a magnet cannot pick out the iron. Solutions and colloids (aerosols, emulsions, and foams) are all mixtures.

▲ *An aerosol is a cloud of fine liquid or solid particles suspended in a gas.*

▼ *Milk and paint are emulsions; colloids of two liquids whose particles are evenly mixed without being in solution.*

◀ *Colloids are liquids or gases containing larger particles than solutions. Butter, foam, and toothpaste are all examples of colloids.*

THE THREE GAS LAWS

1 Gas pressure doubles if its volume is halved and its temperature stays constant (unchanged). The molecules are squashed together.
2 Gas volume doubles if its temperature doubles and the pressure stays the same. The molecules move faster but in more space.
3 Gas pressure doubles if the volume stays the same and the temperature doubles. The molecules move faster in the same space.

▼ *Most liquids contract as they cool and freeze. Water is unusual in contracting until it reaches 39°F (4°C), and then expanding as it freezes, at 32°F (0°C).*

DISTILLATION

A solution is a liquid mixture. One substance, called the solute, dissolves in a liquid, the solvent. When a solution (for example, salt and water) is boiled in a flask, vapor is given off.

▼ *In a condensing vessel, the vapor cools and becomes a liquid. Salt crystals are left in the flask. This apparatus is called a still; the process is distillation.*

Energy

Energy is all around us. We can sometimes see it as light, hear it as sound, and feel it as heat. All these are forms of energy. Energy is the ability to do work. The spring in a toy car provides energy to drive the car. A flashlight battery gives the energy needed to light the bulb. All living things on the Earth depend on the Sun for their energy. Energy comes from matter. It is everywhere in the universe. It can be neither created nor destroyed; it can only be changed into a different form.

▼ *The Sun is powered by the energy from nuclear fusion in which hydrogen atoms combine to form helium.*

THE ENERGY CYCLE

Energy from the Sun is absorbed by the sea and land. Plants use light energy to grow, and animals that eat plants convert the energy for their own use. Decaying plants and animals became fossil fuels; we use the energy from these fuels for light and power.

We turn the energy from food into the energy of movement

A

B

▲ *Kinetic energy is the energy of movement. Any moving body, such as a child on a swing, has kinetic (moving) energy. By the time she has reached position* A *the child has maximum potential (stored) energy; gravity then swings her to* B, *the point at which she has maximum kinetic energy.*

JOULES

The joule is the SI unit (see p.88) of work or energy. One joule of work is done when a force of one newton moves through a distance of one meter. The joule is named after the English scientist James Joule (1818–1889).

1 meter

1 joule effort

1 newton

► *Coal, oil, and natural gas store energy that came originally from the Sun. Coal stores the chemical energy of the ancient plants that lived and died in the prehistoric coal-forests; oil and gas store energy from the bodies of tiny dead sea creatures.*

Energy and matter are never destroyed. Energy released into space may one day help form a new star.

THE CONSERVATION OF ENERGY

The law of conservation of energy says that energy is never created or destroyed. Energy is constantly changing its state, from one form to another. Every time it is used, energy is converted. For example, steam from water heated in a boiler can be released as the kinetic energy of a spinning turbine, which can be changed into electrical energy in a generator; this energy, in turn, can power devices that produce sound, heat, and light.

Electric bell (sound energy)

Electric light (light energy)

Spinning turbine (electrical energy)

Storage battery (chemical energy)

Steam (kinetic energy)

Heat (thermal energy)

Chemical energy

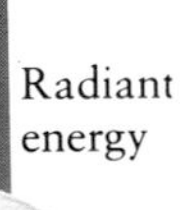

Radiant energy

ver
ions
n coal,
verting heat energy
electrical energy

► Potential energy is stored energy, as in a pendulum weight at the top of its swing. Gravity makes it swing, work is done, and potential energy becomes kinetic energy. A simple pendulum takes the same time to swing, no matter how heavy its weight. A basketball about to be thrown has potential energy. When the player passes, the ball has kinetic energy.

◄ Chemical energy is released in a chemical reaction, such as burning. Our digested food releases chemical energy for use by the body. Heat is thermal energy. The hotter an object is, the faster its molecules move around. Heat is also a form of kinetic energy. Radiant energy is transmitted as electromagnetic radiation, such as light from the Sun or a light bulb.

FACTS ABOUT ENERGY

- The Sun is an immense source of energy. Scientists calculate that every year the Earth receives from the Sun an amount of energy equal to burning 227 trillion tons of coal.
- Energy always ends up as heat energy. A bouncing ball (with kinetic energy) slowly rolls to a standstill. But the energy that made it bounce has not been lost. Kinetic energy changes to heat energy as the ball rubs against the air and the ground (friction).
- Most energy loss is through heat escaping. Even the most efficient car engine wastes about 60 percent of the energy in the fuel it burns. An electric lamp is even less efficient; it turns only about 20 percent of the electricity it uses into light. The rest is "lost" as heat.
- Every year we consume energy equivalent to burning two billion tons of coal. Fossil fuels (coal, oil, gas) will eventually run out. Alternative energy sources (wind, tide, solar power) may not be able to replace them. Just 135 tons of deuterium (found in the oceans), one of the hydrogen-like fuels used in nuclear fusion, would give the same amount of energy as two billion tons of coal.

Heat

Scientists once believed that heat was an invisible fluid. Not until the late 1700s did Benjamin Thompson show that heat is a form of energy. To make atoms move faster you have to add energy, often in the form of heat. A large pan of boiling water contains more heat than another pan with only a little boiling water, though the water in each is at the same temperature: 212°F (100°C). Heat can move from one substance to another in one of three ways: by convection, by radiation, and by conduction.

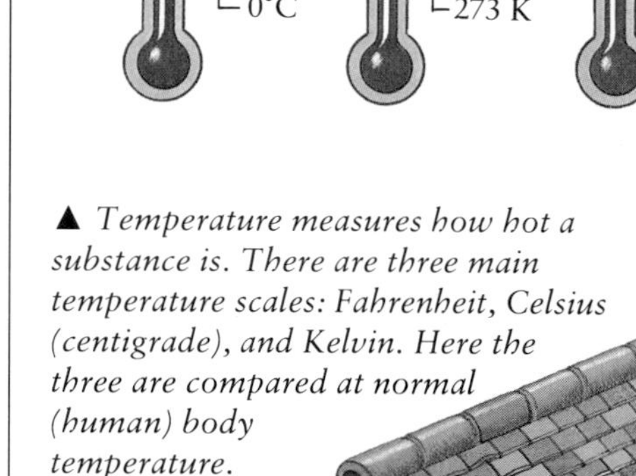

▲ *Temperature measures how hot a substance is. There are three main temperature scales: Fahrenheit, Celsius (centigrade), and Kelvin. Here the three are compared at normal (human) body temperature.*

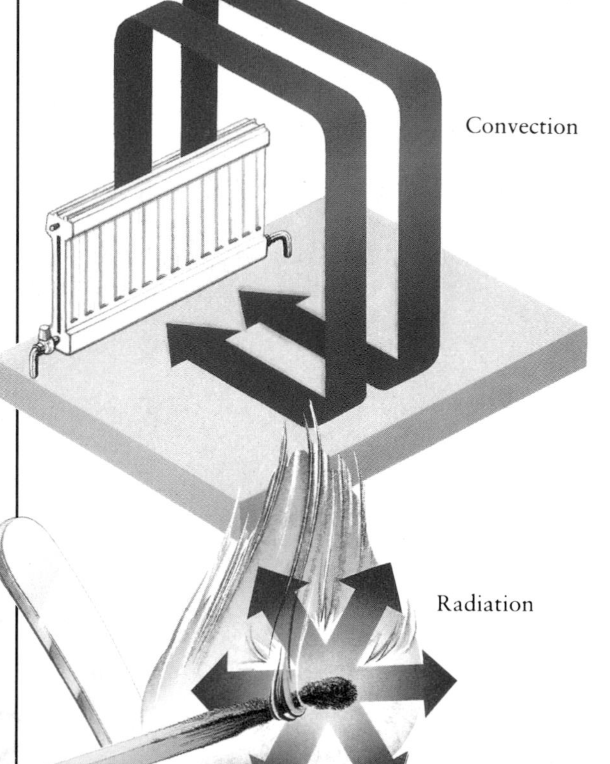

◄ *Convection takes place only in liquids (for example, water in a kettle) and gases (for example, air warmed by a heat source such as a fire or radiator). When a liquid or gas is heated, it expands and becomes less dense. Warm air above the radiator rises and cool air moves in to take its place, creating a convection current.*

◄ *Radiation is the movement of heat through the air. Heat from a match sets molecules of air moving and rays of heat spread out around the heat source.*

◄ *Conduction occurs in solids such as metals. The handle of a metal spoon left in hot liquid warms up as molecules at the heated end move faster and collide with their neighbors, setting them moving. The heat travels through the metal, which is a good conductor of heat.*

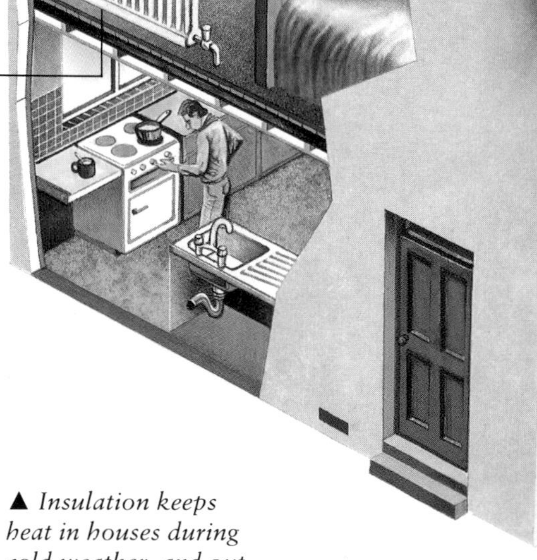

▲ *Insulation keeps heat in houses during cold weather, and out during hot weather. Plastic, wood, and air-filled spaces are good insulators.*

HEAT REFLECTION

Shiny surfaces such as mirrors reflect heat. Shiny or pale-colored walls reflect heat from the Sun. Dark surfaces absorb more heat than pale or shiny ones.

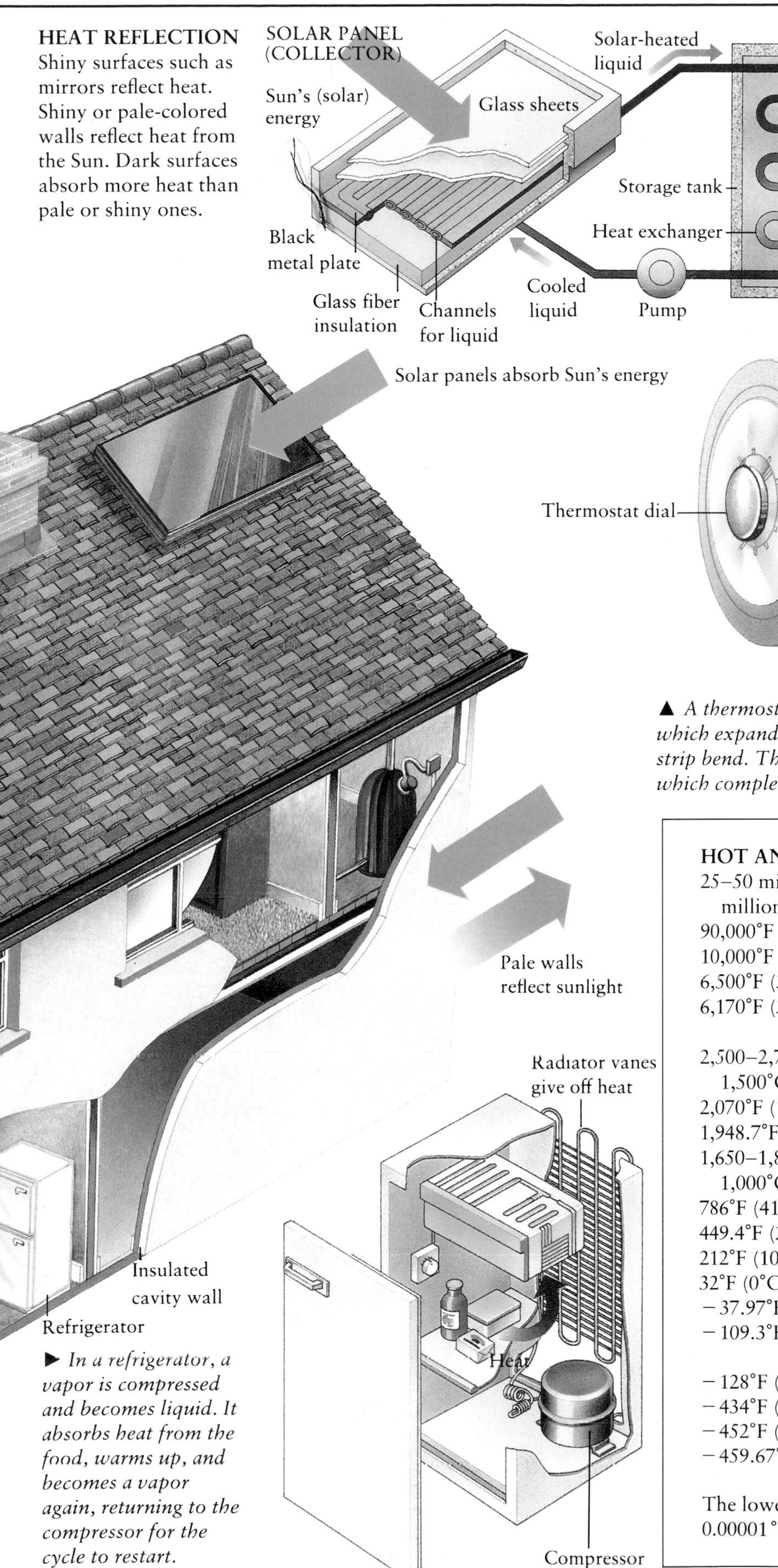

◀ *Solar panels absorb energy from the Sun. The collector is on the roof. Inside is a metal plate that absorbs sunlight. The hot plate heats a liquid in the collector which flows through a heat exchanger to heat the domestic water system.*

▶ *In a refrigerator, a vapor is compressed and becomes liquid. It absorbs heat from the food, warms up, and becomes a vapor again, returning to the compressor for the cycle to restart.*

Tube of mercury

Thermostat dial

Strip of two metals

▲ *A thermostat has a strip made from two different metals, which expand at different rates when heated, making the strip bend. This tilts a tube containing a blob of mercury, which completes the electrical circuit, switching it on or off.*

HOT AND COLD

25–50 million °F (15–30 million °C)	inside Sun
90,000°F (50,000°C)	hydrogen bomb explosion
10,000°F (5,500°C)	temperature at Sun's surface
6,500°F (3,500°C)	carbon sublimes (solid to gas)
6,170°F (3,410°C)	highest melting point of any metal (tungsten)
2,500–2,700°F (1,400–1,500°C)	iron and steel melt
2,070°F (1,132°C)	uranium melts
1,948.7°F (1,064.8°C)	gold melts
1,650–1,830°F (900–1,000°C)	brass melts
786°F (419°C)	zinc melts
449.4°F (231.9°C)	tin melts
212°F (100°C)	water boils
32°F (0°C)	water freezes
−37.97°F (−38.87°C)	mercury melts
−109.3°F (−78.5°C)	carbon dioxide gas becomes solid; "dry ice"
−128°F (−89°C)	coldest place on Earth
−434°F (−259°C)	solid hydrogen melts
−452°F (−268.9°C)	helium gas liquefies
−459.67°F (−273.15°C)	absolute zero

The lowest temperature so far recorded is about 0.00001 °C above absolute zero.

Fuels

Fuels are materials with energy stored in them. Burning the fuel releases the energy as heat. This heat is used to drive machines. Burning (combustion) is a chemical action: carbon and hydrogen in the fuel combine with oxygen in the air, releasing energy as heat and light. The fuels we burn include coal, oil, gas, and wood. Their energy was originally absorbed from the Sun. Nuclear reactors use a different fuel—by splitting atoms of unstable substances such as uranium and plutonium, energy is released.

FACTS ABOUT FUELS

- Gasoline is a blend of chemical compounds made up of hydrogen and carbon, called hydrocarbons. Different gasoline-based fuels contain different blends of hydrocarbons.
- The barrel is the unit used to measure crude oil. A barrel = 42 gallons (159 liters).
- Future oil supplies may come from oil shale, rock layers containing a substance called kerogen that can be turned into oil.
- A coal seam may be as thin as 1 inch (25 mm) or more than 300 ft. (100 m) thick.
- In countries such as Scotland and Ireland, people often use the natural resource of peat as fuel. Peat is decayed plant matter.
- The first people to mine coal were the Chinese, who were digging out surface deposits of coal almost 2,000 years ago.
- Coal can be made into liquid fuel by combining it with oil and hydrogen gas at high temperature and pressure.

▼ *Until the 1950s many millions of people, particularly in Europe, burned coal on household fires. Coal is no longer such an important domestic fuel. Most of the world's coal is now burned in power plants that generate electricity.*

▼ *Coal mining is a worldwide industry. Most coal is either dug from close to the surface (drift or open-pit mining) or brought up from deep seams, or layers, hundreds of feet under the ground.*

▼ *An offshore oil platform. Oil and gas are found beneath the desert, under ice, and beneath the ocean floor. Oil reserves may last for only 30 years, gas for 40 to 50 years.*

▲ *Burning waste could be an alternative energy source. A power station in California generates electricity by burning old tires. The station's fuel supply is 40 million tires, enough to provide heat to drive the generators for ten years.*

OIL, or petroleum, is found as crude oil, which is refined to produce fuels such as gasoline and diesel oil. Also the raw material for petrochemicals and plastics.

COAL occurs as lignite, bituminous coal, sub-bituminous coal, and anthracite. Bituminous is used most. Anthracite gives most heat but burns slowly.

GAS: natural gas is mostly methane but also contains butane and propane. These gases become liquid when pressurized and are usually stored in liquid form.

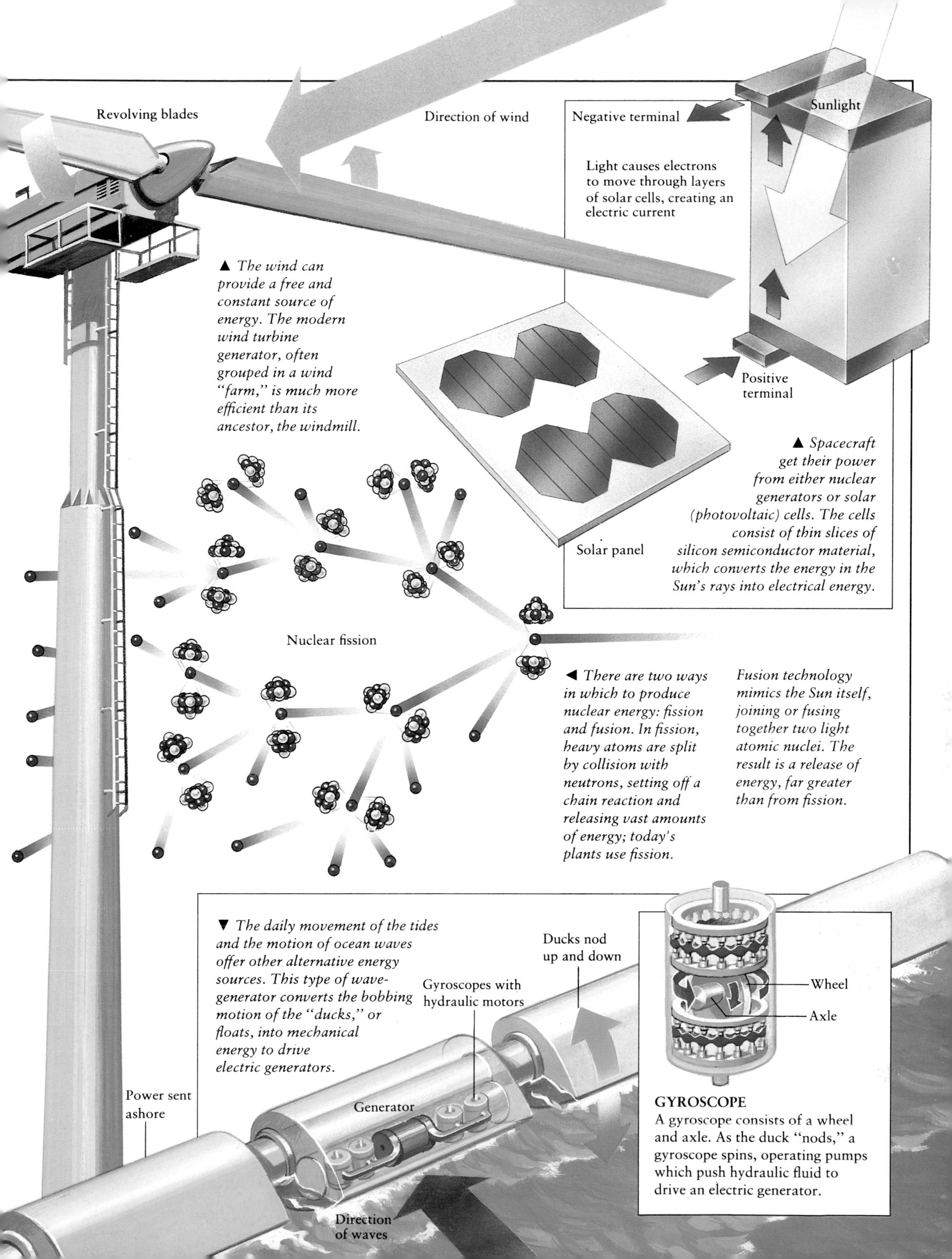

▲ *The wind can provide a free and constant source of energy. The modern wind turbine generator, often grouped in a wind "farm," is much more efficient than its ancestor, the windmill.*

▲ *Spacecraft get their power from either nuclear generators or solar (photovoltaic) cells. The cells consist of thin slices of silicon semiconductor material, which converts the energy in the Sun's rays into electrical energy.*

◀ *There are two ways in which to produce nuclear energy: fission and fusion. In fission, heavy atoms are split by collision with neutrons, setting off a chain reaction and releasing vast amounts of energy; today's plants use fission.*

Fusion technology mimics the Sun itself, joining or fusing together two light atomic nuclei. The result is a release of energy, far greater than from fission.

▼ *The daily movement of the tides and the motion of ocean waves offer other alternative energy sources. This type of wave-generator converts the bobbing motion of the "ducks," or floats, into mechanical energy to drive electric generators.*

GYROSCOPE

A gyroscope consists of a wheel and axle. As the duck "nods," a gyroscope spins, operating pumps which push hydraulic fluid to drive an electric generator.

FORCE AND MOTION

Gravity and Mass

There are four basic forces that control how everything in the universe behaves. They are gravity, electromagnetism, and two subatomic forces (strong and weak) that control the atoms of which all matter is made. Everything in the universe is either at rest or in motion. Motion is change of position, and a force is needed either to start an object moving or to stop it. A force pushes or pulls an object in a particular direction. The most obvious forces, known as mechanical forces, act directly on an object (for example, a person pushing a barrow, or a nail holding up a picture). Other forces, such as magnetism and gravity—the force that pulls objects toward the Earth—act at a distance. The force that presses evenly on a surface is pressure; the air exerts a pressure of 14.7 pounds per square inch (1 kg per sq. cm) on the Earth's surface.

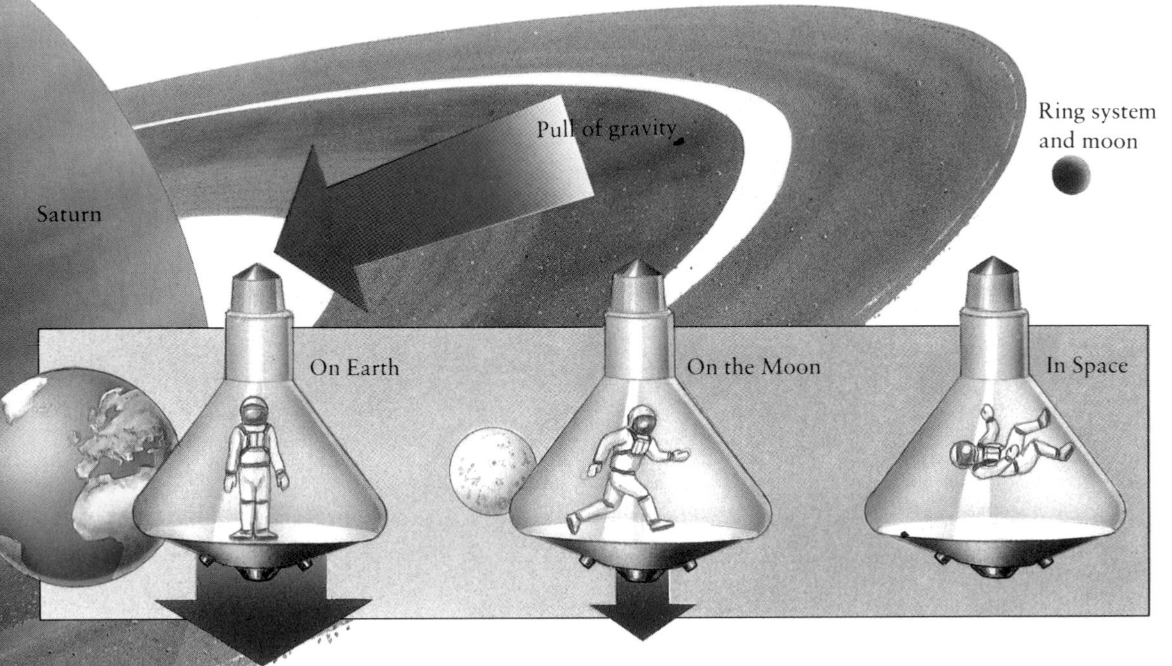

◄ *Gravity is the force that pulls everything on Earth toward the ground and makes things feel heavy. Gravity keeps the rings of Saturn in orbit around the planet.*

◄ *As they escape the influence of the Earth's gravity, astronauts become weightless. On the Moon the pull of gravity is weaker than on Earth, so an astronaut there would seem to weigh less.*

FACTS ABOUT GRAVITY AND MASS

- When the Moon and Sun pull together on the same side of the Earth, ocean tides are highest (spring tides).
- The mass of an object is the amount of material it contains. A body with greater mass has more inertia; it needs a greater force to accelerate. That is why a bus is harder to push than a car.
- Weight depends on the force of gravity, but mass does not.

► *When an object spins around another (for example, a satellite orbiting the Earth) it is pushed outward. Two forces are at work here: centrifugal (pushing outward) and centripetal (pulling inward). If you whirl a ball around you on a string, you pull it inward (centripetal force). The ball seems to pull outward (centrifugal force) and if released will fly off in a straight line.*

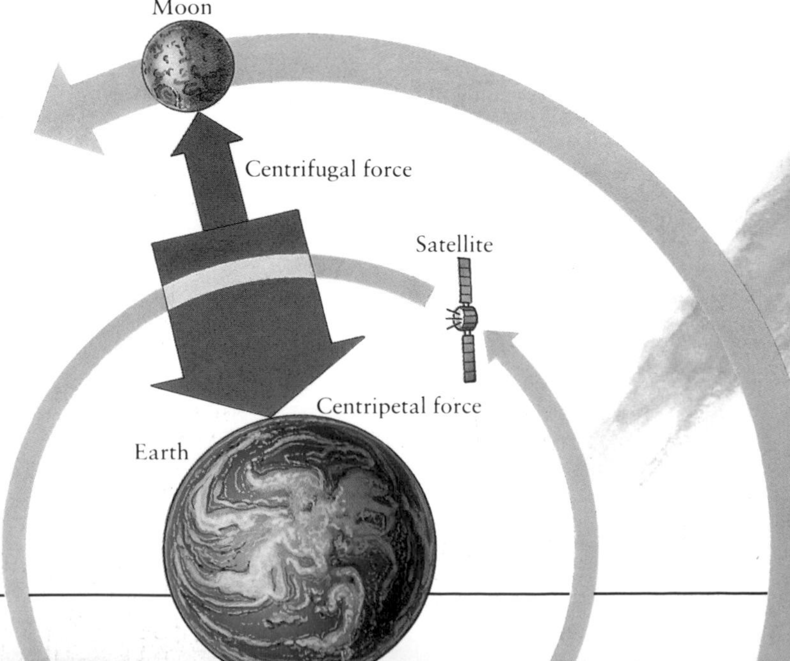

NEWTON'S LAWS

In 1687, the English scientist Isaac Newton *(left)* published three rules about force and motion. He realized that forces act on one another. His laws of motion are set out below.

1 Every object remains stopped or goes on moving at a steady rate in a straight line unless acted upon by another force. This is the inertia principle.

2 The amount of force needed to make an object change its speed depends on the mass of the object and the amount of acceleration or deceleration required.

3 To every action there is an equal and opposite reaction. When a body is pushed one way by a force, another force pushes back with equal strength.

► *Acceleration means increasing speed. Gravity produces a standard acceleration of 1G. Jet fighters and spacecraft subject pilots to greater acceleration. At 2G pilots feels twice as heavy as normal. At 5G they may pass out.*

▲ *A rocket motor is an example of Newton's third law of motion in action. The burning fuel forces gas backward (action). The spacecraft is forced in the opposite direction (reaction).*

▲ *A spacecraft accelerates to a high speed after its launch. To escape the Earth's gravity a spacecraft must reach a speed of about 25,000 mph (40,000 km/h). This is called the "escape velocity."*

► *Ocean tides are caused by the gravitational pull of the Sun and Moon on the Earth. The land is also pulled, but the water movement is more apparent.*

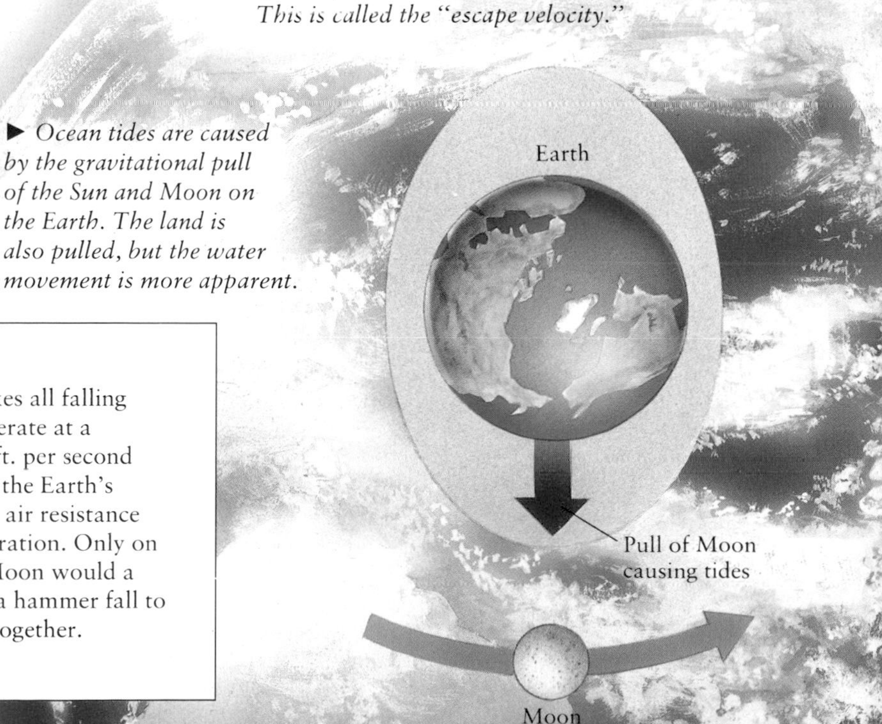

GRAVITY

Gravity makes all falling bodies accelerate at a constant 32 ft. per second (9.8 m/s). In the Earth's atmosphere, air resistance slows acceleration. Only on the airless Moon would a feather and a hammer fall to the ground together.

Motion

All motion is relative. So when a car passes a house it is not strictly true to say that only the car is moving; the house is moving too, as the Earth itself moves. The car is moving relative to the Earth. The Earth is moving relative to the Sun. To move at all, the car must overcome the forces of gravity (holding it down) and friction—the force that resists the movement of surfaces in contact. Only in space, where there is no air, is there no friction to slow down a moving object.

INERTIA
A body with a large mass is harder to start and also harder to stop. A heavy truck traveling at 30 mph (50 km/h) needs more powerful brakes to stop its motion than a car traveling at the same speed. Inertia is the tendency of an object either to stay still or to move steadily in a straight line, unless another force—such as the brick wall stopping the vehicle in the illustration below—makes it behave differently.

▼ *Wind resistance is a form of friction. Any vehicle must push aside the air in front of it. Car designers aim for a smooth, or streamlined, body shape so that airflow created by the forward motion slips easily over and around the vehicle. Streamlined vehicles use less fuel.*

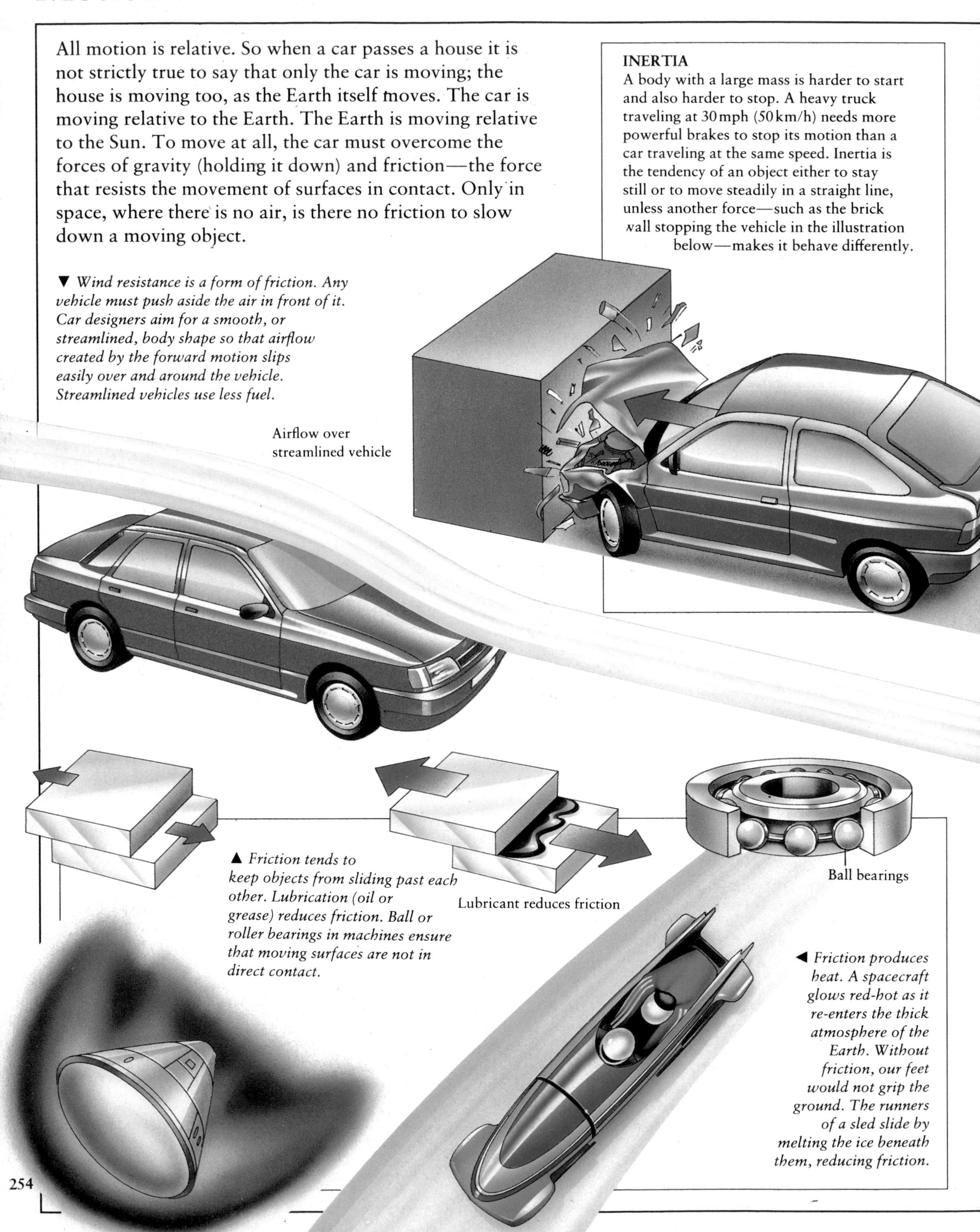

▲ *Friction tends to keep objects from sliding past each other. Lubrication (oil or grease) reduces friction. Ball or roller bearings in machines ensure that moving surfaces are not in direct contact.*

◀ *Friction produces heat. A spacecraft glows red-hot as it re-enters the thick atmosphere of the Earth. Without friction, our feet would not grip the ground. The runners of a sled slide by melting the ice beneath them, reducing friction.*

► Speed is the rate at which a moving object changes position —how far it moves in a fixed time. Velocity is speed in a particular direction. If either speed or direction changes, velocity changes too. Riders on a fairground ride travel at a constant speed, but their velocity changes continually because they keep pointing in a different direction.

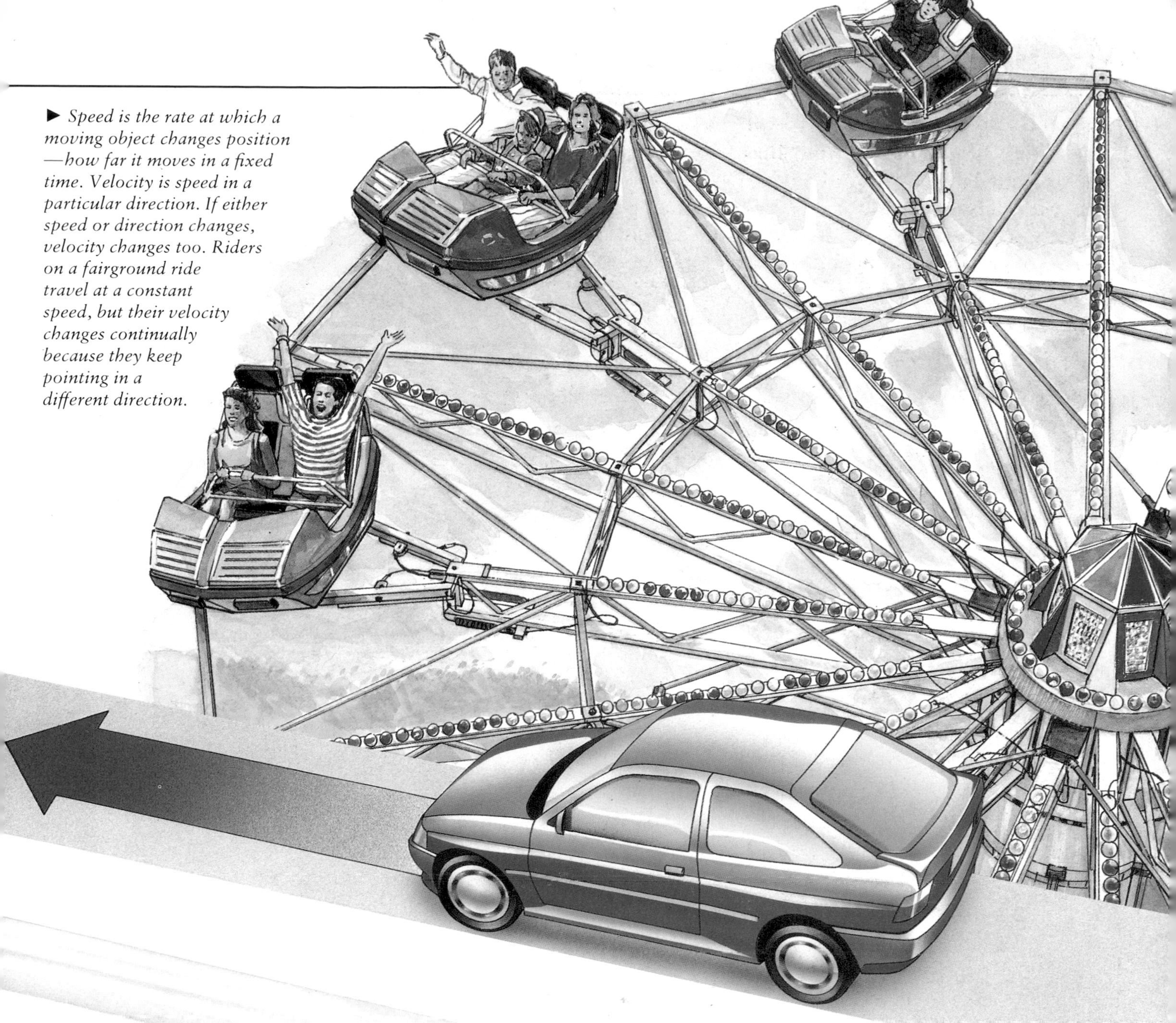

FACTS ABOUT MOTION

- The study of the effects of moving objects on gases, liquids or solids is called mechanics.
- The greatest possible velocity is the velocity of light, about 186,000 miles per second (300 million meters per second).
- A car traveling at 30 mph (50 km/h) needs about 75 ft. (23 m) to come to a stop. A car traveling at 50 mph (80 km/h) needs 175 ft (53 m) to stop. The faster car has greater momentum.
- A perpetual motion machine is supposed to run forever. So far, no one has invented one that works. All machines need energy input to keep them working and to overcome friction. Even an orbiting satellite in space (where there is no friction) is eventually brought down from orbit by the pull of the Earth's gravity.

► Momentum is equal to the mass of a moving object (its weight on Earth) multiplied by its velocity (speed and direction). You may hardly feel a softball thrown gently at you; but a ball hit hard by a softball bat moves faster and gains momentum—so you feel it! One pool ball hitting another transfers momentum to the target ball and sets it moving.

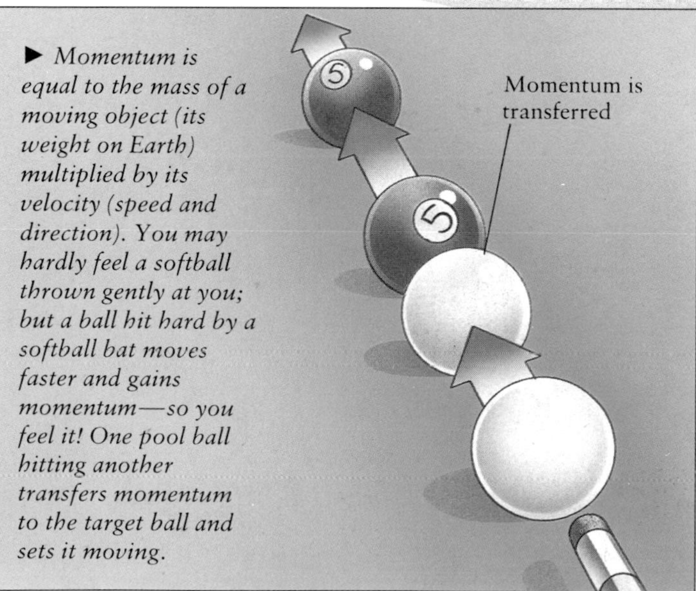

Machines

Machines are devices that make work easier, either by reducing the effort needed, or making it simpler to use effort. A machine need not have moving parts—an earth ramp is a machine, and so is a wedge used to split logs. The oldest known machines are the simplest. There are six basic kinds. The lever, the pulley, and the wheel and axle form one group; the inclined or sloping plane, the wedge, and the screw make up the other group. These were the keys that unlocked the marvels of technology.

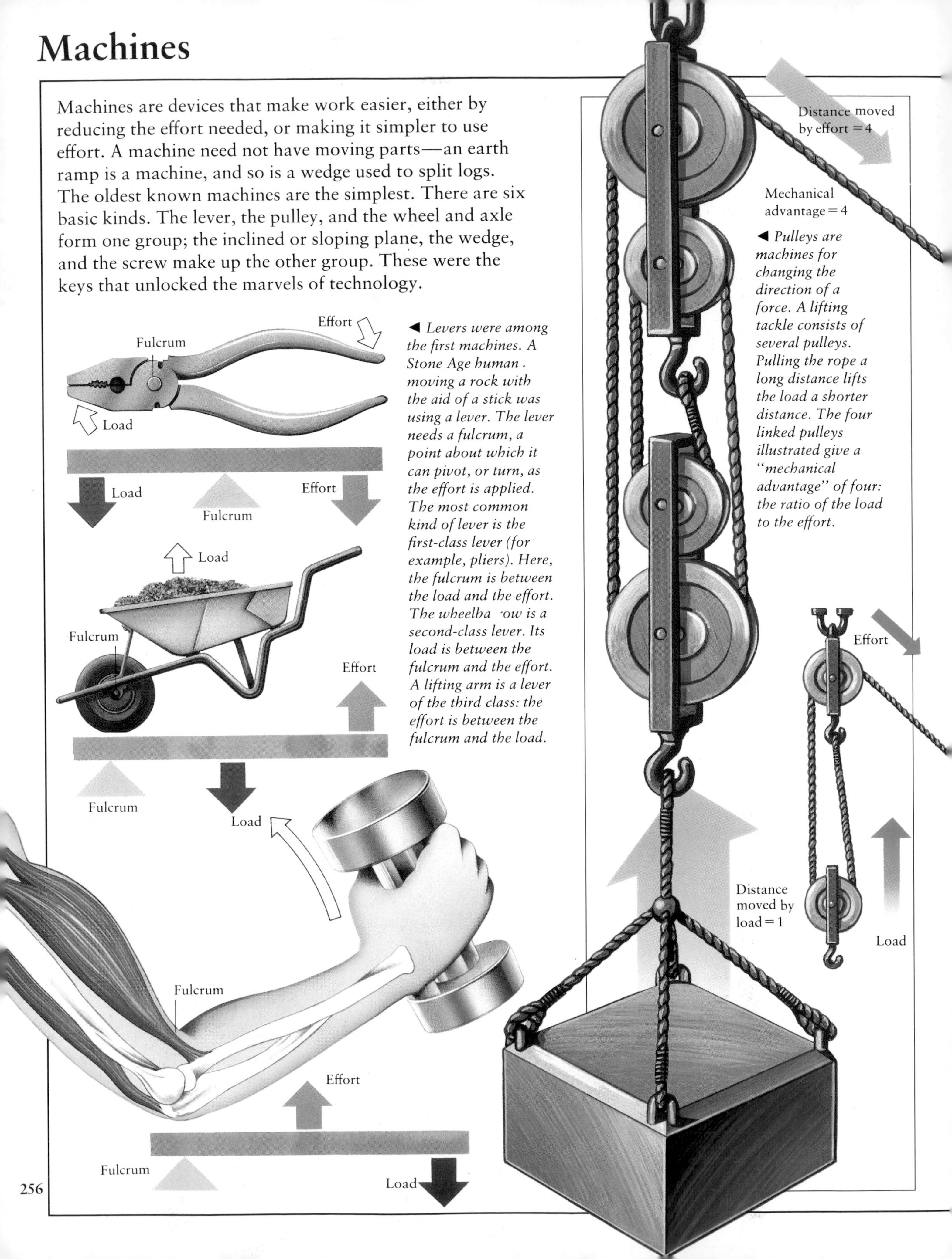

◀ *Levers were among the first machines. A Stone Age human . moving a rock with the aid of a stick was using a lever. The lever needs a fulcrum, a point about which it can pivot, or turn, as the effort is applied. The most common kind of lever is the first-class lever (for example, pliers). Here, the fulcrum is between the load and the effort. The wheelba ·ow is a second-class lever. Its load is between the fulcrum and the effort. A lifting arm is a lever of the third class: the effort is between the fulcrum and the load.*

◀ *Pulleys are machines for changing the direction of a force. A lifting tackle consists of several pulleys. Pulling the rope a long distance lifts the load a shorter distance. The four linked pulleys illustrated give a "mechanical advantage" of four: the ratio of the load to the effort.*

INTERNAL COMBUSTION ENGINE

Typical four-stroke cycle:
1 Induction: piston moves down and fuel mixture is drawn in.
2 Compression: piston moves up
3 Power: combustion forces piston down
4 Exhaust: piston goes up, gases expelled

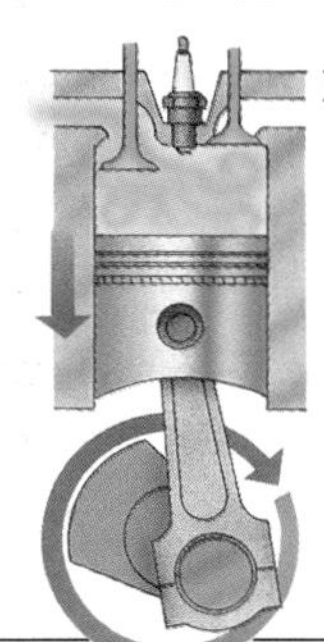

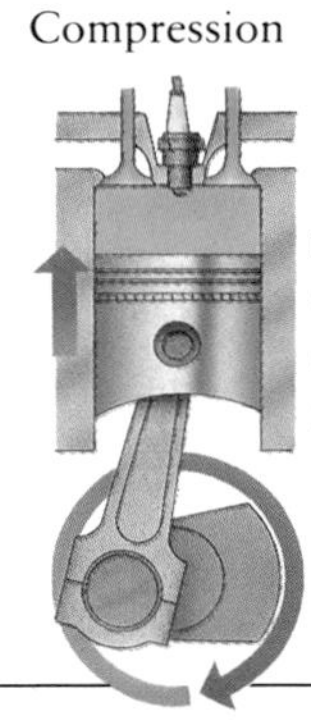

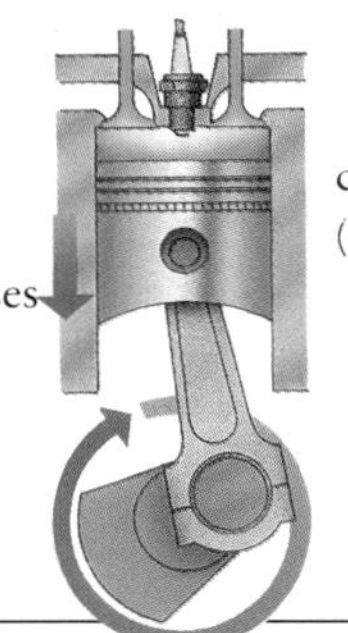

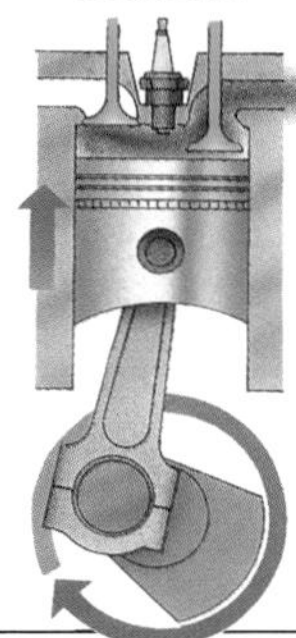

Gearbox

GEARS

Machines have gears (toothed wheels) to make one shaft turn another. Gears can change the direction of movement, or alter the speed and power of a machine. A car has a main gearbox and differential gearing (*see page 293*).

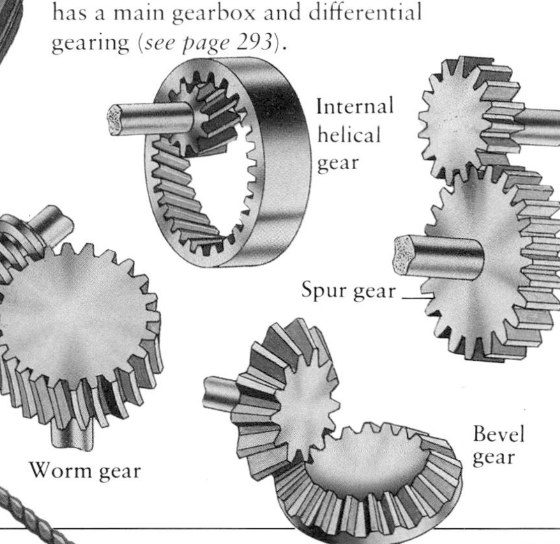

► *Hydraulic machines, such as a forklift truck, use liquid pressure to transmit power, because the liquid cannot be compressed. Inside a hydraulic jack, a large movement of the smaller piston causes a small movement of the large piston.*

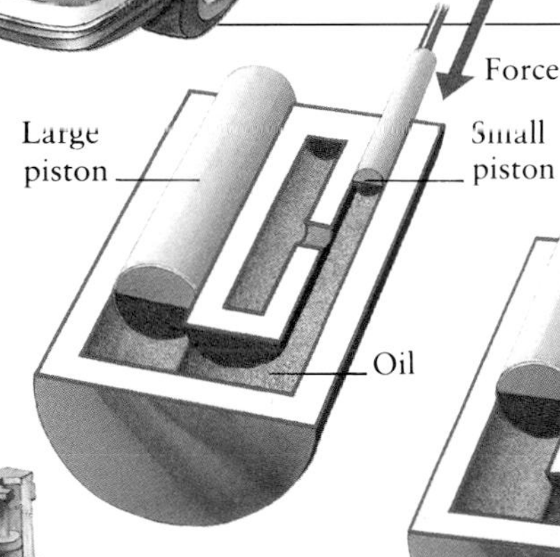

HYDRAULICS

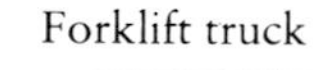

Forklift truck

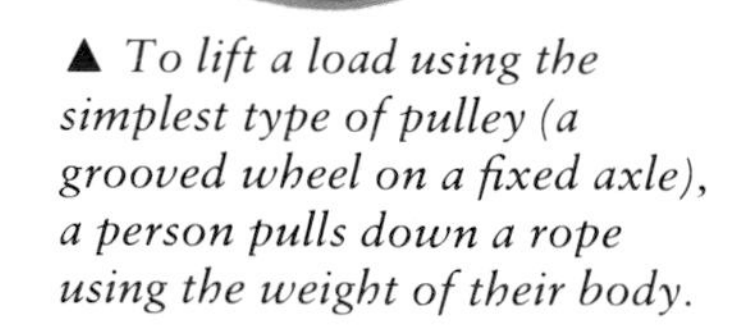

▲ *To lift a load using the simplest type of pulley (a grooved wheel on a fixed axle), a person pulls down a rope using the weight of their body.*

Flying and Floating

A balloon rises if the gas inside it is less dense than the air around it. A bird and an airliner are both heavier than air. To fly, they produce an upward force, called lift, to overcome their weight. They also need to generate a forward thrust to overcome the drag or resistance of the air. A ship floats in water because (being hollow and filled with air) it has a lower density than the water. Any floating object displaces its own weight of the fluid it is floating in.

FACTS ABOUT FLYING

- Bernoulli's principle (the faster air flows, the lower its pressure) explains why an airfoil shape produces lift.
- Drag causes problems. To reduce drag, fast jets are smooth-shaped.
- Jet planes fly faster than propeller planes. Propellers do not work properly at speeds above about 500 mph (800 km/h).

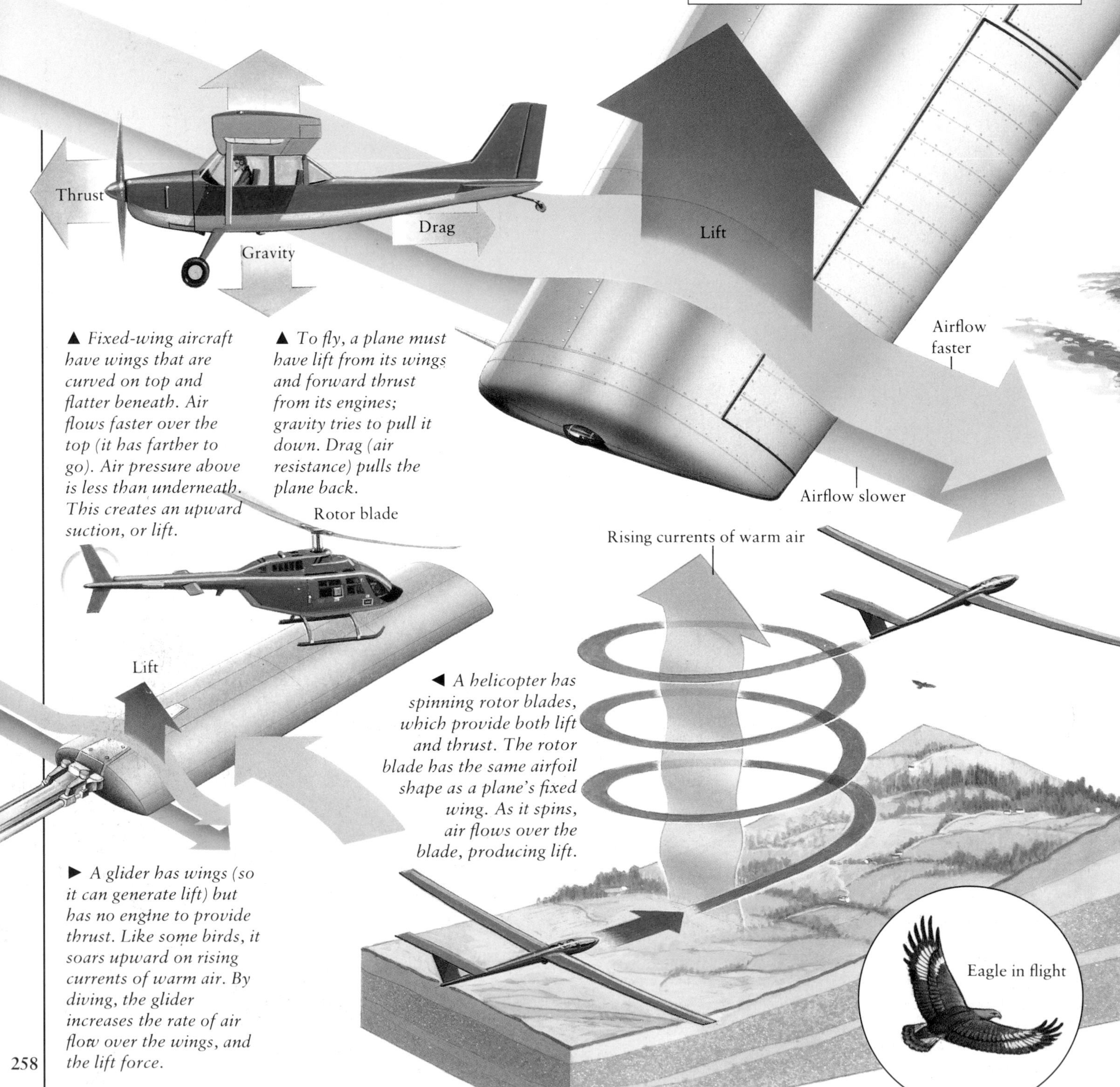

▲ *Fixed-wing aircraft have wings that are curved on top and flatter beneath. Air flows faster over the top (it has farther to go). Air pressure above is less than underneath. This creates an upward suction, or lift.*

▲ *To fly, a plane must have lift from its wings and forward thrust from its engines; gravity tries to pull it down. Drag (air resistance) pulls the plane back.*

◄ *A helicopter has spinning rotor blades, which provide both lift and thrust. The rotor blade has the same airfoil shape as a plane's fixed wing. As it spins, air flows over the blade, producing lift.*

► *A glider has wings (so it can generate lift) but has no engine to provide thrust. Like some birds, it soars upward on rising currents of warm air. By diving, the glider increases the rate of air flow over the wings, and the lift force.*

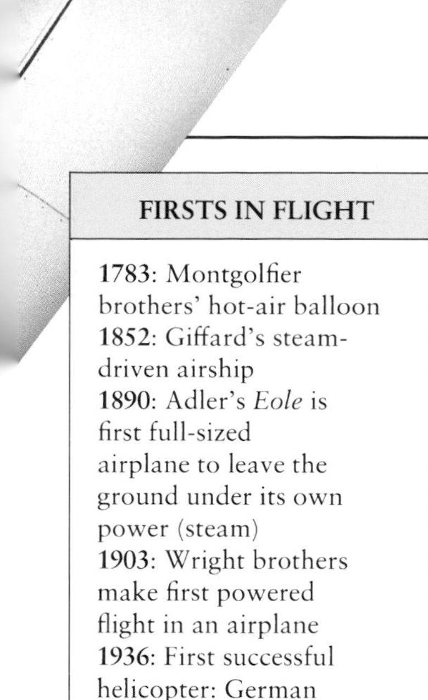

FIRSTS IN FLIGHT

1783: Montgolfier brothers' hot-air balloon
1852: Giffard's steam-driven airship
1890: Adler's *Eole* is first full-sized airplane to leave the ground under its own power (steam)
1903: Wright brothers make first powered flight in an airplane
1936: First successful helicopter: German Focke-Wulf 61
1939: First jet plane: German Heinkel He 178

▲ *Airships are power-driven balloons filled with a lighter-than-air gas such as helium. They are slow but can lift considerable loads.*

► *There are two kinds of balloon: filled with a lighter-than-air gas, or with hot air. In a hot-air balloon, a gas burner warms the air inside the balloon. The hot air is less dense than the surrounding air, so the balloon rises.*

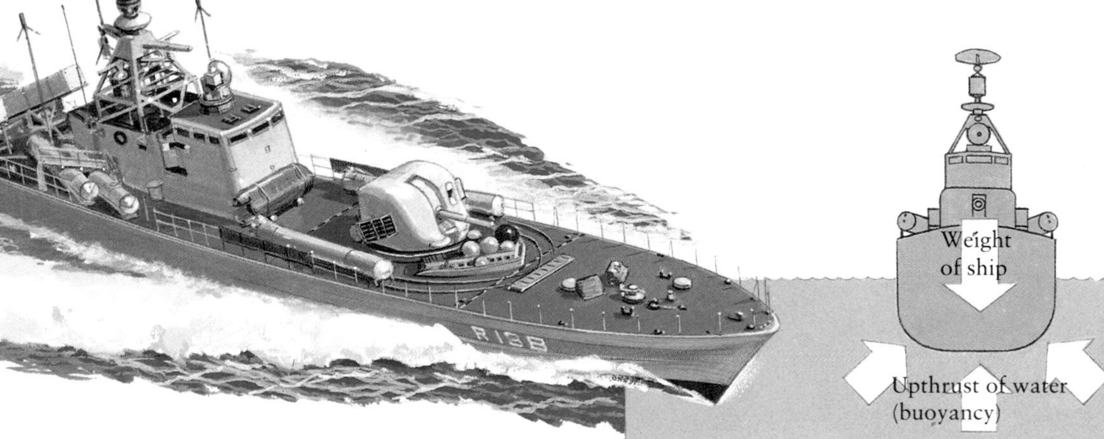

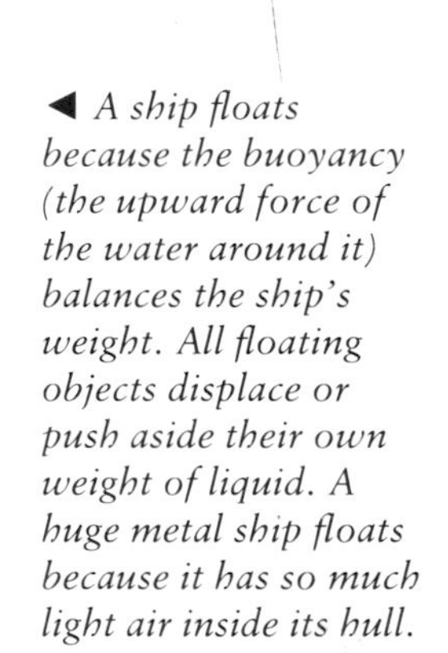

◄ *A ship floats because the buoyancy (the upward force of the water around it) balances the ship's weight. All floating objects displace or push aside their own weight of liquid. A huge metal ship floats because it has so much light air inside its hull.*

Submarine on surface

Compressed air blown into tanks

Submarine rises

Water let into tanks; air pushed out

Submarine sinks

Submarine on sea floor

► *A submarine on the surface maintains positive buoyancy with air-filled tanks. To dive, water is let into the tanks, pushing out the air. As the tanks fill with water, the submarine becomes heavier. It becomes neutrally buoyant and is able to stay under water. To surface, compressed air is blown into the tanks. The air forces the water out through valves, and the submarine is able to rise to the surface.*

SPACE AND TIME

Counting and Measuring

Human curiosity about space and time began when our earliest human ancestors gazed at the stars and wondered about the daily miracle of sunrise. Cave paintings made by prehistoric hunters show animals hit many times by spears. The artist knew the difference between "one" and "many." Keeping count of things became important when people became farmers. From counting on fingers and toes, people progressed to tally-sticks (sticks with notches cut in them) and eventually to number systems, with which people could record measurements and calculate. The numbers we use today came from the Arabs, who in turn developed them from the Hindus of India. The system for writing numbers that we take for granted, based on powers of 10, is known as the base 10 or decimal system and did not become widespread in Europe until after about A.D. 1100.

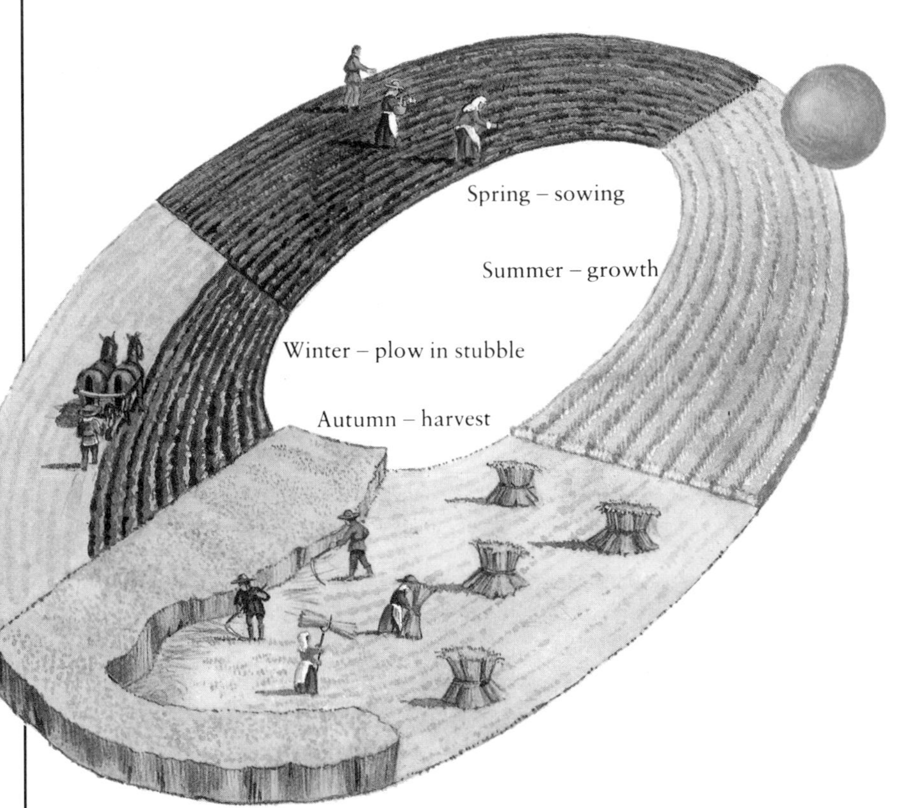

◀ *Early farmers recorded the changing seasons, and in this way made the first calendars, based on either the seasons or the phases of the Moon.*

DEVELOPMENTS IN COUNTING
4000 B.C.: Babylonians wrote numbers from left to right, grouping them in tens
Before 3000 B.C.: Egyptians used measuring rods and balance scales for weighing things
1400 B.C.: Decimals used in China
300 B.C.: Euclid's *Elements* summarized Greek geometry
A.D. 600: Zero (0) used in China and India
700: First mechanical clocks in China
1500s: Negative (−) numbers first used in Europe
1789: France adopted the metric system
1960: Modern SI units adopted

▼ *The Babylonians (who lived in what is now Iraq) used these standard weights and notched tally-sticks (for counting). As trade developed, people needed accurate measurements to make sure that buyer and seller agreed. Surprisingly accurate scales were used for weighing.*

MEASURING
Over 5,000 years ago, the builders of the pyramids in Egypt had to measure length, to know how many stones they needed, and how to drive shafts accurately through the huge structures. The Egyptians also used delicate weighing scales, to weigh gold and precious stones.

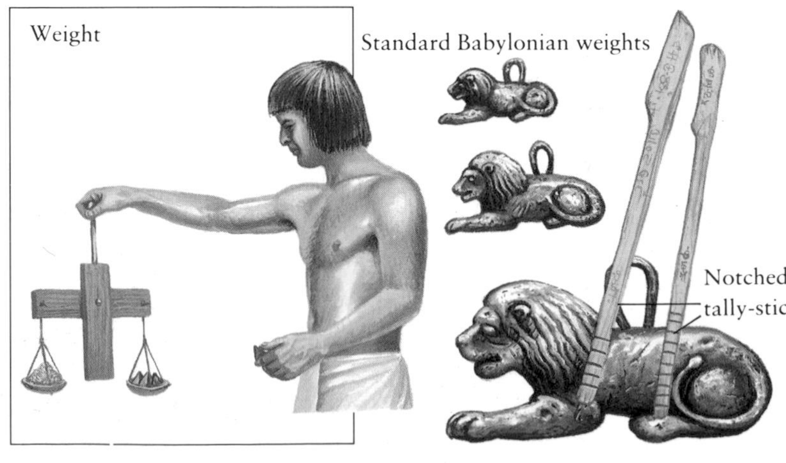

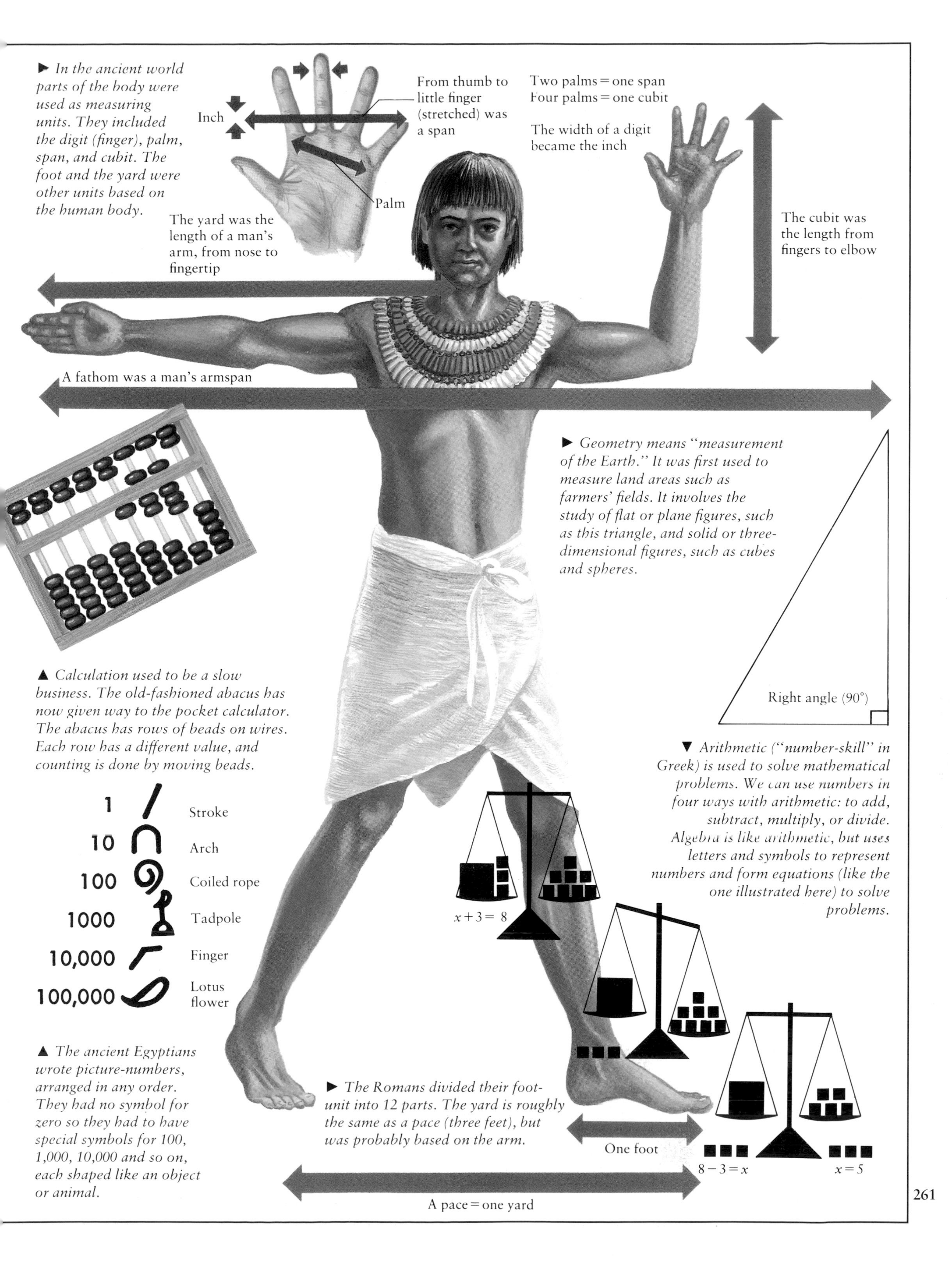

► *In the ancient world parts of the body were used as measuring units. They included the digit (finger), palm, span, and cubit. The foot and the yard were other units based on the human body.*

► *Geometry means "measurement of the Earth." It was first used to measure land areas such as farmers' fields. It involves the study of flat or plane figures, such as this triangle, and solid or three-dimensional figures, such as cubes and spheres.*

▲ *Calculation used to be a slow business. The old-fashioned abacus has now given way to the pocket calculator. The abacus has rows of beads on wires. Each row has a different value, and counting is done by moving beads.*

▼ *Arithmetic ("number-skill" in Greek) is used to solve mathematical problems. We can use numbers in four ways with arithmetic: to add, subtract, multiply, or divide. Algebra is like arithmetic, but uses letters and symbols to represent numbers and form equations (like the one illustrated here) to solve problems.*

▲ *The ancient Egyptians wrote picture-numbers, arranged in any order. They had no symbol for zero so they had to have special symbols for 100, 1,000, 10,000 and so on, each shaped like an object or animal.*

► *The Romans divided their foot-unit into 12 parts. The yard is roughly the same as a pace (three feet), but was probably based on the arm.*

Mathematics

People have found mathematics useful for thousands of years, ever since the Egyptians and Babylonians used it to figure out a calendar (as a guide to planting crops). The peoples of Asia were the first masters of mathematics, developing key branches of mathematics such as arithmetic, geometry, and algebra. "Pure" math has to do with theory; "applied" math deals with problems in engineering, science, business, or any other activity when we need to calculate, measure, and predict.

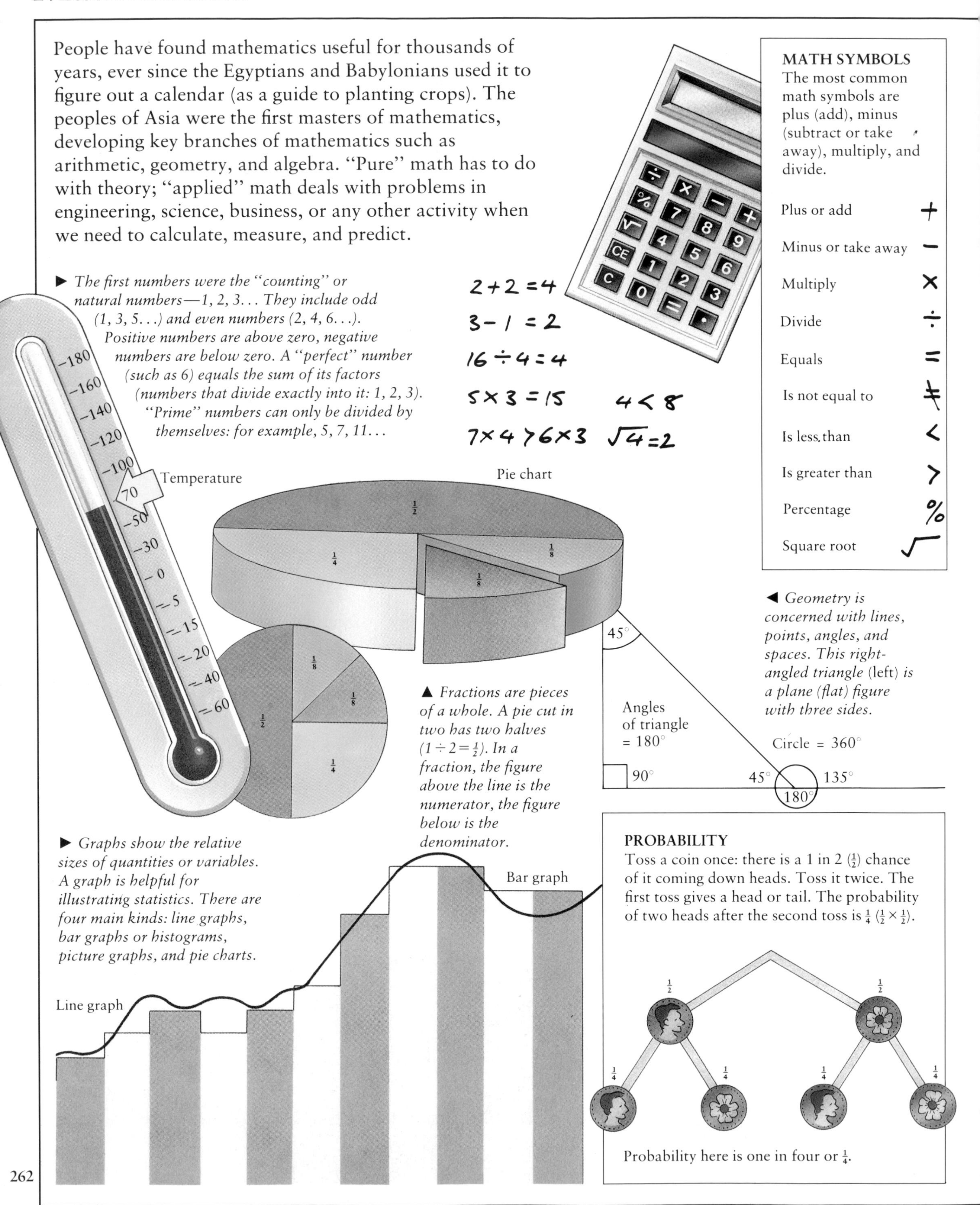

MATH SYMBOLS

The most common math symbols are plus (add), minus (subtract or take away), multiply, and divide.

Plus or add	+
Minus or take away	−
Multiply	×
Divide	÷
Equals	=
Is not equal to	≠
Is less than	<
Is greater than	>
Percentage	%
Square root	√

► *The first numbers were the "counting" or natural numbers—1, 2, 3... They include odd (1, 3, 5...) and even numbers (2, 4, 6...). Positive numbers are above zero, negative numbers are below zero. A "perfect" number (such as 6) equals the sum of its factors (numbers that divide exactly into it: 1, 2, 3). "Prime" numbers can only be divided by themselves: for example, 5, 7, 11...*

▲ *Fractions are pieces of a whole. A pie cut in two has two halves ($1 \div 2 = \frac{1}{2}$). In a fraction, the figure above the line is the numerator, the figure below is the denominator.*

◄ *Geometry is concerned with lines, points, angles, and spaces. This right-angled triangle (left) is a plane (flat) figure with three sides.*

► *Graphs show the relative sizes of quantities or variables. A graph is helpful for illustrating statistics. There are four main kinds: line graphs, bar graphs or histograms, picture graphs, and pie charts.*

PROBABILITY

Toss a coin once: there is a 1 in 2 ($\frac{1}{2}$) chance of it coming down heads. Toss it twice. The first toss gives a head or tail. The probability of two heads after the second toss is $\frac{1}{4}$ ($\frac{1}{2} \times \frac{1}{2}$).

Probability here is one in four or $\frac{1}{4}$.

THE FIBONACCI SERIES

The greatest medieval mathematician, Fibonacci of Pisa, Italy, helped introduce Arabic-Hindu numerals into Europe. He discovered a special sequence of numbers, beginning: 0+1=1, 1+1=2, 1+2=3, 2+3=5, 3+5=8, which gives the sequence: 0 1 1 2 3 5 8 13 21 34 55 89 144... and so on (each number is the sum of the previous two).

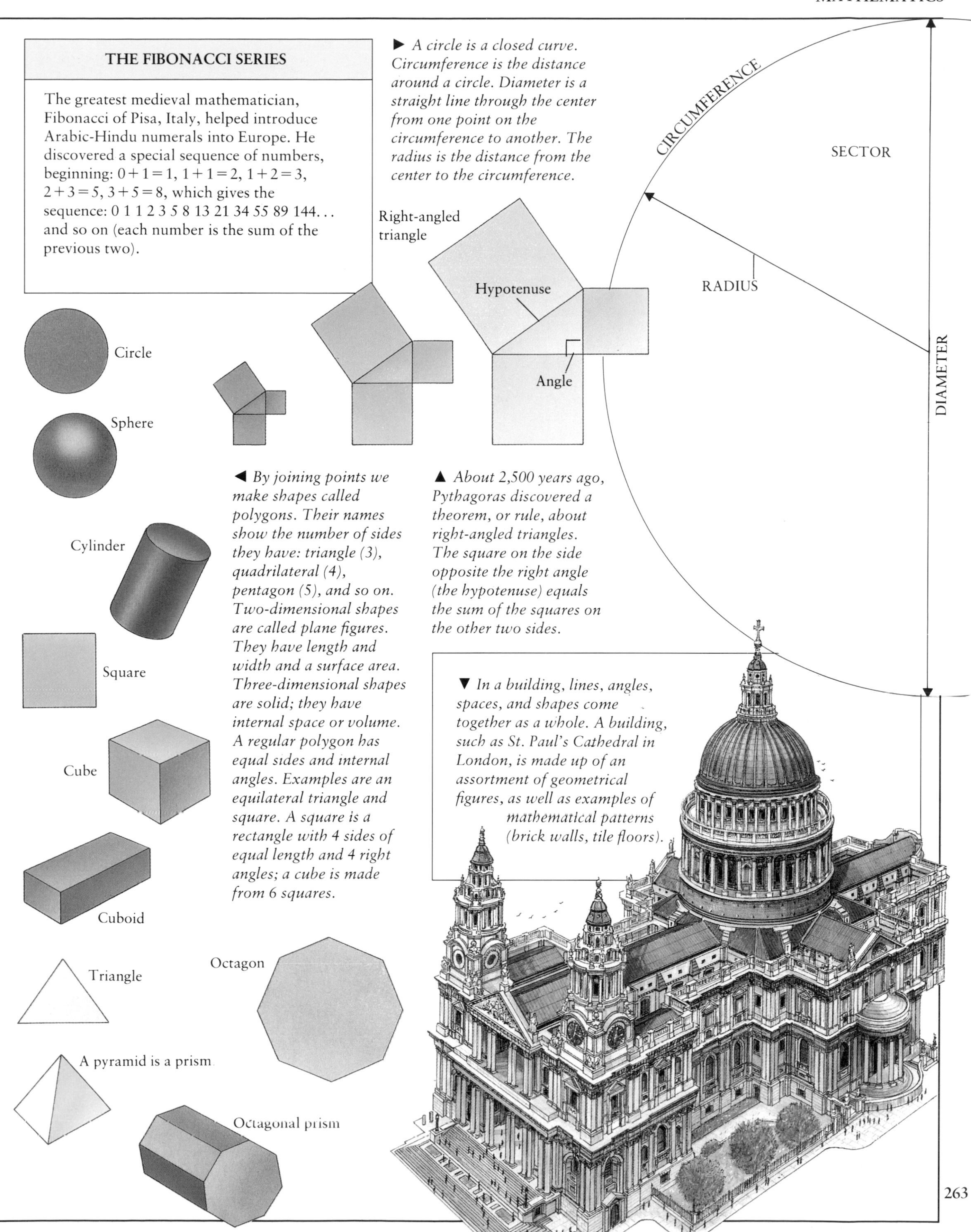

► *A circle is a closed curve. Circumference is the distance around a circle. Diameter is a straight line through the center from one point on the circumference to another. The radius is the distance from the center to the circumference.*

◄ *By joining points we make shapes called polygons. Their names show the number of sides they have: triangle (3), quadrilateral (4), pentagon (5), and so on. Two-dimensional shapes are called plane figures. They have length and width and a surface area. Three-dimensional shapes are solid; they have internal space or volume. A regular polygon has equal sides and internal angles. Examples are an equilateral triangle and square. A square is a rectangle with 4 sides of equal length and 4 right angles; a cube is made from 6 squares.*

▲ *About 2,500 years ago, Pythagoras discovered a theorem, or rule, about right-angled triangles. The square on the side opposite the right angle (the hypotenuse) equals the sum of the squares on the other two sides.*

▼ *In a building, lines, angles, spaces, and shapes come together as a whole. A building, such as St. Paul's Cathedral in London, is made up of an assortment of geometrical figures, as well as examples of mathematical patterns (brick walls, tile floors).*

Measuring Time

Long ago, people counted the days from sunrise to sunset, either as "suns" or "nights." They measured the phases of the Sun, and the Babylonians figured out that in a year there were twelve equal parts. The Babylonians and Egyptians then calculated the year's length at 365 days and 6 hours, astonishingly accurate, since it is in fact 365 days 6 hours 41 minutes and 59 seconds! The first clocks were not very precise. However, the latest atomic clock is accurate to a second in 1.6 million years.

CALENDAR CHANGES
In 46 B.C., the Roman ruler Julius Caesar ordered a 366-day leap year every fourth year to correct previous calendar errors. By the 1500s this too was wrong. So in 1582 Pope Gregory XIII decreed a new calendar in which 11 days were "lost." The Gregorian calendar (adopted by England in 1752) is the calendar we use today.

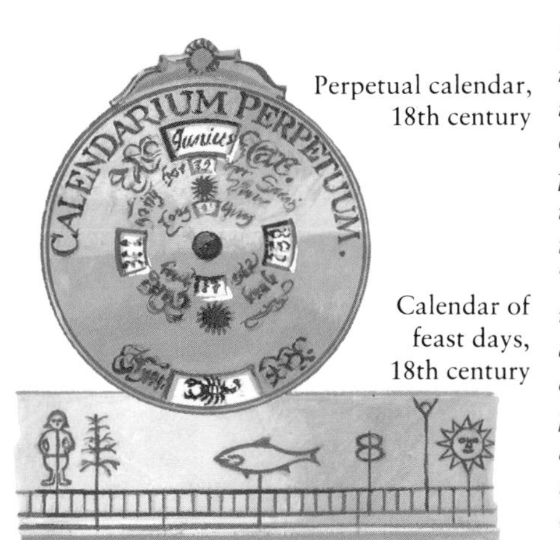

Perpetual calendar, 18th century

Calendar of feast days, 18th century

▲ *Calendars are systems by which the length of the year is fixed. Perpetual calendars show the day of the week for any year that may be wanted.*

► *A solar day starts at midnight and lasts for the time it takes the Earth to spin once on its axis. The time when our part of the Earth faces the Sun we call "day"; the time it is turned away from the Sun is "night." The length of day and night varies; at the North Pole a midwinter day is 24 hours of darkness. When the North Pole faces the Sun the South Pole is always dark. A month is the time it takes for the Moon to orbit the Earth. A year is the time it takes the Earth to travel around the Sun.*

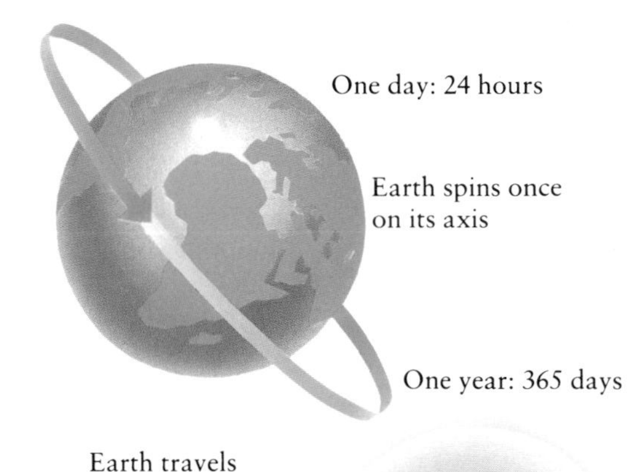

One day: 24 hours

Earth spins once on its axis

One year: 365 days

Earth travels once around the Sun

▼ *The interval between one new Moon and the next is 29.5 days. The new Moon waxes (gets larger) and halfway through its cycle, we see the whole face of the Moon lit by sunlight.*

Full Moon

Gibbous

Last quarter

New Moon

First quarter

Waxing gibbous

Full Moon

Phases of the Moon

One lunar month: average 29.5 days

Moon travels once around the Earth

▼ *Candle clocks were common in the Middle Ages. The candle was marked down its side to show the passing of the hours.*

▼ *In a water clock of ancient Egypt, water dripped slowly from one vessel to another. Some water clocks had floats fixed to pointers.*

▼ *The sundial, which uses the Sun's motion across the sky, was first used to indicate the time of day in ancient Egypt. In Babylon, where astronomers divided the circle into 360 parts, or degrees, the dial was marked into 12 hours.*

▼ *The sandglass, sometimes called an "hourglass," was a popular type of clock in the Middle Ages, before mechanical clocks were invented. Sandglasses were used to measure short periods of time.*

Water clock

Candle clock

Sundial

Sandglass

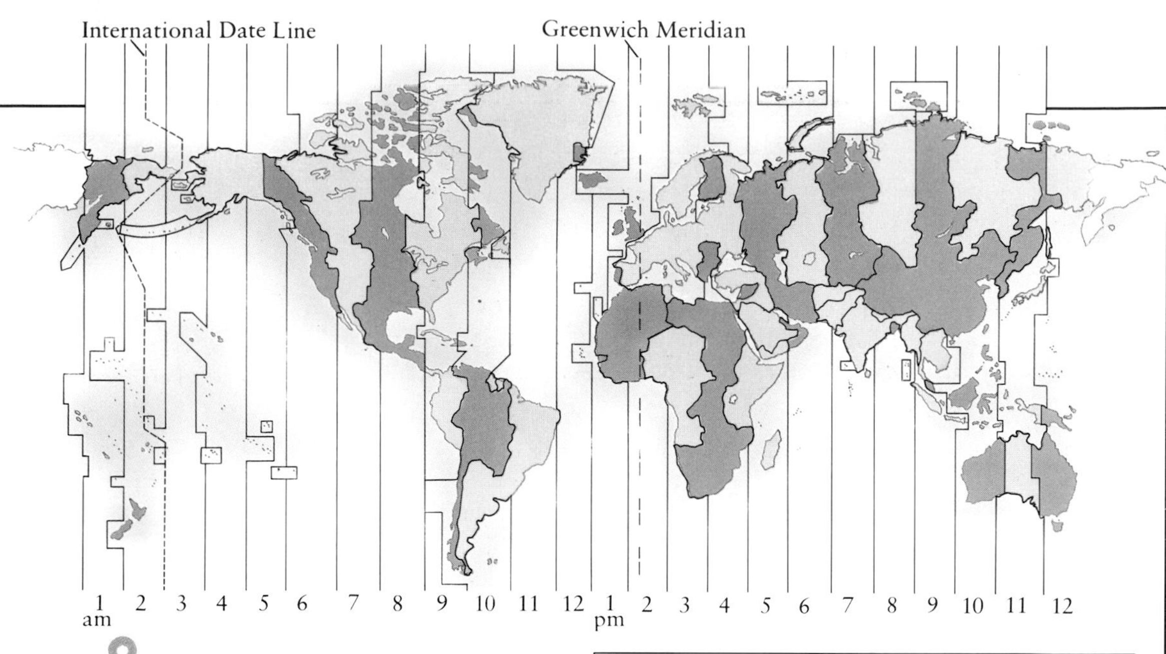

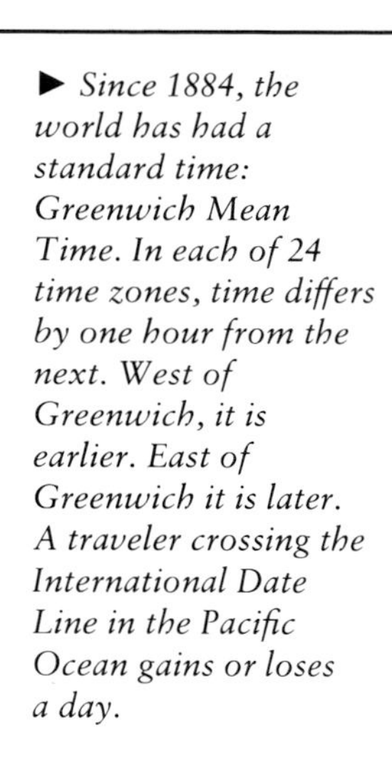

► *Since 1884, the world has had a standard time: Greenwich Mean Time. In each of 24 time zones, time differs by one hour from the next. West of Greenwich, it is earlier. East of Greenwich it is later. A traveler crossing the International Date Line in the Pacific Ocean gains or loses a day.*

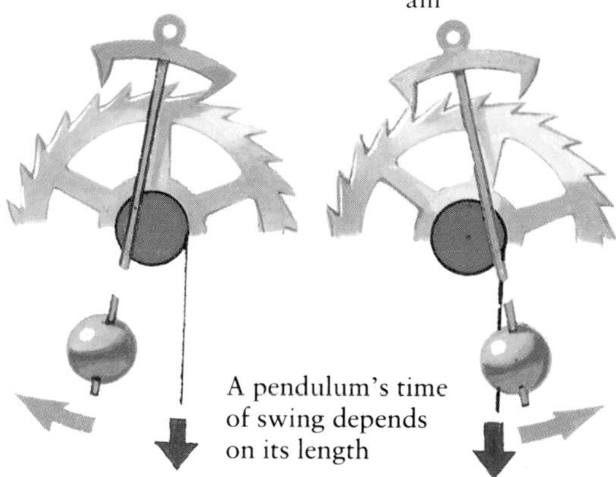

A pendulum's time of swing depends on its length

◄ *The first accurate pendulum clock was made by Christian Huygens of Holland in 1656. The pendulum takes the same time for each swing, and so regulates the clock movement.*

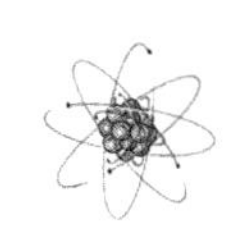

Physical time: a quantity that can be measured (like length or mass), in seconds, minutes, and hours.

Geological time: measured in billions of years, from the origin of the Solar System and beyond.

Biological time: plants and animals follow natural rhythms or cycles (wake-sleep, night-day).

NAVAL CLOCKS

A ship's navigator needs an accurate timekeeper to know where the ship is and how fast it is traveling. The Sun and stars help to find latitude (north-south position). But finding east-west (longitude) is difficult unless you know the exact time. To keep time on a ship, a chronometer or clock had to be accurate without a pendulum. In the 1700s, an English inventor, John Harrison, devised a slowly unwinding spring.

Harrison's chronometer

◄ *Without this, Cook might not have found his way to Australia.*

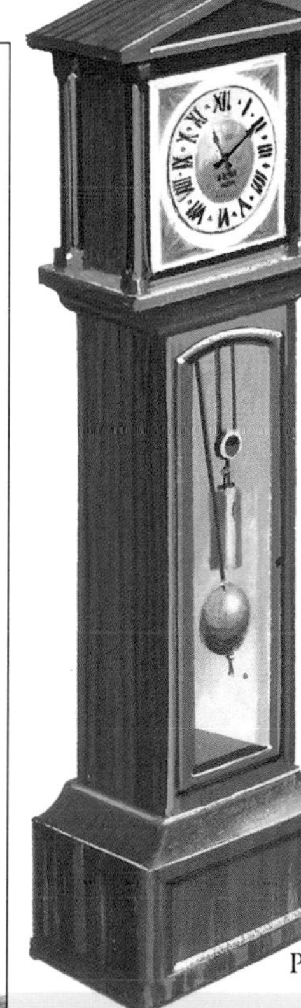

Pendulum clock

▲ *The pendulum of this 19th-century clock controls its speed.*

► *An "atomic" clock counts the vibrations of the light given off by atoms. Communications networks, navigation satellites, and astronomers rely on these very accurate clocks, which lose less than a second in a million years.*

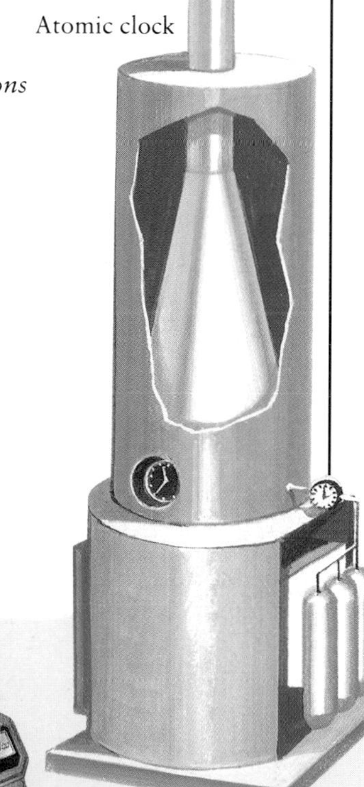

Atomic clock

▼ *A quartz watch works by recording the regular vibrations of a quartz crystal.*

Quartz watch

LIGHT AND SOUND

The Spectrum

We live in a world in which light and sound constantly convey information to our senses. Light and sound are both forms of energy that travel in waves. Light comes originally from the Sun. It can travel through space, whereas sound can only travel through a substance that has molecules which are able to move around. Sound travels far more slowly than light—between 1,083 and 1,115 ft. (330 and 340 m) a second in air. Light travels very quickly, at 186,000 miles (300,000 km) per second. Light rays are a form of electromagnetic radiation. Other rays making up the electromagnetic spectrum are invisible. At the red end of the visible spectrum are infrared rays, microwaves, radar, television, and radio waves. At the other (violet) end are ultraviolet, X-rays, gamma rays, and cosmic rays.

ELECTROMAGNETIC SPECTRUM

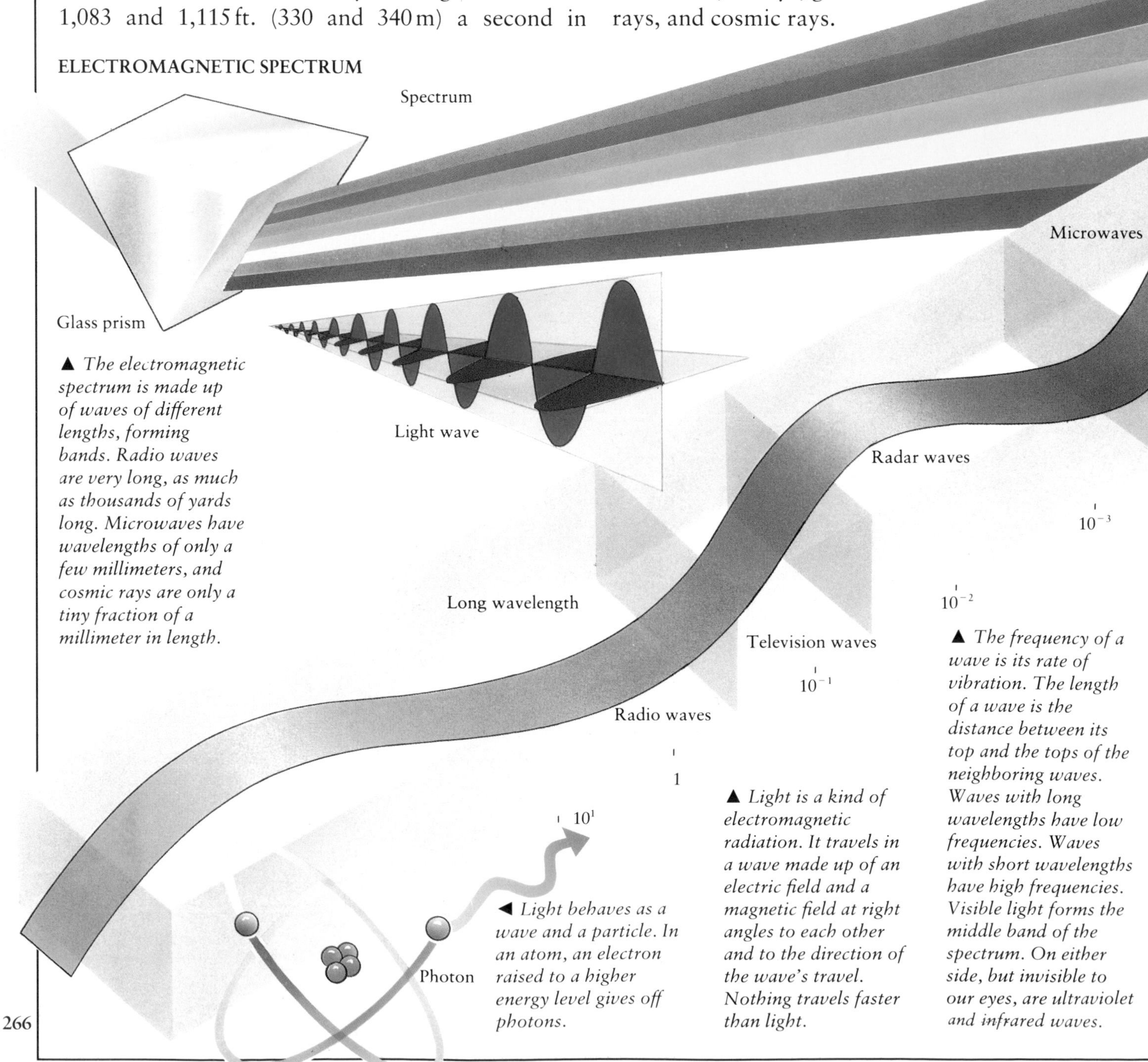

▲ *The electromagnetic spectrum is made up of waves of different lengths, forming bands. Radio waves are very long, as much as thousands of yards long. Microwaves have wavelengths of only a few millimeters, and cosmic rays are only a tiny fraction of a millimeter in length.*

◀ *Light behaves as a wave and a particle. In an atom, an electron raised to a higher energy level gives off photons.*

▲ *Light is a kind of electromagnetic radiation. It travels in a wave made up of an electric field and a magnetic field at right angles to each other and to the direction of the wave's travel. Nothing travels faster than light.*

▲ *The frequency of a wave is its rate of vibration. The length of a wave is the distance between its top and the tops of the neighboring waves. Waves with long wavelengths have low frequencies. Waves with short wavelengths have high frequencies. Visible light forms the middle band of the spectrum. On either side, but invisible to our eyes, are ultraviolet and infrared waves.*

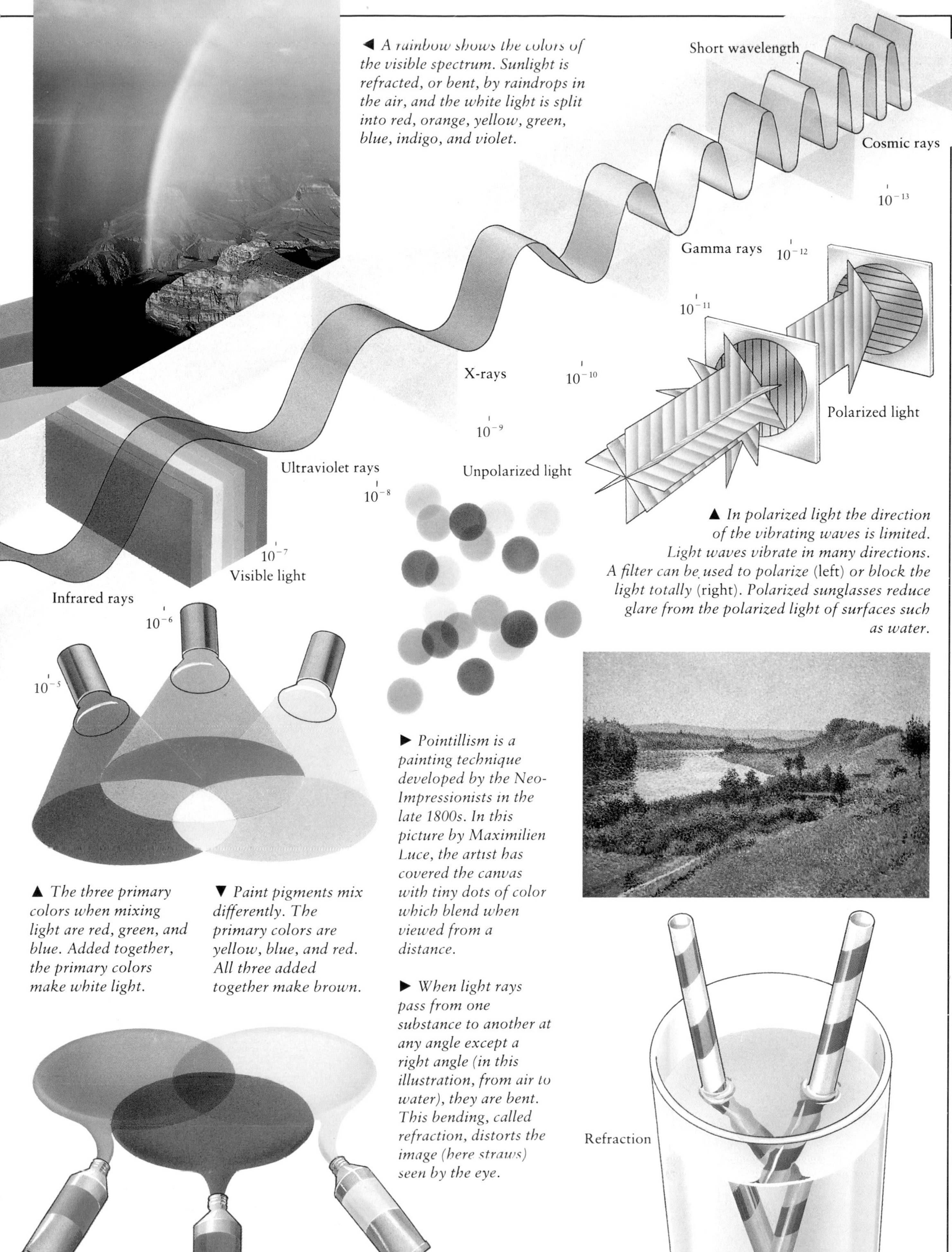

◀ *A rainbow shows the colors of the visible spectrum. Sunlight is refracted, or bent, by raindrops in the air, and the white light is split into red, orange, yellow, green, blue, indigo, and violet.*

▲ *In polarized light the direction of the vibrating waves is limited. Light waves vibrate in many directions. A filter can be used to polarize* (left) *or block the light totally* (right). *Polarized sunglasses reduce glare from the polarized light of surfaces such as water.*

▲ *The three primary colors when mixing light are red, green, and blue. Added together, the primary colors make white light.*

▼ *Paint pigments mix differently. The primary colors are yellow, blue, and red. All three added together make brown.*

▶ *Pointillism is a painting technique developed by the Neo-Impressionists in the late 1800s. In this picture by Maximilien Luce, the artist has covered the canvas with tiny dots of color which blend when viewed from a distance.*

▶ *When light rays pass from one substance to another at any angle except a right angle (in this illustration, from air to water), they are bent. This bending, called refraction, distorts the image (here straws) seen by the eye.*

Light

Light is as vital to our existence on Earth as oxygen or water. Without light we could not see. Without sunlight the Earth would be a frozen and lifeless planet. Without light, there would be no green plants. The energy in the fuels we burn came originally from sunlight. Ever since people discovered how to make fire, they have found ways to make their own light—candles, oil lamps, gas and electric lights. Today, light is used in areas as diverse as entertainment, surgery, and telecommunications.

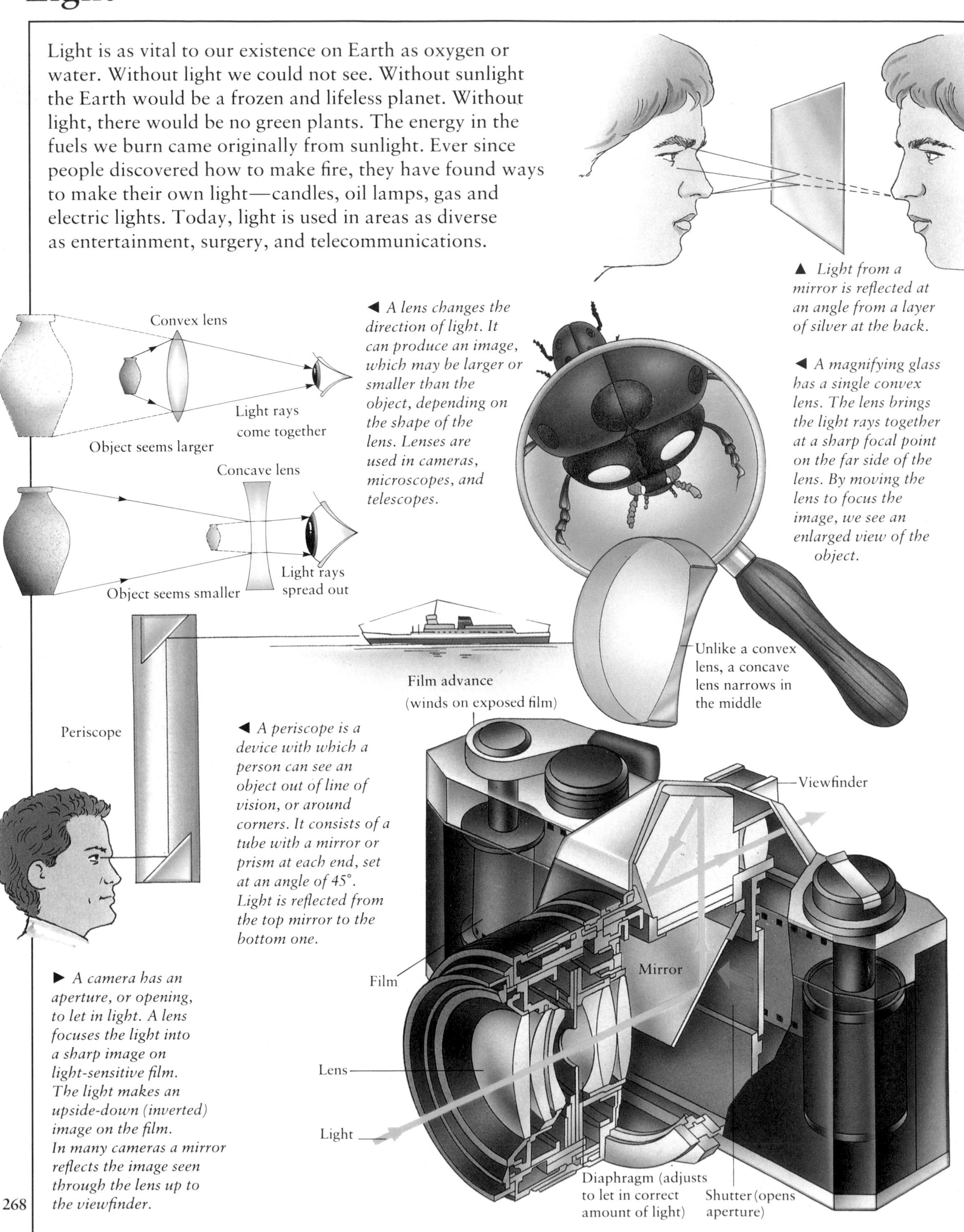

▲ *Light from a mirror is reflected at an angle from a layer of silver at the back.*

◀ *A lens changes the direction of light. It can produce an image, which may be larger or smaller than the object, depending on the shape of the lens. Lenses are used in cameras, microscopes, and telescopes.*

◀ *A magnifying glass has a single convex lens. The lens brings the light rays together at a sharp focal point on the far side of the lens. By moving the lens to focus the image, we see an enlarged view of the object.*

◀ *A periscope is a device with which a person can see an object out of line of vision, or around corners. It consists of a tube with a mirror or prism at each end, set at an angle of 45°. Light is reflected from the top mirror to the bottom one.*

▶ *A camera has an aperture, or opening, to let in light. A lens focuses the light into a sharp image on light-sensitive film. The light makes an upside-down (inverted) image on the film. In many cameras a mirror reflects the image seen through the lens up to the viewfinder.*

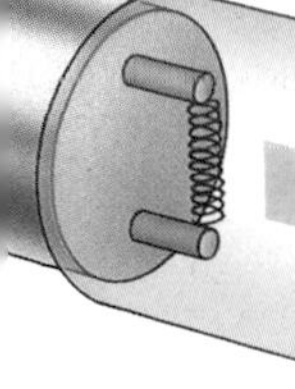

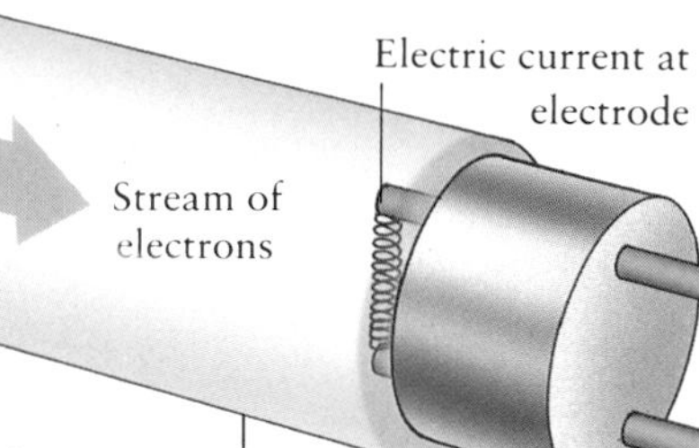

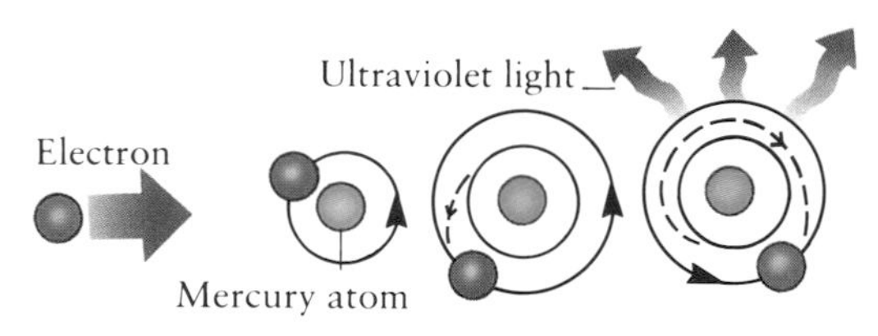

▲ *Inside a fluorescent tube is a gas (argon) plus mercury vapor; when an electric current passes through this, ultraviolet rays are released. They strike the coating of the tube, causing it to glow.*

▲ *Most objects look the color they are because of the way their structure reflects light. An object that reflects light of all wavelengths in equal amounts looks white. An object that absorbs shorter wavelengths of light but reflects longer waves looks red.*

Grass appears green because it reflects green light

▼ *An object that absorbs light of all wavelengths in equal amounts looks black. In dim light, colored objects often look gray to the human eye.*

LASER LIGHT

A laser beam is a beam of pure light, with waves that are all the same length. The waves move in step, giving a narrow but very powerful beam. Laser light hardly spreads at all, unlike the light from a flashlight.

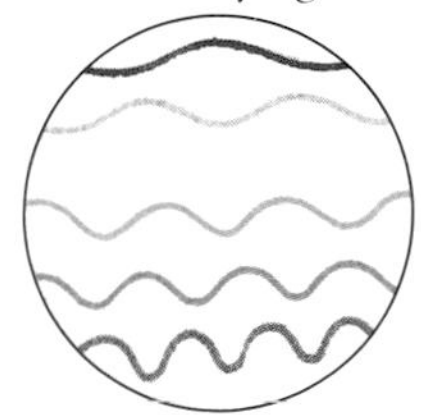

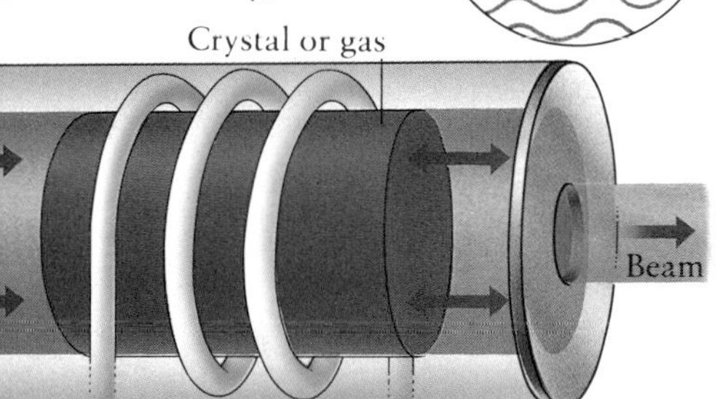

◀ *A laser has three main parts: the material that produces the beam, a power source, and a reflective resonator (usually mirrors).*

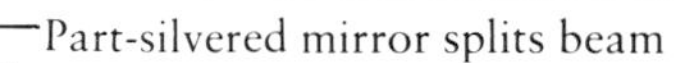

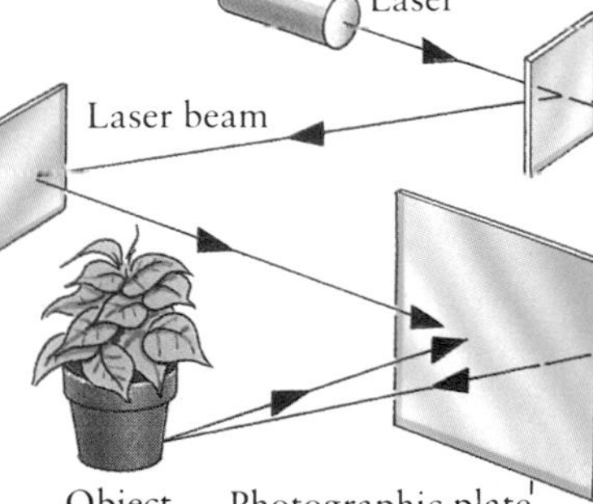

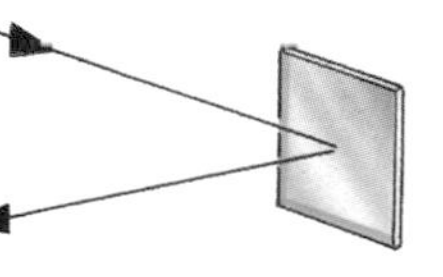

▶ *A hologram is a "solid" picture. A laser beam is split: one part reflects off the object onto the film; the other part lights the photographic plate. The interference between the two beams makes a 3-D picture viewed by shining another laser onto the hologram.*

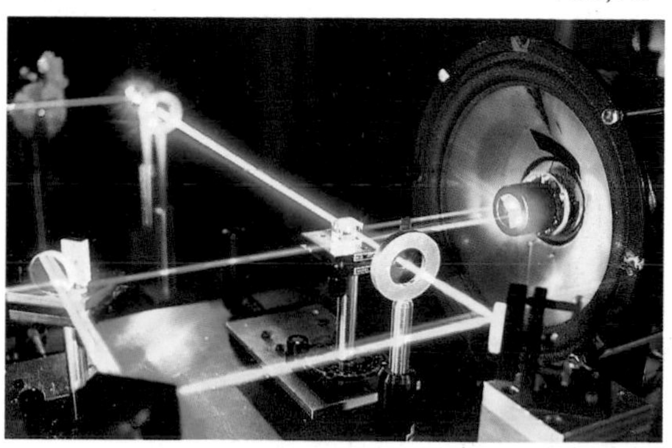

◀ *An experiment using a high-speed laser beam. Laser light can be bent around corners. Because it doesn't disperse, it can be used for exact measurement—for example, to check a deviation in a jet engine's fan blades.*

FACTS ABOUT LASERS

- Lasers do not generate light, they amplify (strengthen) it.
- Laser beams were used to measure the distance between the Earth and the Moon to within 6 in. (15 cm).
- Surgeons can now use lasers to reattach a damaged eye retina.
- Lasers can cut through concrete or steel. Industrial uses for lasers include cutting teeth in saws or drilling eyes in needles.
- "Laser" stands for *l*ight *a*mplification by *s*timulated *e*mission of *r*adiation.

Sound

All sounds are made by vibrations of objects. When an object moves backward and forward, or vibrates, it produces sound. The molecules around it vibrate and set up a sound wave. Sound can travel through air, water, metal, or other materials. Unlike light, sound waves cannot travel through a vacuum (airless space), so there is no sound in space or on the Moon. Sound travels faster through water 4,600 ft./s (1,400 m/s) than through the air (around 1,100 ft./s (330–340 m/s)), because water is more dense.

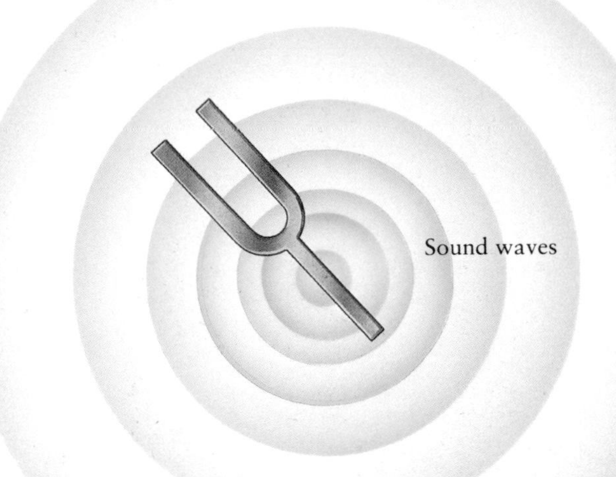

▲ *The air around a vibrating tuning fork is first pushed outward; then the fork moves inward. This vibration produces a series of ripplelike sound waves.*

▼ *Pitch (how high or low a sound is) depends on the frequency of waves: the number of air vibrations per second.*

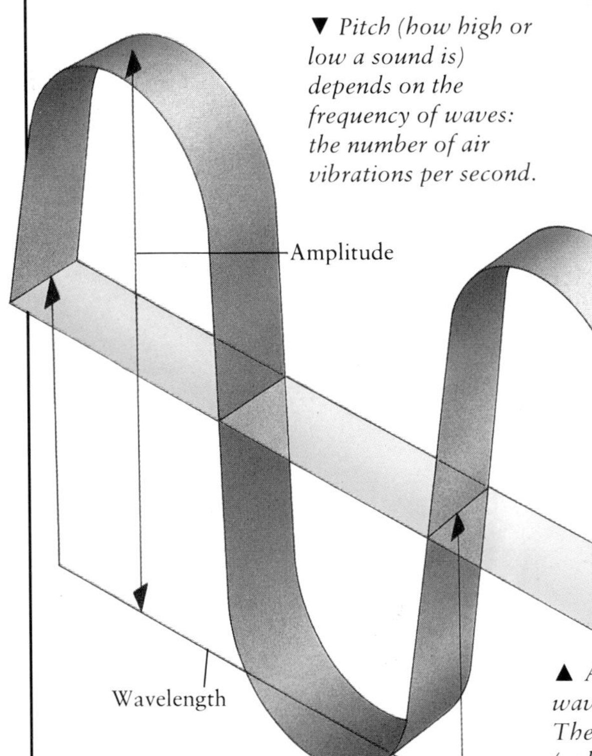

▲ *As frequency increases, wavelength decreases. The loudness of a sound (volume) depends on the height of the wave (amplitude).*

THE SONIC BOOM

An aircraft flying at the speed of sound (Mach 1) flies inside a pattern of pressure waves caused by its movement through the air. Above Mach 1 a supersonic jet overtakes the pressure waves, creating a cone-shaped shock wave.

Shock wave

Sonic boom heard

SONAR

Sounds bounce back from solid objects. Sonar (echo-sounding) devices on ships send sound waves toward the seabed. The waves are reflected back from any obstacles in their way. The time it takes for the sound to come back indicates the depth of the obstacle.

► *Fishing vessels use sonar to locate fish. Survey ships use sonar to chart the seabed.*

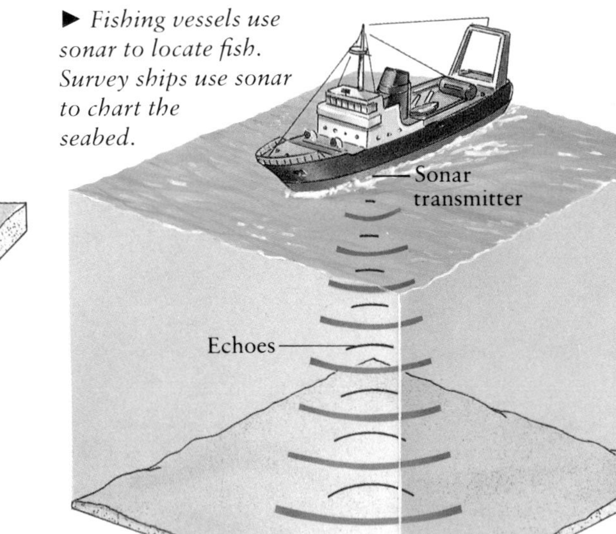

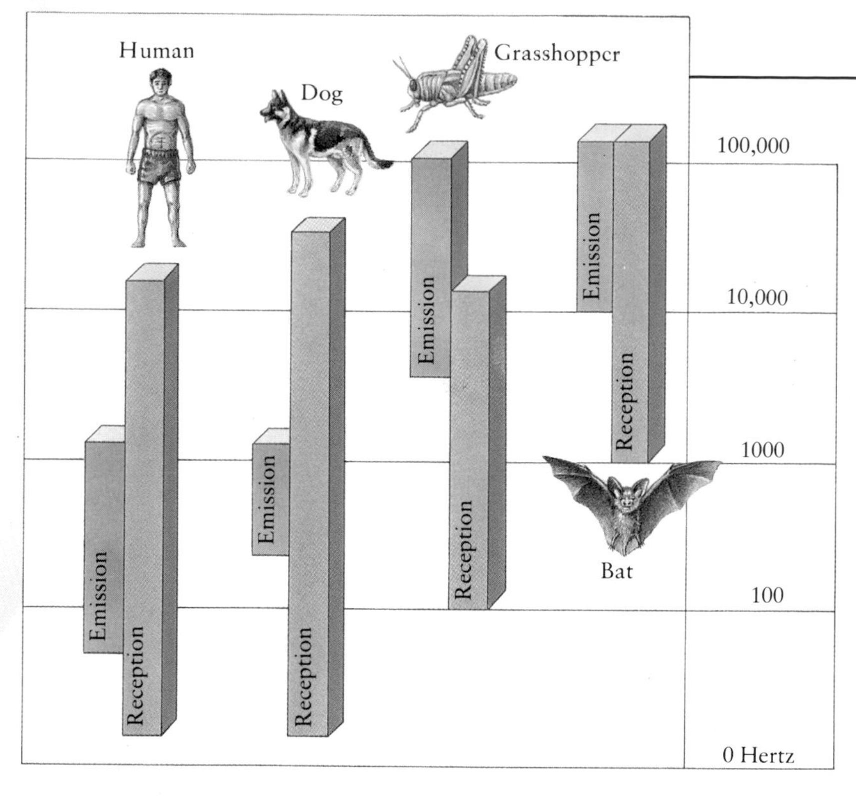

HOW WE HEAR SOUNDS

We can hear sounds with frequencies between about 20 hertz (Hz), or cycles per second, and 20,000 Hz. As we get older, we lose our ability to hear across a wide range of frequencies. Many animals emit (give out) and receive frequencies far above those we can hear. Bats hear and produce ultrasounds up to 100,000 Hz. Ultrasound waves are used in medical scanners and dental drills.

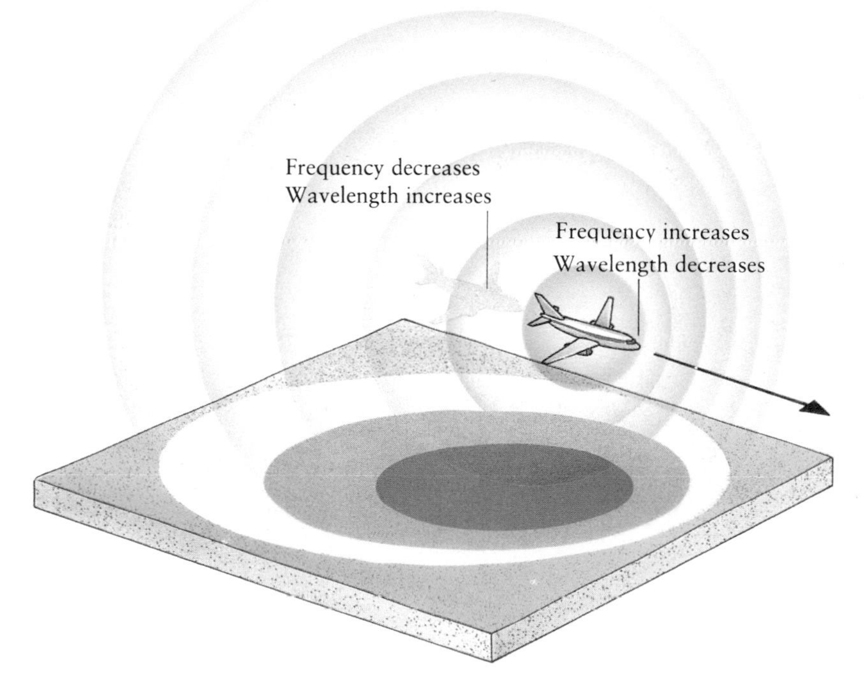

▲ *If a plane is flying toward you, the pitch (frequency) of its sound appears to increase. The wavelength of the engine noise gets shorter because the plane is moving with the sound waves. Likewise, sounds from an object moving away appear lower. This apparent change is called the Doppler effect.*

COMMON FREQUENCY CHANGES

The loudness of sound is how strong it seems when the sound waves hit our ears. At any frequency, the more intense the sound the louder it seems. Loudness is measured in decibels (dB). The chart shows some everyday sounds, loud and soft. Loud noises can damage our ears. Noise louder than about 140 dB can cause pain.

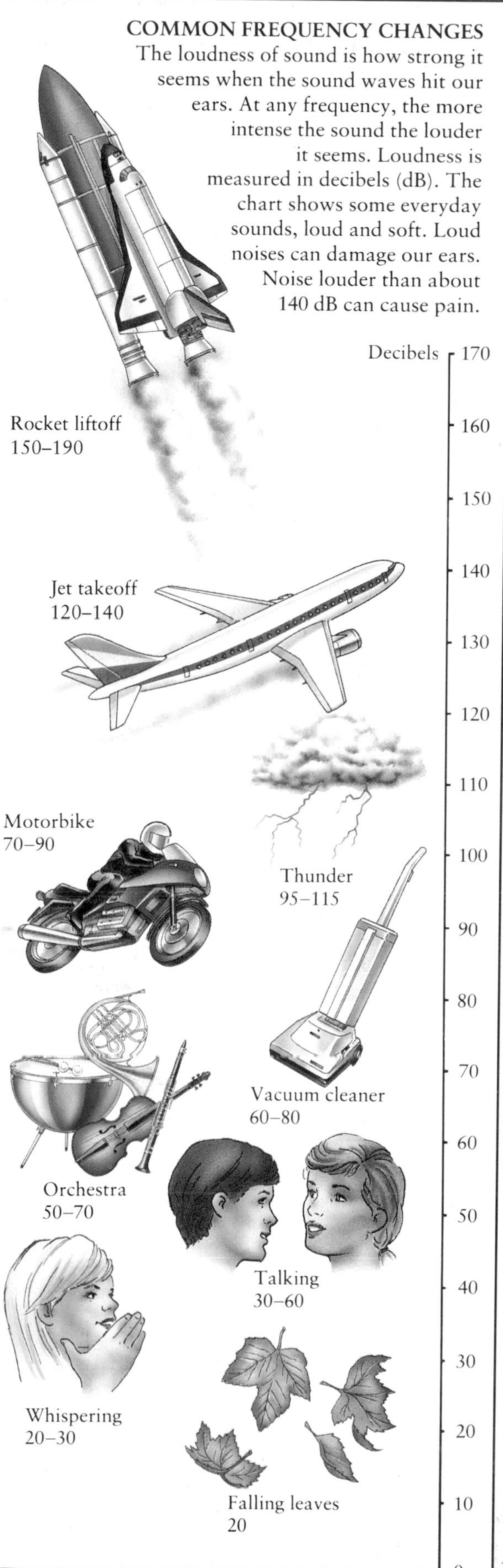

Sound Applications

Sound has many uses—as entertainment, as a means of communication, and as a tool in industry and scientific research. From ancient times, music and the different kinds of instruments used to make music have played a key role in human expression. From the 1870s it became possible to record sounds, and from the early 1900s sound broadcasting opened up new worlds of communication. The microphone, radio, tape player, and compact disc have become part of our everyday lives.

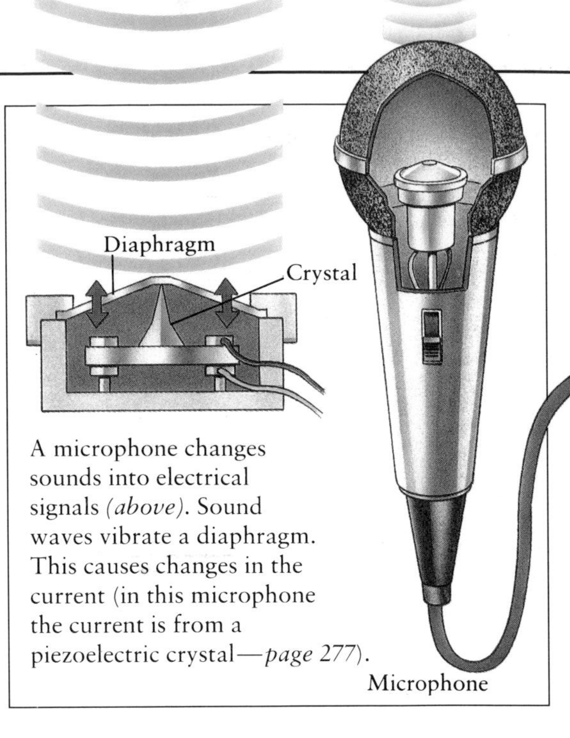

A microphone changes sounds into electrical signals *(above)*. Sound waves vibrate a diaphragm. This causes changes in the current (in this microphone the current is from a piezoelectric crystal—*page 277*).

DIGITAL AND ANALOG
Digital recording is done by a very rapid measurement of sound waves, changing the signals into off-on electrical pulses. Analog recording builds up a continuous sound "image" of variations in the electrical signals from the microphone.

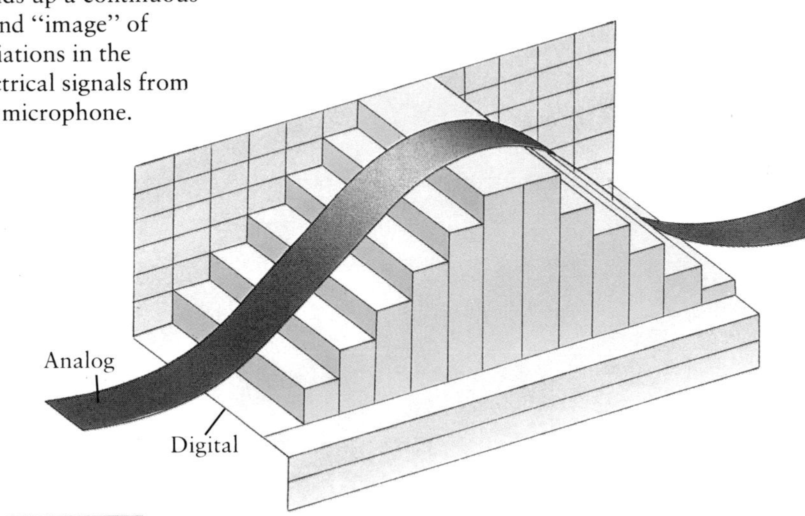

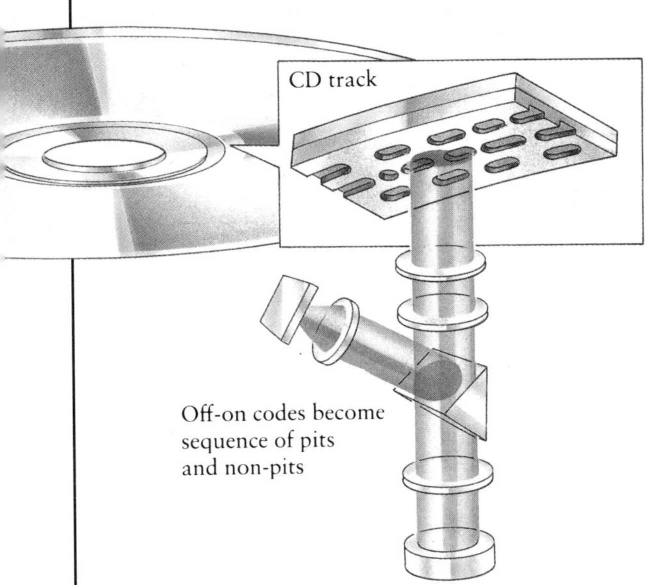

COMPACT DISC
The track on a compact disc contains a sequence of pits, storing sounds in coded form. As the disc spins in the player, a laser reading device scans the track (inside to outside) and converts the code signals into electrical pulses. These are changed back into reproductions of the original sounds.

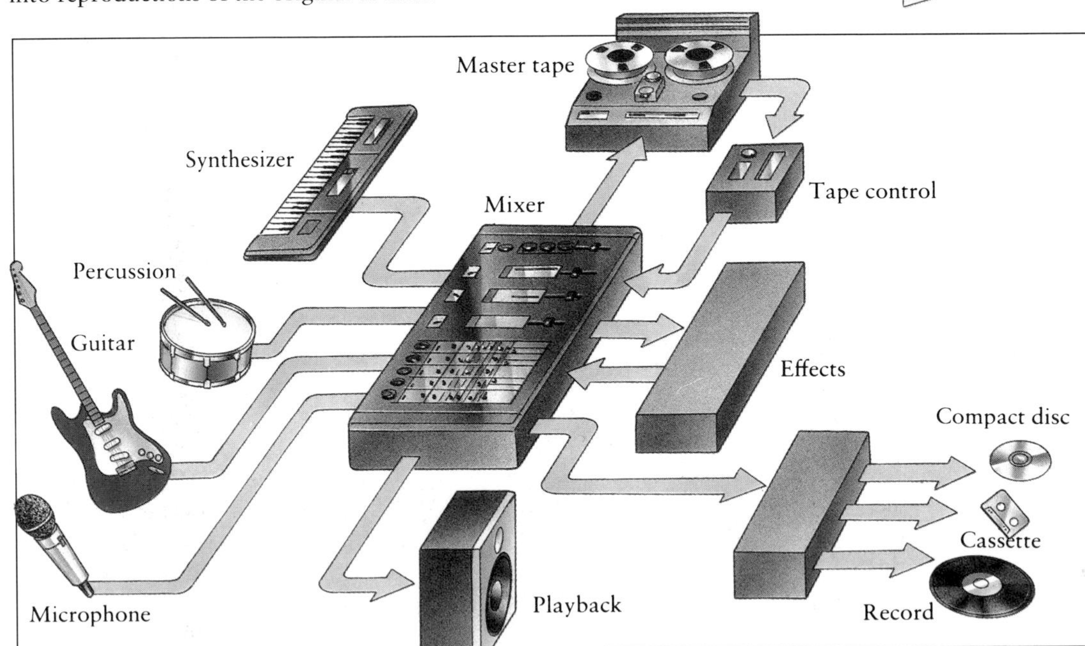

RECORDING
Optical digital recording cuts out the electrical noise, or background "hiss," in analog recording. Compact discs (CDs) and digital compact cassettes (DCCs) have better quality sound than standard vinyl discs and magnetic tapes. In the recording studio, sounds can be mixed and remixed electronically. Stereophonic sound is reproduced using two independent sound channels. Most movies have four-channel sound.

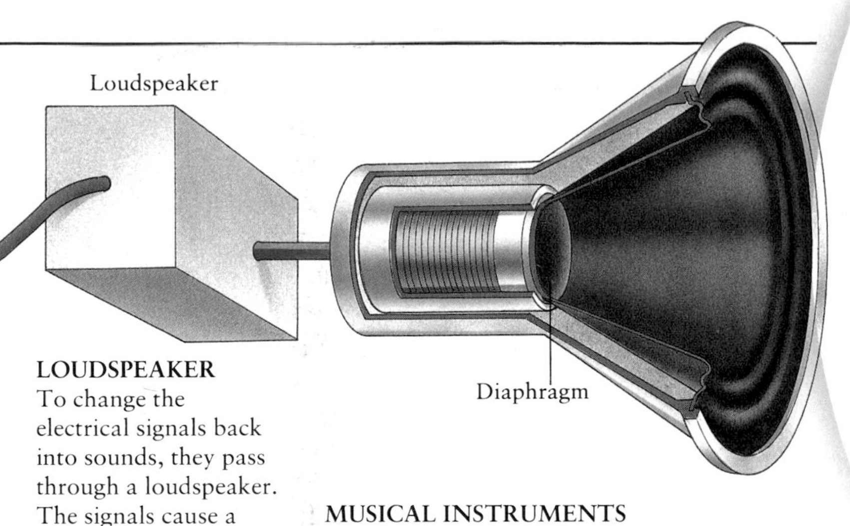

LOUDSPEAKER
To change the electrical signals back into sounds, they pass through a loudspeaker. The signals cause a diaphragm in the loudspeaker to vibrate, and this motion reproduces the original sounds.

MUSICAL INSTRUMENTS
Musical instruments make sounds by vibrating the air in different ways. A violin's strings are bowed or plucked. As the string vibrates, the hollow body and the air around the instrument vibrate. We hear a musical tone.

Holes

◀ *In a woodwind instrument, the longer the column of air inside, the lower the note. Fingering the holes alters the note.*

Valves

◀ *A trumpet has valves that change the length of the air column. The player pushes a valve open to open up extra loops of tube.*

Sound waves

◀ *When a drummer taps the drum skin with a stick, the skin vibrates. The resulting sound resonates (increases) inside the hollow drum body.*

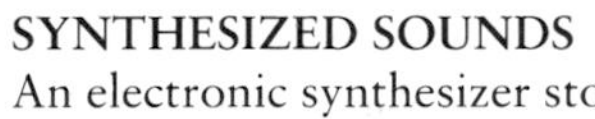

SYNTHESIZED SOUNDS
An electronic synthesizer stores and reproduces sounds digitally. The keyboard controls the frequency of the signals. The sounds are made by circuits called oscillators, and blended or shaped by other circuits called filters. Synthesized sound emerges from a loudspeaker. A synthesizer can produce an amazing range of musical and other sounds, including special effects.

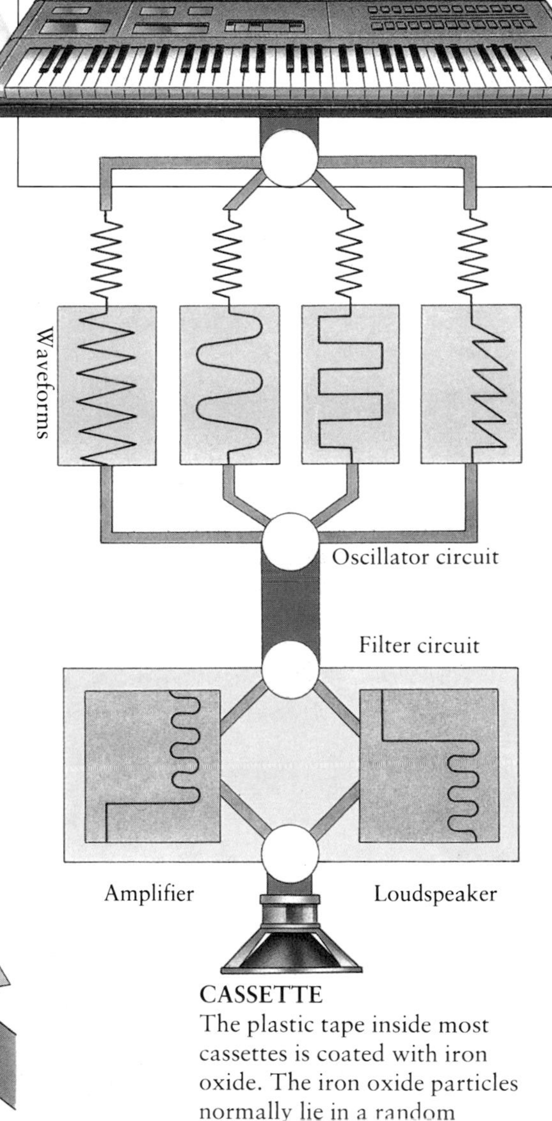

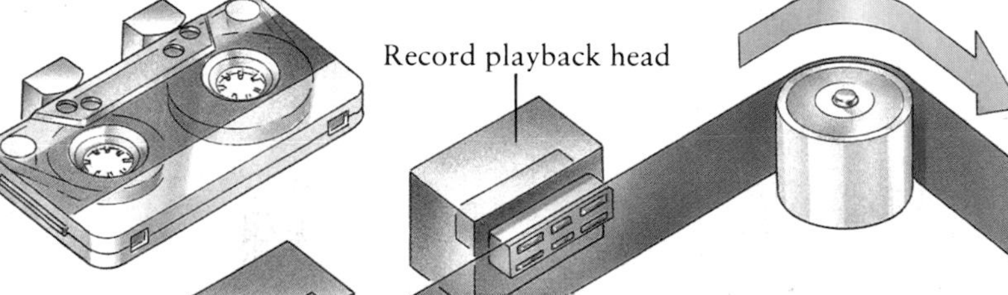

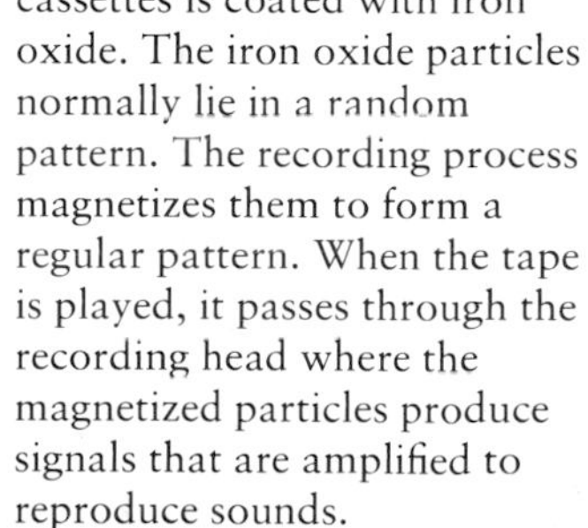

CASSETTE
The plastic tape inside most cassettes is coated with iron oxide. The iron oxide particles normally lie in a random pattern. The recording process magnetizes them to form a regular pattern. When the tape is played, it passes through the recording head where the magnetized particles produce signals that are amplified to reproduce sounds.

ELECTRICITY

Electromagnetism

Electromagnetism is one of the most influential forces shaping our world. Human life changed dramatically from the late 1700s when science opened the way to the widespread use of the enormous power of electricity. Magnetism had been known about for thousands of years. But it was not until 1820 that the Danish scientist Hans Christian Oersted discovered what we now call the magnetic effect—that magnetism and electricity are closely connected and that one can produce the other. This discovery made possible the development of electric generators and motors. Later, radio and television communication and the use of X-rays became possible. Devices to control the motion of electrons (the elementary particles which make up electrical current) brought about the modern electronics revolution in both the home and at work.

▼ *Electromagnetic radiation travels as waves. A wave has two main properties: amplitude, or height, and length (the distance between two neighboring waves). The number of waves that pass a certain point in a certain time gives the frequency.*

► *Radio and TV signals are transmitted on "carrier" waves. In AM (amplitude modulation) the amplified signal from a microphone is made to modulate, or vary, the amplitude of a carrier wave. In FM (frequency modulation) the voice signal modulates the frequency of the carrier wave.*

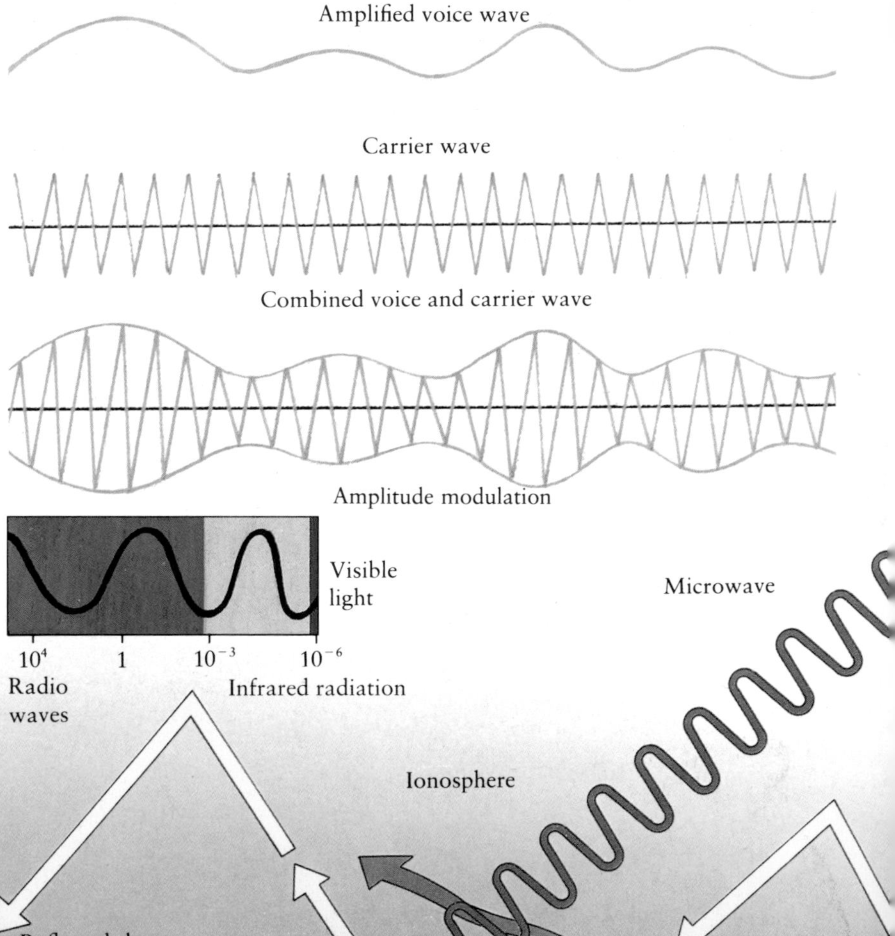

RADIO WAVES
Radio waves with very short wavelengths penetrate the atmosphere and are used to communicate with spacecraft. Longer wave signals are reflected back from the ionosphere. In this way, radio waves can be "bounced" around the world.

◀ *A lightning flash—an example of electricity in nature—is a huge electric spark in the atmosphere. A flash produces about 100 million volts of electricity.*

▼ *X-rays have a very high frequency, and a very short wavelength. Wilhelm Roentgen discovered X-rays in 1895. The rays pass easily through most living tissue, but not through material such as bone which contains heavier atoms. It is important to control doses of X-rays, since too many can cause cell damage.*

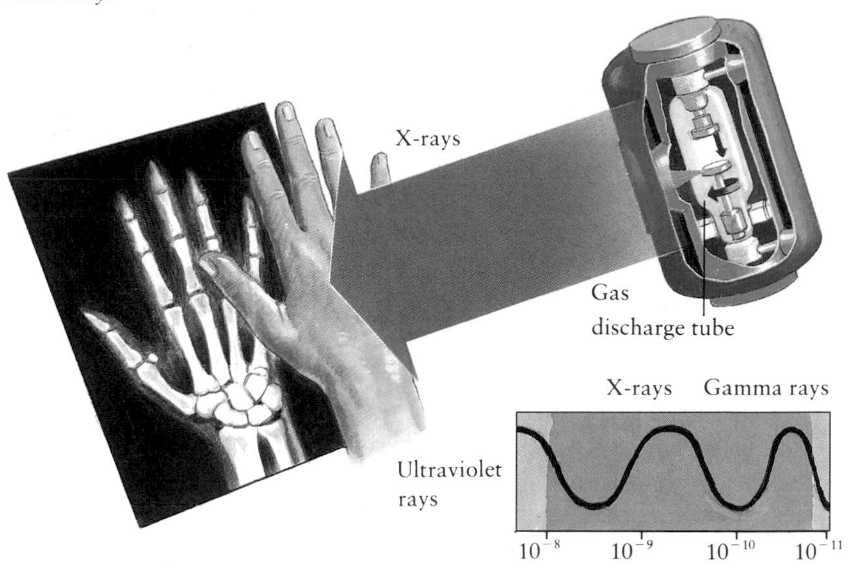

Communications satellite

10^{-1} Radar waves

10^{-2} Microwave waves

10^{-3} Infrared waves

Ionosphere

Microwave

Reflected sky wave

Microwave transmitter

MICROWAVES

Microwaves are a form of shortwave electromagnetic radiation. They are used for communications, since they are conveniently easy to direct. They are also used in microwave ovens, in which they are generated from a tube called a magnetron.

FACTS ABOUT ELECTROMAGNETISM

- In 1752 Benjamin Franklin proved that lightning is electric by flying a kite in a thunderstorm. Attached to the string was a metal key. As lightning struck the kite, a spark flashed as electricity passed down the wet string to the key.
- Wilhelm Roentgen, a German physicist, was awarded the first Nobel Prize in physics in 1901 for his discovery of X-rays. He photographed the bones of his wife's hand.

▼ *Ultraviolet radiation has a higher frequency than light we can see. Some insects can see ultraviolet light. A bee, for example, sees "target" marks which attract it to the center of the flower where the nectar is. Ultraviolet light can burn human skin. The ozone layer in the atmosphere absorbs most ultraviolet radiation in sunlight.*

Visible light

X-rays

10^{-6} 10^{-7} 10^{-8} 10^{-9}

Ultraviolet light

Electricity in Action

Electricity was long regarded as a mystery, because it is an invisible form of energy. It is produced when electrons, tiny subatomic particles, move from one atom to another. Electrons carry negative electric charges; the nucleus of an atom carries a positive charge. An electric current is a flow of electric charges, pushed along by a force. There are two kinds of current. Direct current (DC) flows in only one direction. Alternating current (AC) changes its direction of flow at regular intervals—one hundred times a second.

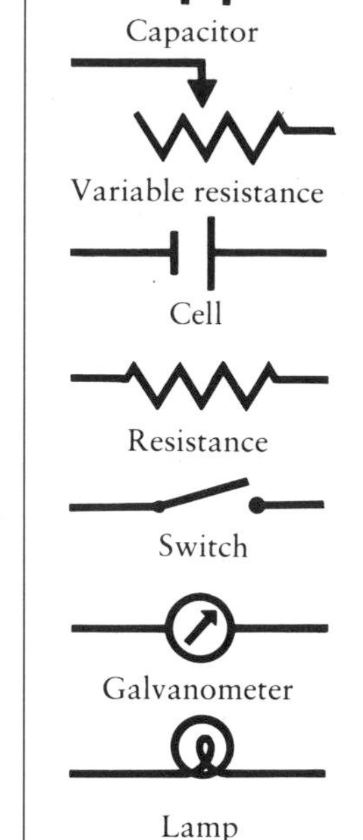

◀ *All circuits need a force to make the current flow. This force, the electromotive force (emf) has to overcome electrical resistance to make the current flow. Circuits with batteries work with direct current. Mechanical generators usually produce alternating current. Shown here are some symbols used when drawing a circuit.*

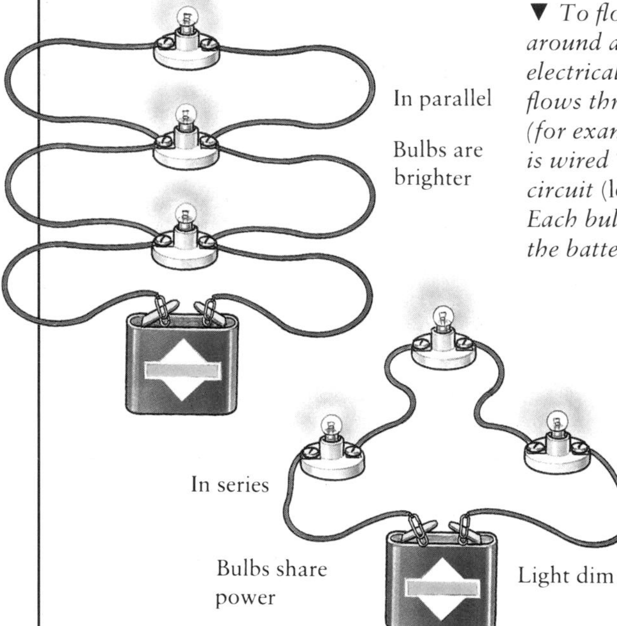

▼ *To flow, a current must move around a circuit—a loop of electrical conductors. If the current flows through several resistances (for example, light bulbs) the circuit is wired "in series." The bulbs in the circuit* (left) *are wired "in parallel." Each bulb has its own connection to the battery.*

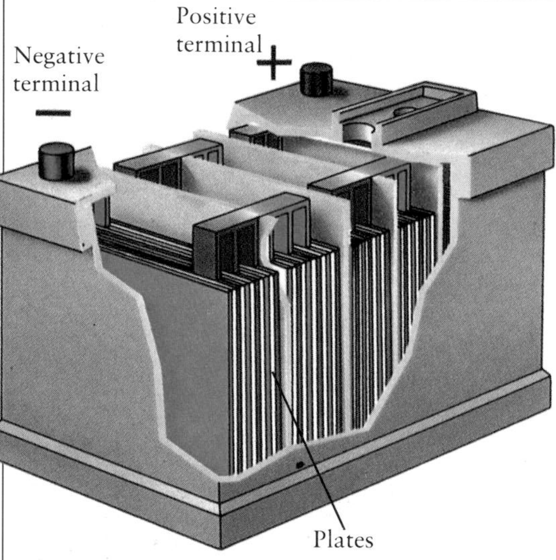

WET BATTERY
A battery uses chemical energy to produce electricity. A car battery is a "wet" battery (its metal plates are in acid). These lead-acid storage batteries store or accumulate electricity and can be recharged by a battery charger when exhausted. This type of battery can also provide vital emergency electricity in hospitals.

▼ *A high-voltage performance at a rock concert. The electricity from a power station is generated at high voltage and low current. The voltage is the force that pushes the current along. Transformers reduce the voltage to a level suitable for general use.*

DRY BATTERY
A dry battery (the kind of battery used in a flashlight) contains a dry pastelike or jellylike chemical. The acid in a lemon may produce enough current to light a small bulb if wired to suitable terminals, such as copper and zinc wires.

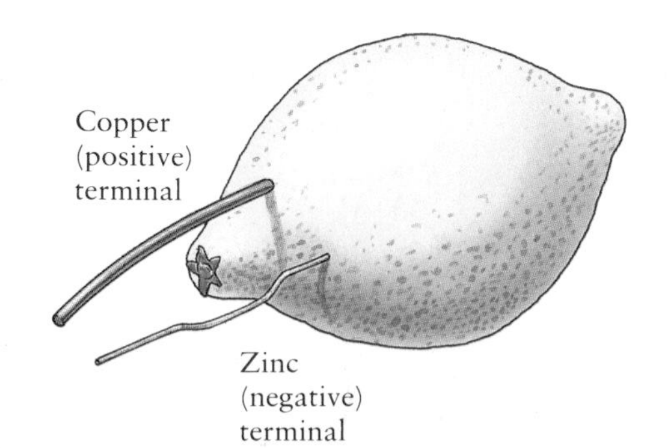

GENERATORS

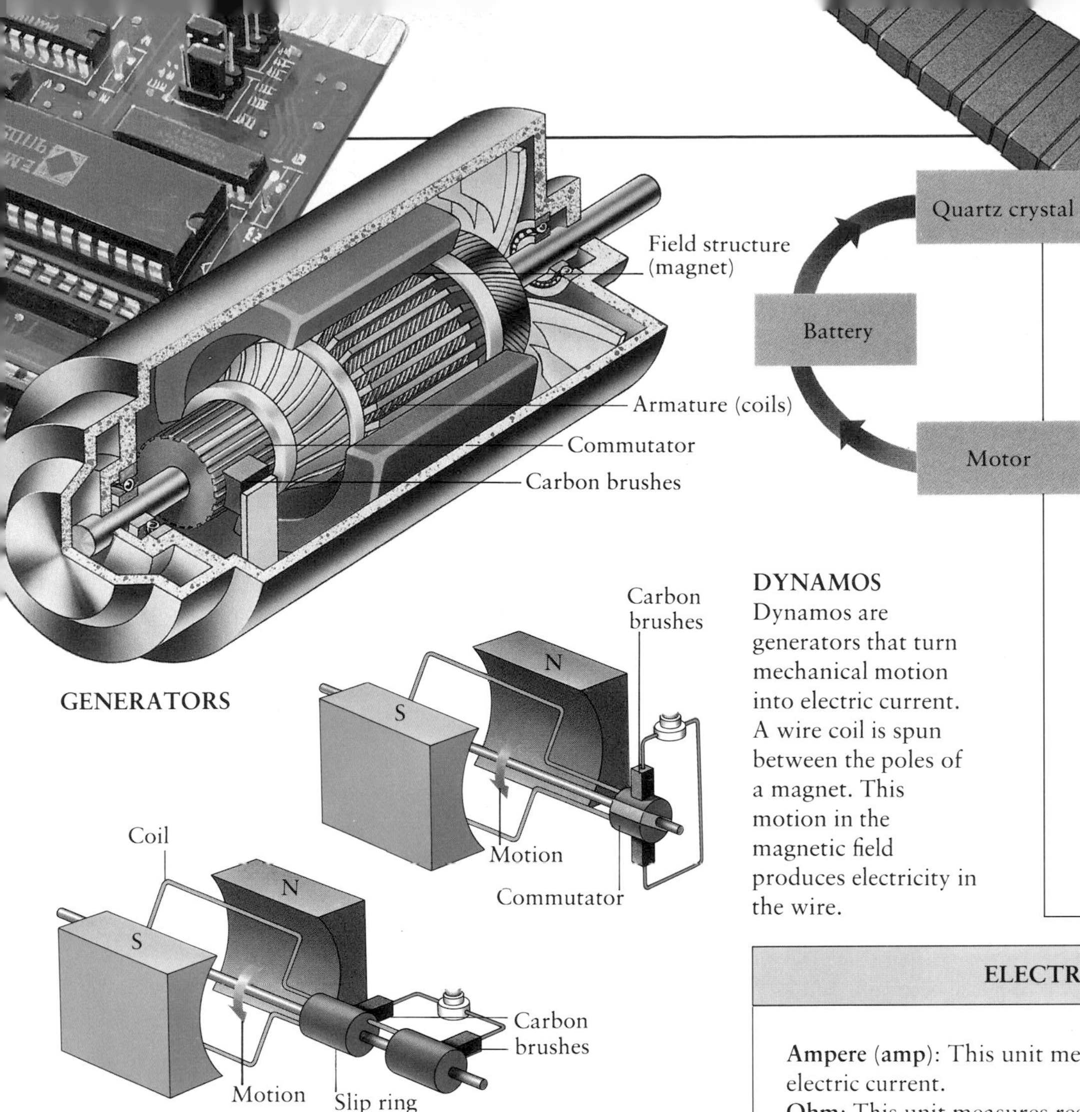

▲ *In an alternating current (AC) generator (also called an alternator), a slip ring, or collector ring, and fixed pieces of carbon called brushes transfer the current to the rest of the circuit.*

▲ *In a direct current (DC) generator, the coils are connected to a device called a commutator that keeps the electric current output constant and flowing in one direction only.*

DYNAMOS

Dynamos are generators that turn mechanical motion into electric current. A wire coil is spun between the poles of a magnet. This motion in the magnetic field produces electricity in the wire.

PIEZOELECTRICITY

A quartz watch has a crystal in it which when squashed or stretched produces a voltage. In reverse, an electrical signal makes the crystal vibrate at a definite frequency, which keeps the watch accurate. This is known as the piezoelectric effect.

ELECTRICAL UNITS

Ampere (amp): This unit measures the strength of an electric current.
Ohm: This unit measures resistance in a circuit.
Volt: Voltage, or electromotive force, is the "potential difference" in energy of an electric charge at two points. One volt across a resistance of one ohm produces a current of one amp.
Watt: This is the unit of power. A 100-watt light bulb turns 100 joules of electrical energy into heat and light every second.

Electromagnet used to move metal objects in a scrapyard

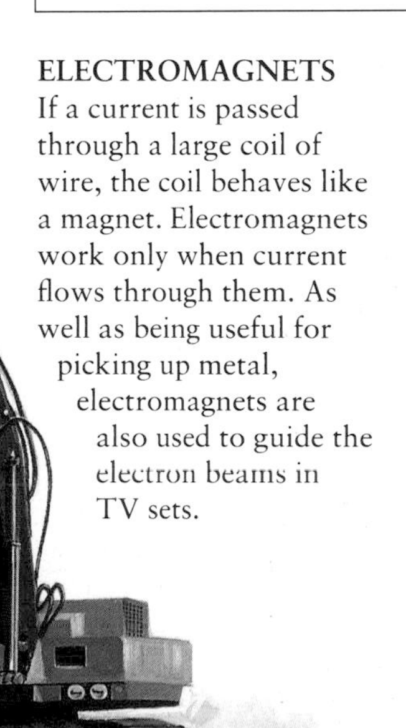

ELECTROMAGNETS

If a current is passed through a large coil of wire, the coil behaves like a magnet. Electromagnets work only when current flows through them. As well as being useful for picking up metal, electromagnets are also used to guide the electron beams in TV sets.

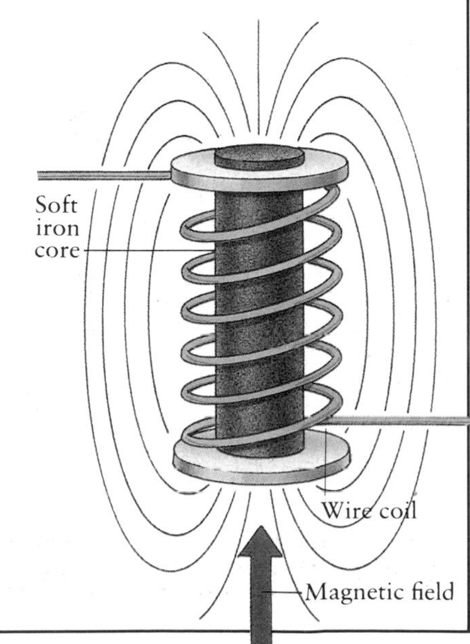

Electronics

In our daily lives, we rely constantly on devices controlled by electronic circuits. These circuits are made up of components (such as diodes and resistors). Electronic components affect the way electrons, and therefore electric currents, flow through them. Complete electronic circuits can now be made small enough to fit on a single silicon chip (a wafer of silicon semiconductor material). Components, including silicon chips, are fitted together to make printed circuits.

FACTS ABOUT ELECTRONICS

- The circuit on a silicon chip may contain up to several hundred thousand components.
- The first integrated circuit was made in 1958.
- Semiconductors are materials that conduct electricity, though not as well as metals.
- Microchips can work at almost the speed of light—performing more than a million operations every second.
- Outside of the world of science fiction, few robots look like people. Human-like robots, or androids, could be built, but there is little advantage in making a machine look exactly like a person.
- The most popular home video format, VHS (Video Home System), was developed in Japan in the mid-1970s.

◀ *Vacuum tubes were the first devices for controlling the flow of an electric current. The diode (1904) and triode (1906) made modern radio and TV possible. Tubes were replaced by the transistor, invented in 1948.*

▼ *Transistors are smaller and use less power than tubes. They are made of semiconducting material. A microchip is an electronic circuit mounted on a single piece of silicon. It works at great speed.*

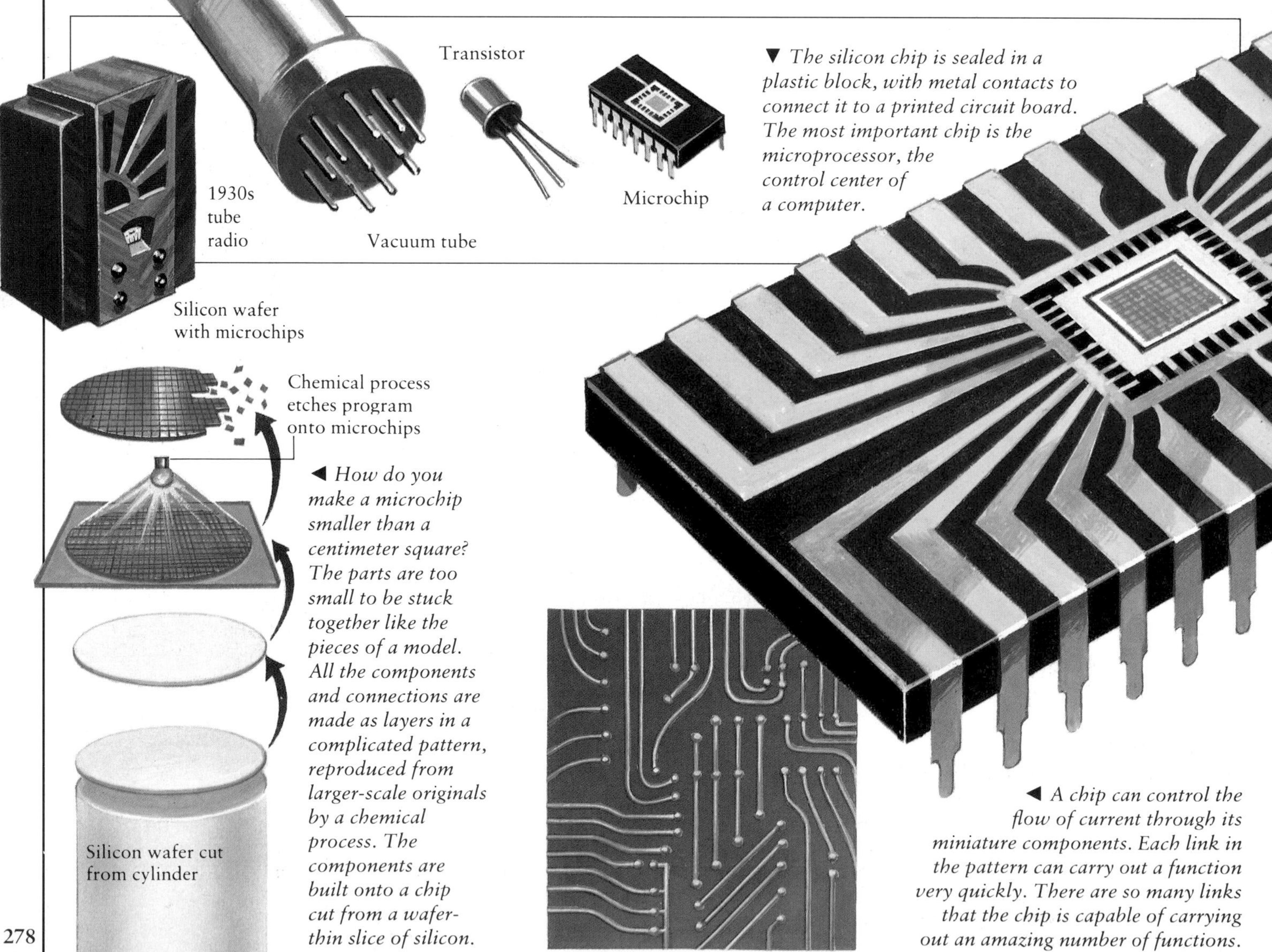

▼ *The silicon chip is sealed in a plastic block, with metal contacts to connect it to a printed circuit board. The most important chip is the microprocessor, the control center of a computer.*

◀ *How do you make a microchip smaller than a centimeter square? The parts are too small to be stuck together like the pieces of a model. All the components and connections are made as layers in a complicated pattern, reproduced from larger-scale originals by a chemical process. The components are built onto a chip cut from a wafer-thin slice of silicon.*

◀ *A chip can control the flow of current through its miniature components. Each link in the pattern can carry out a function very quickly. There are so many links that the chip is capable of carrying out an amazing number of functions.*

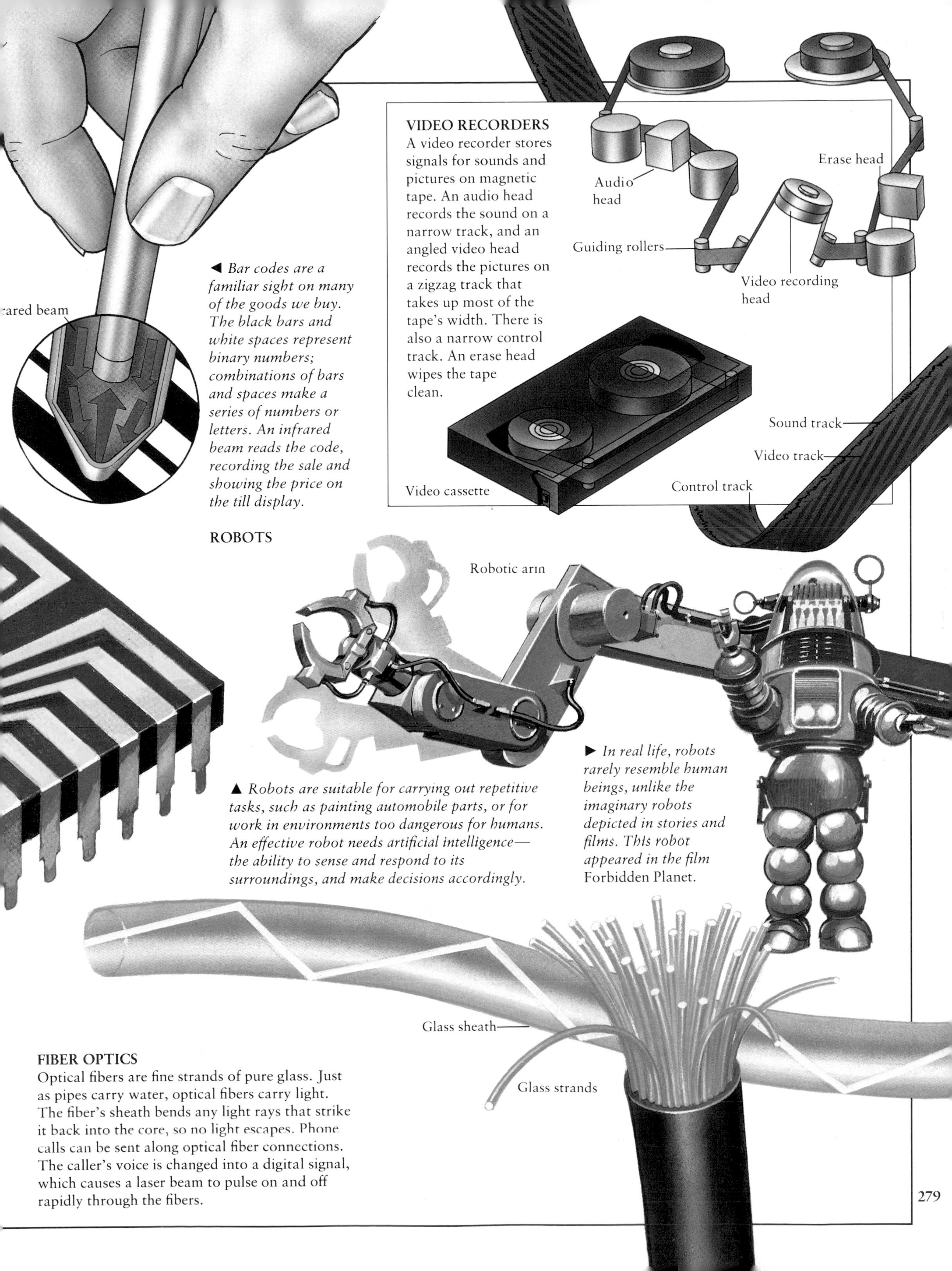

◀ *Bar codes are a familiar sight on many of the goods we buy. The black bars and white spaces represent binary numbers; combinations of bars and spaces make a series of numbers or letters. An infrared beam reads the code, recording the sale and showing the price on the till display.*

VIDEO RECORDERS

A video recorder stores signals for sounds and pictures on magnetic tape. An audio head records the sound on a narrow track, and an angled video head records the pictures on a zigzag track that takes up most of the tape's width. There is also a narrow control track. An erase head wipes the tape clean.

ROBOTS

▲ *Robots are suitable for carrying out repetitive tasks, such as painting automobile parts, or for work in environments too dangerous for humans. An effective robot needs artificial intelligence—the ability to sense and respond to its surroundings, and make decisions accordingly.*

▶ *In real life, robots rarely resemble human beings, unlike the imaginary robots depicted in stories and films. This robot appeared in the film* Forbidden Planet.

FIBER OPTICS

Optical fibers are fine strands of pure glass. Just as pipes carry water, optical fibers carry light. The fiber's sheath bends any light rays that strike it back into the core, so no light escapes. Phone calls can be sent along optical fiber connections. The caller's voice is changed into a digital signal, which causes a laser beam to pulse on and off rapidly through the fibers.

Computers

The fastest computers can do billions of calculations every second. Their main ability is to do a lot of basic tasks quickly and accurately. We use computers at work, school, and play, even if we never tap a keyboard. When we buy goods in a supermarket, draw money from a bank, or ask for information about a vacation, we are making transactions that are now done with the aid of computers. A revolution that began in the 1940s with room-sized machines has today given us the laptop personal computer.

The English mathematician Charles Babbage (1792–1871) designed a large calculating machine which, in theory, could be programmed, like a computer. But the electronic age had not yet dawned, and there was no way to give Babbage's Analytical Engine the power it needed.

PERSONAL COMPUTERS
Small personal computers can be found in offices, schools, and homes all over the world. The machinery and electronics (screen, keyboard, disk drives, printer) are the hardware. Software programs stored on disks tell the computer what to do—in this illustration, a graphics program is being used to design a building.

Printer

Disk

Disk drive

Screen

Keyboard

Mouse

◄ *The output from a computer can be in the form of a visual display on a screen, printed paper, a modem link, or a disk.*

◄ *A keyboard, a mouse, a modem linked to other computers, and a floppy disk input data and instructions.*

◄ *A mouse is a simple hand-controller. By moving the mouse, you direct the cursor around the screen to give the computer commands.*

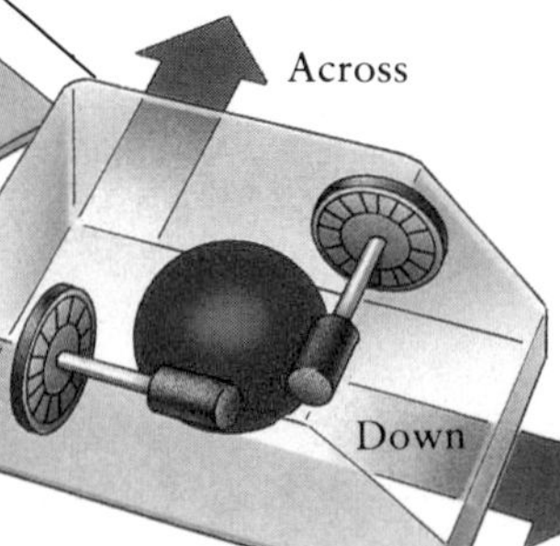

INSIDE A COMPUTER
A computer has four basic parts: input, processing, memory, and output. It also needs a program—instructions telling the processing unit how to do different tasks.

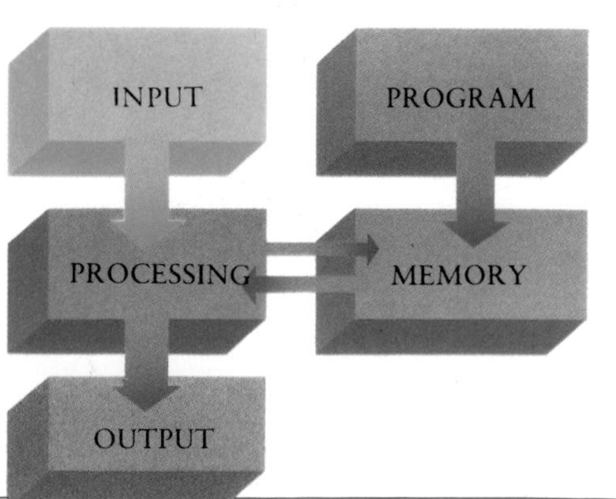

▲ *Computers are now used in almost every business. For the financial dealer, computers are vital for storing business data and analyzing movements on the world's money markets, 24 hours a day. Computers in offices and laboratories all around the world are constantly exchanging information.*

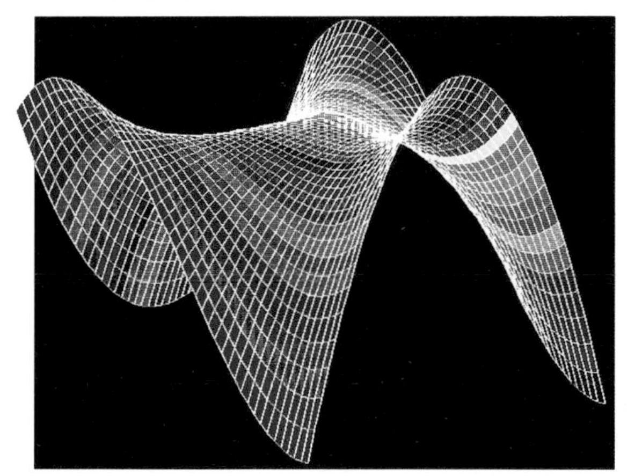

LARGE AND SMALL

The first fully electronic computers were the British Colossus (completed in 1943) and the American ENIAC—*E*lectronic *N*umerical *I*ntegrator *A*nd *C*omputer—(completed in 1946). Both needed teams of operators. These computers, called mainframes, were at first so big that they needed whole rooms to contain them. In the 1950s, smaller transistorized computers were developed. By the 1970s the miniaturization of electronic equipment led to advances in computer technology such as the small, less expensive personal computer (pc).

Mainframe computer

Laptop computer

Computer software (program)

Computer notebook

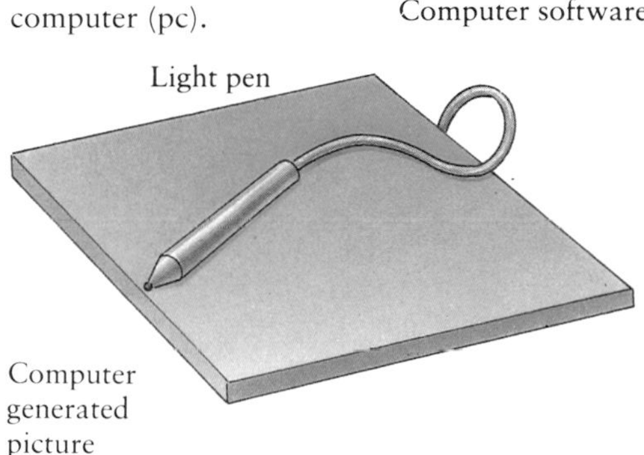

◀ *Computer graphics are made by converting information into pictures. Computer-aided design programs allow engineers, architects, and scientists to study and modify designs on screen. The operator can use a lightpen to alter the computer-made picture.*

USEFUL TERMS

Bit: Binary digit; a numeral in binary notation (0 or 1)
Byte: Space in a computer's memory occupied by one letter or numeral
Cursor: Highlighted area on screen
Data: Information processed by a computer
Hardware: The physical parts of a computer system
Network: Several computers connected, e.g. by a modem (telephone)
Software: Programs for a computer

VIRTUAL REALITY

Graphics programs are now so sophisticated that computers can interact with people to create "virtual reality." The computer generates sounds and images, creating a "landscape," heard and seen inside a special helmet. The effect is to construct a seemingly real world that the person feels a part of. The possibilities for entertainment and education are endless.

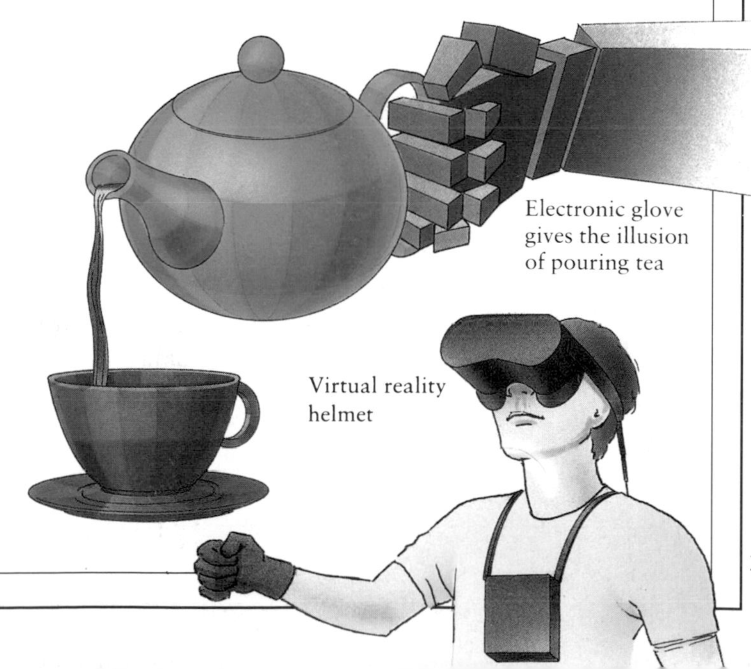

TECHNOLOGY

Engineering

Science is knowledge. Technology is the process of applying knowledge to make work easier, to make our lives more comfortable, and to prevent and cure disease. Technology began when human beings first shaped tools and used fire. It remained very simple until about 10,000 years ago, when farming began. Key inventions quickly followed this shift to a settled way of life — for example, the wheel, metalworking, pottery, and weaving. Industrial technology began in the late 1700s with the steam engine and powered machinery. Engineers mastered new technologies — steel-making, electricity, and the manufacturing of automobiles and airplanes. Technology has given us the power to control the environment. It has brought faster travel and communications. But it has also created problems which today's technology must solve for the sake of tomorrow.

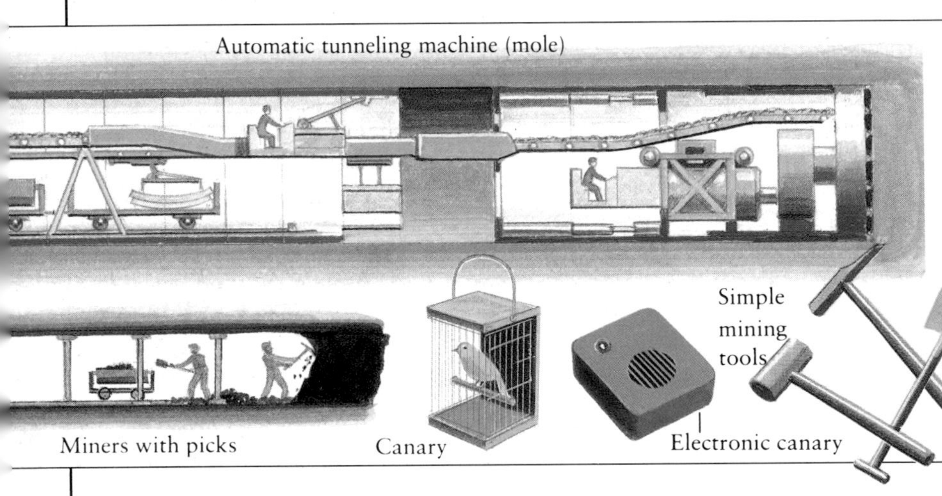

DIGGING AND DRILLING

Some of the most advanced technology is used in mining, drilling, and tunneling. From Stone Age times, miners were pioneer technologists. Today, explosives, pneumatic drills, and the automatic tunneling machine, or mole, have replaced the pick-ax and human muscle-power of the past.

▼ Production line assembly was an important step toward automation — the use of machines to do work previously done by people. Today, computer-controlled robots are used for routine assembly jobs. Before the assembly line was introduced (in the early 1900s, for car-making), most goods were made one at a time, just as potters have made pots by hand for thousands of years.

FACTS ABOUT TECHNOLOGY

- Inventions can come too early. A spinning steam device was designed by an ancient Greek called Hero of Alexandria. It was thought of as a toy.
- Leonardo da Vinci's flying machine and Babbage's computer were also ahead of their time. But they were unworkable without a power source.

▲ Machine tools are powered tools used for shaping metal or other materials, by drilling holes, chiseling, grinding, pressing, or cutting. Often the material (the workpiece) is moved while the tool stays still.

CLASSIC INVENTIONS
Commonplace but extremely useful inventions are often minor miracles of engineering. We take the zipper and the ballpoint pen for granted — both were unknown 100 years ago. The chemical or foam fire-extinguisher is not particularly sophisticated, yet its technology would have baffled an 18th-century firefighter.

▲ *With submersibles and bathyscaphes, we can explore the depths of the oceans.*

▲ *Space technology has developed from the modest experiments of the 1930s to interplanetary probes.*

▼ *Technology has given us the power to fight disasters, natural and human-made.*

▼ *Technology keeps a deep sea diver alive underwater, able to observe and work.*

▼ *Early people slowly mastered technological skills such as house-building, clothes- and fire-making, food preservation, and water storage. They were using technology to adapt their environment and make their lives easier. Domestic technology is still developing, though at different rates in different parts of the world.*

▼ *The modern Western house is far removed from the smoky huts of our ancestors. It has TV, radio, and telephones, as well as electricity, gas, water, and drainage.*

▼ *Modern technology can create a clean, comfortable and energy-efficient home. The house of tomorrow may well be computer-controlled, with all its systems (heating, lighting, water supply, security, and communications) run by a small computer.*

Buildings: Construction

The builders of the past used materials such as wood, stone, and brick. Most of the work was done by hand. The construction of massive structures such as the pyramids of Egypt and Central America required thousands of workers and sophisticated mathematics. There were even some machines on ancient building sites—the Romans had cranes, for example. But not until the 1800s did the new steel and concrete construction techniques make it possible to build a skyscraper.

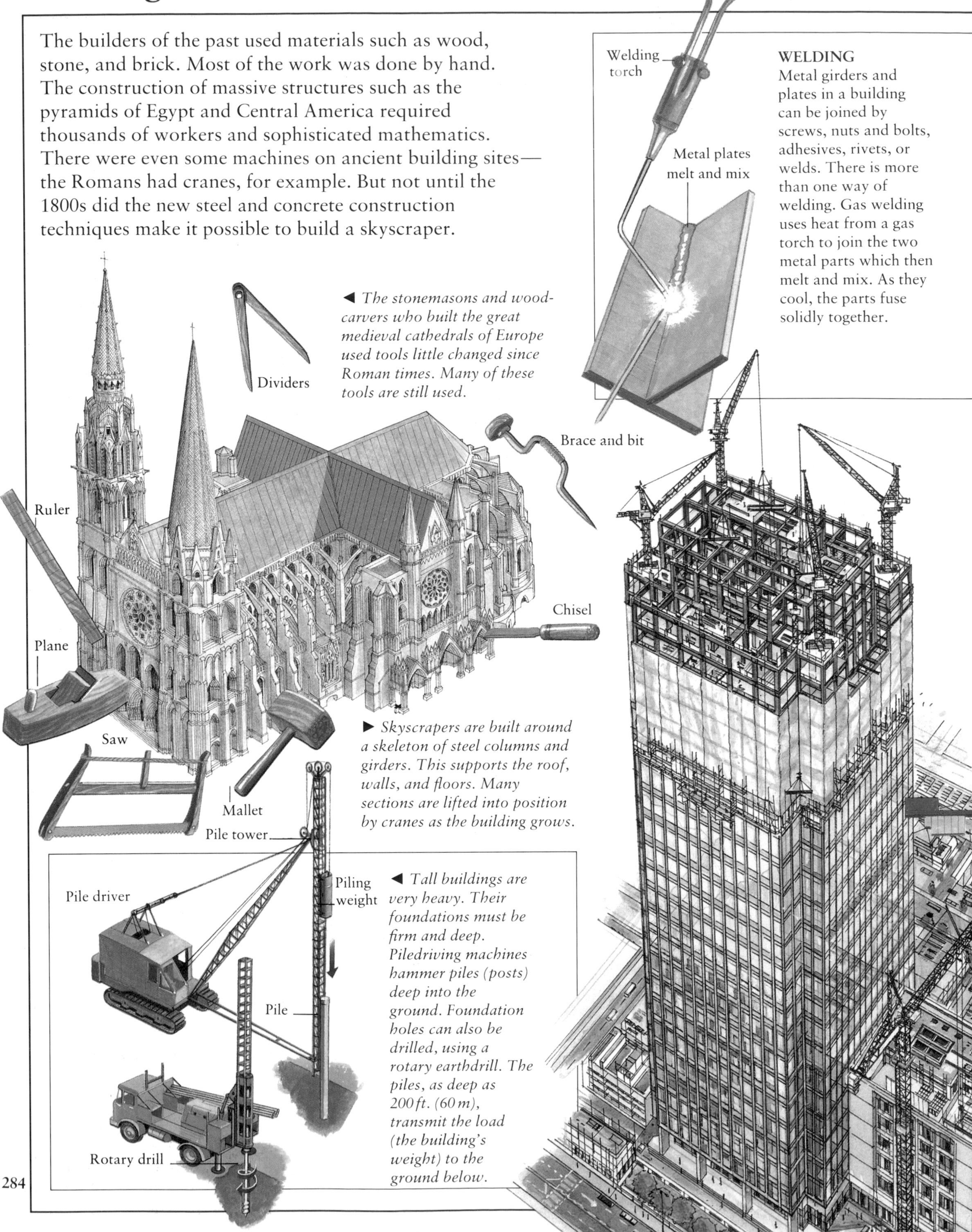

WELDING
Metal girders and plates in a building can be joined by screws, nuts and bolts, adhesives, rivets, or welds. There is more than one way of welding. Gas welding uses heat from a gas torch to join the two metal parts which then melt and mix. As they cool, the parts fuse solidly together.

◀ *The stonemasons and woodcarvers who built the great medieval cathedrals of Europe used tools little changed since Roman times. Many of these tools are still used.*

▶ *Skyscrapers are built around a skeleton of steel columns and girders. This supports the roof, walls, and floors. Many sections are lifted into position by cranes as the building grows.*

◀ *Tall buildings are very heavy. Their foundations must be firm and deep. Piledriving machines hammer piles (posts) deep into the ground. Foundation holes can also be drilled, using a rotary earthdrill. The piles, as deep as 200 ft. (60 m), transmit the load (the building's weight) to the ground below.*

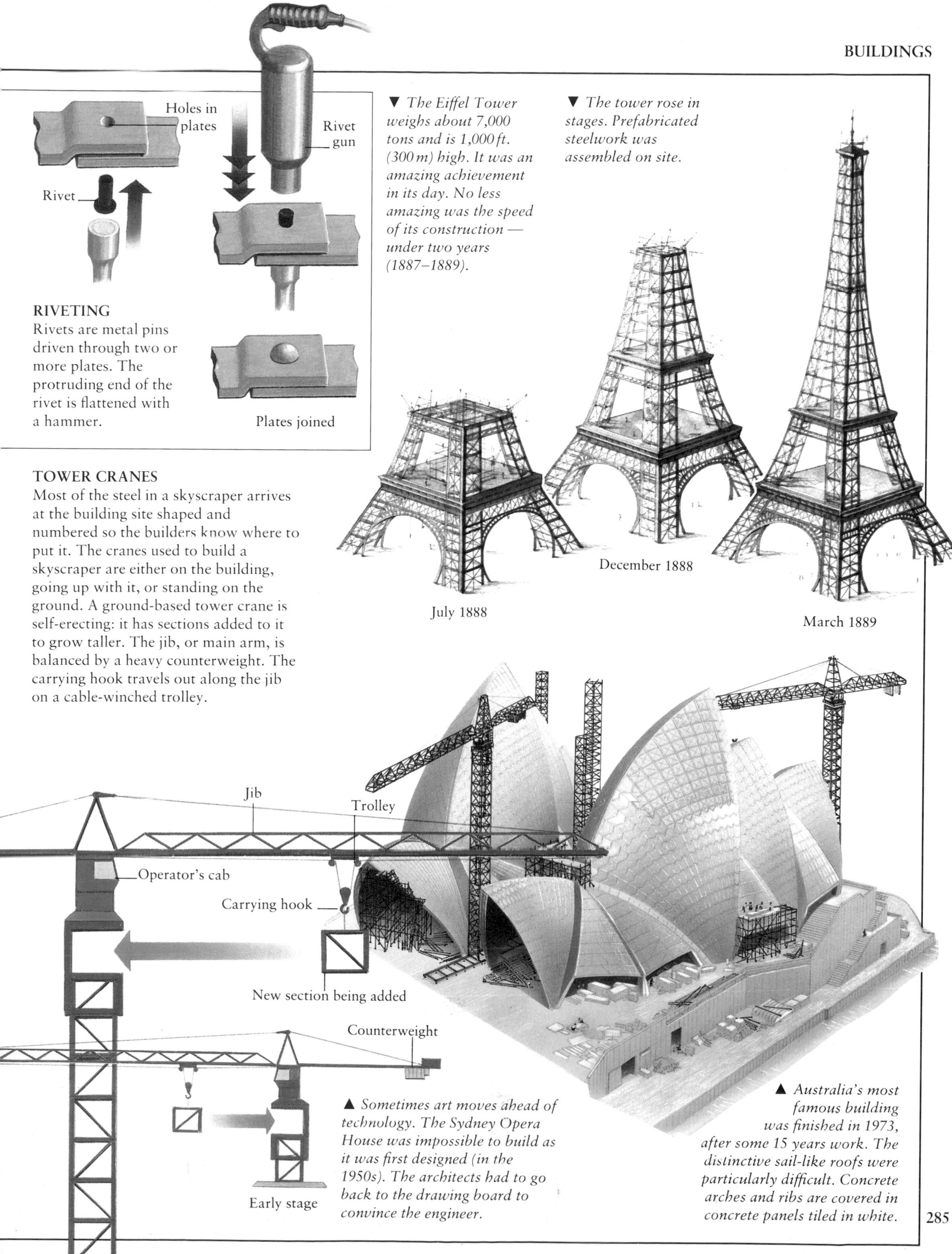

RIVETING
Rivets are metal pins driven through two or more plates. The protruding end of the rivet is flattened with a hammer.

TOWER CRANES
Most of the steel in a skyscraper arrives at the building site shaped and numbered so the builders know where to put it. The cranes used to build a skyscraper are either on the building, going up with it, or standing on the ground. A ground-based tower crane is self-erecting: it has sections added to it to grow taller. The jib, or main arm, is balanced by a heavy counterweight. The carrying hook travels out along the jib on a cable-winched trolley.

▼ *The Eiffel Tower weighs about 7,000 tons and is 1,000 ft. (300 m) high. It was an amazing achievement in its day. No less amazing was the speed of its construction — under two years (1887–1889).*

▼ *The tower rose in stages. Prefabricated steelwork was assembled on site.*

▲ *Sometimes art moves ahead of technology. The Sydney Opera House was impossible to build as it was first designed (in the 1950s). The architects had to go back to the drawing board to convince the engineer.*

▲ *Australia's most famous building was finished in 1973, after some 15 years work. The distinctive sail-like roofs were particularly difficult. Concrete arches and ribs are covered in concrete panels tiled in white.*

Buildings: Interiors

A building is an artificial environment. Its outside may look traditional or futuristic, but for the people who live and work in it, the inside is just as important. A building's internal systems must work efficiently for the comfort and safety of the people in it. The systems installed in many modern buildings include heating, air-conditioning, fire sprinklers, power and water supplies, waste disposal, and communications. In multistory buildings, elevators and escalators are also important components.

▲ *Modern building techniques can create huge interior spaces. In the Hong Kong and Shanghai Bank in Hong Kong (47 stories, opened 1986), the floors hang from steel masts. This unusual design creates a light and airy open space, called an atrium, inside the building.*

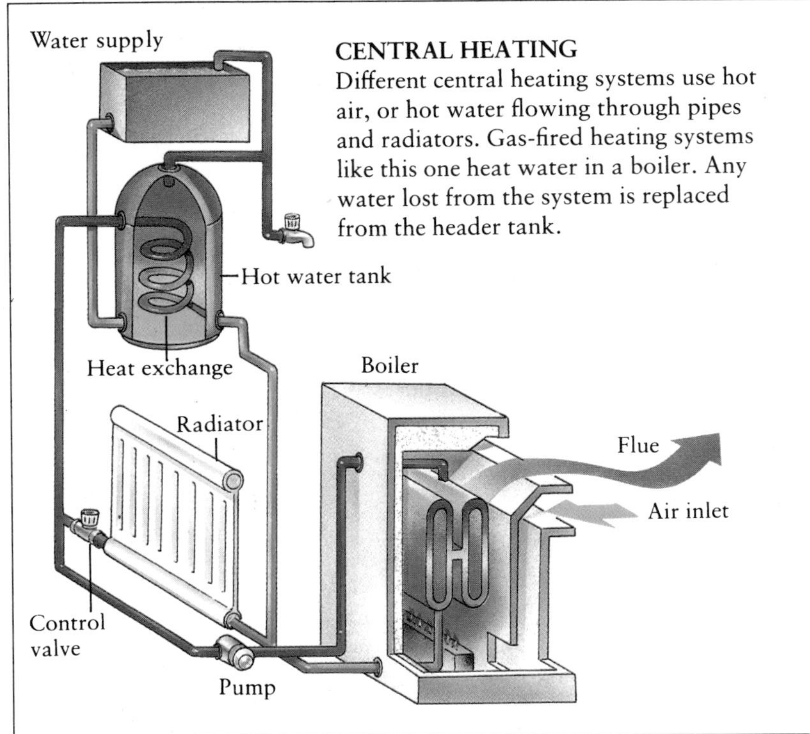

CENTRAL HEATING
Different central heating systems use hot air, or hot water flowing through pipes and radiators. Gas-fired heating systems like this one heat water in a boiler. Any water lost from the system is replaced from the header tank.

▼ *The Pompidou Arts Center in Paris has its "insides" on the outside. The pipes and tubes that carry essential services (the blue pipes are air-conditioning) are hidden in most buildings, but here they are on the outside. The elevators are also on the outside. This controversial design gave easy access to the services for maintenance and more room inside the building.*

ROMAN CENTRAL HEATING
The Romans devised a method of heating their homes with hot air. This system was called a "hypocaust." The floors of their homes were raised on piers. The air was heated by burning fuel in a furnace. It then flowed through tiled flues (air channels) in the walls into the space beneath the floors.

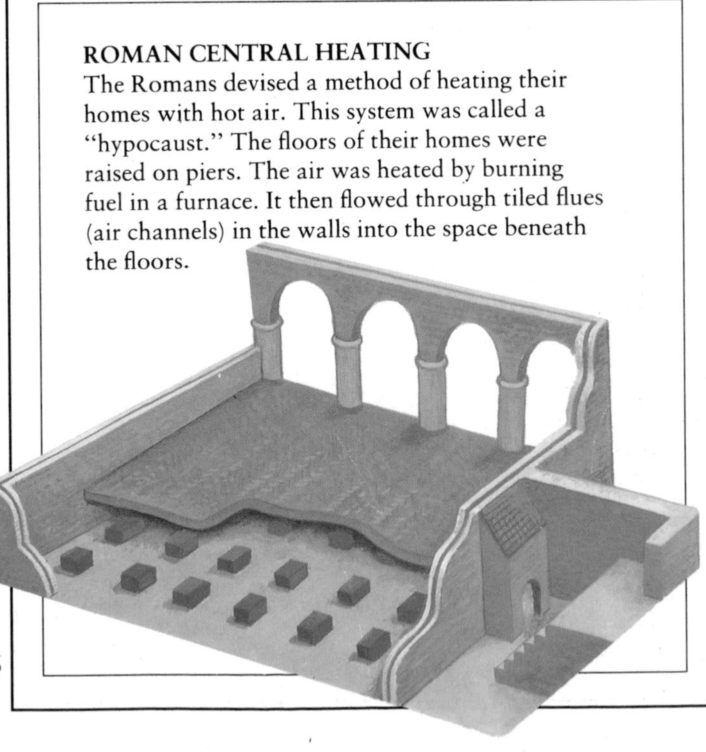

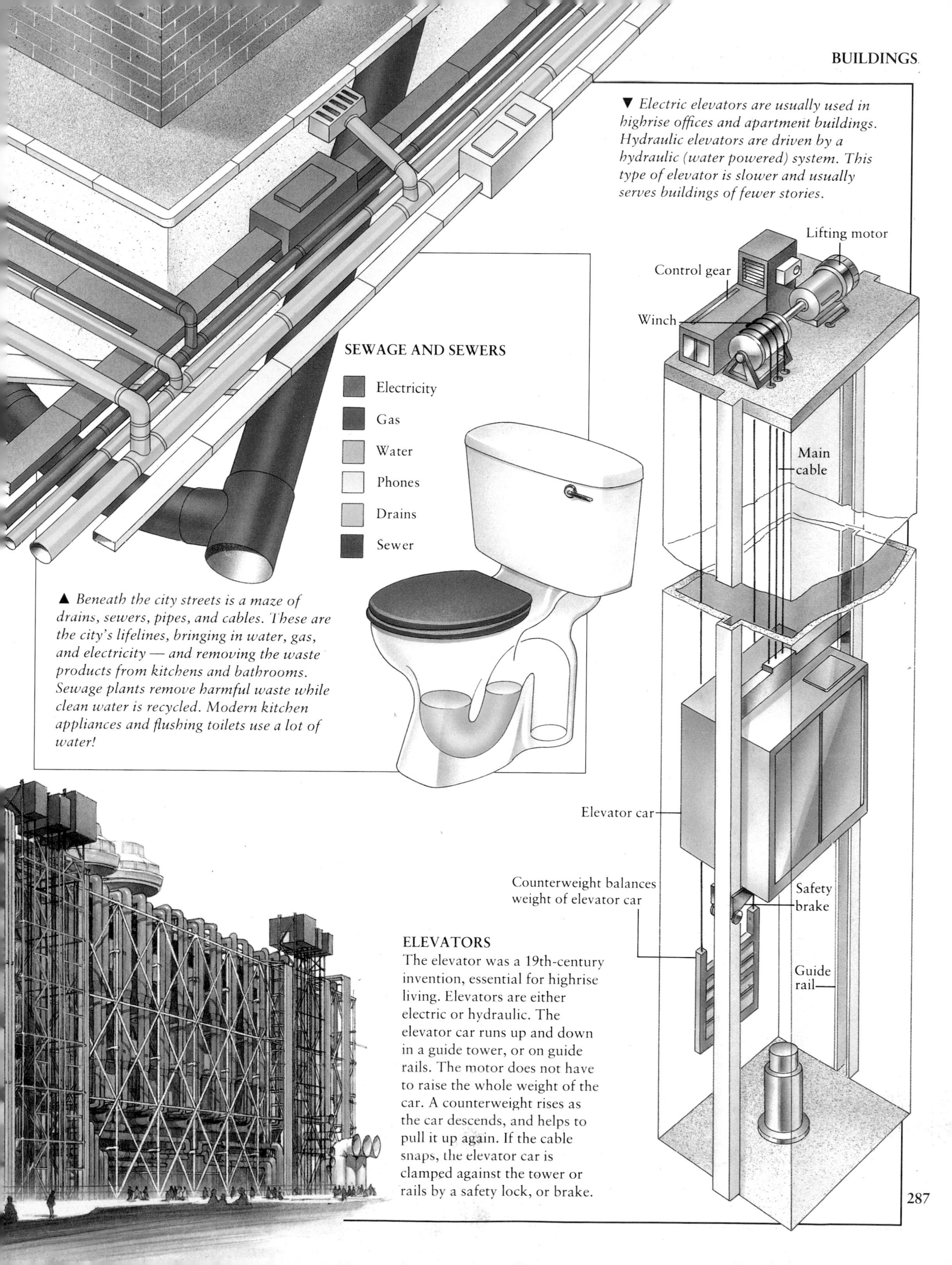

▼ *Electric elevators are usually used in highrise offices and apartment buildings. Hydraulic elevators are driven by a hydraulic (water powered) system. This type of elevator is slower and usually serves buildings of fewer stories.*

▲ *Beneath the city streets is a maze of drains, sewers, pipes, and cables. These are the city's lifelines, bringing in water, gas, and electricity — and removing the waste products from kitchens and bathrooms. Sewage plants remove harmful waste while clean water is recycled. Modern kitchen appliances and flushing toilets use a lot of water!*

ELEVATORS

The elevator was a 19th-century invention, essential for highrise living. Elevators are either electric or hydraulic. The elevator car runs up and down in a guide tower, or on guide rails. The motor does not have to raise the whole weight of the car. A counterweight rises as the car descends, and helps to pull it up again. If the cable snaps, the elevator car is clamped against the tower or rails by a safety lock, or brake.

Bridges, Tunnels, Dams, and Roads

Bridges were among the earliest structures. The simplest were log or rope bridges across rivers. The Romans spanned chasms with stone arches and built some of the most durable roads the world has seen. Dams were built 5,000 years ago to control river waters, and tunnels have a similarly long history. Methods of building were revolutionized in the 1800s by the use of steel-reinforced concrete, and by a more exact knowledge of the stresses and loads that such massive structures must withstand.

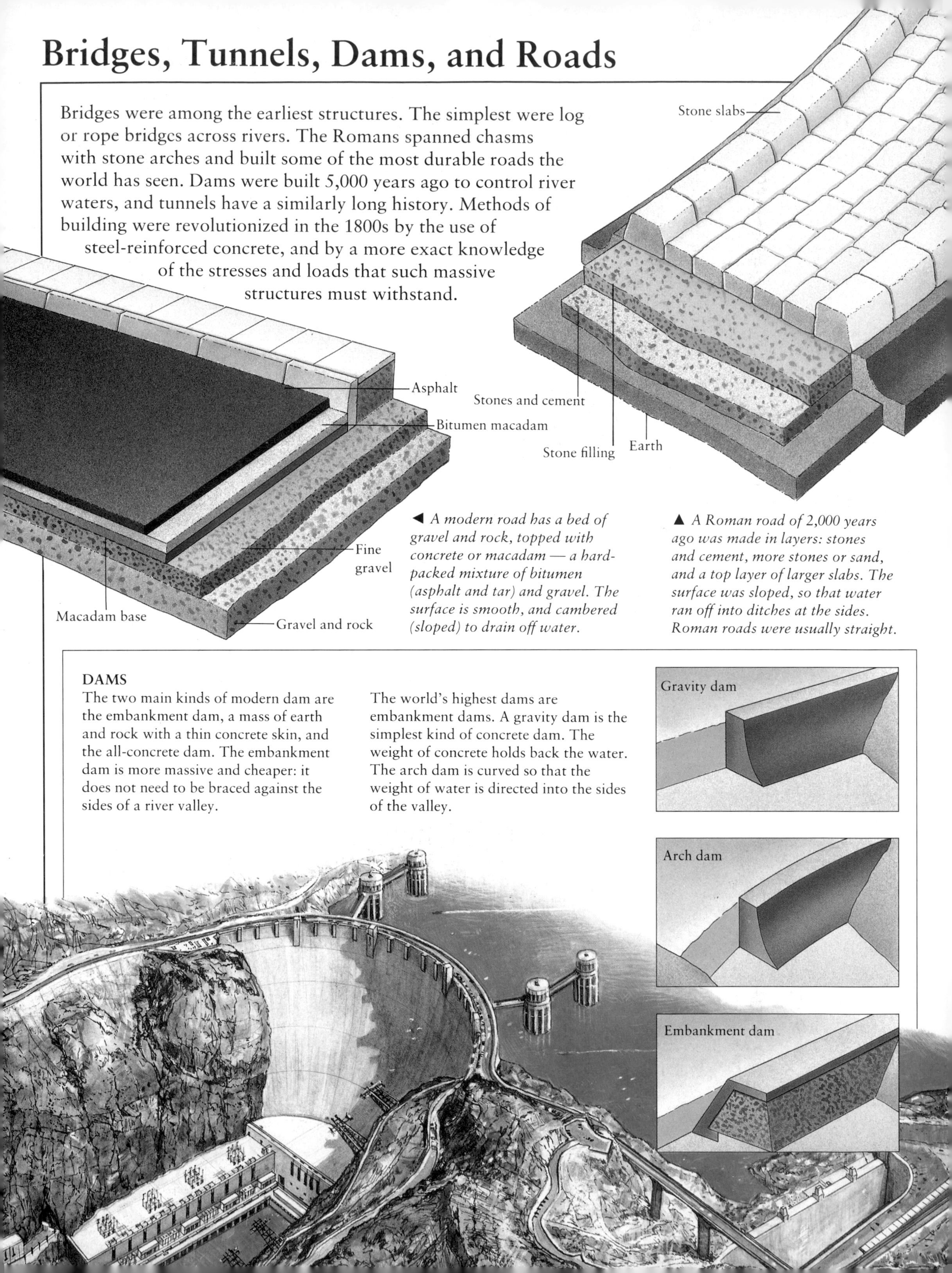

◀ A modern road has a bed of gravel and rock, topped with concrete or macadam — a hard-packed mixture of bitumen (asphalt and tar) and gravel. The surface is smooth, and cambered (sloped) to drain off water.

▲ A Roman road of 2,000 years ago was made in layers: stones and cement, more stones or sand, and a top layer of larger slabs. The surface was sloped, so that water ran off into ditches at the sides. Roman roads were usually straight.

DAMS

The two main kinds of modern dam are the embankment dam, a mass of earth and rock with a thin concrete skin, and the all-concrete dam. The embankment dam is more massive and cheaper: it does not need to be braced against the sides of a river valley.

The world's highest dams are embankment dams. A gravity dam is the simplest kind of concrete dam. The weight of concrete holds back the water. The arch dam is curved so that the weight of water is directed into the sides of the valley.

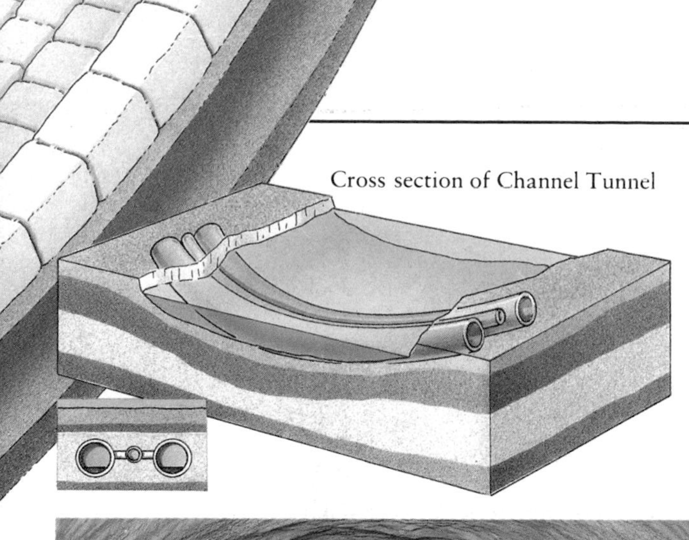

▲ *The Channel Tunnel, a rail link between England and France* (top), *was bored by tunneling machines weighing 1,300 tons each. Lasers kept the machines on course. The rotating cutting heads* (above) *were moved forward by hydraulic jacks.*

▼ *A bridge must support its own weight, the weight of the traffic crossing it, and withstand stresses caused by high winds. A steel girder cantilever design was used for the rail bridge across the Firth of Forth in Scotland (completed in 1890).*

TYPES OF BRIDGES

► *A beam bridge is basically a giant "log" placed across a stream. The beam (the roadway) stretches from bank to bank and can be supported by pillars buried deep in the riverbed.*

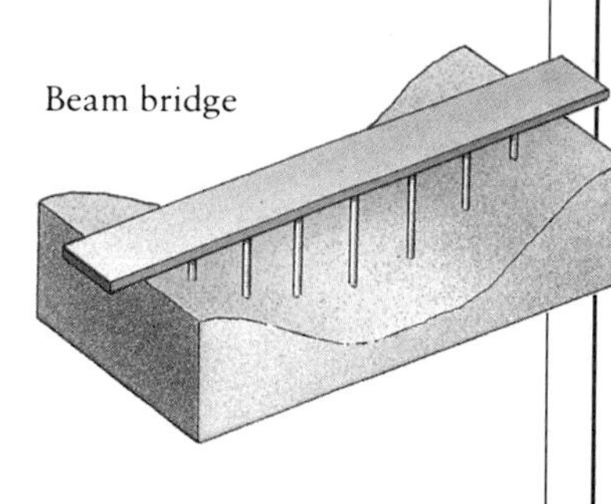

► *Arch bridges may have one or more arches. The arch is immensely strong, so long as its ends are fixed. The road is built on top of the arch, which takes the weight of the whole structure.*

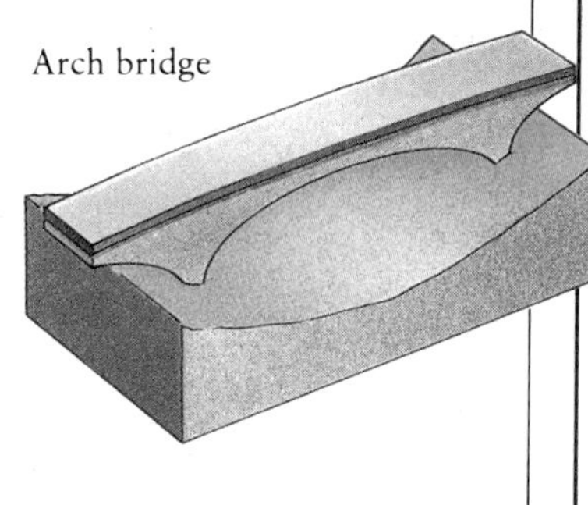

► *Suspension bridges are hung by steel cables from tall towers. The cables can carry huge loads and a single span (distance from tower to tower) over 3,300 ft. (1,000 m) is possible.*

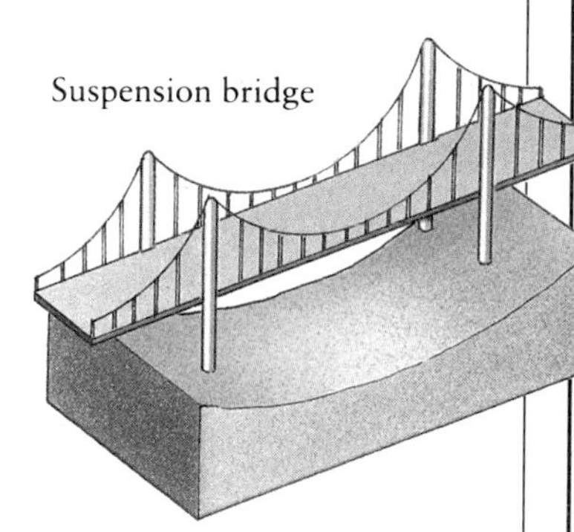

Transportation

Transportation developments have speeded every advance in civilization. Stone Age hunters dragged food home on wooden sleds or floated across streams on log rafts. By 3000 B.C. people had domesticated animals, fitted the wheel to the cart, and built the first seagoing ships. The horse and the sailing ship remained supreme until the late 1800s, when the railroad, steamship, and car brought undreamed-of speed and power. In the past 100 years, airplanes and spacecraft have transformed our world.

▼ *In 1804 Richard Trevithick developed the first steam locomotive.*

▲ *In the 1800s, people took coaches for long journeys.*

► *A powerfully-engined GMC "dump" truck.*

ROAD TRANSPORTATION

Trains now match some airplanes for speed. But most goods are carried by road in diesel trucks, and most people travel by car. The pedal bicycle (1839) offered cheap personal mobility. In the 1880s the car and motorcycle brought in the auto age, and a social and technological revolution. Horse cabs and horse-drawn carts lingered on into the early 1900s. By then, electric streetcars running on rails laid in city streets were popular. Buses replaced streetcars by the 1950s, but the "environmentally friendly" streetcar is now making a comeback in some cities.

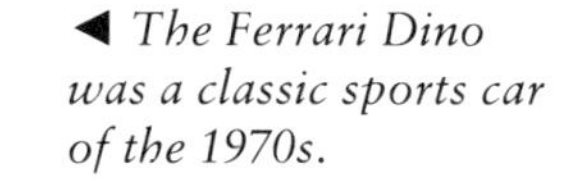

◄ *The Ferrari Dino was a classic sports car of the 1970s.*

OCEAN TRANSPORTATION

The shape of sails was a key to the development of ships. The fully rigged sailing ship — galleons and clippers — used a combination of square and triangular sails. Steam power was used first to drive paddlewheels, then screws, or propellers.

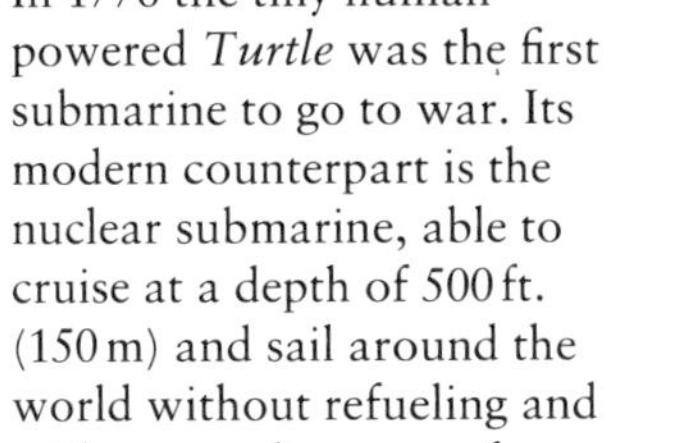

In 1776 the tiny human-powered *Turtle* was the first submarine to go to war. Its modern counterpart is the nuclear submarine, able to cruise at a depth of 500 ft. (150 m) and sail around the world without refueling and without needing to surface.

▼ *In ancient Egypt, ships were driven by oars and a single sail.*

▼ *Both the Americans and the British built fleets of clippers, the superfast merchant ships of the mid-1800s.*

▼ *American David Bushnell's submarine of 1775 was operated by hand-cranked screws.*

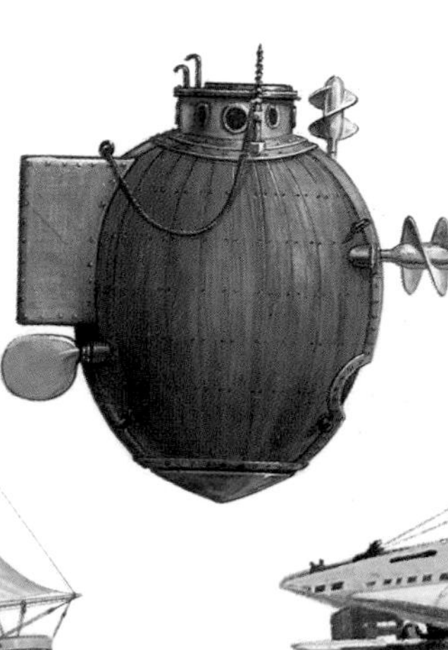

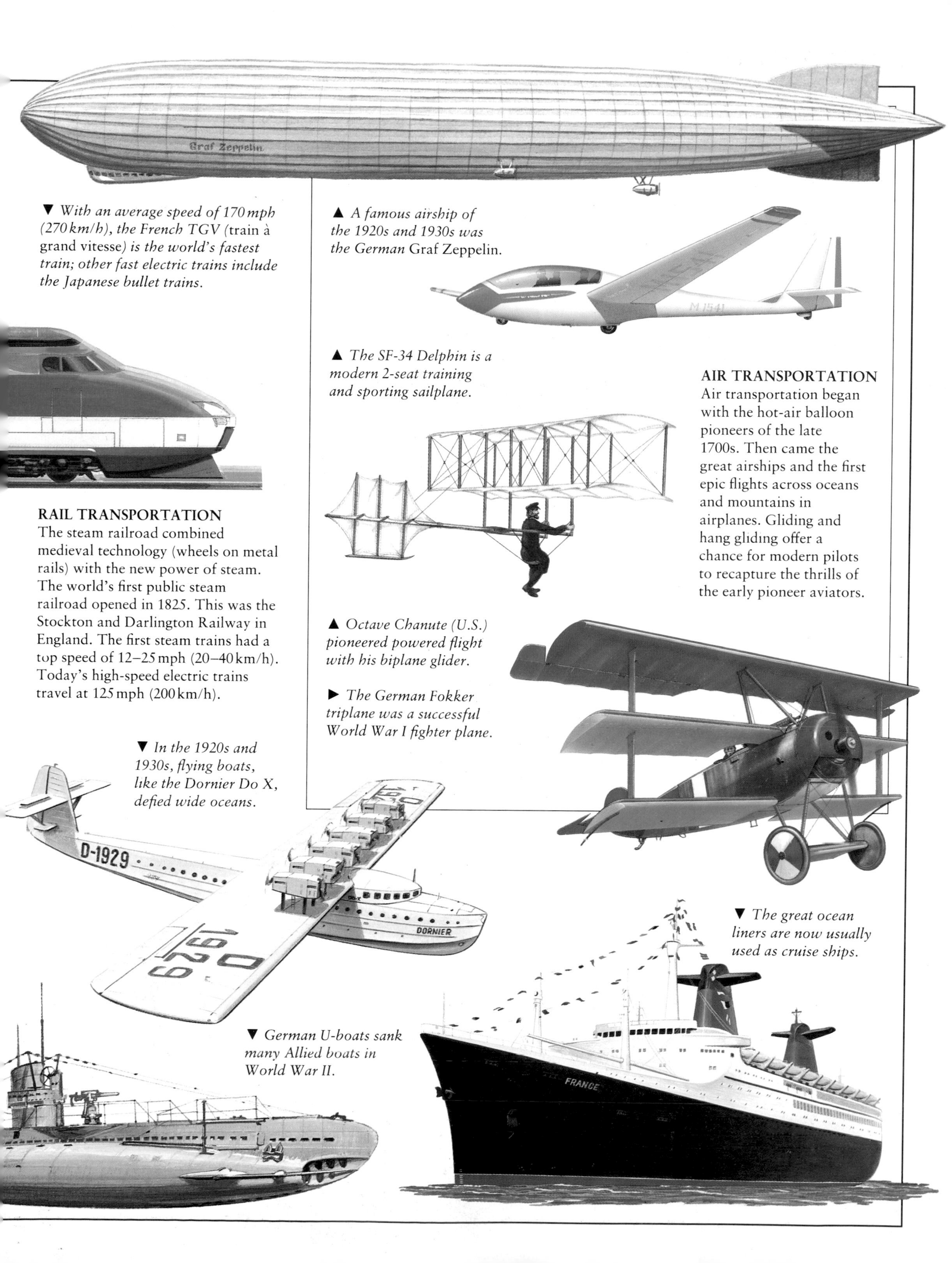

▼ *With an average speed of 170 mph (270 km/h), the French TGV* (train à grand vitesse) *is the world's fastest train; other fast electric trains include the Japanese bullet trains.*

RAIL TRANSPORTATION

The steam railroad combined medieval technology (wheels on metal rails) with the new power of steam. The world's first public steam railroad opened in 1825. This was the Stockton and Darlington Railway in England. The first steam trains had a top speed of 12–25 mph (20–40 km/h). Today's high-speed electric trains travel at 125 mph (200 km/h).

▼ *In the 1920s and 1930s, flying boats, like the Dornier Do X, defied wide oceans.*

▼ *German U-boats sank many Allied boats in World War II.*

▲ *A famous airship of the 1920s and 1930s was the German* Graf Zeppelin.

▲ *The SF-34 Delphin is a modern 2-seat training and sporting sailplane.*

▲ *Octave Chanute (U.S.) pioneered powered flight with his biplane glider.*

► *The German Fokker triplane was a successful World War I fighter plane.*

AIR TRANSPORTATION

Air transportation began with the hot-air balloon pioneers of the late 1700s. Then came the great airships and the first epic flights across oceans and mountains in airplanes. Gliding and hang gliding offer a chance for modern pilots to recapture the thrills of the early pioneer aviators.

▼ *The great ocean liners are now usually used as cruise ships.*

Land Transportation

The wheel was probably the single most important invention in the history of transportation. Its unknown inventor may have first used log rollers to move heavy stones. Together with the speed and muscle power of the horse, the wheel gave people new mobility. But for thousands of years, roads were usually terrible and few people actually traveled very far. Today, millions take to the roads every day. Cars and trucks fill the roads, causing traffic jams. The wheel's success has created a modern-day problem.

◀ *A wheel and axle together are a rotating lever, which increases force. A short movement of the axle produces a greater movement of the wheel. The wheel makes movement easier by reducing friction. Yet it needs friction, to push against the ground. Otherwise it would just spin around.*

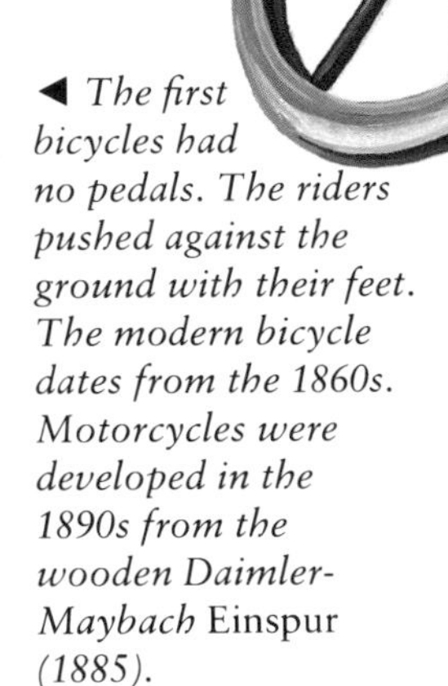

WHEEL

The wheel was invented in the Middle East over 5,000 years ago. The earliest land vehicles were four-wheeled carts. The wheels were made from wooden boards fixed together and roughly rounded. Spoked wheels appeared by 1500 B.C. Wheel design changed little until the 1800s when pneumatic (air-filled) tires were developed.

◀ *The first bicycles had no pedals. The riders pushed against the ground with their feet. The modern bicycle dates from the 1860s. Motorcycles were developed in the 1890s from the wooden Daimler-Maybach* Einspur *(1885).*

▼ *Spoked wheels were fitted to light chariots from about 1500 B.C. The horse collar (invented c.A.D. 800), which did not choke the animal, allowed horses to pull much heavier wagons.*

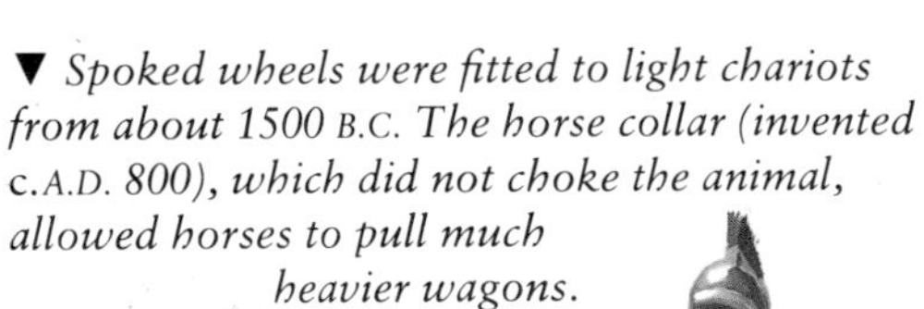

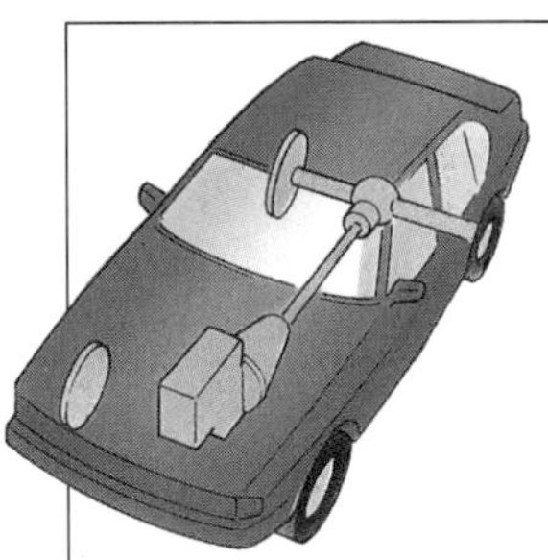

DISC BRAKE

A car brake works by friction. There are two kinds of car brake, disc brakes and drum brakes. In a disc brake, pads press against both sides of a disc on the axle. Pressure from the brake pedal is increased by fluid hydraulics or by air pressure.

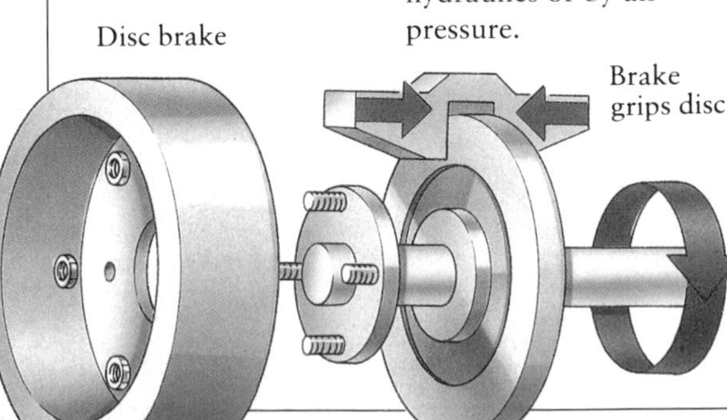

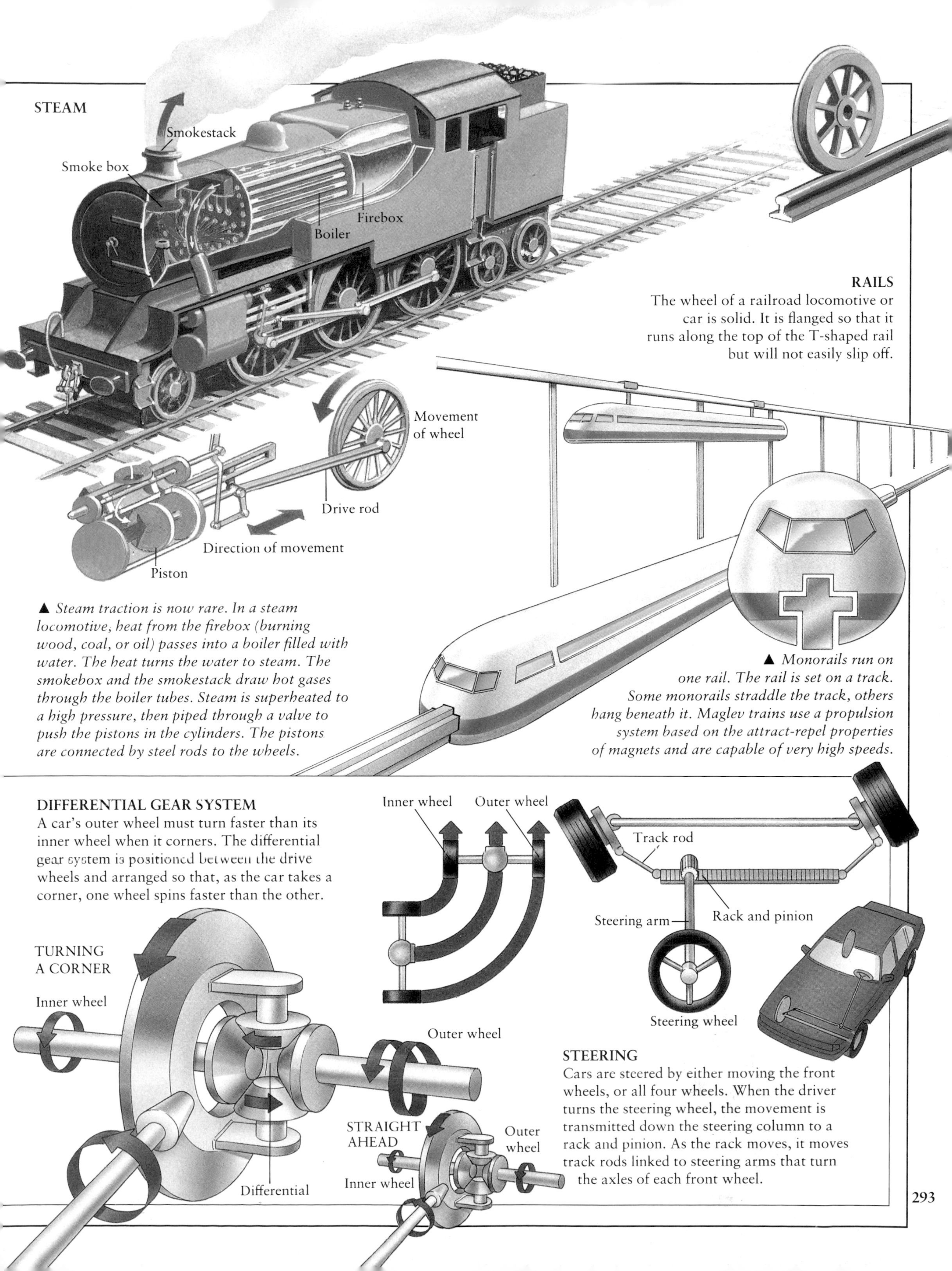

STEAM

▲ *Steam traction is now rare. In a steam locomotive, heat from the firebox (burning wood, coal, or oil) passes into a boiler filled with water. The heat turns the water to steam. The smokebox and the smokestack draw hot gases through the boiler tubes. Steam is superheated to a high pressure, then piped through a valve to push the pistons in the cylinders. The pistons are connected by steel rods to the wheels.*

RAILS

The wheel of a railroad locomotive or car is solid. It is flanged so that it runs along the top of the T-shaped rail but will not easily slip off.

▲ *Monorails run on one rail. The rail is set on a track. Some monorails straddle the track, others hang beneath it. Maglev trains use a propulsion system based on the attract-repel properties of magnets and are capable of very high speeds.*

DIFFERENTIAL GEAR SYSTEM

A car's outer wheel must turn faster than its inner wheel when it corners. The differential gear system is positioned between the drive wheels and arranged so that, as the car takes a corner, one wheel spins faster than the other.

STEERING

Cars are steered by either moving the front wheels, or all four wheels. When the driver turns the steering wheel, the movement is transmitted down the steering column to a rack and pinion. As the rack moves, it moves track rods linked to steering arms that turn the axles of each front wheel.

Sea Transportation

The first boats were floating logs or rafts, driven by paddles. The development of sail took thousands of years, with sailors in different lands borrowing ideas from each other. The stern rudder, for example, came from China. Paddle-driven steamships were developed in the early 1800s, at about the same time as iron hulls began to be used instead of wooden ones. From the 1830s, the screw propeller replaced the paddle wheel. Ships grew larger and today the biggest are 500,000-ton supertankers.

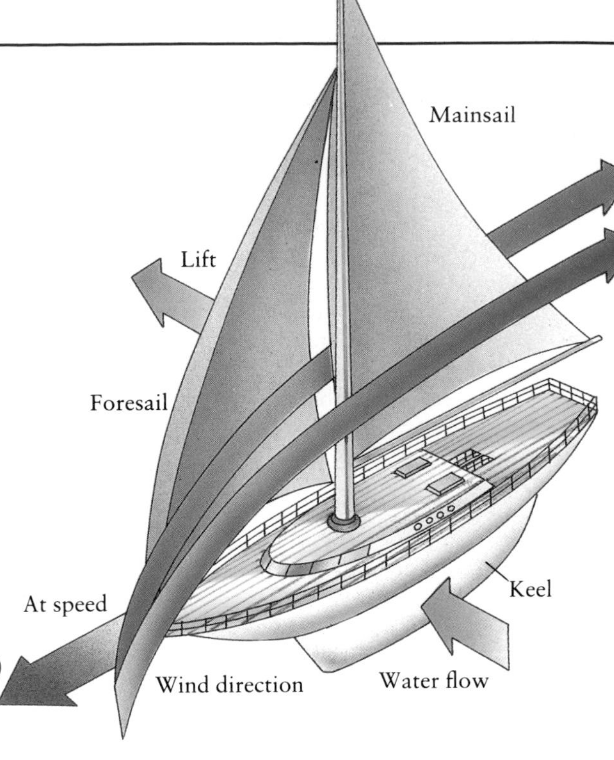

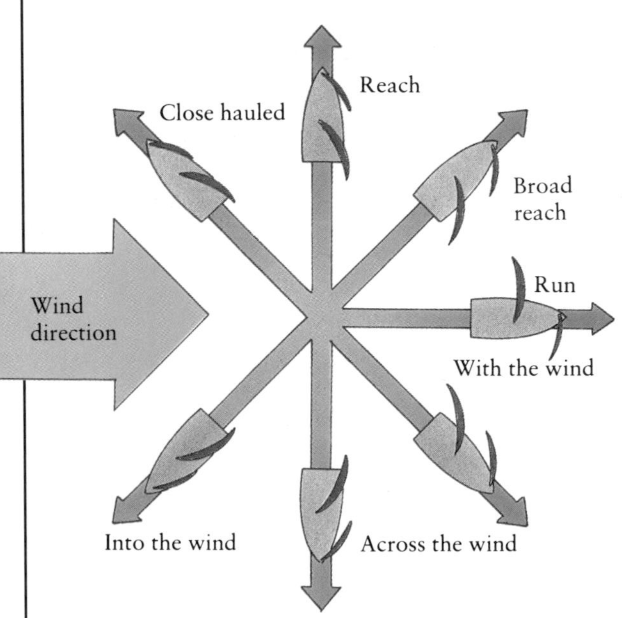

SAILING CRAFT

A sailing craft, such as a yacht, is dependent on the wind, but it can make headway in any direction — though at different rates. Sailing with the wind (running) is not the fastest. Sailing across the wind (reaching) is faster. To sail into the wind, a boat must tack or zigzag at an angle of 45° to the wind direction. Sailing directly into the wind causes the sails to flap uselessly.

SAILS

When sailing into the wind, the sails act like a slotted wing, and a strong suction force (similar to an airfoil's lift) is produced. This operates in two directions: a forward thrust and a sideways push. The keel beneath the boat counteracts this sideways push and keeps the boat from capsizing.

HOVERCRAFT

Air-cushion vehicles (ACVs), or hovercraft, ride on a cushion of air, blown downward by fans. The air is trapped inside the skirt of the craft. ACVs can travel over both land and water.

Air cushion

MODERN SHIPS

Most modern ships are powered either by steam turbines or diesel engines. Both need oil as fuel. Nuclear power plants have been tried in naval ships but not adopted for general use. The biggest ships afloat are oil tankers, bulk carriers, and aircraft carriers. A big tanker may need 5 miles (8 km) to slow to a halt; at normal speed this can take as long as 20 minutes.

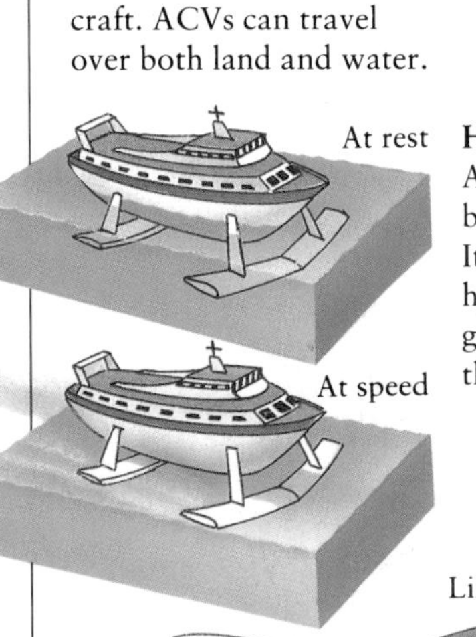

HYDROFOIL

A hydrofoil is faster than ordinary craft because its design reduces water resistance. It has wings beneath its hull. At rest, the hydrofoil floats low in the water. As it gains speed, the wings lift the craft so that it skims across the water.

SHIP DEVELOPMENT

4000 B.C.: Egyptians build reed boats
500 B.C.: Oared trireme galley
A.D. 1200s: Stern rudder replaces steering oar
1500: First galleons
1800: Fulton's submarine *Nautilus*
1802: First working steamship, Symington's *Charlotte Dundas*
1807: Fulton's steamboat *Clermont* first carries passengers
1818: First all-iron sailing ship
1830s: Screw propeller
1838: Steamer *Sirius* crosses Atlantic
1840s: First clippers
1872: First oil tanker with engines in the stern (as now)
1894: First turbine ship
1906: First hydrofoil
1954: First nuclear submarine

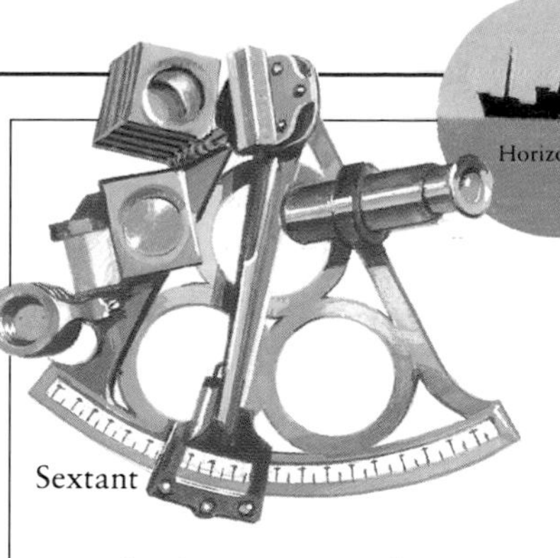

Horizon in view

Mirrors adjusted so horizon and Sun align; altitude measured

NAVIGATION

The sextant is an optical instrument with mirrors and a telescope. It measures the altitude of the Sun above the horizon, in order to find the ship's latitude.

► *The ship's compass has a compass card, with magnets beneath, floating in liquid inside a bowl. Navigators also use radar scanners and direction-finders, to receive signals from other ships, shore beacons, and satellites. Echo-sounders measure the depth of water beneath the ship.*

PROPELLER BLADES

The blades of a ship's screw propeller force water backward as they are turned by the engines. The water pushes back, with equal force, against the ship. This reaction force moves the ship forward. The propeller blades are curved, like airfoils, producing additional forward suction.

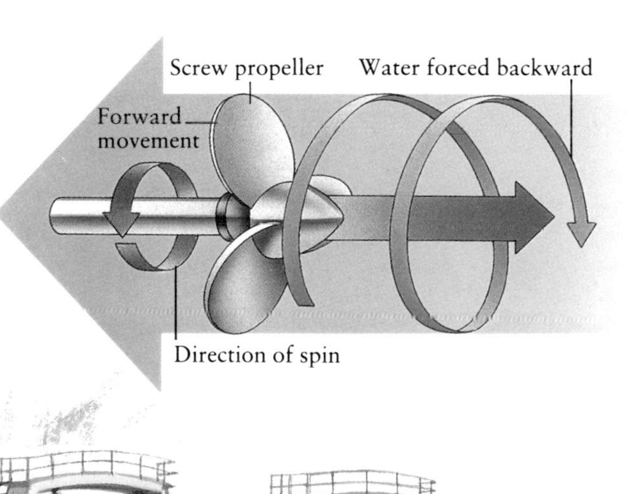

STEERING

The ship's rudder hangs in the water flow produced by the ship's forward motion. In straight motion, the water flows past the rudder. Moving the rudder left or right pushes water aside. The water pushes back with equal and opposite force. The effect of this reaction is to push the stern of the boat in the opposite direction to the rudder's movement.

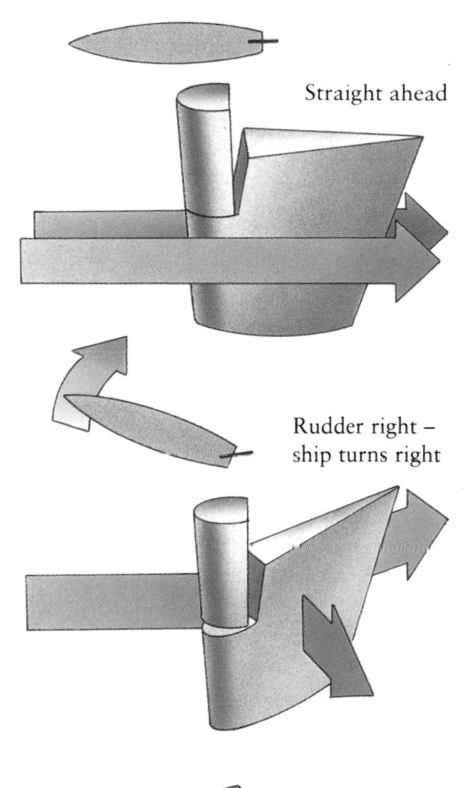

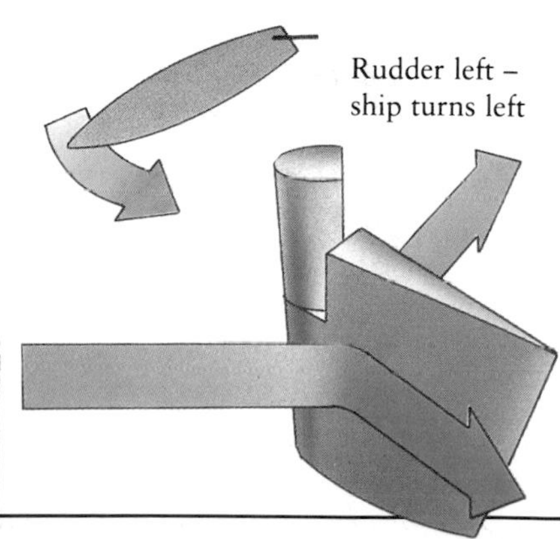

▼ *A ship floats if it weighs less than the weight of water it displaces. Only warships are measured by displacement. Others are measured in deadweight tonnage (total weight) or gross tonnage (internal volume).*

paration of fish
uel tanks
Diesel engines

Air Transportation

The pioneers of air travel knew little about the forces acting on an object in flight. The Montgolfier brothers' hot-air balloons of the late 1700s were rumored to contain a mysterious "rising air." During the 1800s, scientists began to figure out the forces involved in flight, through experiments with gliders and airships. From 1903, when the Wright brothers flew their first flimsy airplane, to the jumbos and supersonic jets of today, air transportation has developed with astonishing speed.

INSIDE AN AIRLINER
A modern airliner, such as the Airbus, has a wide body able to carry 300 or more passengers. The body is known as the fuselage, and houses crew, passengers, and cargo. Major sections of the plane (wings, tail, engines) are often built in different factories.

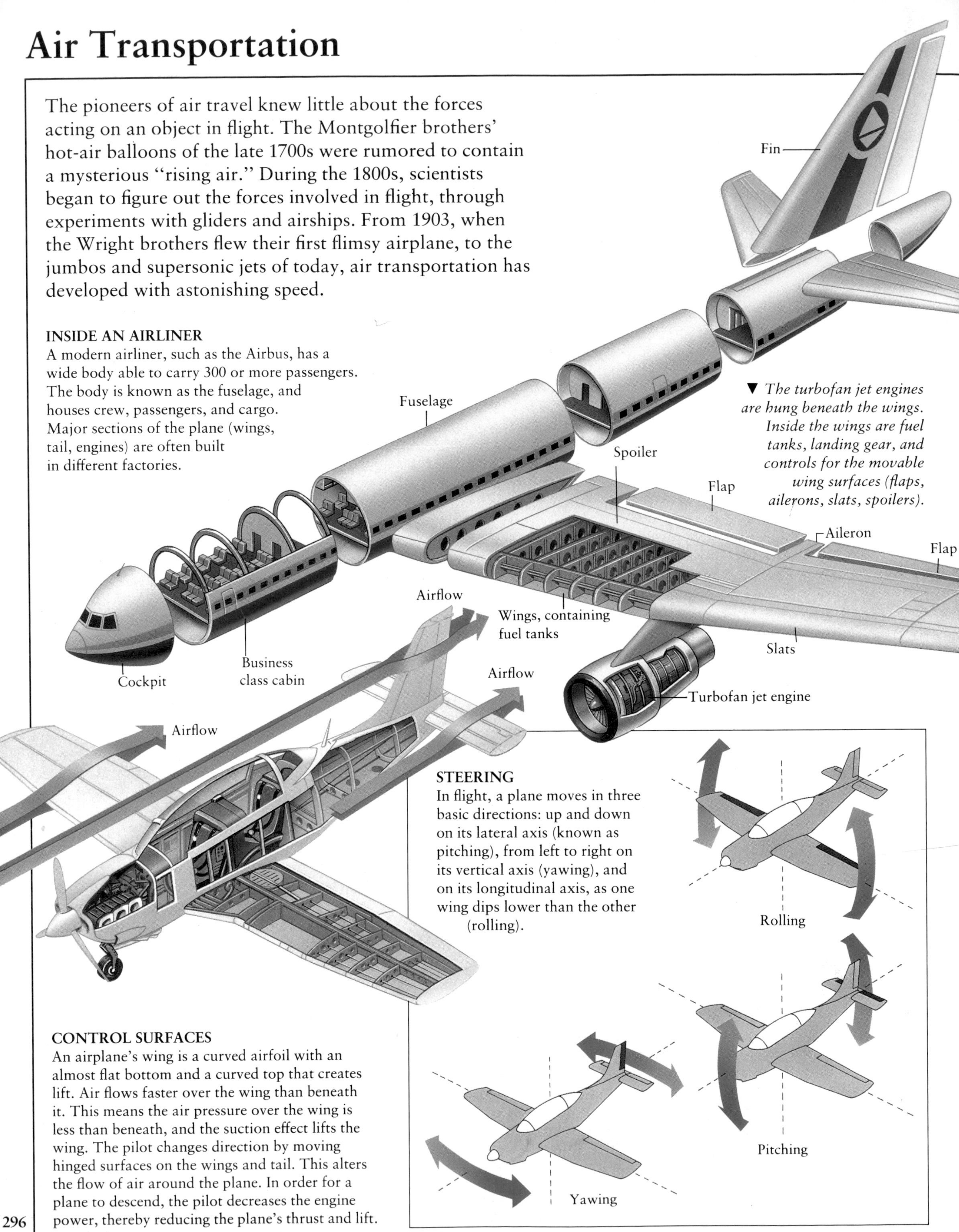

▼ *The turbofan jet engines are hung beneath the wings. Inside the wings are fuel tanks, landing gear, and controls for the movable wing surfaces (flaps, ailerons, slats, spoilers).*

STEERING
In flight, a plane moves in three basic directions: up and down on its lateral axis (known as pitching), from left to right on its vertical axis (yawing), and on its longitudinal axis, as one wing dips lower than the other (rolling).

CONTROL SURFACES
An airplane's wing is a curved airfoil with an almost flat bottom and a curved top that creates lift. Air flows faster over the wing than beneath it. This means the air pressure over the wing is less than beneath, and the suction effect lifts the wing. The pilot changes direction by moving hinged surfaces on the wings and tail. This alters the flow of air around the plane. In order for a plane to descend, the pilot decreases the engine power, thereby reducing the plane's thrust and lift.

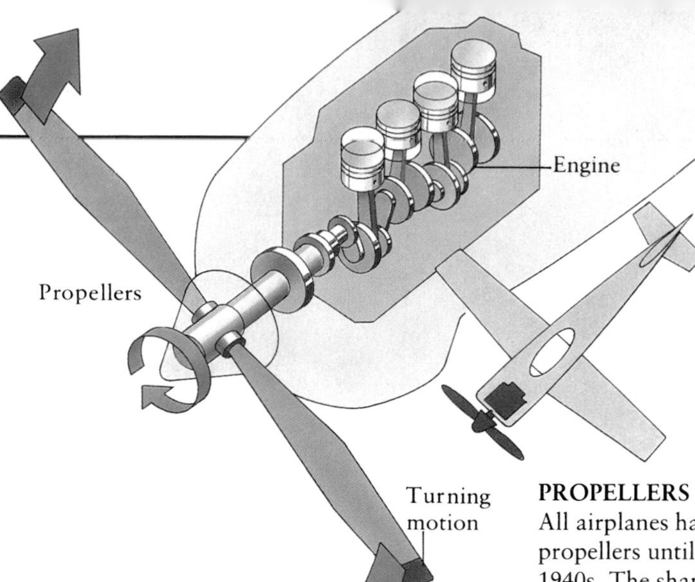

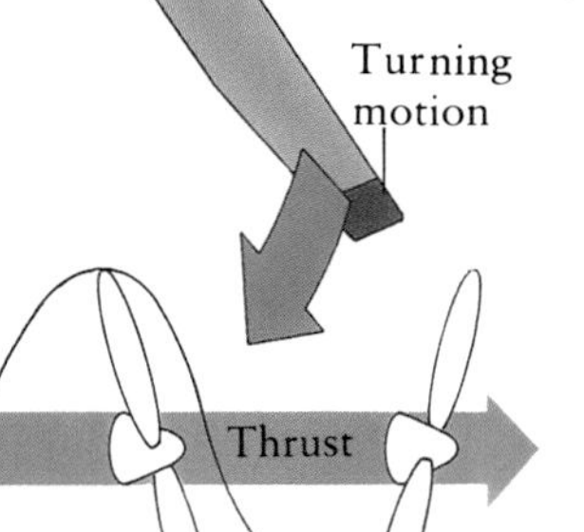

PROPELLERS
All airplanes had propellers until the 1940s. The shape of a propeller, or airscrew, is an airfoil, similar to the wing. As it is spun by the engine, it bites into the air to produce thrust.

JET ENGINES
Jet engines were invented in the 1930s and first used in aircraft during World War II. In a simple jet engine, air is sucked in at the front by compressors into a combustion chamber where fuel is burned. Hot exhaust gases shoot out backward, pushing the plane forward. Modern turbofan engines have longer blades at the front of the compressor, to push in extra air.

Higher speed

Lower speed

▲ *A fast jet has a swept-back wing, to minimize drag. But at low speeds, straight wings fly better. The swing-wing airplane has movable wings. They stick out for takeoff and landing but are swept back for high speed.*

FACTS ABOUT FLIGHT

- The first jet-engine aircraft was the German Heinkel He 178; its first flight was in 1939.
- The first jet airliners were the De Havilland Comet and the Boeing 707, both introduced in the 1950s.
- The first plane to fly faster than sound and break the "sound barrier" was the U.S. Bell XS-1 in 1947.
- The world's fastest plane was the U.S. X-15 rocket plane, which had a top speed of 4,534 mph (7,297 km/h). It was used for research in the 1960s.

HELICOPTERS
Helicopters are the most successful vertical takeoff and landing aircraft (VTOL) and the most versatile of all air transportation. Helicopters obtain lift and propulsion from engine-driven rotors. As they are able to hover at nearly zero ground speed, they can play vital roles in transporting ill people, air-sea rescue, sowing crops, or as military transportation. The CH-47 Chinook *(left)* is a medium-sized helicopter with a tandem-rotor; it was originally developed for the U.S. Army.

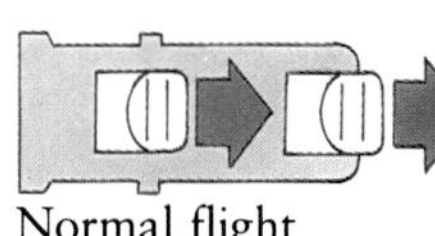
Normal flight

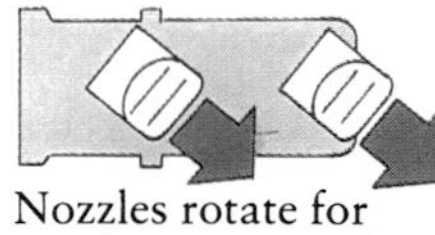
Nozzles rotate for acceleration

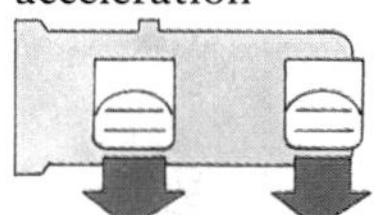
Thrust engines blast downward for takeoff

VERTICAL FLIGHT
The Harrier is one of the few fixed-wing planes able to rise straight into the air and hover in midair. It has four swiveling nozzles, which direct exhaust gases from the jet engines. When the nozzles direct thrust downward, the plane takes off or lands vertically, or hovers.

Materials

The materials we use are natural (wood, cotton, wool) or synthetic (glass, steel, plastic), or sometimes a mixture of the two. The first materials people used were natural. They learned to shape stones, to weave fibers, and to make pots and bricks from clay. They discovered how to extract metals from ores in the ground, and how to shape the metal. In the Bronze and Iron Ages, the Earth's resources seemed limitless. Now we must learn to conserve and recycle, and to make the best use of all materials.

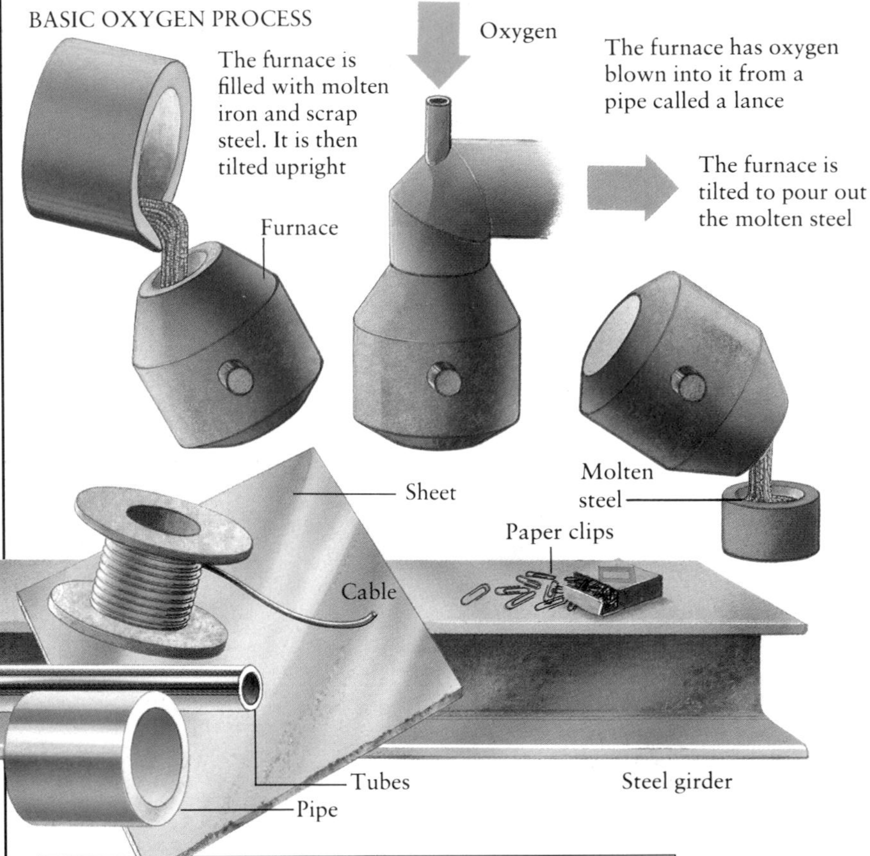

STEEL
More than half the world's steel is made by the basic oxygen process. The raw material is about three parts melted iron and one part scrap steel. Blowing oxygen into the melted iron raises the temperature and gets rid of impurities. Two other processes (electric arc and open hearth) are also common.

USES OF STEEL
Solid lumps of steel (known as blooms, billets, or slabs) are finished in various ways: for example, by rolling them into sheets or bars, or by drawing them into thin wires. Rust can be prevented by coating the steel with zinc (a process called "galvanizing").

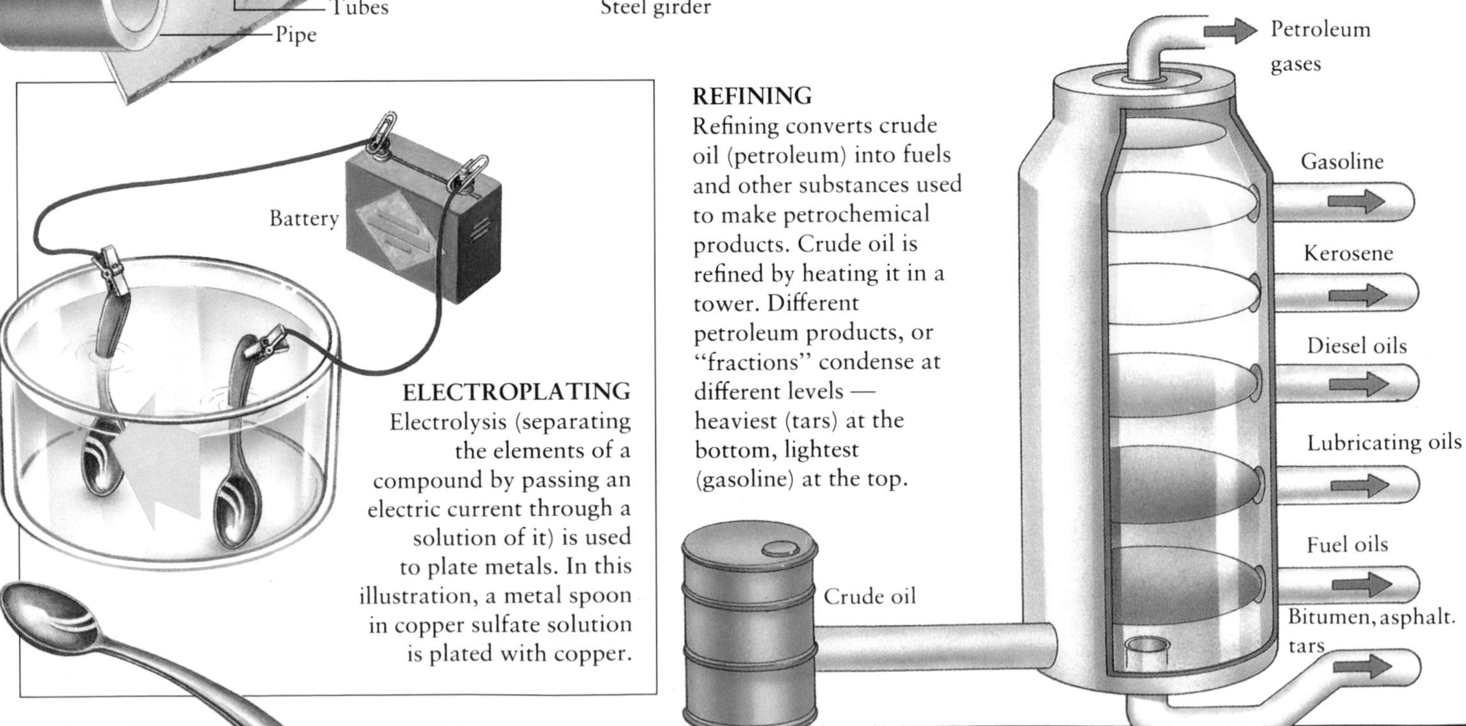

ELECTROPLATING
Electrolysis (separating the elements of a compound by passing an electric current through a solution of it) is used to plate metals. In this illustration, a metal spoon in copper sulfate solution is plated with copper.

REFINING
Refining converts crude oil (petroleum) into fuels and other substances used to make petrochemical products. Crude oil is refined by heating it in a tower. Different petroleum products, or "fractions" condense at different levels — heaviest (tars) at the bottom, lightest (gasoline) at the top.

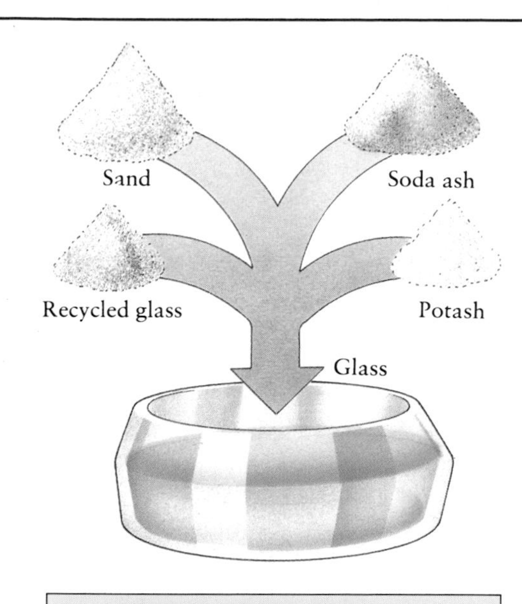

GLASS

Glass is made by mixing and heating sand, limestone, and soda ash. When these ingredients melt they turn into glass, which hardens as it cools. Glass is in fact not a solid but a "supercooled" liquid. It can be shaped by blowing, pressing, drawing, casting into molds, rolling, and floating across molten tin, to make large sheets.

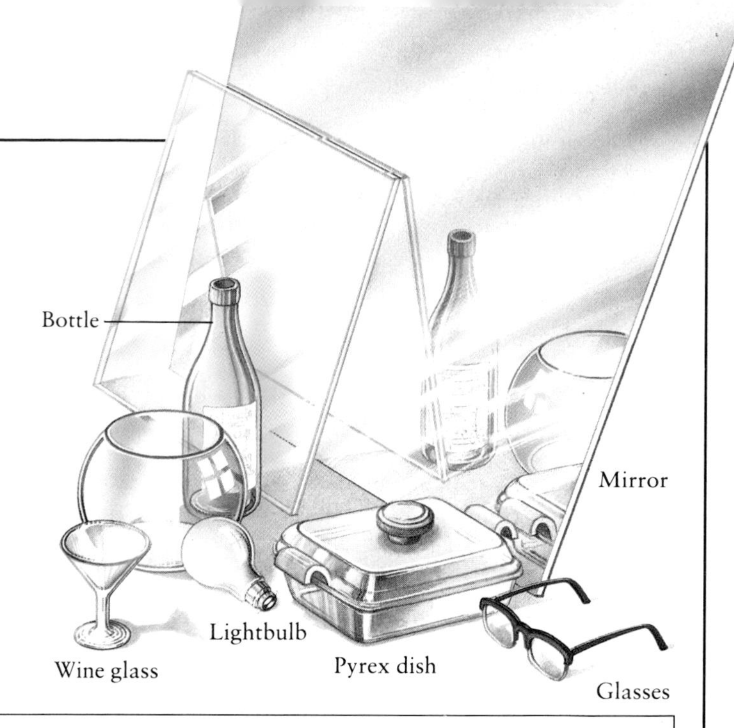

MATERIAL BREAKTHROUGHS

- Stainless steel is an alloy of steel with chromium or nickel. It was first made by accident.
- All plastics are chemical compounds called polymers. The first synthetic plastic was Bakelite (1908). Other important plastic breakthroughs were cellophane (1912), polystyrene (1930s) and PVC (1940s).
- The first synthetic fiber was nylon, developed in the 1930s as a cheap alternative to silk.
- Carbon fibers were first used by Edison as filaments in his light bulb (1879).

CERAMICS

Ceramic objects, such as pottery and porcelain, electrical insulators, bricks, and roof tiles, are all made from clay. The clay is shaped or molded when wet and soft, and heated or "fired" in a kiln until it hardens.

TEXTILES

A textile is any cloth made from woven fibers. People first wove textiles nearly 8,000 years ago. The first synthetic fibers, such as acrylic and polyester, were made from cellulose in the 1930s and 1940s. Other synthetics, like nylon, come from oil. Most synthetic fibers are stronger than natural fibers.

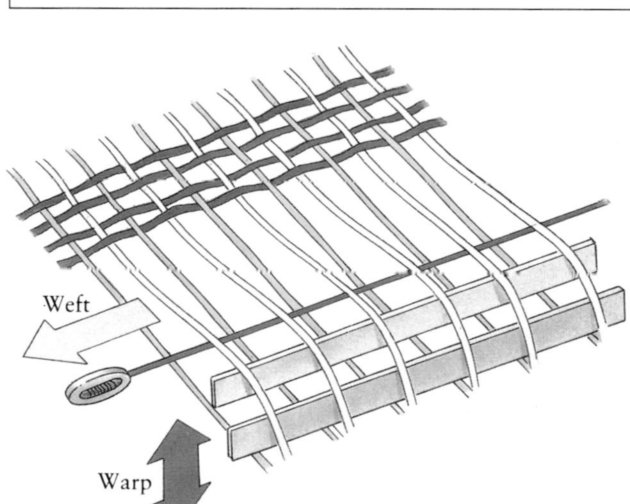
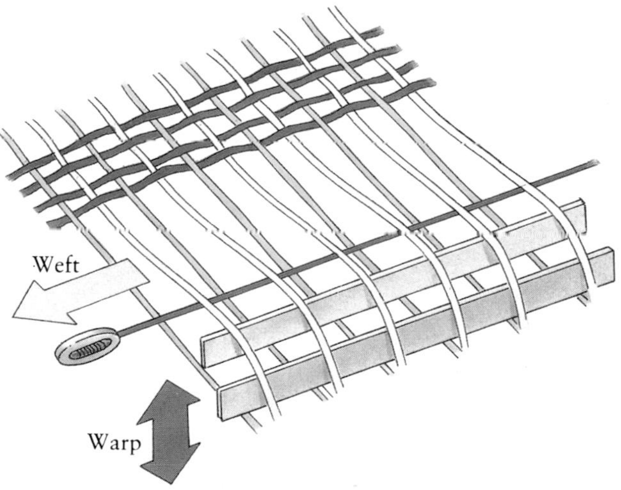

WEAVING

In weaving, threads are joined in a criss-cross pattern to make cloth. Fibers from plants (flax, cotton) or animals (wool) are first spun (twisted) into thread. Then the threads are woven on a loom. Lengthwise threads make up the warp; crosswise threads are the weft.

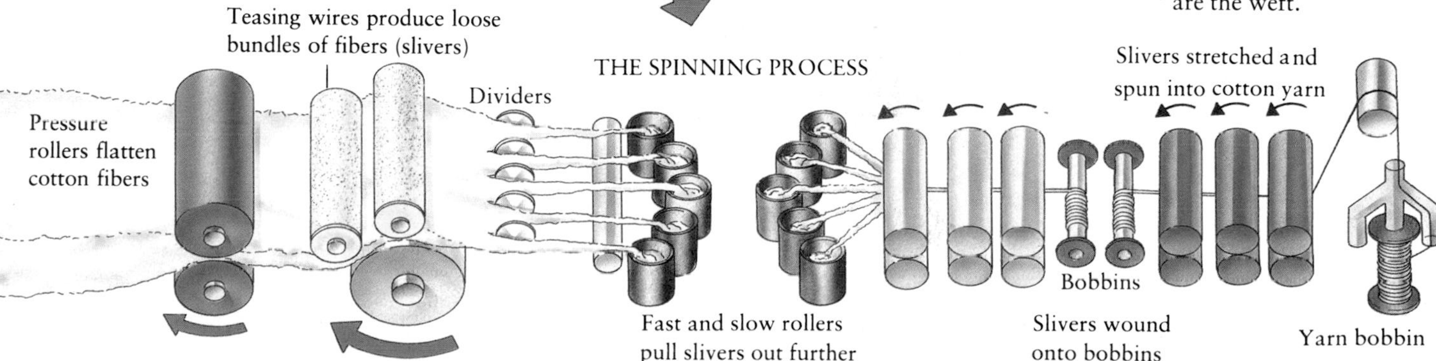

Farming

Farming began about 10,000 years ago, probably in the Middle East. Farming technology developed slowly as iron tools replaced wooden sticks and flint sickles. Later, heavier plows, drawn by animals, came into use, and crop rotation was developed. After the 1700s farms began to be mechanized. Two kinds of farming are now practiced worldwide: subsistence (the farmer plants crops to eat) and cash (the farmer plants crops to sell). Science has helped boost crop yields and reduce losses from pests and disease.

CHANGES IN FARMING

8000–7000 B.C.: Sowing of cereals, domestication of animals.
4000 B.C.: Irrigation of crops.
500 B.C.: Iron tools, ox-drawn plow.
100 B.C.: Romans practice rotation of crops.
1400s: Farmers in Europe begin to enclose (fence) their land. Enclosure meant bigger farms and improved methods.
1500s: New plants (corn, tomato, potato) brought from America to Europe.
1600s: European farmers start to grow clover and turnips to feed their animals in winter. Before this, most pigs and cattle were killed in the autumn.
1700s: Scientific farming begins. Improved breeds of livestock. Tull's seed drill (1701).
1800s: Threshing and reaping machines. McCormick's reaper-binder (1873). Steam power. Food canning and refrigeration.
1900s: Tractors replace horses on many farms. Fewer but larger farms means fewer farmworkers. Chemical fertilizers, weedkillers, and pesticides. Green revolution boosts crop yield. Genetically engineered disease-free plants. Intensive rearing of livestock.

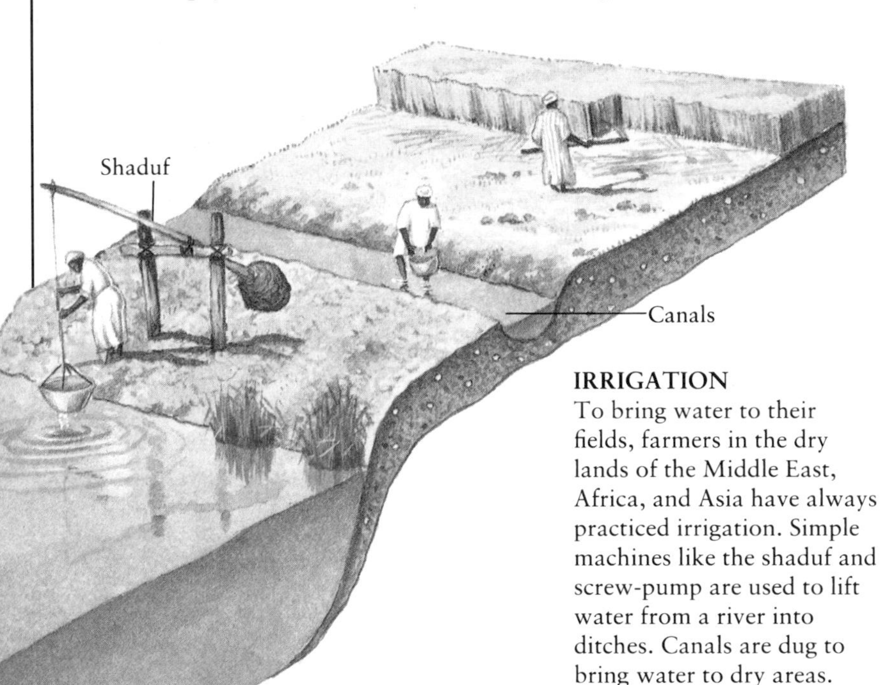

IRRIGATION

To bring water to their fields, farmers in the dry lands of the Middle East, Africa, and Asia have always practiced irrigation. Simple machines like the shaduf and screw-pump are used to lift water from a river into ditches. Canals are dug to bring water to dry areas.

▼ *Aerial crop spraying defends field crops against pests.*

PEST CONTROL

Anything that eats crops or causes disease is a pest. Insect pests (for example, locusts) are devastating. Chemicals can destroy pests, but biological control (for example, the sterilization of males, or finding a natural predator) does less harm to other living things.

THE PLOW

The modern tractor plow has a number of blades, or bottoms. In a typical plow, the share cuts a furrow through the soil; the moldboard then turns the soil and breaks it up. There are also disk and rotary plows. But in order to prevent soil erosion, a serious problem, "no till" agriculture is now being practiced by many farmers.

Early plows

Modern plow

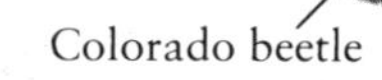

Colorado beetle

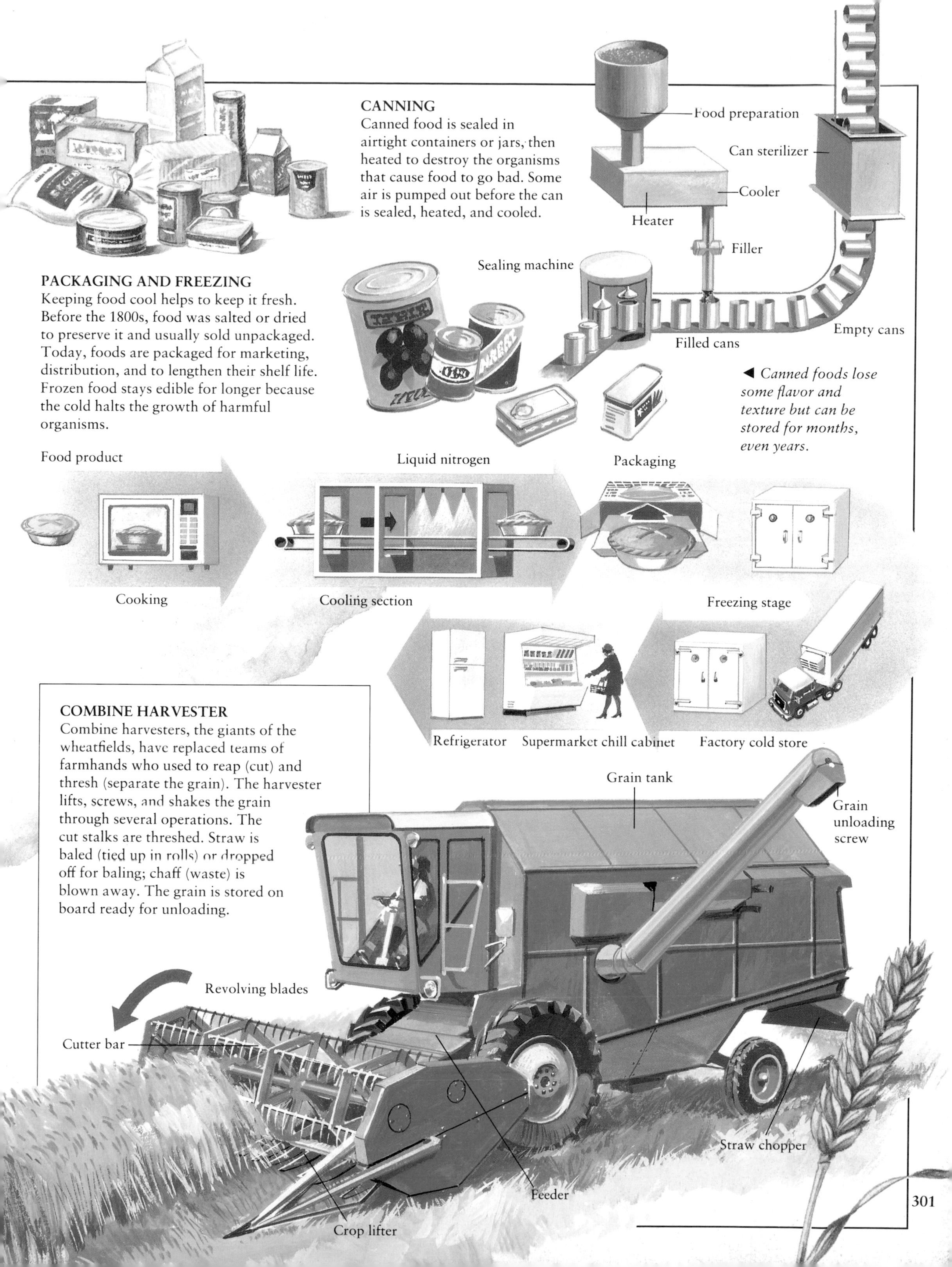

CANNING

Canned food is sealed in airtight containers or jars, then heated to destroy the organisms that cause food to go bad. Some air is pumped out before the can is sealed, heated, and cooled.

PACKAGING AND FREEZING

Keeping food cool helps to keep it fresh. Before the 1800s, food was salted or dried to preserve it and usually sold unpackaged. Today, foods are packaged for marketing, distribution, and to lengthen their shelf life. Frozen food stays edible for longer because the cold halts the growth of harmful organisms.

◀ *Canned foods lose some flavor and texture but can be stored for months, even years.*

COMBINE HARVESTER

Combine harvesters, the giants of the wheatfields, have replaced teams of farmhands who used to reap (cut) and thresh (separate the grain). The harvester lifts, screws, and shakes the grain through several operations. The cut stalks are threshed. Straw is baled (tied up in rolls) or dropped off for baling; chaff (waste) is blown away. The grain is stored on board ready for unloading.

Medicine

Doctors of the ancient world combined folklore, observation, and philosophy to treat illness. They carried out surgery without anesthetics, using techniques such as acupuncture. Scientific medicine began in the 1600s, with the discovery of microscopic bacteria and with advances in anatomical knowledge. Surgery was made safer in the 1800s with anesthetics and antiseptics. Today, doctors use antibiotics to kill bacteria. Surgeons can transplant organs, and scientists probe the secrets of the gene.

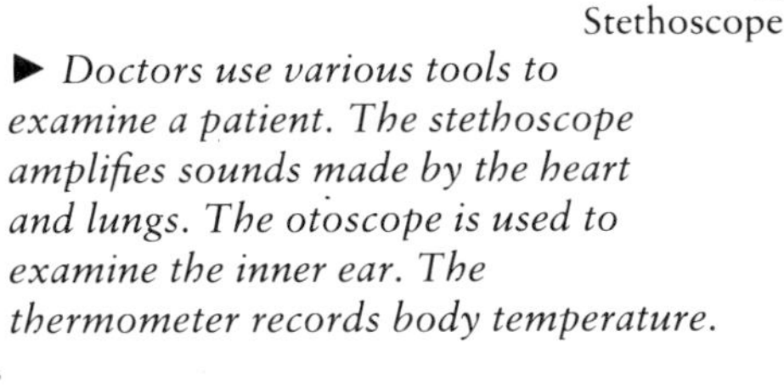

► *Doctors use various tools to examine a patient. The stethoscope amplifies sounds made by the heart and lungs. The otoscope is used to examine the inner ear. The thermometer records body temperature.*

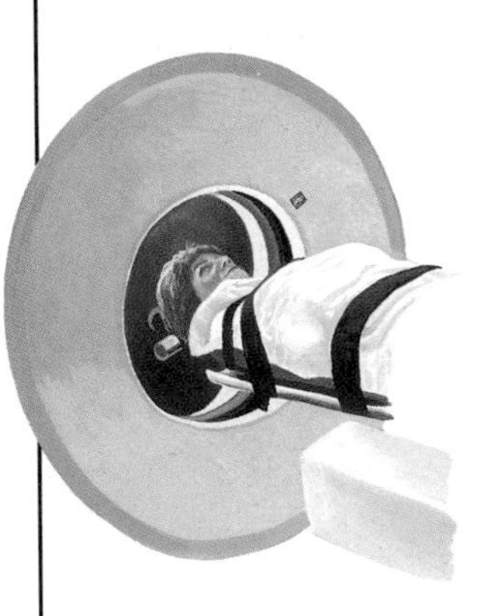

▲ *A patient having a body scan. CAT (computerized axial tomography) takes X-ray pictures, computerized into "slice" images.*

MICROSCOPE

The medical microscope dates from the mid 1600s, when bacteria were seen through lenses for the first time. The optical microscope, and the more powerful electron microscope (1930s), have enabled scientists to unravel the secrets of the cell.

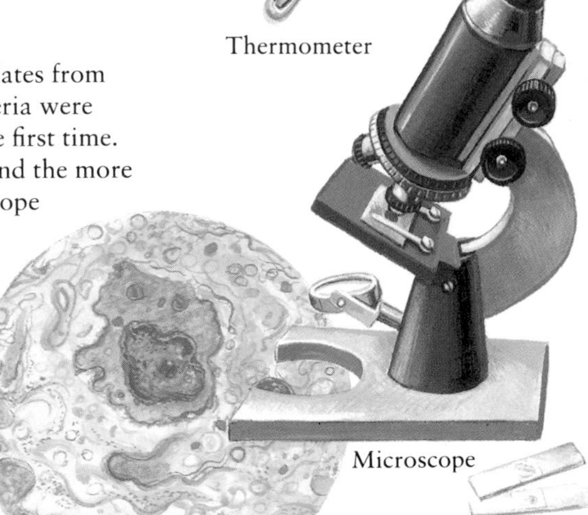

► *Ultrasound scanners are used in the treatment of kidney or heart disease. A computer changes the data received from a transducer into an image of the organ.*

▼ *This HIV virus causes AIDS (acquired immune deficiency syndrome). Viruses live in other living things. They invade cells, causing disease. Drugs such as antibiotics do not work against viruses.*

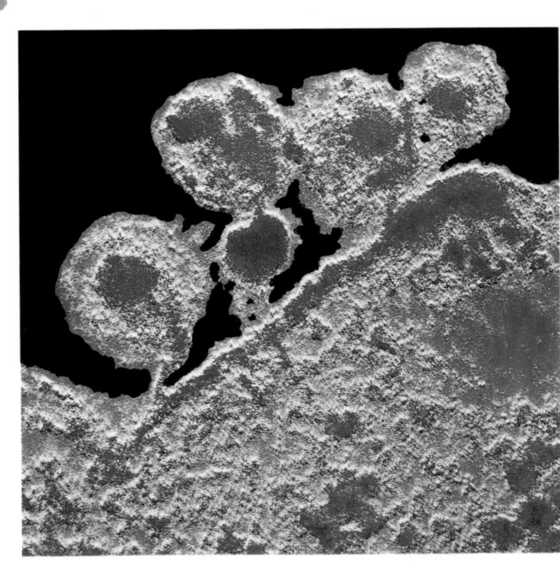

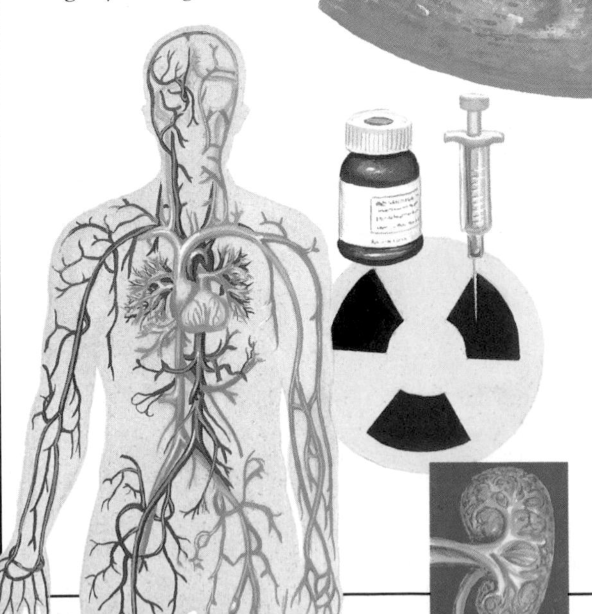

◄ *Radioisotopes are used in medical detective work. A radioactive trace (a radioisotope attached to a carrier substance) is put into the patient's body. It is carried around the bloodstream to the target organ. As the radioisotope decays, it gives off radiation that forms a scanner image.*

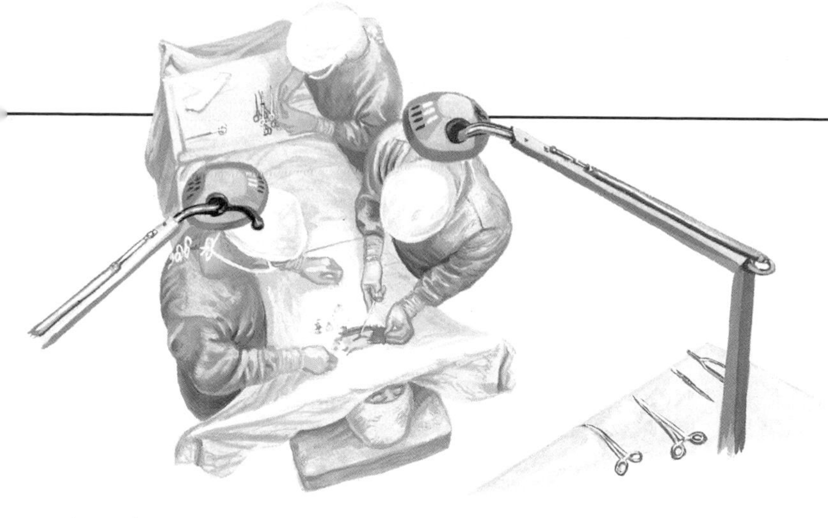

▲ *Modern surgeons use many instruments apart from the scalpel (knife). They can remove diseased organs and replace them with healthy or artificial ones.*

▼ *Modern drugs are chemicals that kill bacteria and fight infection (antibiotics), help to deaden pain (anesthetics), and calm the mind (tranquilizers).*

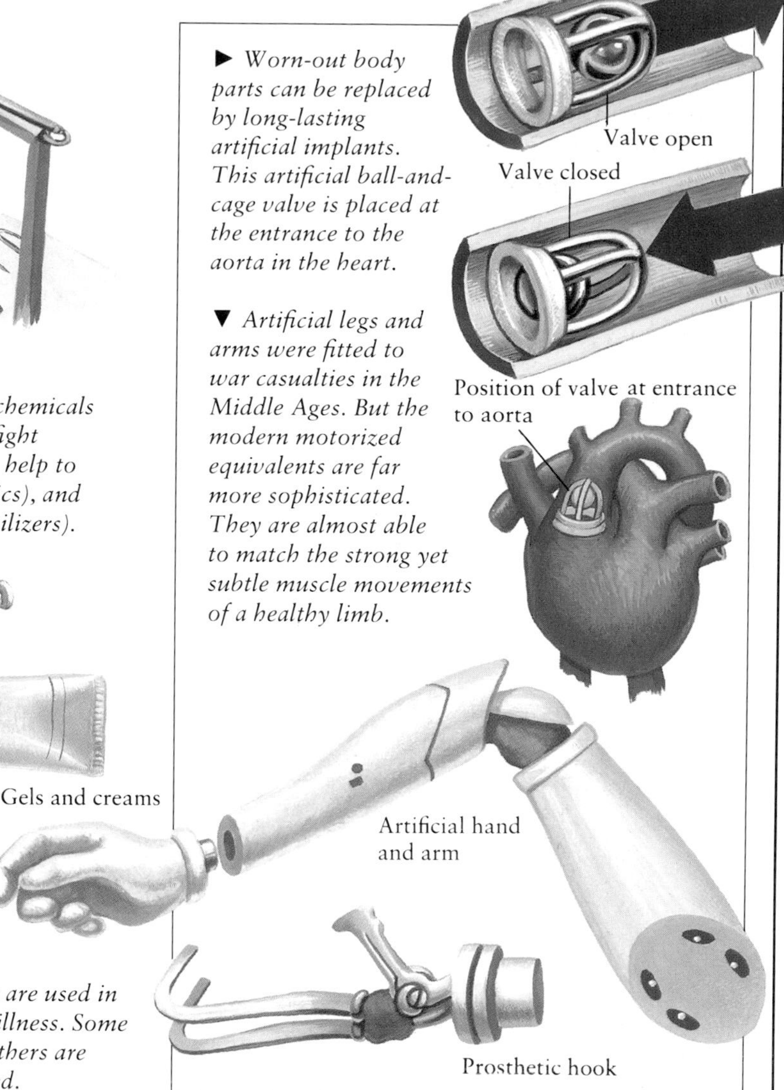

► *Worn-out body parts can be replaced by long-lasting artificial implants. This artificial ball-and-cage valve is placed at the entrance to the aorta in the heart.*

▼ *Artificial legs and arms were fitted to war casualties in the Middle Ages. But the modern motorized equivalents are far more sophisticated. They are almost able to match the strong yet subtle muscle movements of a healthy limb.*

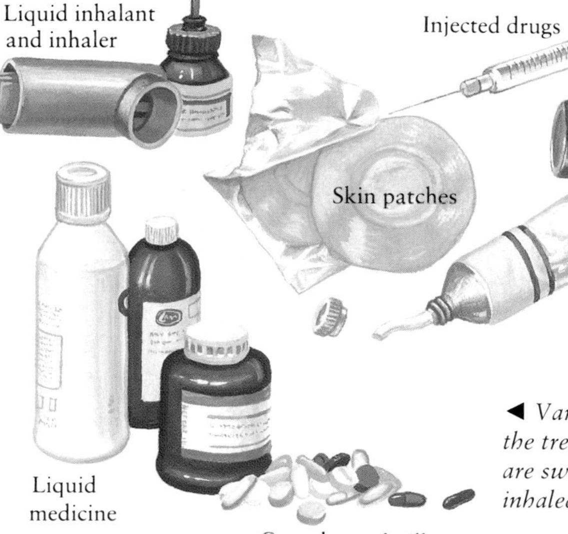

◄ *Various drugs are used in the treatment of illness. Some are swallowed, others are inhaled or injected.*

VACCINATION
The Arabs first discovered that catching a mild dose of a disease gave a person immunity from a more serious attack. In the 1790s Edward Jenner gave a boy an injection of cowpox, to see if "vaccination" would ward off smallpox. Louis Pasteur developed immunization against rabies in the mid-1800s. Today there are vaccines against infectious diseases such as diphtheria and polio.

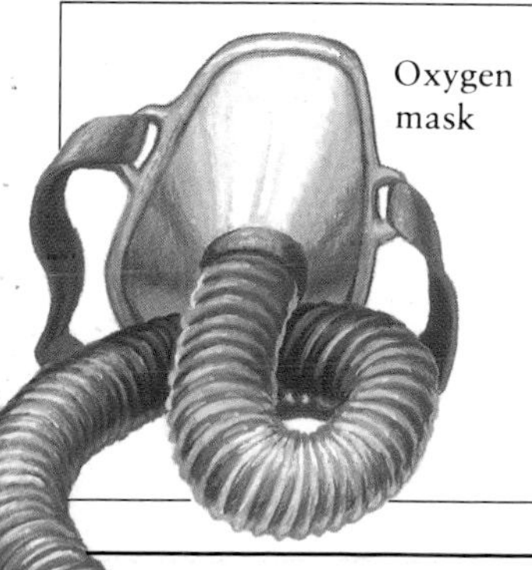

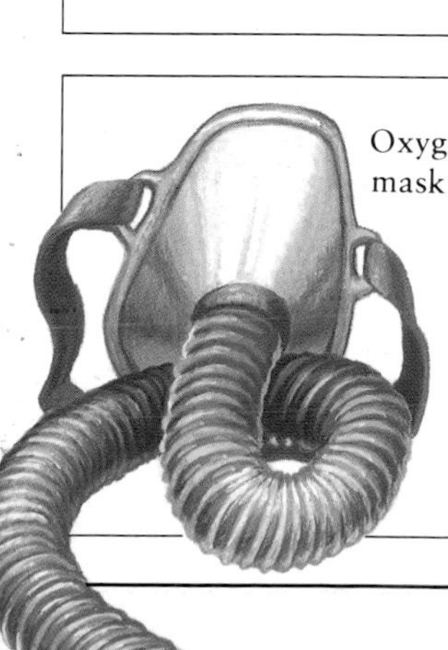

ANESTHETICS
Anesthetics can be breathed in or injected. A general anesthetic makes the patient unconscious. A local anesthetic deadens pain around a particular area.

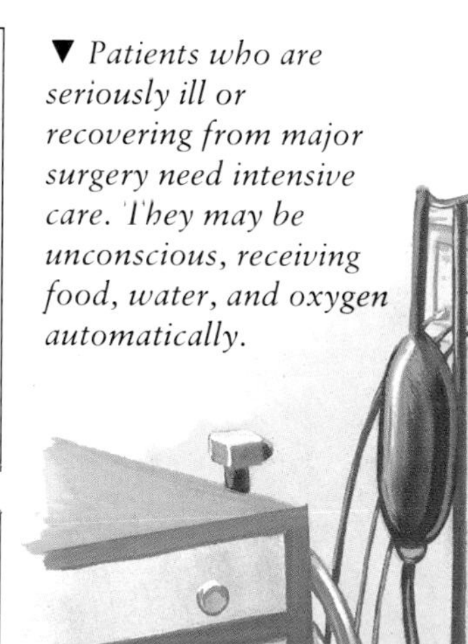

▼ *Patients who are seriously ill or recovering from major surgery need intensive care. They may be unconscious, receiving food, water, and oxygen automatically.*

▼ *Electronic devices monitor (record) heart beat, breathing, and other body functions.*

Communications

Communication is the exchange of ideas and information. It has developed from the beginnings of language in the Stone Age to modern methods of message-sending and storing, such as radio, television, recording, and computers. The communications revolution began with the development of printing in the 1400s, allowing information to be copied and distributed easily for the first time. In this century, electronic telecommunications have given us almost instant communications with people and machines far apart.

▼ Modems are devices that enable computers to send data to one another, as signals passed through the telephone network.

▼ A fax machine sends words and pictures across long distances. It codes images into signals sent through the telephone system.

COMMUNICATIONS TECHNOLOGY
The media — TV, radio, and newspapers — now have access to an array of technologies. Books, film, TV, radio — all make use of the marvels of electronics. Pictures are broadcast by satellite TV. In print, images and words are stored electronically and exchanged.

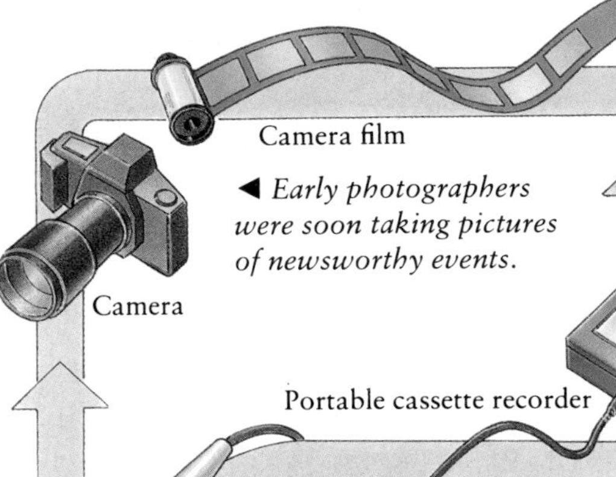

◀ Early photographers were soon taking pictures of newsworthy events.

◀ Journalists now use tape recorders and cameras as well as pencils and notebooks. Stories are relayed to the computerized editorial systems.

▼ Videotape provides instant storage of filmed events, from a hand-held camera.

◀ The images taken by a video camera can be viewed on a TV set.

◀ Radio was the first non-print medium to win a mass audience, and an outside broadcast over the airwaves still has great impact.

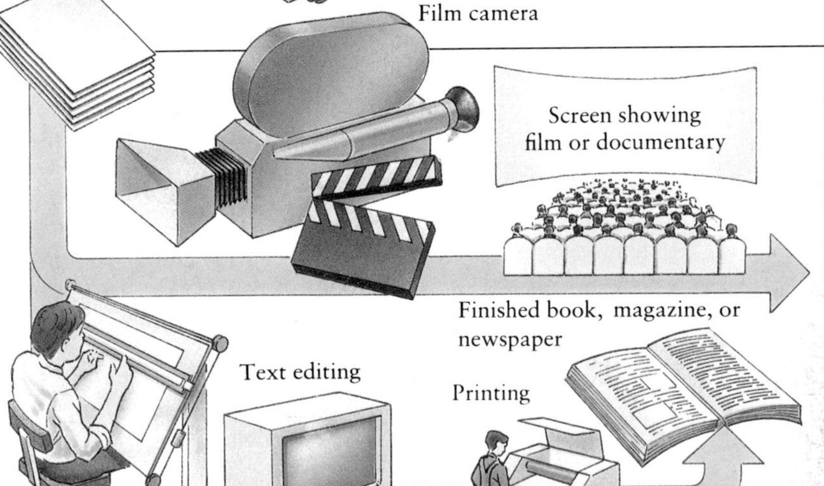

FILM
The art of filmmaking (cinematography) was invented in the late 1800s. A movie is a long strip of thousands of still photographs, projected so fast (24 frames a second) that they appear to the eye to make moving pictures. Movies with soundtracks were introduced in the 1920s. Color came in the 1930s.

PUBLISHING
The first newspapers and magazines in the 1600s offered cheap daily or weekly news. In the 1930s book publishers began to produce paperback books more cheaply than hardbacks. Today designers, editors, and writers use electronics to process words and pictures onto the page.

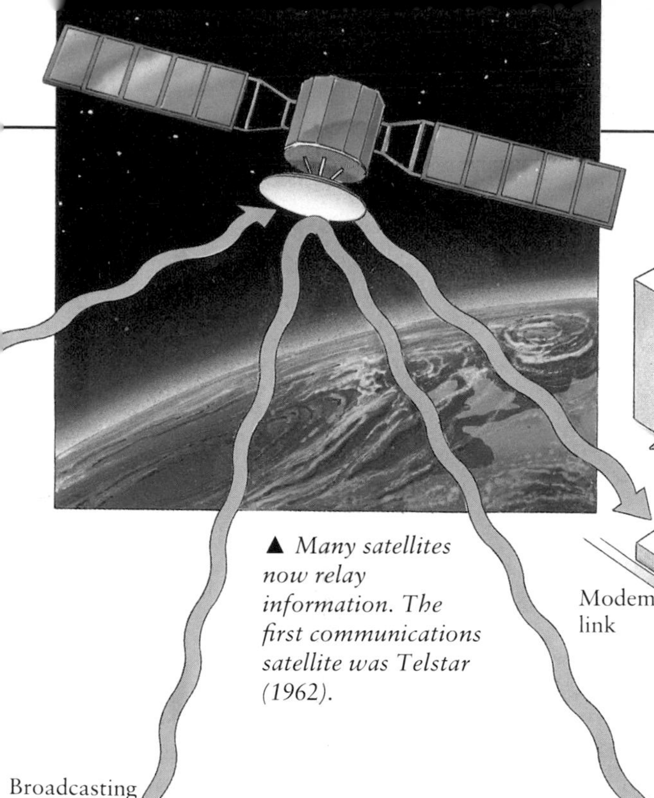

▲ *Many satellites now relay information. The first communications satellite was Telstar (1962).*

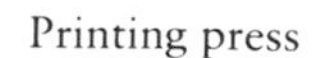

◀ *A writer or editor can use a desktop publishing computer system to make up the pages of a book. It can either become an electronic "book" on disk, or be printed.*

► *Some of today's newspapers depend on new technology that can lay out news pages on video display terminals, or uses lasers to make printing plates.*

Broadcasting center

◀ *Television pictures may be broadcast to a receiver antenna. They can also be transmitted to subscribers through cables or from a satellite beaming signals to small dish receivers.*

Home receiver dish

Television

◀ *Videotape has changed the way we watch TV and make home movies.*

Videotape

◀ *The modern radio is much smaller than the bulky tube-sets of the 1930s and 1940s.*

CHANGES IN COMMUNICATIONS

1820s: First experiments with photographs
1837. Morse's electric telegraph
1867: Typewriter invented
1877: Sound recording (phonograph)
1895: Wireless (radio) demonstrated
1895: First public motion picture shows
1899: First tape recorder
1929: Electronic television system
1936: First public television service, in Britain
1940s: First electronic computers
1956: Invention of videotape
1960: Invention of laser
1962: First communications satellite
1971: First microprocessor
1980s: Compact discs, lightweight camcorders, and small personal computers
1990s: Satellite TV expands, offering multi-channel broadcasts

ELECTRONIC EDUCATION

Computers can store vast amounts of information, and video graphics are becoming more realistic. CD-ROM disks store electronic books which are playable through a TV and illustrated with sounds and moving pictures.

INVENTIONS

7000 B.C.: Pottery (Mesopotamia)
4000 B.C.: Iron smelting (origin unknown)
4000–3000 B.C.: Bricks (Egypt and Assyria)
*c.*3000 B.C.: Wheel (Asia)
*c.*3000 B.C.: Plow (Egypt and Mesopotamia)
3000 B.C.: Glass (Egypt)
2600 B.C.: Geometry (Egypt)
700 B.C.: Calendar (Babylon)
*c.*500 B.C.: Abacus (China)
200s B.C.: Screw (for raising water), Archimedes (Greece)

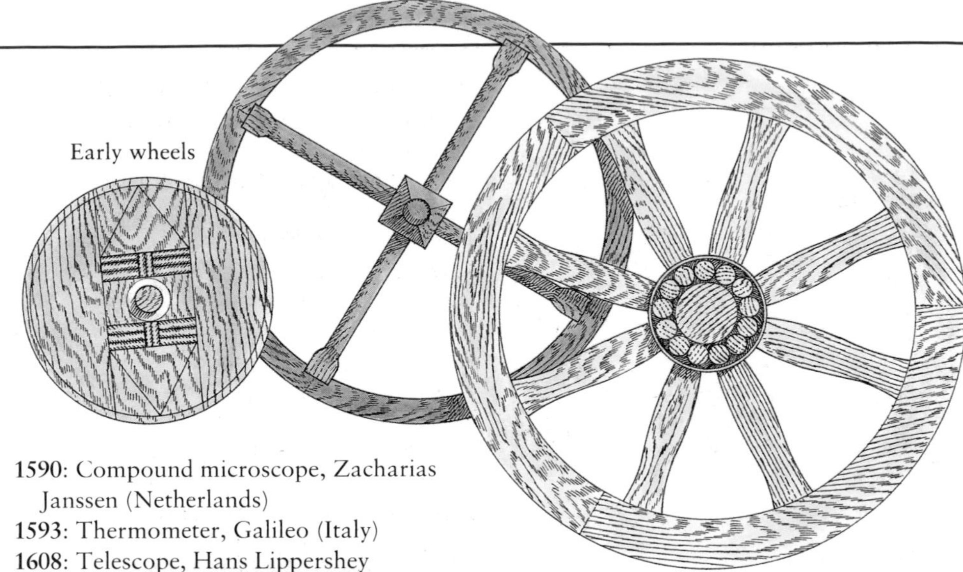
Early wheels

1590: Compound microscope, Zacharias Janssen (Netherlands)
1593: Thermometer, Galileo (Italy)
1608: Telescope, Hans Lippershey (Netherlands)

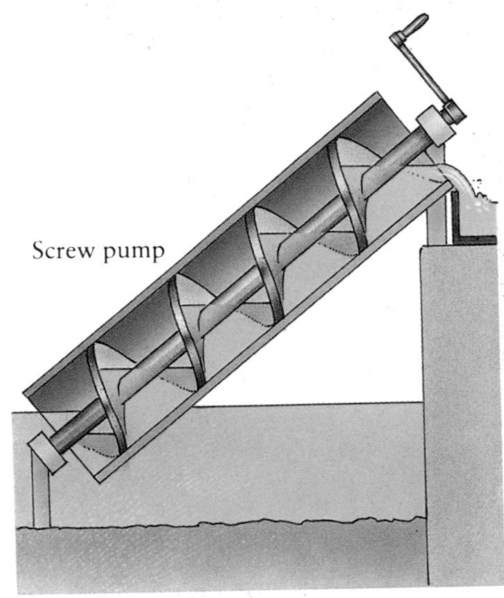
Screw pump

Lippershey's telescope

A.D. 105: Paper (from pulp), Ts'ai Lun (China)
250: Algebra, Diophantus (Greece)
*c.*1000: Gunpowder (China)
1100: Magnetic compass (China)
1100: Rocket (China)
1250: Cannon (China)
1440: Printing press (movable type), Johannes Gutenberg (Germany)

Early printing press

1520: Rifle, Joseph Kotter (Germany)
1589: Knitting machine, William Lee (England)

1614: Logarithms, John Napier (Scotland)
1636: Micrometer, William Gascoigne (England)
1637: Coordinate geometry, René Descartes (France)
1640: Theory of numbers, Pierre de Fermat (France)
1642: Calculating machine, Blaise Pascal (France)
1643: Barometer, Evangelista Torricelli (Italy)
1650: Air pump, Otto von Guericke (Germany)
1656: Pendulum clock, Christian Huygens (Netherlands)
1665–75: Calculus, Sir Isaac Newton (England) and Gottfried Leibniz (Germany), independently
1675: Pressure cooker, Denis Papin (France)
1698: Steam pump, Thomas Savery (England)
1712: Steam engine, Thomas Newcomen (England)
1714: Mercury thermometer, Gabriel Fahrenheit (Germany)
1725: Stereotyping (printing), William Ged (Scotland)

1733: Flying shuttle, John Kay (England)
1735: Chronometer, John Harrison (England)
1752: Lightning rod, Benjamin Franklin (U.S.)
1764: Spinning jenny, James Hargreaves (England)
1765: Condensing steam engine, James Watt (Scotland)
1768: Hydrometer, Antoine Baume (France)
1783: Parachute, Louis Lenormand (France)
1783: Hot-air balloon: Montgolfier brothers (France)

First hot-air balloon

1785: Power loom, Edmund Cartwright (England)
1793: Cotton gin, Eli Whitney (U.S.)
1796: Lithography, Aloys Senefelder (Germany)
1800: Electric battery, Alessandro Volta (Italy)
1800: Lathe, Henry Maudslay (England)
1804: Steam locomotive, Richard Trevithick (England)

Early steam locomotive

1815: Miner's safety lamp, Sir Humphry Davy (England)
1816: Metronome, Johann Malzel (Germany)
1816: Bicycle, Karl von Sauerbronn (Germany)
1817: Kaleidoscope, David Brewster (Scotland)
1822: Camera, Joseph Niepce (France)
1823: Digital calculating machine, Charles Babbage (England)

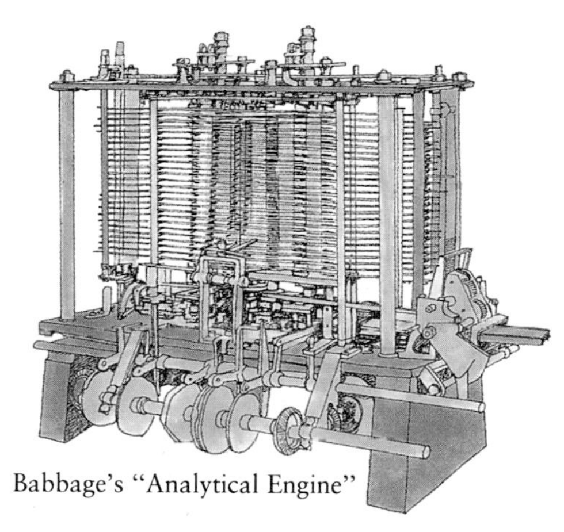
Babbage's "Analytical Engine"

1823: Waterproof cloth, Charles Macintosh (Scotland)
1824: Portland cement, Joseph Aspdin (England)
1825: Electromagnet, William Sturgeon (England)
1826: Photography, Joseph Niepce (France)
1827: Friction match, John Walker (England)
1828: Blast furnace, James Neilson (Scotland)
1831: Dynamo, Michael Faraday (England)
1834: Reaping machine, Cyrus McCormick (U.S.)
1836: Revolver, Samuel Colt (U.S.)
1837: Telegraph, Samuel Morse (U.S.)
1839: Vulcanized rubber, Charles Goodyear (U.S.)
1844: Safety match, Gustave Pasch (Sweden)
1846: Sewing machine, Elias Howe (U.S.)
1849: Safety pin, Walter Hunt (U.S.)
1852: Gyroscope, Léon Foucault (France)
1852: Passenger elevator, Elisha Otis (U.S.)
1855: Celluloid, Alexander Parkes (England)
1855: Bessemer converter, Henry Bessemer (England)
1855: Bunsen burner, Robert Bunsen (Germany)
1858: Refrigerator, Ferdinand Carré (France)

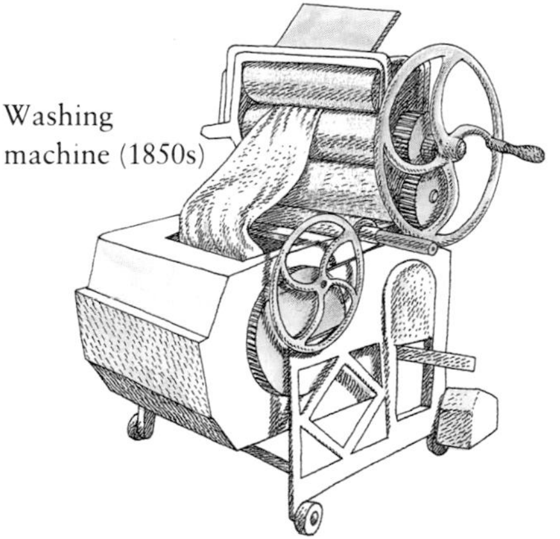
Washing machine (1850s)

1858: Washing machine, Hamilton Smith (U.S.)
1859: Internal-combustion engine, Jean-Joseph-Etienne Lenoir (France)
1859: Storage battery, which can be recharged again and again, Gaston Planté (France)
1861: Linoleum, Frederick Walton (England)
1862: Rapid-fire gun, Richard Gatling (U.S.)

Gatling gun

Early typewriter

1865: Cylinder lock, Linus Yale Jr. (U.S.)
1866: Dynamite, Alfred Nobel (Sweden)
1867: Typewriter, Christopher Sholes (U.S.)
1868: Motorized bicycle, Michaux brothers (France)
1870: Margarine, Hippolyte Mège-mouriès (France)
1873: Barbed wire, Joseph Glidden (U.S.)
1876: Telephone, Alexander Graham Bell (Scotland/U.S.)

Bell's telephone

1876: Carpet sweeper, Melville Bissell (U.S.)
1877: Phonograph, Thomas Edison (U.S.)
1878: Microphone, David Edward Hughes (England/U.S.)
1879: Incandescent lamp, Thomas Edison (U.S.)
1879: Cash register, James Ritty (U.S.)
1884: Fountain pen, Lewis Waterman (U.S.)
1884: Linotype, Ottmar Mergenthaler (U.S.)
1885: Motorcycle, Edward Butler (England)

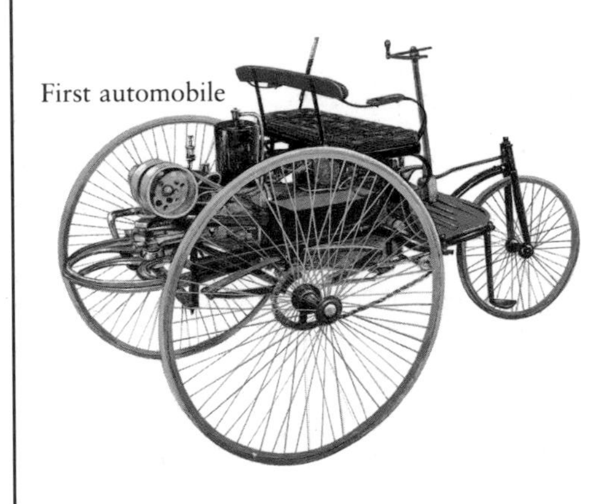
First automobile

1885: Automobile engine, Gottlieb Daimler and Karl Benz (Germany), independently
1885: Vacuum bottle, James Dewar (Scotland)
1885: Electric transformer, William Stanley (U.S.)
1886: Electric fan, Schuyler Wheeler (U.S.)
1886: Halftone engraving, Frederick Ives (U.S.)
1887: Gramophone, Emile Berliner (Germany/U.S.)
1887: Monotype, Tolbert Lanston (U.S.)
1888: Pneumatic tire, John Boyd Dunlop (Scotland)
1888: Kodak camera, George Eastman (U.S.)
1890: Rotogravure printing, Karl Klie (Czechoslovakia)
1892: Diesel engine, Rudolf Diesel (Germany)
1892: Zipper, Whitcomb Judson (U.S.)

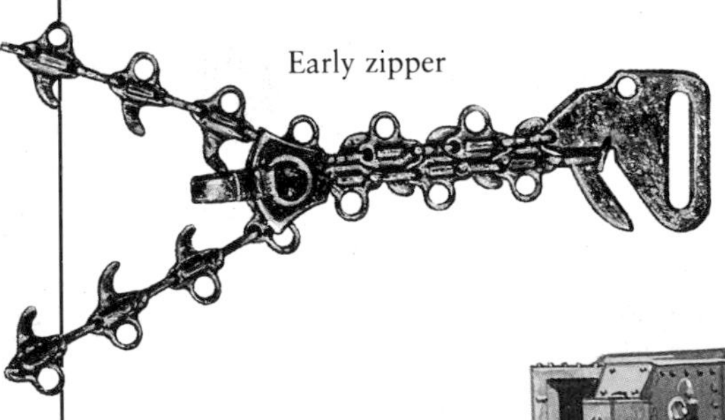
Early zipper

1895: Wireless, Guglielmo Marconi (Italy)
1895: Photoelectric cell, Julius Elster and Hans Geitel (Germany)
1895: Safety razor, King C. Gillette (U.S.)
1898: Submarine, John P. Holland (Ireland/U.S.)
1899: Tape recorder, Valdemar Poulsen (Denmark)
1901: Vacuum cleaner, Hubert Booth (England)
1902: Radiotelephone, Reginald Fessenden (U.S.)
1902: Air conditioner, Willis H. Carrier (U.S.)
1903: Airplane, Wilbur and Orville Wright (U.S.)

Early Marcon
radio set

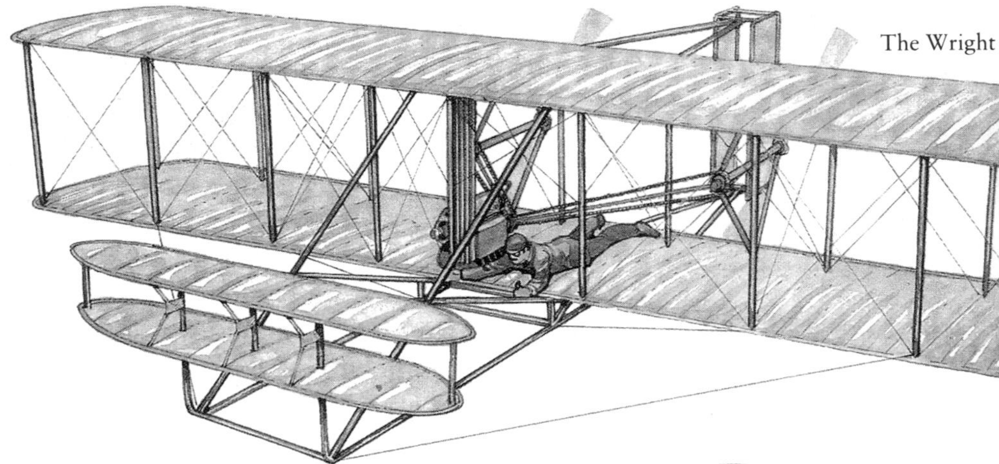
The Wright

1904: Diode, John Fleming (England)
1906: Triode, Lee De Forest (U.S.)
1908: Bakelite, Leo Baekeland (Belgium/U.S.)
1908: Cellophane, Jacques Brandenberger (Switzerland)
1911: Combine harvester, Benjamin Holt (U.S.)
1913: Geiger counter, Hans Geiger (England)
1914: Tank, Ernest Swinton (England)

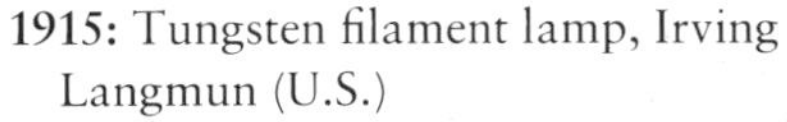
Baird's televisor (1930)

First tank

1915: Tungsten filament lamp, Irving Langmun (U.S.)
1918: Automatic rifle, John Browning (U.S.)
1925: Television (mechanical system), John Logie Baird (Scotland) and others
1925: Frozen food process, Clarence Birdseye (U.S.)
1926: Rocket (liquid fuel), Robert H. Goddard (U.S.)
1927: First talking picture, *The Jazz Singer* (U.S.)

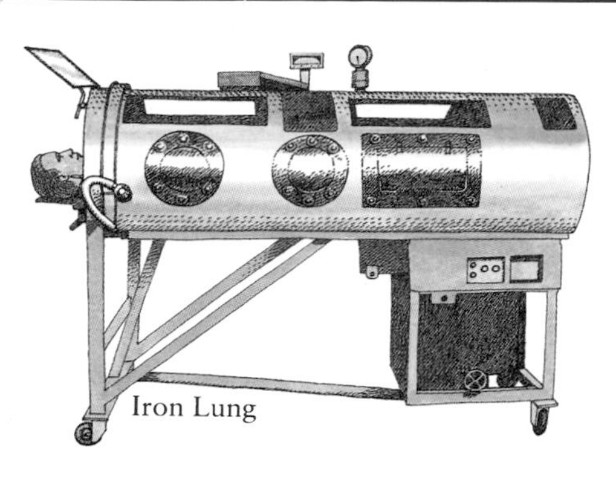
Iron Lung

1927: Iron lung, Philip Drinker and Louis Shaw (U.S.)
1928: Electric razor, Jacob Schick (U.S.)
1929: Television (electronic system, adopted as standard), Vladimir Zworykin (U.S.)
1930: Jet engine, Frank Whittle (England)
1931: Cyclotron, Ernest Lawrence (U.S.)
1935: Nylon, Wallace Carothers (U.S.)
1935: Parking meter, Carlton Magee (U.S.)
1936: First practical helicopter, Heinrich Focke (Germany)
1938: Ballpoint pen, Ladislao and Georg Biro (Hungary)
1939: Electron microscope, Vladimir Zworykin and others (U.S.)
1940: Penicillin as an antibiotic, Howard Florey (Australia) and Ernst Chain (Germany/England)
1940: Xerography photocopier, Chester Carlson (U.S.)
1942: Nuclear reactor, Enrico Fermi (Italy) and others, in Chicago (U.S.)
1944: Automatic digital computer, Howard Aiken (U.S.)
1946: Electronic computer, J. Presper Eckert and John W. Mauchly (U.S.)
1947: Polaroid camera, Edwin Land (U.S.)

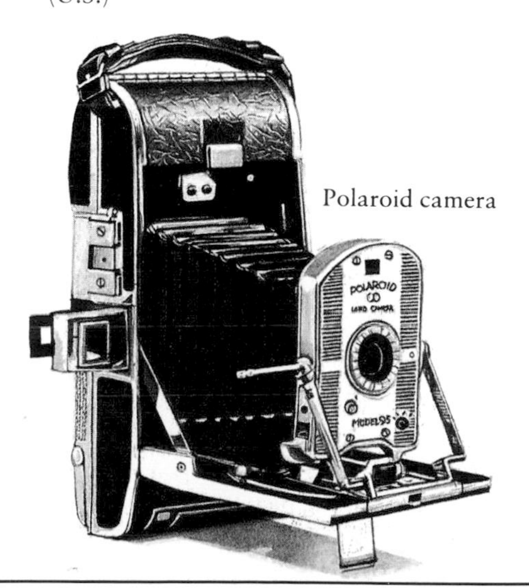

Polaroid camera

1948: Transistor, John Bardeen, Walter Brattain, and William Shockley (U.S.)
1948: Long-playing record, Peter Goldmark (U.S.)
1954: Maser, Charles Townes (U.S.)
1954: Solar battery, D. Pearson, C. Fuller, G. Pearson (U.S.)
1955: Hovercraft, Christopher Cockerell (England)
1955: Contraceptive pill, Gregory Pincus and others (U.S.)
1956: Rotary engine (car), Felix Wankel (Germany)

The Gloster E 28/39 aircraft first flew with a Whittle jet engine in May 1941.

Sputnik 1, the first artificial Earth satellite

1956: Videotape recording, A. Poniatoff (U.S.)
1957: Artificial Earth satellite (U.S.S.R.)
1959: Fuel cell, Francis Bacon (England)
1959: Microchip, Fairchild Semiconductors Corp (U.S.)
1960: Laser, Theodore Maiman (U.S.)
1962: Industrial robot, Unimation (U.S.)

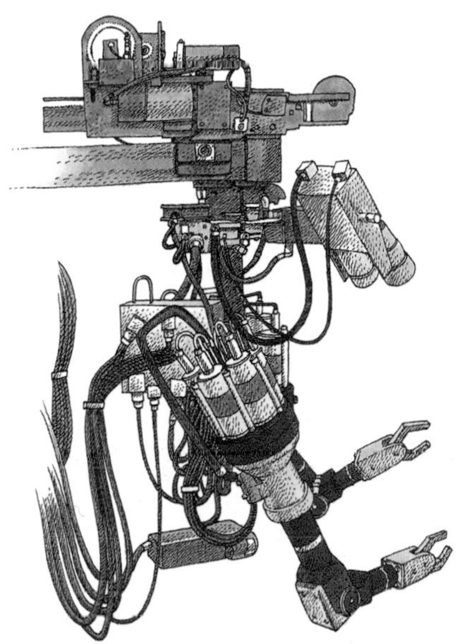
Japanese automated "arm"

1965: Holography (an idea conceived in 1947 and subsequently developed using laser), Dennis Gabor (Hungary/England)
1971: EMI-Scanner, Godfrey Hounsfield (England); developed from his invention of computerized tomography in 1967

A computer, aiding design

1971: Microprocessor, Intel (U.S.)
1973: Teletext, British Broadcasting Corporation and Independent Television (UK)
1979: Compact disc, Sony (Japan) and Philips (Netherlands)
1981: Space shuttle (U.S.)
1984: Macintosh PC, Apple Corp (U.S.)

INDEX

Page numbers in *italic* type refer to illustrations.
Page number in **bold** type refer to key topics.

A

Abacus 237, *237*, 261, *261*
Absolute magnitude 209
Acceleration 253, *253*
Acid 241
Acid rain 77, *77*, 241, *241*
Adler 259
Advection fog 55, *55*
Aerial telescopes 216, *216*
Aerosols 245, *245*
Africa 300; continental drift 21; grasslands 66, 67, *67*; Great Rift Valley 27, *27*, 46; plate tectonics 23, *23*, *27*
Aging (human) 154–155, *154–155*
Agriculture *61*, 71, 78, *78*, **300–301**, *300–301*
AIDS *302*
Air 49, *49*; *see also* Atmosphere
Air-cushion vehicles (ACVs) 294–*294*
Airfoils 294, 296, *296*, 297
Airliners 296, *296*
Air mass 51
Air pollution *15*, **76–77**, *76–77*
Airships 259, *259*, 291, *291*, 296
Air transportation 258, 270, *270*, 291, *291*, **296–297**, *296–297*
Alaska 62, 63, *63*, 79, *79*
Albatrosses 124, *124*, 125
Alchemy 236
Aldrin, Buzz 222
Algae 84, *85*, 87, 90, 92, *92*, 93, *93*
Algebra 261, 262
Alligator gar 116, *116*
Alligators 120, 121, *121*, 137
Alloys 241, *241*
Alpine habitat 89, *89*, 99, *99*
Alpine swift *123*
Alps 24, 25, *25*
Alternating current (AC) 276
Altocumulus clouds 54, *57*
Altostratus clouds 54
Aluminum 13, 30, 74, 80
AM (amplitude modulation) 274, *274*
Amazon River 44
Amber *32*
Ammonites *32–33*, 33
Amperes 277
Amphibians *84*, 118–119, *118–119*
Anacondas 121, *121*
Analog recordings 272, *272*
Analysis, chemical 241, *241*
Andes Mountains 24, 25, 38, 44, 175
Andromeda galaxy 199, 212, 216, 219
Anemometers 52, 58
Anesthetics 302, 303, *303*
Anglerfish 117, *117*
Animals: carbon cycle 15; continental drift 20; fossils **32–33**, *32–33*; grasslands *67*; hearing 271, *271*; mountainous areas 25; ocean life 41, *41*; oxygen cycle 15; temperate woodland 64; tundra *62*
Annelids 108, *108*
Annular eclipse 167, *167*
Antarctic *14*, 21, 23, 37, 52, 62–63, *63*, 69, *76*, *77*
Antarctic Ocean *38*
Antares 213, *213*
Anteaters 129, *129*
Anthers 100, *100*
Antibiotics 302, 303
Ants 115, *115*, 135
Apollo asteroid *192*
Apollo program 176, 177, 179, 222–223, *222–223*, 228
Appalachian Mountains 25, 64
Apples 106, *106*
Aquifers 69, *69*
Arabic numerals 260, 263
Arachnids 109, 112, *112*, 113, *113*, 130, 132, 135, 137
Arctic *14*, 51, 52, 55, 59, 62–63, *63*, 76
Arctic Ocean *38*
Arctic habitat *88*, 89
Arecibo radio telescope 219, 232
Argon 49, 243
Ariel 187, *187*
Aries *211*, *212*
Aristarchus 215
Aristotle 8
Arithmetic 261, 262
Armadillos 127, *127*
Armstrong, Neil 222
Arthropods 109, *109*, 112–113, *112–113*
Asia 64, 66, 262, 300
Asteroids 163, 180, **192–193**, *192–193*, 226, 233, *233 see also* minor planets
Asthenosphere *12*, 20, 22, *22*
Astronauts 222, *222–223*, **228–229**, *228–229*, 252, *252*, 283
Astronomers: discoveries 186, *186*, 188, *188*, 190, 194, 198, 205, 215, 216, *216*, 219; in history 212, **214–215**, *214–215*
Astronomy 161, 236
birth of **214–215**, *214–215*; from space **220–221**, *220–221*; optical **216–217**, *216–217*; radio telescopes **218–219**, *218–219*, 232
Atlantic Ocean 20, 21, *21*, 38, *38*; hurricanes 53; plate tectonics 22, 23, *23*; Sargasso Sea 41
Atmosphere 168; Earth 174, *174*; Earth's Moon 177, 179; Jupiter 182, *182*; Mars 180, *180*; Mercury 171, *171*; Neptune 188, *188*; Pluto 190, *190*; Saturn 184, *184*; Titan 185, *185*; Uranus 186, *186*; Venus 172, *172*
Atmosphere 48–49, *48–49*; balance **14–15**; climate 50; pollution **76–77**, *76–77*; radiation from Sun *11*, 15; water vapor 38
"Atomic" clocks 264, 265, *265*
Atomic number 238, *238*, 242
Atoms 160, 166, 207, 237, *237*, **238–241**, *238–241*, 252
Augite 30, *30*
Aurorae *13*, *48*
Australia 21, *21*, 23, 53, 64, 66, 67, *67*, *75*
Australia Telescope 219, 227
Automation 282, *282*
Axles 256, 292, *292*
Axolotls 119, *119*

B

Babylon 260, *260*, 262, 264, *264*
Babylonians 212, 214, *214*, *215*
Background radiation 161, 219
Bacteria 84, *85*, 87, 90, 92, *92*, 106, 302
Balloons 258, 259, *259*, 291, 296
Balloons, weather *59*
Ballpoint pens 283, *283*
Bamboos 91, *91*
Banyan trees 105, *105*
Baobabs 105, *105*
Bar codes *279*
Bathyscaphs *283*
Barchan dunes 69, *69*
Barn owl *123*
Barographs 58, *58*
Barometers 58
Barred spiral galaxies 196, 198, *198*
Barycenter 191, *191*
Basalt *12*, 23, 28, *28*
Bats 109, *109*, 127, *127*, 129, *129*, 130, *130*; hearing 271, *271*
Batteries 276, *276*
Bauxite 74, *74*
Beaches 43, *43*
Beans 98, 102, *102*
Bearded saki 138, *138*
Bears 134, *134*
Beavers 130, *130*
Bee-eaters 130, *130*
Bees 101, 114, *114*, 136, *136*
Beetles 115, *115*
Bellatrix *208*
Bernoulli's principle 258
Berries 103, *103*
Betelgeuse *208*, *209*
Bicycles 290, 292
Big Bang 160–161, 200, 219, 236
Big Crunch 200, 201
Big Dipper 210, *210*, *214*
Binary stars 204, 205, *205*
Binary system (numbers) 237, 281
Biological time 265
Biological pest control 81
Biomes 60 *see also* habitats
Biosphere reserves 80, *81*
Bird-eating spider 113, *113*
Birds 81, *84*, 122–125, *122–125*, *133*, *134*, 137, *138*, 139, 141
Bits 281
Bivalves 111, *111*
Blackberries 103, *103*
Black dwarf 203, *203*
Black holes 198, *198*, 201, 203, *203*, **206–207**, *206–207*
Black swan *122*
Black widow spider 113, *113*
Bladderwrack 92, *92*
Blood (human) 142, *142*, *143*, 148, *148*
Blue giant stars 203, *203*, 208
Blue whale 109, *109*
Bluebell 98, *98*
Boats *see* Ships
Body scans 302, *302*
Bogs 47, *47*
Bones (human) 142, *142*, 144, *144*, 145, *145*
Bonsai 105, *105*
Books 304, *304*
Bornhardts *67*
Bosons 239
Brachiopods 84
Brain (human) *146*, 146–147
Brakes 292, *292*
Breathing *see* respiration
Bridges 288–289, *289*
Brightness, stars **208–209**, *208–209*
Bristlecone pines 105, *105*
Broadcasting 272

Broadleaved trees 64–65, *65*
Bronze Age 298
Brown dwarf 203, *203*
Buildings 284–287, *284–287*
Bulb plants 98, 100, *100*, 102, 107
Bumblebee bats 109, *109*
Buoyancy 259, *259*
Burning 250
Buses 290
Butterflies 114–115, *114–115*, *133*, *138–139*
Butterfly fish 117, *117*
Bytes 281

C

Cacti 91, *91*, 99, *99*
Caecilians 118, 119, *119*
Caimans 121, *121*
Calcium 13, 30
Calculation 261, *261*
Calculators 280
Caldera *18*
Calendars 260, 262, 264, *264*
California Current 51
California Nebula *197*
Callisto 183, *183*
Caloris Basin 171, *171*
Cambrian Period *28–29*
Camels 136, *136*
Cameras 268, *268*
Canada 55, 62, *63*, 64
Canals 300
Cancer *211*, *213*
Candle clocks 264, *264*
Canned food 301, *301*
Capricorn *211*, *213*
Carbon *31*, 72, 161, 202, 238, *238*, 240, *240*
Carbon cycle 15, *15*
Carbon dioxide 11, 14, 15, 34, 49, *49*, 76, 77, 172, 174, 180, 242
Carbon fibers 299
Carboniferous Period *28–29*
Carnivores *128–129*, 129
Carnivorous plants 101, *101*
Carrots 98, 102, *102*
Cars 290, *290*; brakes 292, *292*; gears 293, *293*; steering 293, *293*
Caspian Sea 46, *47*
Cassegrain telescopes 217, *217*
Cassettes 273, *273*
Cassini Division 184, *184*
Cassini space probe 185, *185*
Caterpillars 114, *114*
Catfish 117, *117*
Cathedrals 284, *284*
Cats 128, 135, 137, 138, *138*, 141, *141*
Cattle 136, *136*
Cauliflowers 102, *102*
Cave paintings 260
Caves and caverns 35, *35*
CD-ROM disks 305, *305*
Celestial sphere 210, *210*, 211, 212, *212*
Cell division 155, *155*
Cellophane 299
Celsius scale 248, *248*
Cenozoic Era *28–29*
Centipedes 109, *109*, 112, *112*
Central America 284
Central heating 286, *286*
Centrifugal force 252, *252*
Centripetal force 252, *252*
Cepheid variable stars 219
Ceramics 299, *299*
Cereals 98, 102
Ceres 192, *192*
Chalk 28, *28*
Chameleons 120, *120*
Chaos theory 59
Charged particles 239
Chariots 292, *292*
Charon 190, *190*, 191, *191*
Chemical energy 247, *247*
Chemicals: analysis 241, *241* compounds 239, *239*, 240, 245; reactions 243, *243*
Chemistry 236
Chestnuts *103*
Chicago *57*
Chile 64
Chimpanzees 129, *129*
China 64, 260, 294
China, ancient 214, *214*, *215*, 223
Chinook wind 52
Chlorine 76
Chlorofluorocarbons (CFCs) 76, 77, 167
Chlorophyll 100, 104, 106
Chordates 109, *109*
Chronometers 265, *265*
Chrysalis 114, *114*
Chub 117, *117*
Cichlids 117, *117*
Cinchona tree 106, *106*
Circles 263, *263*
Circuits, electric 276, *276*
Circumference 263, *263*
Cirques *37*
Cirrocumulus clouds *54*
Cirrostratus clouds *53*, *54*
Cirrus clouds 51, *53*, *54*
Clams 110, *110*, *111*, 135
Classes (classification) 86, 87; of mammals 127
Classification 86–87, *86–87*, 90
Clay 28; ceramics 299, *299*
Cleaner wrasse 117, *117*
Cliffs 42, 43
Climate 11, 42, 49, **50–51** *50–51*, 60, 77, *77*
Clipper ships *290*
Clocks 264, *264–265*
Cloth 299
Clouds *8*, *48*, 49, 53, **54–55**, *54–55*, *56–57*, 59
Club mosses 91, *91*, 94, *95*
Clusters: galaxies **198–199**, *198–199*; stars 208, *208*
Coaches 290, *290*
Coal, as fuel 246, *246*, 250, *250*
Coal 32, *32*, 63, 72, *72*, 73
Coast redwood *97*
Coasts 42–43, *42–43*
Cobras 120, *120*
Cockroaches 114
Cockscomb *92*
Coelacanths 116, *116*
Coelenterates *85*
Colloids 245, *245*
Colombia *71*
Colonies, animal 130
Color, red shift 200, *200*
Color: rainbows 57, *57*; skies 58, *58*
Colors 267, *267*, 269, *269*
Colors of stars 209, *209*
Combine harvesters 301, *301*
Combustion 250
Comets 193, **194–195**, *194–195*, 226
Communications 304–305, *304–305*
Compact discs (CDs) 272, *272*, 305
Compasses 13, *13*, 295, *295*
Composite flowers 100, *100*
Compounds 239, *239*, 240, 245
Computers 237, *237*, **280–281**, *280–281*; communications technology 304, *305*; domestic *283*; modems *304*
Concrete 288
Concrete dams 288
Condensation 244, *244*
Conduction 248, *248*
Cone Nebula *197*
Coniferous forests 64–65, *64*
Conifers *85*, 91, 96–97, 99, 105
Conservation 80–81, *80–81*
Conservation of energy 247, *247*
Constellations 208, 210, *210*, *211*, **212–213**, *212–213*, 214; Northern Hemisphere 212–213, *213*; Southern Hemisphere 213, *213*
Continental plates 20, *20*, *23*
Continental drift 20–21, *20–21*, 22
Continental shelf 38
Convection 248, *248*
Convection, Sun 165
Cook, Captain James 265
Copernicus, Nicolaus 215, 216, 236
view of universe 215, *215*
Copper 74, *74*, *75*
Coral 110, *110*, 117
Coral polyps 108, *108*
Coral reefs 39, *39*
Coral snake *120*
Core, Earth's 13
Coriolis effect 51, *51*, 52
Cork 106, *106*
Corms 100, *100*
Corn 102, *103*
Corn bunting 124, *124*
Corona, Sun 164, *164*, 167
Cosmic rays 266, *266*
Cosmology 200–201, *200–201*; 236
Cosmonauts 231
Cotton 106, *106*
Counting 260
Courtship 119, 125, 134
Covalent bonding 239, *239*
Cows 129, 136, *136*
Crabs 109, *109*, 112, *112*, *113*
Crane, black crowned *122*
Cranes 284, *284*, 285, *285*
Crater Cloud *199*
Craters: Last bombardment 169; Mars 181, *181*; Mercury 169; Caloris Basin 171, *171*; Moon 169, 176, *176*, 216; formation 177, *177*
Cretaceous Period *28–29*
Crocodiles 120, 121, *121*, 137, *137*, 141, *141*
Cross-breeding of plants 106
Cross-pollination 100
Crossbills 124, *124*
Crust, Earth's 12, *12*, 13, 23, *23*, 24
Crustaceans *84*, 109, *109*, 112, *112*, *113*
Crux constellation 210, 212, *213*
Cryptic mantid 115, *115*
Crystals 30, *31*, 56, *56*, 239, *277*
Cubes 263, *263*
Cuckoos *122*, 123
Cumulonimbus clouds *53*, *54*
Cumulus clouds *54*, *55*
Current, electric 276, *276*
Currents, ocean 40, *40*, 41, 51
Cursor 281
Cuttlefish 110, *110*
Cycads *85*, 91, *91*, 96, *96*
Cyclones *53*
Cygnus *212*
Cypresses *97*, *105*

D

Dams *73*, **288**, *288*
Data, computer 281
Dating rocks 33

Davida *192*
Days 10
Dead Sea *30*, 40
Death Valley 69
Decibels 271
Deciduous forests 64–65, *64*
Deciduous trees 104, *104*
Decimal system 260
Deep Space Network (DSN) 227, *227*
Deer 127, *127*, 131, *131*
Deforestation 15, 65, 71, 78, *78*
Deimos 181, *181*, 185
Depressions *51*
Desalination 75, *75*
Desdemona 187
Desertification 68, 78, *78–79*
Deserts *34*, 60, *60*, 66, **68–69**, *68–71*, 88, *88*, 99, *99*
Desktop publishing 305, *305*
Detergents 243, *243*
Devonian Period *28–29*
Diameter 263
Diamond 30, *31*, 240, *240*
Diatoms 92, *92*
Diatryma 141, *141*
Dicotyledons 91, *91*, 98–101, *98–101*
Differential gear systems 293, *293*
Digestive system (human) 143, *143*, 150–151, *151*
Digital recordings 272
Dinosaurs 33, 140, *140*, 141, *141*
Diodes 278
Direct current (DC) 276
Distances in space 168, **208–209**, *208–209*
Diseases 115, 137, 157, 302–303
Distillation 245, *245*
DNA 237, 240; molecule 152, *152*
Doctors 302–303
Dodder 101, *101*
Dodo 138, *138*
Dogs 129, *129*, *135*, 136, 137, *137*
Doldrums 52
Dolomite 30, *30*
Dolphins *126*, 127, 129, *129*
Domestic animals 136, *136*
Doppler effect 271, *271*
Dormouse 128, *128*
Doves 123, *123*
Dragonflies 114, 115, 133, *133*
Drilling 282
Drinking water 75, *75*
Drizzle 56
Drugs 106, 157, 303, *303*
Drums 273, *273*
Drupes 103, *103*
Dry batteries 276, *276*
Duckbilled platypus 135, *135*
Ducks 124, *124*, 125, *125*, 136, *136*
Dugong *126*
Dunes 43, 69, *69*
"Dust Bowl" 67
Dynamos 277

E

Eagles *122*, 123
Ear (human) 147, *147*
Ears, hearing 271, *271*
Earthquakes 12, **16–17**, *16–17*, 22, 35
"Earth Summit" 81
Earth 161, 168, *168*, **174–175**, *174–175*; artificial satellites **224–225**, *224–225*; asteroids 192; atmosphere 174, *174*, 220, *220*; climate 165, *165*; density 168; eclipses, seen from 167, *167*; historical views of 214, 215, *215*; meteoroids 193, *193*; Moon **176–179**, *176–179*; orbit 163, *163*, 175; Solar System 162
Earthworms 108, 111
Echidna *126*, 127
Echinoderms *84*, 109, *109*
Echo-sounders 295
Eclipses 166–167, *166–167*, 179, *179*
Ecology 88
Edison, Thomas Alva 299
Eels 116, 117, *117*
Effelsburg Observatory 218, 219
Egrets 134, *134*, 138, *138*
Egypt, ancient 214, *214*
Egypt: calendar 262, 264; clocks 264, *264*; measurements 260, 260, 261, 261; pyramids 237, 260, 284; ships 290, 295
Eider ducks 125, *125*, 132, *132*
Eiffel Tower 285, *285*
Einstein, Albert 160, 219
Electricity 73, *73*
Electricity 247, 250, *250*
Electrolysis 298
Electromagnetism 252, **274–275**, *274–275;* radiant energy 247; spectrum 266, *266–267*
Electromagnets 277, *277*
Electromotive force 276, *276*
Electron microscopes 268, 302
Electronics 274, **278–279** computers **280–281**, *280–281* telecommunications 304
Electrons 207, 232, 238, *238*, *239*; electricity 276; electronics 274; periodic table 242
Electroplating 298, *298*
Elements 30, **242–243**, *242–243*
Elephants *126*, 127, 134, *134*, 135, *135*
Ellipses 163, *163*, 215, *215*
Elliptical galaxies 198, *198*
Emeralds 30, 31
Emulsions 245, *245*
Emus 122, *123*
Energiya rocket 223, *223*
Energy 11, **72–73**, *72–73*, 239, **246–247**, *246–247*; fuels **250–251**, *250–251*; heat **248–249**, *248–249*
Engineering 282–283, *282–283*
Energy: Big Bang 160; radiation 218, *218*; solar energy collectors 233, *233;* Sun 164, 165, *165*, 166
Engines, rockets 222
Engines: internal-combustion 257, *257* jet 296, 297, *297*
Environmental problems 80–83, *80–83*, **78–81**, *78–81*
Eocene Period *28–29*
Equator 8, 9, 10, *40*, 49, *49*, 50, 51, *51*, 52, 60, 61, 70
Equator, celestial 210, *210*, 211, *211*
Eratosthenes, astronomer 215
Erosion 34–35, *34–35*, 42, *43*, 44, *44*, 60, 68, *68*, 71, 78, *79*
Erratics, glacial *37*
Eruptions, volcanoes 18–19, *18–19*
Escape velocity 253, *253*
Ethiopian Highlands 25
Eucalyptus, Australian 91, *91*
Euclid 260
Eunumia *192*
Europa 183, *183*
Europe 51, 64, 65, 66, 67, 76
European Space Agency 223, 230
Evening star 172, *172*
Everest, Mount 25, 34, 39, *39*
Evolution 84, 140
Exploding stars 204–205, *204–205*
Extinction, animal 138, 141
Exxon Valdez 79, *79*
Eyes (mammal) 128, *128*; (human) 143, *143*, 146, *146*
Eyre, Lake 47

F

Fabrics 299
Factories 282, *282*
Faculae, Sun 164, *164*
Fahrenheit scale 248, *248*
Fallow deer *127*
False acacia *104*
Families (classification) 86, *86*, 87, *87*
Farming *61*, 71, 78, *78*, **300–301**, *300–301*
Faults 26, *26*, *27*, *46*
Fax machines 304, *304*
Feathers 124, *124*
Ferns *85*, 91, *91*, 94, 95, *95*
Ferrel cells *50*
Fertilization *see* reproduction
Fiber optics 279, *279*
Fibonacci series 263
Films 304, *304*
Fireballs 193, *193*
Fire extinguishers 283, *283*
Fire salamander 119, *119*
Firs 96, *105*
Fish *41*, *84*, 116–117, *116–117*, 133, *133*, 135, *135*, 140, *140*, 156
Fitness (human) 156–157, *156–157*
Fjords *43*
Flame test 241, *241*
Flares, Sun 164, *164*
Flatworms 108, *108*, 111, *111*, 133, *133*
Fleas 115, *115*
Flies 115, *115*, 135, 137, *137*
Floating **258**, 259, 295, *295*
Floods *57*
Florida Current 41
Flower, parts of 100; types of 100, *100*
Flowering plants 98–101, *98–101*, 102
Flukes 108, *108*, 111
Fluorescent lighting 269, *269*
Flying 258–259, *258–259*
Flying fish 116, *116*
Flying lemur 127
Flying squirrel 133, *133*
FM (frequency modulation) 274, *274*
Fog 55, *55*
Fogbows *57*
Föhn wind 52
Folds, rock 26, *26*, *27*
Food, preservation 301, *301*
Foods 156, *156*; food plants 102–103, *102–103*, 106, *106*
Forces 252–253
Forecasting weather 58–59, *58–59*
Forests 64–65, *64–65*, 66, **70–71**, *70–71*, 78, *78*, 88, *88*, 89, *89*, 98, *98*, 99, *99*, 138

Forget-me-not, alpine 99
Fossil fuels *15*, 32, 72, *72*, *77*, 246, *246*, 247
Fossils 20, **32–33**, *32–33*, 85, *85*, 140
Foundations, buildings 284, *284*
Foxes 86, *86*, 135, *135*
Foxgloves 98, 106, 107
Fractions 262, *262*
France 260
Franklin, Benjamin 275
Frequency 266, *266*, 270, *270*, 271
Friction 247, 254, *254*, 255, 292
Frogs 118, *118*, 119, *119*
Fronts 51, *51*, 54, 58
Frost 34
Frozen food 301, *301*
Fruits 102–103, *102–103*
Fuels 72–73, *72–73*, 76, **250–251**, *250–251*
Fumaroles *19*
Fundy, Bay of *11*, *73*
Fungi *85*, 87, 90, *90*, 92–93, *92–93*
Fur trade 138
Furnaces 298, *298*

G

Gabbro *12*, 23
Gagarin, Yuri 228, *228*
Gaia hypothesis 14
Galaxies: Big Bang 200, *200*; clusters and superclusters **198–199**, *198–199*; formation 160, 196, *196*; light from 161; Milky Way **196–197**, *196–197*, 219; types 198, *198*
Galilean satellites 183, *183*
Galileo Galilei 183, 216
Galileo space probe 183, *183*, 192, 226
Galle ring, Neptune 188, *188*
Gamma rays 220, 243, *243*, 266
Ganymede 170, 183, *183*
Gaping Ghyll 35
Gas, natural 32, 72, *72*, 73
Gas clouds 162, *162*
Gases 238, **244–245**, *244–245*; convection *248*; natural gas *250*; noble gases 243
Gases, air 49, *49*
Gasoline 250, 298
Gaspra 192, 226
Gastropods 110, *110*, 111, *111*, 132, *132*
Gavials 121, *121*
Gazelles 132, *132*
Gears 257, *257*, 293, *293*
Geckos 120, *120*
Gemini *211*, *213*
Gems 30–31, *30–31*
Genera (genus) (classification) 86, 87
Generators 277, *277*
Genetic engineering 106
Genetics 237
Geographical poles 9, 13, *13*
Geological time 265
Geological timescale *28–29*
Geometry 237, *237*, 260, 261, 262, *262*
Geostationary orbit 224, *224*
Geothermal power 73
Geysers *19*, *23*, *73*
Gharials 121, *121*
Giant clam 110, 111
Giant planets 168, *168 see also* individual planets
Giant snail 110, *110*
Giant squid 109, *109*
Giant tortoise 121, *121*
Giant waterlily 99
Gibbons *127*, *135*
Gila monster (lizard) 120, *120*
Ginger 103, *103*
Ginkgos 91, *91*, 96, *96*
Giraffes 88, 109, *109*, 127, *135*
Glaciers 36–37, *36–37*, 44, 63
Glass 299, *299*
Glass recycling 80, *80*
Gliders 258, *258*, 291, 296
Global warming 77, *77*
Gobi Desert 68
Gold 30, *31*, 74, *74*, 75, 242, *242*
Goliath beetles 115, *115*
Gondwanaland 20
Graben *22*, 23
Graf Zeppelin (airship) 291, *291*
Grand Banks *55*
Grand Canyon 39, *39*
Grand Erg Oriental 69
Granite 23, 28, *28*, 67
Graphics, computer 281, *281*
Graphite *31*, 240, *240*
Graphs 262, *262*
Grasses 91, *91*, 98, 107
Grasshoppers 114, *114*
Grasslands 66–67, *66–67*, 88, *89*, 98, *98*, *131*
Gravity 10–11, 35, 160, 247, *247*, **252–253**, *252–253*; Big Bang 200; black holes *206*; comets 195; galaxies 198; orbits 191, *191*; planets 163, 188; rockets 222; space probes 227, *227;* weightlessness 228, 229, *229*
Great auk 138, *138*
Great Dark Spot, Neptune 188, *188*, 189, *189*
Great Dividing Range *25*
Great Glen *23*, *46*
Great Lakes 46
Great Nebula 197, *197*
Great Red Spot, Jupiter 182, *182*
Great Rift Valley 27, *27*, *46*
Great Salt Lake *47*
Grebes 122, *123*, 125, *125*
Greece 260
Green Bank Observatory *219*
Greenhouse effect 42, 77, *77*, 174
Greenland 36, 37, 62, 63, *63*
Greenwich Mean Time 265, *265*
Greenwich meridian 9
Growth (human) 154–155, *154–155*
Gulf Stream *40*, *50*
Gust 202
Guyots 39
Gymnosperms 91, *91*, 96, *96*
Gyres 40
Gyroscopes 251, *251*

H

Habitats 81, *81*, 88–89, *88–89*, 138
Hadley cell 50
Hailstones 56, 57, *57*
Half-life, radioactivity 243
Halley's Comet 194, *194*, 195, *195*, 226
Halo, Sun 57, *57*
Hang gliders 291
Hardware, computer *281*
Hardwoods 107, *107*
Hares 127, *127*
Harmattan wind 52
Harrier aircraft 297, *297*
Hawaii *19*, 39
Hawaiian eruptions *18*
Hawks 123, 124, *124*
Health (human) 156–157, *156–157*
Hearing 129, 147, 271, *271*
Heart (human) 148, *148*
Heat 247, **248–249**,*248–249*
Heat, Sun 165, *165*
Heating systems 286, *286*
Hedgehog 127, *127*
Helicopters 258, *258*, 259, 297, *297*
Heliosphere 166, *166*
Helium 160, 165, 182, 242, *242*, 243, 259, *259*
Hell's Gate *27*
Herbivores 129, *129*
Hercules *212*, 232
Herds 131, *131*
Hermes 192
Hermit crab 113, *113*
Herons 122, *123*
Herring, saltwater 116, *116*
Hertzsprung-Russell Diagram 208, *208*
Hibernation 128, *128*
Hidalgo *192*
Hillary, Edmund *25*
Himalayas 23, 24, *24*, 25, 34, *47*
Hipparchus 215
HIV virus 302
Holograms 269, *269*
Honeybees 114, *114*
Hong Kong and Shanghai Bank 286, *286*
Hornets 114, *114*
Horse-drawn transportation 292, *292*
Horses 128, *128*, 132, 136, *136*, 137, *137*, 141
Horseshoe bat *127*
Horseshoe crab *112*
Horsetails *85*, 90, *90*, 94, 95, *95*
Horst *26*
Hourglasses 264, *264*
Housing 283, *283*
Hovercraft 294, *294*
Huang He River 44
Hubble, Edwin 216, 219, 230
Hubble Space Telescope (HST) 190, 219, 220, 221, *221*, 230
Hudson Bay 39
Human body 142–157, *142–157*; digestion and respiration 150–151, *150–151*; growth and aging 154–155, *154–155*; heart, blood, and skin 148–149, *148–149*; looking after your body 156–157, *156–157*; nervous system 146–147, *146–147*; reproduction 152–153, *152–153*; skeleton and muscles 144–145, *144–145*; systems of the body 142–143, *142–143*
Humboldt Current *40*
Hummingbirds 101, *101*, 125
Hurricanes 53, *53*, *59*
Huygens, Christiaan 265
Hydra constellation 212, *213*
Hydraulic machines 257, *257*
Hydrocarbons 240, *240*, 250
Hydroelectric power 73, 78
Hydrofoils 294, *294*
Hydrogen 160, 162, 165, 182, 238, *238*, 240, 241, 242
Hydrosphere *12*, 38
Hyenas *129*
Hygiene 157, *157*
Hyrax *126*, 127

I

Ice 36–37, *36–37*, 38, 46, 56, *57*, *63*
Ice Ages 36, 37, 42, 43, 64, 175, *175*

Ice, comets 194, 195
Icebergs 36, 37, 65, *65*
Iceland 23, *23*, *63*
Igneous rocks 28, *28*, 30
Ikeya-Seki, Comet 195
Immune system 149, *149*, 157
Inclined planes 256
India 23, 51, *51*, 66, 67, 260
Indian Ocean *38*, 51, *51*, 53
Inertia 254, *254*
Infrared Astronomy Satellite (IRAS) 221, *221*
Infrared radiation 218, *218*, 220
Inmarsat satellite 225, *225*
Infrared rays 266, *266*
Insect pests 300, *300*
Insects *84*, 109, 114–115, *114–115*
Insects, fossils *32*
Inselbergs *71*
Insulation 248, *248*
Intelsat satellite 225, *225*, *221*
Intensive care 303, *303*
Internal-combustion engines 257, *257*
International Date Line 265, *265*
International Ultraviolet Explorer (IUE) 221, *221*
Invertebrates *109*, 110–111, *110–111*
Invisible material *see* Missing Mass
Io 183, *183*
Ionosphere 49, *49*
Ionosphere 274
Ions 239, *239*
IRAS-Araki-Alcock, Comet 195, 221
Iron 13, 30, 74, *74*
Iron Age 298
Irregular galaxies 198, *198*
Irrigation *78*, 300, *300*
Islands 19, *19*, 39, *39*, *43*
Isobars 58, *58*

J

Japan 17, 75
Jellyfish 108, *108*, 110, *110*, 111, *111*
Jet aircraft 258, 259, 296, *296*, 297, *297*
Jet streams 50, *51*
Joule, James 246
Joules 246
Journalism 304, *304*
Judas tree *104*
Juniper *97*
Jupiter 9, 168, *168*, 169, **182–183**, *182–183*; asteroid belt 192, *192*; atmosphere 182, *182*; composition 168; orbit 163, *163*; ring 169, *182*; satellites 183, *183*, 216; space probes 226
Jurassic Period *28–29*

K

Kangaroos 135, *135*
Kaolin 74
Kelp *92*
Kelvin scale 248, *248*
Kepler, Johannes 214, 215, *215*
Kestrels 124, *124*
Keyboards, computers 280, *280*
Kidney fern *95*
Kidneys (human) 151, *151*
Kinetic energy 246, *246*, 247, *247*
Kingdoms (classification) 86, 87
Kingfishers 123, *123*
King penguin 135, *135*
Kiwi 122, *122*
Kjolen Mountains *25*
Koala *126*, 127
Kon-Tiki 40
Kopjes *67*
Kwanto plain 17

L

La Palma observatory 217, *217*
La Rance estuary 73
Labrador Current *63*
Laburnum seeds *107*
Ladybugs 115, *115*
Lagoon Nebula *197*
Lagoons *43*
Laika 225
Lakes 37, *37*, 38, 45, *45*, **46–47**, *46–47*
Lambert Glacier 37
Lampreys 116, *116*
Land resources 74, *75*
Land use 61
Landscape preservation 80, *80–81*
Landscapes 60–61, *60–61*
Landslides *35*
Land transportation 292–293, *292–293*
Larch *105*
Laser light 269, *269*, 279
Latitude, lines of 9, *9*
Laurasia 20
Lava 18, 28
Lava plains, Moon 176, *176*
Lavatories *287*
Lead 74, *74*, 75
Leap years 10, 264
Leaves 100, *100*, 102, 104, *104*
Leeches 108, *108*
Leeks *102*
Lenses 268, *268*
Lenticular clouds 55
Leo *211*, *213*
Leo Cloud *199*
Leonardo da Vinci 237, *237*, 282
Levers 256, *256*
Libra *211*, *213*
Lichens 90, *90*, 93, *93*
Life in universe 161, 174
Life span, animal 134, *134*; human 155
Lifts 286, 287, *287*
Light 266–269, *266–269*; fiber optics 279, *279*; red shift 200, *200*, 201; spectrum **266–267**, *266–267*; speed of 161, 196, 266; Sun 165, *165*
Light bulbs 276, *276*
Lightning 52, *52*, 161, 275, *275*
Light-years 168
Limbs, artificial 303, *303*
Limestone *28*, 29, 32, 34, 35, *35*, 42
Limpets 110, *110*
Linear dunes 69, *69*
Liners 291, *291*
Lions *126*, 129, 132, *135*
Liquids 238, **244–245**, *244–245*, 248, *248*
Lithosphere *12*, 16, 20, 22, *22*
Liver (human) 151, *151*
Liverworts 90, *90*, 94, *94*
Lizards 120, *120*, 141, *141*
Lobsters 109, *109*
Local Group 198, 199, *199*
Local Supercluster 198, 199, *199*
Locator maps: Local Group 199, *199*; Solar System 197, *197*; Supercluster 199, *199*
London, England 17, *57*, *76*
Long March rockets 223, *223*
Longitude, lines of 9, *9*
Longshore drift 43, *43*
Looms 299, *299*
Loris 128, *128*
Loudspeakers 273, *273*
Lubrication 254, *254*
Luce, Maximilien 267, *267*
Lugworms 108, *108*
Luminosity, stars 208
Luna space probes 226, *226*
Lungfish 117, *117*
Lungs (human) 150, *150*
Lunokhods 226, *226*
Lymph system 149, *149*
Lyrebirds 125, *125*

M

Maat Mons, Venus 173, *173*
Macaque 135, *135*
Macaw *122*
Machine tools 282, *282*
Machines 256–257, *256–257*
Madagascar 21, *71*
Magellan space probe 173, *173*
Magellanic Clouds 198, 199, *199*
Maglev trains 293, *293*
Magma 18, 28, 30
Magnesium 13, 30, 74
Magnetic fields, planets 166, *166*
Magnetism 13, *13*, 23, 252, 274 *see also* Electromagnetism
Magnetosphere 13, *166*
Magnetotails 166, *166*
Magnifying glasses 268, *268*
Magnitude, stars 208–209, *209*
Maidenhair spleenwort *95*
Maidenhair tree 96, *96*
Main sequence stars 202, *202*, 208
Mammals *84*, *86*, 126–127, *126–127*, 134–35, *134–35*
Mammoth Cave National Park *35*
Mammoths 141, *141*
Manganese 74, *74*
Mangrove swamps 71, *71*
Manned Maneuvering Units (MMUs) 229, *229*
Mantle 12, *12*, 13, 20, 23, 24
Maps: Mars *180*; Moon *176*; stars *210*, *211*, *212*; Venus 173, *173*; weather 58, *58*
Marble 28, 29, *29*
Mardalsfossen Falls *44*
Margay 138, *138*
Marianas Trench 39, *39*
Marine invertebrates 110–111, *110–111*
Mariner space probes *160*, 170, 171, *226*
Marine turtles 121, *121*
Mars 9, 165, *168*, **180–181**, *180–181*; asteroid belt 192, *192*; atmosphere 180, *180*; composition 168; orbit 163, *163*, *215*; satellites 181, *181*; space probes 226
Marshes *43*, *47*
Marsh horsetail *95*
Marsupials 20, 135, *135*
Mass 252
Materials 298–299, *298–299*
Mathematics 237, 261, **262–263**, *262–263*, **238–241**, *238–241*
Matter 160
Matter: atoms and molecules 238–241, *238–241*; periodic table

242–243, *242–243*; solids, liquids and gases **244–245**, *244–245*
Mauna Kea 39, *39*
Maxwell Montes 173
Mayfly 135, *135*
Meanders 45, *45*
Measurements 260–261, *260–261*
time **264–265**
Medicine 302–303, *302–303*
Mediterranean 17, 64
Mendeleyev, Dmitri 243
Menstruation 154, *154*
Mercalli scale, earthquakes 17
Mercury (metal) 76
Mercury (planet) 9, *168*, 169, **170–171**, *170–171*; composition 168; craters 169, *170*, 171, *171*; orbit 163, *163*, 170; space probes 226
Mesosphere 48, *48*, 174, *174*
Mesozoic Era *28–29*
Messages, coded 232, *232*
Metals 74–75, *74–75*; alloys 241, *241*; electroplating 298, *298*; periodic table 242; welding 284, *284*
Metals, asteroids 233, *233*
Metamorphic rocks 28, 29, *29*, 30
Meteorites 193, *193*
Meteoroids 174, 193, *193*
Meteorologists 58, 59
Meteors 193, *193*
Meteor showers *193*
Meteosat satellite 225, *225*
Methane 14, 77, 185, 188, *188*, 190, 242, *250*
Metric system 260
Mice 135, 157, *157*
Microchips 278, *278*
Microphones 272, *272*
Microprocessors 278, *278*, 305
Microscopes 268, 302, *302*
Microwaves 266, *266*, 275
Middle East 300
Midwife toad 118, *118*
Migration 125, *125*, 133, *133*
Milky Way 196–197, 198, 199, 212, 219
Millipedes 112, *112*
Minerals 30–31, 30–31, 74, *74*
Mining 75, *75*, 250, *250*, 282, *282*
Minor planets 192–193, *192–193*; *see also* asteroids
Mint 103, *103*
Mintaka *208*, *209*
Miocene Period *28–29*
Miranda 187, *187*
Mir space station 228, 231, *231*
Mirrors 268, *268*
Mirrors, telescopes *217*
Missing mass 200
Mistal 52
Mistletoe 101, *101*, 107
Mixtures 245
Modems 304, *304*
Mohole Project 12
Mohorovicić discontinuity 12
Mojave Desert 69
Molecules 238, 239, *239*, **240–241**, 244, *244*
Moles 128, *128*
Mollusks *84*, 109, *109*, 110, *110*, 111, *111*
Momentum 255, *255*
Monarch butterfly 133, *133*
Monkey puzzle tree *105*
Monkeys 135, *135*, 138, *138*
Monocotyledons 91, *91*, 98–101, *98–101*
Monorails 293, *293*
Monotremes 135, *135*
Monsoon 51, *51*
Montgolfier brothers 259, 296, *306*
Months 264, *264*
Moon 11, *49* *57*, **176–177**, *176–177*, 252, *252*, 264; astronauts on 222, *222–223*, 228, 232, 233; craters 169, *176*, 177, *177*, 216; and Earth **178–179**, *178–179*; eclipses 167, *167*, 179, *179*; formation *178*; orbit *179*; space probes 226, *226*
Moons *see* satellites
Moraine *36*
Morse, Samuel 305
Mosquitoes 115, *115*
Mosses *85*, 90, *90*, 94, *94*, 95, *95*
Moths 114, *114*
Motion 252–253, *253*, **254–255**, *254–255*
Motorcycles 290, 292, *292*
Mountaineering *21*
Mountain habitats 89, *89*, 99, *99*
Mount Hopkins 217
Mount Palomar 217, *217*
Mount Wilson telescope 216, *216*
Mountains 24–25, 24–25, 34, 36, 39, *39*, 55, 175, *176*, *181*
Mouse, computers 280, *280*
Mousebird 123, *123*
Movies 304, *304*
Mudflows *35*
Multiple Mirror Telescope 217, *217*
Muscles (human) 142, *142*, 143, 144, *144*, 145, *145*
Mushrooms 92–93, *92–93*, 107, *107*
Music 272, 273, *273*

N

Nacreous clouds 49, 55
Nails (human) 149, *149*
Nam Tso, Lake 47
National Radio Astronomy Observatory 219
Natural gas 246, *246*, 250, *250*
Natural regions, world 88–89, *88–89*
Natural resources 72–75, *72–75*
Navigation 13, 210, 265, *265*, 295
Neagh, Lough 47
Neap tides *11*
Nebulae 162, *162*, 196, *197*, 198, 202, 205, *205*
Nematodes 108, *108*, 111
Neon 242, *242*, 243
Neptune 9, 169, *169*, **188–189**, *188–189*, 216; atmosphere 188, *188*; composition 168; orbit 163, *163*, 190, 191; rings 169, 188, *188*; satellites 188, 189, *189*; space probes 226
Nereid 189
Nervous system, human 142, *142*, 143, 146–147, *146–147*
Ness, Loch *47*
Nests 130–131, *130–131*
Netherlands *75*
Network, computers 281
Neutrinos 239
Neutron stars 203, *203*, 205, *205*, **206–207**, *206–207*
Neutrons 166, 238, *238*
Névé *36*
Newspapers 304, *305*
Newton, Isaac 217, 253, *253*
Newts 118, 119, *119*
Nickel 13, 74, 75
Night and day 264, *264*
Nile River 44
Nimbostratus clouds *54*
Nitrogen 14, 49, 65, 65
Nitrogen oxides 76, *76*, 77
Noble gases 243
Noctilucent clouds 49
Noise 271
North America 46, *46*, 51, 64, *64*, 66, 67, 68, 69
North American Nebula *197*
North Atlantic Drift 51
North Pole 9, 13, 36, 49, 50, 62, *264*
Northern Hemisphere 52
Northern Hemisphere, constellations 212, *212*
Norway *43*, *63*
Norway spruce 97, *97*
Nuclear fission 237, 238, 251, *251*
Nuclear fusion 251
Nuclear power *73*, 79
Nuclear reactions, Sun 164; inside stars 202–203
Nuclear reactors 250
Nuclear submarines 290
Nucleus, atoms 238, *238*
Nucleus, comets 194, *194*
Numbers 260, 262, *262*, 263
Nunataks *63*
Nutrition 156, *156*
Nuts 102
Nylon 299

O

Oaks *87*
Oases 69, *69*
Oberon 187
Oberth, Herman 223
Oblique faults *26*
Observatories 217
Oceans 38–39, *38–39*, 88, *88*; coasts **42–43**, *42–43*; currents **40**, *40*, 41, 51; floor 38, *38*; life in **41**, *41*, *63*; minerals *31*, *39*, 74, *74*; plate tectonics 22, *22*, 23; thermocline 41, *41*; tides 11, *11*, 252, 253, *253*; tsunamis *17*; water cycle 14
Octopuses 110, *110*, 132, *132*
Oersted, Hans Christian 274
Ohms 277
Oil 32, *32*, 63, *63*, 72, *72*, *73*, 79, *79*, 250, *250*, 298, *298*
Oil tankers 294, 295
Old Woman meteorite 193, *193*
Oligocene period *28–29*
Olivine 30, *30*
Olympus Mons 181
Onions 102, *102*
Opium poppy 106
Optical astronomy 216–217, *216–217*
Optical digital recording 272
Oranges *103*
Orb-web spider *113*
Orbits: asteroids 192, *192*; ellipses 163; comets 195, *195*; planets 163, *163*; satellites 224, *224*
Orchids 98
Orders (classification) 86, 87; birds 122–123, *122–123*; mammals 126–127, *126–127*
Ordovician period *28–29*
Ores 30, 74, 75
Organic chemistry 240

Orion constellation, 208–209, *208–209*, 210, *212*, *213*
Orion nebula 197
Osprey *124*
Ostriches 122, *123*, 125
Otoscopes 302, *302*
Otters 139, *139*
Ovaries 100, *100*, 143, 153
Owls 123, *123*
Oxbow lakes 45, *45*
Oxen 136, 138
Oxidation 243, *243*
Oxygen 13, 14, 30, 34, 49, 76, 161, 174, 240
Oxygen cycle 15, *15*, 106, *106*
Oystercatchers 124, *124*
Oysters 111, *111*
Ozone layer 15, 76, *76*, 167, *167*, *174*, *275*

P

Pacific Mountain System *25*
Pacific Ocean 38, *38–39*, 39; continental drift 21; earthquakes 17; *Kon-Tiki* expedition *40*; typhoons *53*; volcanoes *19*
Packaging, food 301, *301*
Paddle ships 294
Paint, primary colors 267, *267*
Paleocene period *28–29*
Paleozoic era *28–29*
Palms 91, *91*, 98, 99, *102*
Pampas 66, *66*
Pampas grass *98*
Pan 185
Pandora 185
Pangaea 20, *20*, *26*
Pangolin *126*, 127
Papermaking 107, *107*
Parallax 208, 209, *209*
Parasitic plants 101, *101*
Parkes Observatory 218, *218*
Parrots 137, *137*
Partial eclipse 167, *167*
Partridges 124, *124*
Pasteur, Louis 303
Peaches *104*
Pears *103*
Peat 47, 72
Peat mosses 94, *94*
Pelean eruptions *18*
Pelicans 122, *122*
Pendulums 247, *247*, 265, *265*
Penguins 122, *123*, 135, *135*
Penicillin 106
Pentagons 263, *263*
Penumbra 167, *167*
Penzias, Arno 219
Peppers 102, *103*
Peregrine falcons 125
Perigee, Moon at 179, *179*
Perihelion, Pluto at 190
Period, comet's 195, *195*
Periodic table 242–243, *242–243*
Periscopes 268, *268*
Permafrost 63, *63*
Permian period *28–29*
Personal computers 280, *280*, *281*
Peru *40*, *41*
Pesticides 81
Pests, agriculture 300, *300*
Petals 100, *100*
Petroleum 240, 298, *298*
Pets 137, *137*
Phaethon 192
Phases of Moon 179, *179*
Pheasants 123, *123*
Phloem 104, 105, *105*
Phobos 181, *181*
Phoebe 185
Phoenician juniper 97, *97*
Phosphorus 242, *242*
Photochemical smog 76, *76*
Photography 268, *268*, 304, *304*, 305
Photons 266, *266*
Photosphere *165*
Photosynthesis 15, *41*, *49*, 90, 100, 101, 106
Phyla (phylum) (classification) 86, 87, 90; animals 108–109, *108–109*; plants 90–91, *90–91*
Physical time 265
Piezoelectricity 277, *277*
Pigeons, navigation 13, 210
Pigs 129, *129*, 136, *136*
Piledrivers 284, *284*
Pill bugs 113, *113*
Pines 97, 105, *105*
Pioneer space probes 226, *226*
Pisces *211*, *212*
Pitch, sound 270, *270*
Placental animals 135, *135*
Planck, Max 239
Plane figures 263, *263*
Planes, weather 59
Planetismals 163
Planets 9, *9*, **168–169**, *168–169*, 214; density 168; formation 161, 163, *163*; gravity 188; magnetic fields 166, *166*; minor planets **192–193**, *192–193*; orbits 163, *163*, 215, *215*; space probes 226; *see also* individual planets
Planet X 190, 191, *191*, 219
Plant kingdom, The 90–107, *90–107*; bacteria, algae, lichens and fungi 92–93, *92–93*; fruits or vegetables? 102–103, *102–103*; ginkgos, cycads, and conifers 96–97, *96–97*; liverworts, mosses, horsetails, and ferns 94–95, *94–95*; monocotyledons and dicotyledons 98–101, *98–101*; phyla 90–91, *90–91*; plants and people 106–107, *106–107*; trees 104–105, *104–105*
Plants 60; acid rain damage *77*; carbon cycle 15; continental drift 20; deserts 68; erosion of rocks *34*; fossil fuels 72, *72*; fossils 32; grasses 66–67, *66–67*; mountainous areas 25; ocean life 41; oxygen cycle 15; photosynthesis 15, *41*, *49*; tropical rain forests 70–71, *71*; tundra *62*
Plasma 238
Plastics 240, 241, 299
Plate tectonics 22–23, *22–23*, 42
Platinum 74, 75
Pleiades 208, *208*
Plinian eruptions *18*
Pliocene period *28–29*
Plovers 125, *125*
Plows 300, *300*
Plums *103*
Pluto 169, *169*, 188, **190–191**, *190–191*, 216; atmosphere 190; composition 168; discovery of 190, 219; orbit 163, *163*, 191, *191*; satellites 190, *190*
Pointillism 267, *267*
Poison plants 107, *107*
Polar bears *88*, 89
Polar cells 50
Polaris *212*
Polarized light 267, *267*
Polar regions *62–63*, *62–63*, 68
Polders 75, *75*
Poles: geographical 9, 13, *13*; magnetic 13, *13*, 21, 23; celestial 212, *212*; of Uranus 187, *187*
Pollen, fossils 33
Pollination 100, *100*, 101, *101*; *see also* reproduction
Pollution 76–77, *76–77*, 79, 139
Polygons 263, *263*
Polymers 241, *241*, 299
Polynesian Islands *40*
Polystyrene 299
Pomes *103*
Pompidou Arts Center, Paris *286–287*
Portuguese man-of-war 111, *111*
Positrons 239
Potassium 13, 30
Potatoes 98, 102, *102*, 107
Potential energy 247, *247*
Potholes 35
Pottery 282, *282*, 299, *299*
Power stations 250, *250*, 276
Power stations, solar 233, *233*
Prairies 66, *66*, 67, 98, *98*; prairie dogs 131, *131*
Precambrian eon *28–29*
Precipitation 56
Prehistoric animals 140–141, *140–141*
Pressure 252
Pressure, atmospheric 48, *58*
Prickly pear *99*
Primary colors 267, *267*
Printing 304
Probability 262, *262*
Production line assembly 282, *282*
Programs, computers 280
Prometheus 185
Propellers 294, 295, *295*, 297, *297*
Proteus 189
Protoctists 87, 90, 92, 108, *108*, 111, *111*
Protons 166, 207, 232, 238, *238*
Protozoans 108, 111, *111*
Proxima Centauri 209
Pterosaurs 33
Ptolemy 212, 215, 216
Puberty 154, *154*
Publishing 304
Puffballs 93, *93*
Puffer fish 117, *117*
Pulleys 256, *256–257*
Pulsars 204, 205, *205*, 218, 219
Pulsating variable stars 204, *204*
PVC 299
Pyramids 260, 284
Pyrenees 25
Pythagoras 237, *237*, 263, *263*

Q

Quadrilaterals 262, *263*
Quahog clam 135
Qualitative analysis 241
Quantitative analysis 241
Quantum theory 239, *239*
Quarks 239, *239*
Quartz 30, *30*
Quartz watches 265, *265*, 277, *277*
Quasars 198, *198*, 201, *201*, 218, 219
Quaternary period *28–29*
Queen Alexandra birdwing butterfly *114–115*, 115
Quetzal, resplendent *123*

Quillworts 91, *91*
Quinine 106

R

Rabbits 131, *131*, *135*, 137
Radar 266, 295
Radar mapping 173, *173*
Radiant energy 247
Radiation 160, 218, *218*, 248, *248*; background 161, 219; black holes 206; from Sun 11, *11*, 15, 167, *167*; pulsars 205, *205*
Radio 272, 304, *304*, 305, *305*
Radioactivity 243, *243*
Radioisotopes 302, *302*
Radiosonde balloons *59*
Radio telescopes 218–219, *218–219*, 232
Radio waves 49, *49*, 266, *266*, 274, *274*
Radius 263
Rafflesia 99, 101
Railroads 236, *236*, 290–291, *290–291*, *293*
Rain 10, 14, 34, *48*, 51, 52, 54, **56–57**, *56–57*, 66, 68, 70, 71, 77, *77*
Rain, acid 241, *241*
Rainbows 57, *57*, 267, *267*
Rain forests 70–71, *70–71*, 81, 88, *88*, 99, *99*
Rain shadow *57*
R Aquarii 221, *221*
Raspberries *103*
Rats 129, *129*
Rattlesnakes 120, *120*
Rawinsonde balloons *59*
Rays *84*, 116, *116*
Reactions, chemical 243, *243*
Recordings 272, *272*
Recycling 80, *80*
Red Sea 21, *23*, *27*
Red shift 200, *200*, 201
Red squirrel *127*
Red-throated diver *123*
Reflecting telescopes 216, *216*, 217, *217*
Red dwarf stars 203, *203*
Red giant stars 165, *165*, 202–203, *202–203*, 204, *204*, 205, *205*, 208, 221
Reeds 98, 99, *99*, 107, *107*
Reindeer 136
Refining 298, *298*
Reflection, heat 249, *249*
Refracting telescopes 216, *216*, 217, *217*
Refraction 267, *267*
Refrigerators 249, *249*
Reg deserts *69*
Relativity, General Theory of 219
Renaissance 237
Reproduction; amphibian 118, *118*, 119; bird 125; conifer 97, *97*; fish 117; insect 114, *114*, 115; mammal 134–135, *134–135*, *143*, 152–153, *152–153*; reptile 121; shrimp 112, *112*; *see also* pollination
Reptiles *84*, 88, 120–121, *120–121*
Reservoirs *46*
Resistors 278
Resources 72–75, *72–75*
Respiration (in amphibians) 119, *119*; (in fish) 116, *116*; (in humans) 150, *150*
Reverse faults *26*
Rheas 122, *122*
Rhinoceroses 127, *127*, 138, 139, *139*
Rhizomes 100, *100*
Rhododendrons 98, 107
Rhubarb 107, *107*
Richter scale, earthquakes 17
Ridges, on Mercury 171, *171*
Rift valleys 27, *27*
Rigel 208, *208*
Ring systems: Jupiter 182, *182*; Neptune 188, *188*; Saturn 184, *184*; Uranus 186, *186*
Rivers *34*, 38, **44–45**, *44–45*, 47, 57, 71, 79, 88, *88*, 139
Rivets 285, *285*
Roads 288, *288*, 290, 292
Robber crab 113, *113*
Robins *122*, 125, *125*
Robots 278, 279, *282*
Rockets 222–223, *222–223*, 224, *224*, *229*, 230, *230*, 253, *253*
Rocks 28–29, *28–29*; bending and breaking **26–27**, *26–27*; bornhardts *67*; continental drift **20–21**, *20–21*; deserts *68*; earthquakes 16–17; erosion 44, *44*; fossils **32–33**, *32–33* geothermal power 73; kopjes 67; landscapes **60–61**, *60–61*; mantle 12, 13; minerals and gems **30–31**, *30–31*; mountains **24–25**, *24–25*; plate tectonics **22–23**, *22–23*; seashore **42–43**, *42–43*; weathering and erosion **34–35**, *34–35*
Rocky Mountains 24, *25*, 64
Roentgen, Wilhelm 275
Romans 284, 286, *286*, 288, *288*
Roots 100, *100*, 102, 104, *104*
Rosat satellite 221, *221*
Rosette Nebula *197*
Rousseau, Jean Bernard 35
Rubber 106, *106*
Rudders 295, *295*
Russell, Henry 219
Rust 243, *243*, 298
Rutherford, Ernest 237
Ruwenzori Range *25*

S

Sahara 68, 69, *69*, 78
Sahel 78
Sailfish 109, *109*
Sailing ships 290, *290*, 294, *294*
Salamanders 118, 119, *119*
Salmon 133, *133*
Salt *30*, 40, 47, *47*
Salyut space stations 231, *231*
Sand 43, *43*, 68, 69, *75*, 78, *79*
Sand dunes 69, *69*
Sandglasses 264, *264*
Sandgrouse, pintailed *122*
Sandstone *26*, 28, *28*
San Francisco *17*, *55*
Sarawak Chamber 35
Sargasso Sea 41
Satellites: Earth **176–179**, *176–179*; Jupiter 183, *183*, 216; Mars 181, *181*; Neptune 188, 189, *189*; Pluto 190, *190*; Saturn 185, *185*; Uranus 187, *187*
Satellites, artificial 161, *161*, *220–221*, 221, 222, **224–225**, *224–225*, 238
Satellites, communication 305, *305*
Satellites, weather 59
Saturn 9, 169, 184–185, 184–185; atmosphere 184, 184; composition 168; orbit 163, 163; rings 169, 169, 184, 184, 185, 185; satellites 169, 185, 185; space probes 185, 226
Saturn V rocket 222, *222*
Savannas 66, 67, *67*, 88, *89*, 98, *98*
Schwassmann-Wachmann, Comet 195; orbit *195*
Scorpions 109, *109*, 112, *112*, *132*
Scorpius *211*, *213*
Scotland *23*, *43*, *46*, 47
Screws 256
Sea *see* Oceans
Sea anemones 108, *108*
Sea cucumbers 109, *109*, 110, *110*
Seahorses 117, *117*
Sea lilies 109, *109*
Seals 133, *133*
Seamounts 39
Seashores 42–43, *42–43*
Sea snakes 120
Seasons 10, *10*
Sea transportation 290, *290–291*, **294–295**, *294–295*
Sea urchins 109, *109*, *110*, 111
Seaweeds *41*, *59*, 87, 90, 91, *91*, 92, *92*
Sediment 38, 42, *43*, 44, 45, 47, *47*, 71
Sedimentary rocks 28, *28*, 30, 32, *32–33*
Seeds 96, 98, 103, 105; development of 100, *100*; as food 102
Seif dunes 69
Seismometers 12, 17
Selective breeding 137
Self-pollination 100
Semiconductors 278, *278*
Sepals 100, *100*
Sequoia 91, *91*
Seven Sisters *see* Pleiades
Sewers 287, *287*
Sextants *295*
Shale *26*, 32
Sharks *84*, 116, *116–117*, 137, *137*
Shearwater 122, *123*
Sheep 136, *136*
Sheepdogs 137, *137*
Shepherd moons *185*, 186
Shieldbugs 115, *115*
Ships, weather *59*
Ships: clocks 265, *265*; floating 258, *259*, 295, *295*; sonar 270, *270*; transportation 290, *290*, **294–295**, *294–295*
Shock waves, earthquakes 16, *16*
Shooting stars 193, *193*
SI units 260
Silica *31*, 177
Silicon 13, 60
Silicon chips 278, *278*
Silkworms 136, *136*
Silurian period *28–29*
Silver *31*, 74, *74*, *75*
Silver fir *96*
Silverfish 115, *115*
Single-celled organisms *85*, 87, 92, *92*, 108
Sirius 209
Skeleton (human) 142, *142*, 144, *144*, 145, *145*; (mammal) 128, *128*
Skies, color *58*, *58*
Skin (human) 149, *149*; skin trade 138, *138*
Skylab 228, 231, *231*
Skyscrapers 284, *284*, *285*
Slash and burn *67*
Slate 28, *29*
Sleet *56*

Slugs 111, *111*, 132
Slumps, rock 35, *35*
Smell, sense of 147, *147*
Smilodon 141, *141*
Smog 76, *76*
Snails 110, *110*, 111, *111*, 132, *132*
Snakes 120, *120*, 121, *121*, 132, *132*, 138, *138*, 141, *141*
Snow 36, *36*, **56–57**, *56–57*
Sodium 13, 20
Software, computers 280, 281
Softwoods 107, *107*
Soil 60, 61, *61*, 63, *63*, 66, *66*, 68, 70,*70*, 78, *78*
Soil creep *35*
Solar energy collectors 233, *233*
Solar flares 164, *164*, *225*
Solar Max 225
Solar power 73, *73*, 249, *249*, 251, *251*
Solar sails 232, *232*
Solar system 9, *9*
Solar System: formation **162–163**, *162–163*; minor planets **192–193**, *192–193*; planets **168–191**, *168–191*; Sun **164–167**, *164–167*
Solar wind 13, 166, *166*, 195, 232
Solids 238, **244**, *244*, 248
Solutions 245, *245*
Sonar 270, *270*
Sonic boom 270, *270*
Sound 266, **270–273**, *270–273*
"Sound barrier" 297
South Africa *67*, 74
South America *21*, 23, 64, 66, *66*, *67*
South China Sea 39
Southern Alps *25*
Southern Cross *see* Crux
Southern Hemisphere 52, 64
Southern Hemisphere, constellations 213, *213*
Southern lights *13*
South Pole 9, 36, 49, 50, 57, 62, 264, *264*
Soviet Union 222, 223, 224, 226, 228
Soyuz spacecraft 228
Space and time 260–265
Space cities 233,*233*
Spacecraft 222–223, *222–223*; artificial satellites **224–225**, *224–225*; astronauts **228–229**, *228–229*; future of 232
Spacelab 229
Space planes 232
Space probes 169, 222, **226–227**, *226–227*
Space shuttles 221, 224, *225*, *225*, 228, 229, **230–231**, *230–231*, 232
Space stations 222, 228, **230–231**, *230–231*
Spacesuits 228, *228*
Space technology 251, *251*, 253, *253*, 283, *283*
Space telescopes 220–221, *220–221*
Spain, DSN station 227
Species (classification) 86, *86*, 87, *87*
Spectrographs 200
Spectrum 266–267, *266–267*
Speed 255, *255*
Speed, galaxies 200
Sphagnum mosses 94, *94*
Spicules, Sun 164, *164*
Spiders 109, *109*, 112, 113, *113*, 130, *130*, 132, *132*, 135, 137, *137*
Spinach 102, *102*
Spiral galaxies 198, 198, *198*
Spirogyra 92, *92*
Spits 43, *43*
Sponges *85*, 108, *108*, 111, *111*
Spoonbills 124, *124*
Spores 94, *94*, 95, 95
Springs 44, 69
Spring tides *11*
Spruce *97*, *105*
Squares 263, *263*
Squash 102, *103*
Squids 110, *110*, 132
Squirrels 127, *127*, 133, *133*, 134
St. Helens, Mount 19, *19*
Stags horn moss *95*
Stainless steel 299
Stalactites and stalagmites *35*
Stamen 100, *100*
Starfish 109, *109*, 110, *110*
Starlings *134*
Stars: binary systems 204, 205, *205*; blue giants 203, *203*; brightness **208–209**, *208–209*; clusters 208, *208*; constellations 210, *210*, *211*, **212–213**, *212–213*, 214; distance **208–209**, *208–209*; formation 162–163, *162–163*; galaxies 196, *196–197*; life of 202–203, *202–203*; mapping **210–211**, *210–211*; neutron 203, *203*, 205, *205*, **206–207**, *206–207*; pulsars 204, 205, *205*, 218, 219; red dwarfs 203, *203*; red giants 165, *165*, 202–203, *202–203*, 204, *204*, 208, 221; supernovae 199, *199*, 203, *203*, 204–205, *204–205*, 209; variable 204, *204*, *219*; white dwarfs 165, *165*, 203, *203*, 204, *204*
Statistics 262, *262*
Steam power 236, *236*, 290, *290*, 291, 293, *293*
Steel 298, *298*, 299
Steering 293, *293*, 295, *295*, 296
Steppes 66, 98
Stethoscopes 302, *302*
Stigma 100, *100*
Stingrays 116, *116*
Stone Age 290, 304
Stonehenge 214, *214*, *215*
Stone pine *97*
Storms 43, **52–53**, *52–53*
Stratocumulus clouds *54*
Stratosphere 48, *48*, 76
Stratus clouds *54*, 56
Streams 44
Strike-slip faults 26
Stromatolites 33
Strombolian eruptions *18*
Strong force 252
Sublimation 244
Submarines 259, *259*, *290–291*
Submersibles 283, *283*
Sulfur dioxide 76, *77*
Sun 164–167, *164–167*; and the atmosphere 49, *49*, 50; comets 194, *195*; day and night 264, *264*; Earth's days and years 10; eclipses *166–167*, 167; energy 11, 15, 246, *246*, 247; formation 161, **162–163**, *162–163*; gravity 11; halo *57*, *57*; historical views of 215, *215*; life cycle 165, *165*; light 266, 268; observing 164, *164*; radiation 76, *77*; satellite observation 225; solar power 73, *73*, 249, *249*, 251, *251*; solar wind 13, *13*; space probes 227, *227*; weather *55*
Sundials 264, *264*
Sunspots 164, *164*, 216
Superclusters 198–199, *198–199*
Superior, Lake *46*, 47
Supernovae 199, *199*, 203, *203*, 204, *204–205*, 209
Supersonic aircraft 270, *270*
Surgery 302, 303, *303*
Surinam toad 118, *118*
Swallows 125, *125*
Swamps 47, *47*, 70, 71, *71*
Swans 122, *122*
Swifts 123, *123*, 133, *133*
Swing-wing aircraft 297, *297*
Sycamores *104*
Sydney Opera House 285, *285*
Syenite 28, *28*
Symbiosis 101
Symbols, maths 262, *262*
Symbols, weather *58*
Syncline 27
Synform 27
Synoptic charts 58
Synthesizers 273, *273*
Synthetic fibers 299

T

Tabei, Junko 25
Tadpoles 118, *118*
Tally-sticks 260, *260*
Tape recorders 272, *273*, 304, *304*
Tapeworms 108, *108*, 111
Taproots 100, *100*
Taste, sense of 129; (human) 147, *147*
Taurus *211*, *212*, *215*
Taylor, F.B. 21
Tectonics 22–23, *22–23*, 42
Teeth 129, *129*; (human) 151, *151*, 156, *156*
Telecommunications 304
Telegraph 305
Telescopes 214, 216, *216*, 217; radio **218–219**, *218–219*, 232; space **220–221**, *220–221*
Television 266, 274, 304, *304*, 305, *305*
Tempel-Tuttle, Comet 193
Temperate habitats 89, *89*, 98 *98*
Temperate zones 60, *60*, **64–65**, *64–65*, 66
Temperature 248, *248*
Temperature, stars 208, *209*
Tenzing Norgay *25*
Termites 131, *131*
Terns *122*, 125, *125*
Tertiary Era *28–29*
"Test-tube" babies 155
Tethys Ocean 20, 24
Textiles 299
TGV (*train a grand vitesse*) 291, *291*
Thermocline 41, *41*
Thermographs 58, *58*
Thermometers 58, *58*, 302, *302*
Thermosphere 48, *48*
Thermostats 249, *249*
Thompson, Benjamin 248
Thunder 52, 52, 54, 58
Thunderstorms 275, *275*
Tidal barriers *73*
Tides, ocean 11, *11*, 251, *251*, 252, 253
Tien Shan Mountains *25*
Tigers 138, 139
Timber 107, *107*
Time, measuring **264–265**, *264–265*

Timescale of the universe 160–161, *160–161*, 201
Tin 74
Tinamou 122, *123*
Titan 170, 185, *185*
Titania 187, *187*
Titanium 75
Titicaca, Lake 47
Toads 118, *118*, 119
Toadstools 93, *93*
Tomatoes *103*
Tornadoes 53, *53*
Tortoises 120, 121, *121*
Toucans 123, *123*
Touch, sense of 128; (human) 147, *147*
Tower cranes 285, *285*
Trade winds 50, 51, *51*, 52
Trains 290–291, *290–291*, *293*
Trams 290
Tranquilizers 303
Transantarctic Mountains 62
Transcurrent faults *26*
Transformers 276
Transistors 278, *278*
Transpiration 14, *14*
Transportation 290–297, *290–297*
Trap-door spider *113*
Tree ferns 95
Trees 25, *60*, **64–65**, *64–65*, **70–71**, *70–71*, 78, *78*, 81, *81*, *87*, 91, 96–97, *96–97*, 98, 99, 104–105, *104–105*; timber 107, *107*
Triangles 263, *263*
Triassic period *28–29*
Trifid Nebula *197*
Trilobites 112
Triodes 278
Triton 188, 189, *189*, 190, 191
Trojan asteroids 192, *192*
Tropical rain forests 70–71, *70–71*, 88, *88*, 99, *99*, 138
Tropics 9, 50, 56, 59, 60, 70–71
Troposphere 48, *48*, 174, *174*
Trout 135, *135*
Trucks 290, *290*, 292
Trumpets 273, *273*
Tsetse fly 137, *137*
Tsunamis *17*
Tuatara 120, 141, *141*
Tubers 100, *100*, 102
Tugela Falls *44*
Tull, Jethro 300
Tuna 117, *117*
Tundra 62–63, *62–63*
Tuning forks 270, *270*
Tunnels 282, *288*, 289
Turbofan jet engines 296, *296*, 297
Turtledove *123*
Turtles 120, 121, *121*
Twin stars *see* Binary stars
Typhoons 53

U

U-boats 291, *291*
Ultrasound 271, 302, *302*
Ultraviolet radiation 76, 174, 266, *266*, 275, *275*
Umbra 167, *167*
United Nations 80, 81
United States of America 64, 65, *65*, 67, *77*, 222, 223, 227, 228
Universe: Cosmology **200–201**, *200–201*, *236*; historical views of 215, *215*; timescale of **160–161**, 160–161
Urals 25
Uranium 73, 242, *242*, 243
Uranus 9, 169, **186–187**, *186–187*, 188, 216; composition 168; orbit 163, *163*, 190; rings 169, 186, *186*; satellites 187, *187*; space probes 226; tilt 186, *186*, 187, *187*
Ursa Major *210*, 212, *213*

V

V2 rockets 223, *223*
Vaccination 157, *157*, 303
Vacuums 270
Vacuum tubes 278, *278*
Valles Marineris *189*, 181
Valleys 27, *27*, 36, *36–37*, *43*, 44, 45, *45*
Van Allen belts 13
Vancouver, Canada *57*
Variable stars 204, *204*, 219
Vegetables 102–103, *102–103*, 156, *156*
Vegetation 60, 80 *see also* Plants
Vein, mineral *30*
Velocity 255, *255*
Venus 9, 168, *168*, 169, **172–173**, *172–173*; atmosphere 172, *172*; composition 168; as morning star, evening star 172, *172*; orbit 163, *163*, 216; space probes 226, *227*
Venus's-flytrap 101, *101*
Vertebrates 108, 109, *109*
Vertical take off and landing aircraft (VTOL) 297, *297*
Vibrations, sound *270*
Video recorders 278, *279*, 304, *304*, 305, *305*
Viking space probes 180, *181*, *226*
Vipers 120, *120*
Virgo *211*, *213*
Virgo cluster 199, *199*
Virtual reality 281, *281*
Viruses 302, *302*
Volcanoes 18–19, *18–19*, 22, 23, *23*, 27, 28, 30, *31*, 39, *39*, *46*, 76, 244, *244*; on Io 183, *183*; on Mars 181, *181*
Volts 277
Volvox 92, *92*
Voyager space probes 169, 183, 184, 185, 186, *186*, 187, 188, 189, 226
Vulcanian eruptions *18*

W

Wasps 114, *114*, 115, *115*, 134
Waste disposal 79, *79*, 80
Watches, quartz 265, *265*, 277, *277*
Water: 240, 242, *242*, 300, *300*; aquifers 69, *69*; geysers *19*, 23, 73; hydroelectric power 73; ice 36; irrigation 78; oceans and seas **38–39**, *38–39*; resources 72, 74, 75, *75*; water cycle **14**, *14*; water vapor *48*, 49, 54, 55, *55*; *see also* Lakes; Oceans; Rain; Rivers
Water clocks 264, *264*
Waterfalls *44*
Waterlilies 99
Water spider 130, *130*
Water vapor 175, *175*
Watt, James 236, *236*
Watts 277
Wavelengths, sound 270, *270*, 271, *271*
Waves, electromagnetic radiation 266, *266*, 274, *274*
Waves 42, *43*
Weak force *252*
Weather 11, 41, 49
Weather forecasting 58–59, *58–59*
Weathering 25, **34–35**, *34–35*, 60
Weather satellites 224, *225*
Weaving 299, *299*
Wedges 256
Weeds 81
Wegener, Alfred 20, 21
Weightlessness 228, 229, *229*
Welding 284, *284*
Western pygmy blue 115, *115*
Wet batteries 276, *276*
Wetlands 88, *88*, 99, *99*
Whales 109, *109*, 127
Wheels 256, 292, *292–293*
Whippoorwill *122*
White dwarf stars 165, *165*, 203, *203*, 204, *204*, 208
Wind 40, *40*, *43*, 50, 51, *51*, **52–53**, *52–53*, 58, *68*, *77*; resistance 254, *254*; sailboats 294, *294*; wind power 73, *73*, 251, *251*
Windmills *75*
Wolf-Rayet stars 204, *204*
Wolves 139, *139*
Wood 65
Woodlands, temperate 64–65, *64–65*
Woodlice 113, *113*
Woodpeckers 124, *124*
Woodwind instruments 273, *273*
Worms *85*, 108, *108*, 111, *111*
Wright brothers 259, 296, 308, *308*

X

X-rays 220, 221, *225*, 266, 274, *275*
Xylem 104, 105, *105*

Y

Yellow-billed cuckoo *122*
Yosemite Falls *44*
Yuccas 99, 101

Z

Zebras 129
Ziggurats, Babylonian 214, *214*
Zinc 74, *74*
Zippers 283, *283*, 308, *308*
Zodiac 211, *211*, *214*
Zoos 139, *139*

The publishers would like to thank the following artists
for contributing to the book:

Jonathan Adams, Kevin Addison, Marion Appleton, Andy Archer,
Mike Atkinson (Garden Studios), Owain Bell, Richard Bronson, Peter Bull,
Lynn Chadwick (Garden Studios), Kuo Kang Chen, Richard Coombes,
Joanne Cowne (Garden Studios), Richard Draper, Dean Entwistle,
Michael Fisher (Garden Studios), Eugene Fleury, Chris Forsey,
Mark Franklin, Lee Gibbons (Wild Life Art Agency), Jeremy Gower,
Ray Grinaway, Terry Hadler (Bernard Thornton Artists), Allan Hardcastle,
Alan Harris, J. Haysom, André Hrydziusko (Simon Girling Associates),
Ian Jackson, Ron Jobson (Cathy Jakeman Illustration),
Roger Kent (Garden Studios), Adrian Lascom (Garden Studios),
S. Lings, Jenny Lloyd (The Classroom), Bernhard Long (Temple Rogers),
Chris Lyon, Mainline Design, Alan Male (Linden Artists),
Maltings Partnership, Janos Marffy (Cathy Jakeman Illustration),
Josephine Martin (Garden Studios), Doreen McGuinness (Garden Studios),
David More (Linden Artists), Mullard Radio Astronomy Observatory,
Oxford Illustrators, Bruce Pearson (Wild Life Art Agency),
Sebastian Quigley, Elizabeth Rice (Wild Life Art Agency),
Paul Richardson, John Ridyard, Eric Robson (Garden Studios),
Michael Roffe, David Russell, John Scorey, Nick Shewring (Garden Studios),
Rob Shone, Guy Smith (Mainline Design), Mark Stacey, Roger Stewart,
Lucy Su, Stephen Sweet (Simon Girling Associates),
Myke Taylor (Garden Studios), Kevin Toy (Garden Studios),
Guy Troughton, Phil Weare (Linden Artists), Steve Weston, Keith Woodcock

The publishers wish to thank the following for supplying
photographs for this book:

Page 6 ZEFA; 13 NASA; 20 ZEFA; 26 ZEFA; 27 Ardea; 29 AGE Fotostock; 30 ZEFA; 34 ZEFA;
40 Science Photo Library; 44 Frank Spooner; 49 NASA; 50 Science Photo Library; 53 Frank Spooner
Pictures; 55 ZEFA; 59 Dundee University; 65 ZEFA; 67 Frank Spooner; 71 ZEFA; 73 Spectrum;
79 ZEFA; 93 ZEFA; 96 A–Z Botanical; 101 Science Photo Library; 107 Heather Angel; 110 Tony
Stone; 111 NHPA Spike Walker; 115 NHPA George Bernard; 119 ZEFA; 120 NHPA Tony
Bannister; 125 Frank Lane Picture Agency; 130 Heather Angel; 134 ZEFA; 136 ZEFA;
139 *t* Bruce Coleman, *c* ZEFA; 147 Allsport; 153 Robert Harding; 154 ZEFA;
157 Helene Rogers/TRIP; 158 ZEFA; 161 Science Photo Library; 171 JPL/NASA; 173 Science Photo
Library; 176 Science Photo Library; 177 ZEFA; 179 NASA; 181 NASA; 185 JPL/NASA;
187 JPL/NASA; 189 NASA; 190 NASA; 193 Science Photo Library; 194 Science Photo Library (top);
ESA (bottom); 197 NASA (top); Anglo-Australian Observatory (center right and bottom);
199 Anglo-Australian Observatory; 201 Science Photo Library; 204 Anglo-Australian Observatory;
208 Anglo-Australian Observatory; 209 Science Photo Library; 211 Royal Astronomical Society;
221 NASA; 225 Science Photo Library; 227 NASA; 247 Allsport/Simon Bruty; 250 Science Photo
Library; 259 Goodyear; 267 ZEFA; 269 Science Photo Library; 275 Science Photo Library;
276 Midnight Design Ltd; 286 Sir Norman Foster & Partners; 289 QA Photo Library; 292 Quadrant
Picture Library; 297 Boeing; 300 Science Photo Library